Sustainability in Plant and Crop Protection

Series editor

Aurelio Ciancio, Sezione di Bari, Consiglio Nazionale delle Ricerche Istituto per la Protezione delle Piante, Bari, Italy

More information about this series at http://www.springer.com/series/13031

Waqas Wakil • Jose Romeno Faleiro
Thomas A. Miller
Editors

Sustainable Pest Management in Date Palm: Current Status and Emerging Challenges

Springer

Editors
Waqas Wakil
Department of Entomology
University of Agriculture
Faisalabad, Punjab, Pakistan

Thomas A. Miller
Department of Entomology
University of California
Riverside, CA, USA

Jose Romeno Faleiro
Food and Agriculture Organization
of the United Nations
IPM (Red Palm Weevil)
Al-Ahsa, Saudi Arabia

Sustainability in Plant and Crop Protection
ISBN 978-3-319-24395-5 ISBN 978-3-319-24397-9 (eBook)
DOI 10.1007/978-3-319-24397-9

Library of Congress Control Number: 2015958728

Springer Cham Heidelberg New York Dordrecht London

Printed on acid-free paper

Springer International Publishing AG Switzerland is part of Springer Science+Business Media (www.springer.com)

(*Photo courtesy: Chester N. Roistacher*)

This book is dedicated to John B. Carpenter in recognition of his pioneering work and worldwide reputation as an authority on date palm culture, particularly with regard to pathology and integrated pest management. He was born in 1915 and graduated from the

University of Idaho. He received his PhD in plant pathology from the University of Wisconsin at Madison in 1941. His initial research was on fire blight of apples and Phytophthora diseases of potatoes. After working as Senior Plant Pathologist for the USDA Rubber Disease Investigation Unit in La Hulera, Turrialba, Costa Rica, he returned to USA in 1954 and began a long career as Senior Plant Pathologist at the USDA Date and Citrus Station in Indio, California. During the latter part of his career, Carpenter served as the station superintendent, retiring in 1982 when the Indio Station was closed by an Act of Congress. In the 1980s, Carpenter was appointed as Cooperative Extension Specialist and Professor Emeritus at the University of California, Riverside, USA. He also had an appointment as Research Associate (with faculty privileges) in the Department of Plant Pathology at the University of California, Riverside. In addition to being a prominent citrus pathologist, he oversaw the U.S. Date and Citrus Station's date palm germplasm collection and breeding program after the retirement of J.R. Furr. When the Indio station was closed, he was instrumental in salvaging important date palm germplasm, ensuring its survival and continued existence. After his retirement, Carpenter consulted in many countries regarding date palm culture, resulting in many improvements in date palm culture around the world. Carpenter was the co-author with H.S. Elmer of the book "Pests and Diseases of the Date Palm", which was for many years the standard reference in this field and still contains authoritative information on many date palm pests and diseases.

Foreword

Invasive species are an ineluctable problem in agriculture. Today, with rapidly increasing mobility, commerce and transportation, we have to deal with invasive species occurring at an increasing rate. Moreover, in the modern era, crop plants have been dispersed around the world as market demands create opportunities.

It is estimated that worldwide there are over 100 million date palms, *Phoenix dactylifera* L., cultivated mostly in the arid regions of the world. Increasing global production of dates has significantly improved the livelihood and income security among rural communities of date farmers in several date producing countries, leading to increased internal and external trade of dates and date-based products. The world production of dates has significantly increased from 1.8 million tons in 1962 to nearly 8.0 million in 2012. Climate change, due to global warming, is adversely impacting mankind. This is especially deleterious in arid countries of the world where desertification is a common phenomenon. In this context, the date palm is very useful, for its capability to withstand adverse climate changes; simultaneously providing an environment with a micro-climate in the desert, enabling farmers to grow multiple crops.

Countries with rich agricultural heritage have found it necessary to provide their growers with guidance and advice for agricultural pest and disease problems. This volume on Date Palm Integrated Pest Management (IPM) is a part of an increasing trend to provide such guidance and advice globally. Certainly, any country that harvests dates, and even those using other palms as ornamentals, are in need of advice and information about pest management.

When new pests arrive or finally catch up to date palms that have been cultivated far from their original location, local experts are forced to turn to countries which have dealt with the same pest for many years for primary advice and information. Moreover, foreign exploration for natural enemies of pest insects is an important aspect of IPM.

Increasingly, all countries are engaged in IPM jointly; regardless of the exact agricultural portfolio that they support. The principles employed try to achieve sustainability and support biodiversity. We face common enemies and challenges. The current revolution occurring in electronic communication means that we are in good contact with a greater audience than ever before. This will certainly lead to new opportunities and innovations coming from increased interactions in all fields of science including IPM. Intensification of sustainable crop production has been FAO's strategic objective. In order to achieve this goal, FAO has endorsed the "Ecosystem Approach" in agricultural management, including IPM, combining different management practices to grow high quality crops and minimize pesticide use. In this context, this book comprehensively addresses various issues related to sustainable IPM of date palm pests through the expertise of leading authors worldwide. It will serve as a guide and resource to administrators, researchers, academicians, students and farmers.

University of Agriculture
Faisalabad, Pakistan
April, 2015

Iqrar Ahmad Khan
Vice Chancellor

Preface

Date palm, *Phoenix dactylifera* L. (Arecales: Arecaceae), is important to the agrarian economy of several countries in arid regions of the world. Enhanced monoculture of date palm in several date palm growing countries coupled with global warming, unrestrained use of chemical insecticides, and extensive international trade is likely to have an impact on the species diversity and density of pest complex and related natural enemies in the date palm agro-ecosystem. During the last two decades there has been a significant increase in the production area of date palms especially in the Middle East and North Africa, which are the pre-dominant date palm growing regions worldwide. The Food and Agriculture Organization of the UN estimates that there are over 100 million date palms with an annual world's production of nearly 8 million tons. The development of sustainable pest management strategies in date palm is vital to meet the existing and emerging pest challenges.

The crop is attacked by a wide range of insect pests and phytophagous mites, causing serious losses in yield and is some cases death of palm trees. The increase in monoculture plantations of palms during the last two decades has been accompanied by an increase in arthropod pest challenges. Additionally, climate change and large scale movement of palm species for farming and ornamental gardening has also increased the invasive species events compounding the problems of crop protection.

Over the past 20 years those of us living in the temperate zone have been witness to an unprecedented spread of classic pests of palm trees as invasive species. The effect of red palm weevil, in particular, in the greater Mediterranean area has been catastrophic. Iconic palm trees, used largely as ornamentals, are being protected by extraordinary measures including insecticide drenches. This only underscores how unprepared we were for such invasions. It is perhaps a little more difficult to understand how the same species reached the islands of Curaçao and Aruba in the Americas during 2009. In this context the role of pre- and post-entry quarantine regimes is becoming increasingly important in today's world where regional and even intercontinental shipments of planting material have become common.

The responses to events like these are usually the same. First, the invasive species has to be identified, which reminds us of the importance of systematic taxonomy. Moreover, in the modern era, we have learned that a small change in the gut symbiotic population of a given insect can and has changed the whole biology including host choice. Thus insect identification has reached a more complex level and we are forced to adopt microbiological methods. Second, some form of field identification or trapping has to be devised to get accurate information about the location and spread of the invasive species population. Often this response draws from the powerful field of chemical ecology coupled with insect behaviorists. Population biologists can then describe the dynamics of such invasions and economic impact studies can put the event into a wider perspective. Third, some form of pest management must be supplied based on the expertise of economic entomologists. Most often this involves emergency treatment with neurotoxic insecticides to hold back the threat until the sophistication of more sustainable biological control approaches can be brought to bear. All of this is made more complicated by modern views of organic growing, which restrict the types of materials that can be used.

Ideally, over time, invasive species revert to a dynamic balance where control is achieved and life goes on. We have learned of the extreme difficulty of achieving local elimination once an invasive species has become established. Such programs usually involve mass rearing and releases, such as in Sterile Insect Technique strategies, and thus are very expensive, so that, economic impact is the deciding factor.

Renowned palm scientists have comprehensively addressed issues pertaining to the life cycle, damage, losses, geographical distribution, host range and management of major insect pests of date palm of the orders viz. Coleoptera (red palm weevil, long horn beetle, rhinoceros beetle, frond borer, sap beetles), Hemiptera (dubas bug, scale insects), Lepidoptera (lesser date moth, carob moth, raisin moth) and mites. Furthermore, the role of semiochemicals in date palm pest management, innovative methods for managing storage pests of dates and implications of phytoplasmas and their insect vectors in date palm are also discussed. This book on sustainable date palm IPM is intended for farmers, students, researchers and administrators involved in the date palm industry.

We would like to thank all the authors for contributing excellent chapters to this book. We are greatful to Prof. Dr. Aurelio Ciancio, Bari, Italy for including this volume in his book series "Sustainable Plant and Crop Protection", and for his encouragement and valuable suggestions throughout the project. We extend our thanks to Zuzana Bernhart and Mariska van der Stigchel, Springer Dordrecht, The Netherlands for their facilitation in successful production of this book.

Faisalabad, Punjab, Pakistan — Waqas Wakil
Al-Ahsa, Saudi Arabia — Jose Romeno Faleiro
Riverside, CA, USA — Thomas A. Miller

Preface

This is the second Volume of the Series *Sustainability in Plant and Crop Protection* (SUPP). The series presents each year a topic coverage dedicated to a specific theme, examined in a perspective related to crops sustainability. In particular, it targets issues in plant and crop protection, including not only innovative approaches and data in pest and disease control but also the management of biotic and abiotic stresses. This second volume affords the study of protection in date palm productions and has been produced thanks to the endless efforts and detailed studies provided by its editors, Dr. Waqas Wakil, Dr. Jose R. Faleiro, Dr. Thomas A. Miller and by all the other contributing authors.

Date palm is a challenging crop, not only for the large number of pests that attack dates and stored fruits but also for the peculiar traits of the environment in which productions take place, the dimension of the palm trees and the complexity of the cropping cycles.

It is, furthermore, a very ancient crop. Date palm domestication may be tracked back to the 5th millennium B.C. when, after its likely adoption from wild ancestors, it largely contributed to the success of agriculture in desert or semi-arid environments, otherwise unsuitable for human settlement. Date palm groves provided stable sources of sugars and vitamins and have been integrated with other fruit trees like fig, grape, apple or pomegranate, shaded in various layers under the palm fronds in the cultivated gardens. Cereals, vegetables and pulses were often present in the groves, thus integrating farmers' diet and contributing to the development of agricultural and selection practices.

Date palm is also one early example of irrigated crop, a practice which in turn contributed to the development of ducts, wells or canals, with related machineries and technologies. Irrigation can be dated back to the 3rd millennium to become fully established by the 1st millennium B.C. This crop, hence, largely sustained for several millennia the agriculture and landscape of the desert or peri-fluvial oasis or environments, in a wide territory ranging from North Africa and Middle East to Central Asia. Furthermore, the palm has an overall symbolic meaning, not only because of its intrinsic symmetry and ornamental beauty but also for its historical role in providing food for many societies and settlements, as well as for shaping the art, culture and traditions of many people.

It is hence with interest that the reader may look at this volume. It provides comprehensive reviews of main pests and parasites of date palm, of their impacts and of the possible control solutions adopted in different regions of the globe, including the Americas. Many of the palm threats result by a wide array of insect pests which are listed, described and illustrated in several chapters. The recent outbreak of the red palm weevil, *Rhynchophorus ferrugineus*, in the Mediterranean regions, is just one of the many examples of major threats examined in detail.

Although many measures applied still consider the application of synthetic pesticides, a number of studies and assays dealing with other, more sustainable control practices including biological, physical or non-chemical approaches are also discussed. These options require an increase in the production of knowledge, and in many cases future research guidelines are indicated. In particular, these also consider the substitution of the use of pesticides by information-based and more resource-conservative approaches.

The production of new data on the ecology, taxonomy and genetics of pests as well as on the cultural or management aspects of date palm represents, indeed, a challenge for the years to come. The volume, hence, contributes to present, organize and discuss key aspects of the actual sustainability issues of this crop. The amount of information provided is impressive, with a vast bibliography, updating actual knowledge through the research data produced by the field work and experience of the authors who represent a leading edge in the field, either for their experience or comprehensive contributions.

Aurelio Ciancio

Contents

Contributors

Mohammad Ali Al-Deeb Department of Biology, College of Science, United Arab Emirates University, Al-Ain, United Arab Emirates

Khalid A. Alhudaib Plant Protection Department, College of Agricultural and Food Sciences, King Faisal University, Hofuf, Alhasa, Saudi Arabia

Geoffrey O. Bedford Department of Biological Sciences, Macquarie University, NSW, Australia

Assunta Bertaccini *Alma Mater Studiorum*, University of Bologna, Bologna, Italy

Charles S. Burks USDA-ARS, Commodity Protection and Quality, South Riverbend Avenue, Parlier, CA, USA

Gilberto J. De Moraes Departamento de Entomologia e Acarologia, Escola Superior de Agricultura "Luiz de Queiroz" (ESALQ)-Universidade de São Paulo (USP), Piracicaba, SP, Brazil

Óscar Dembilio Department of Agricultural and Environmental Sciences, Campus del Riu Sec, Universitat Jaume I, Castelló de la Plana, Spain

Ravi Durvasula Center for Global Health, Department of Internal Medicine, UNM School of Medicine, Albuquerque, NM, USA

Hamadttu A.F. El-Shafie Date Palm Research Center of Excellence, King Faisal University, Hofuf, Al-Ahsa, Saudi Arabia

Mevlüt Emekci Department of Plant Protection, Faculty of Agriculture, Ankara University, Ankara, Turkey

Jose Romeno Faleiro Food and Agriculture Organization of the United Nations (UTF/SAU/043/SAU), Date Palm Research Center, Al-Ahsa, Saudi Arabia

Nina Fedoroff Center for Desert Agriculture, King Abdullah University of Science and Technology, Thuwal, Saudi Arabia

Murray J. Fletcher New South Wales Department of Primary Industries, Charles Sturt University, Orange, Australia

David Gopurenko NSW Department of Primary Industries, Wagga Wagga, NSW, Australia

Geoff M. Gurr Fujian Agriculture and Forestry University, Fuzhou, China

Graham Centre Charles Sturt University and NSW Department of Primary Industries, Orange, NSW, Australia

Ally Harari Department of Entomology, Agricultural Research Organization, The Volcani Center, Bet Dagan, Israel

Ali M. Idris Center for Desert Agriculture, King Abdullah University of Science and Technology, Thuwal, Saudi Arabia

Josep A. Jaques Department of Agricultural and Environmental Sciences, Campus del Riu Sec, Universitat Jaume I, Castelló de la Plana, Spain

Mohammed Zaidan Khalaf Integrated Pest Control Research Center, Directorate of Agricultural Research, Ministry of Science and Technology, Baghdad, Iraq

Robert R. Krueger USDA-ARS, National Clonal Germplasm Repository for Citrus and Dates, Riverside, CA, USA

Jian Liu Fujian Agriculture and Forestry University, Fuzhou, China

Thomas A. Miller Department of Entomology, University of California, Riverside, CA, USA

Kazem Mohammadpour Plant Protection Research Department, South Khorasan Agricultural and Natural Resources Research and Education Centre, AREEO, Birjand, Iran

Dave Moore CABI, Bakeham Lane, Egham, Surrey, UK

Hagit Navarro Green Storage Limited, Rishon Lezion, Israel

Shlomo Navarro Green Storage Limited, Rishon Lezion, Israel

Mohamed W. Negm Department of Plant Protection, Faculty of Agriculture, Assiut University, Assiut, Egypt

Jorge E. Peña Department of Entomology and Nematology, Tropical Research and Education Center, University of Florida, Homestead, FL, USA

Thomas M. Perring Department of Entomology, University of California, Riverside, CA, USA

Rasmi Soltani Laboratory of Plant Protection (INRA Tunis), Regional Center of Agriculture Research, Sidi Bouzid, Tunisia

Victoria Soroker Department of Entomology, Agricultural Research Organization, The Volcani Center, Bet Dagan, Israel

Waqas Wakil Department of Entomology, University of Agriculture, Faisalabad, Punjab, Pakistan

Muhammad Yasin Department of Entomology, University of Agriculture, Faisalabad, Punjab, Pakistan

About the Editors

Waqas Wakil is an Assistant Professor in the Department of Entomology, University of Agriculture, Faisalabad (UAF) Pakistan. He obtained PhD from UAF and then Doktor der Naturwissenschaften (Dr. rer. nat.) from Trier University, Trier, Germany. He has worked in many world renowned research laboratories as Visiting Scientist/Professor. His scientific interests are biological control, insect pathology and sustainable integrated pest management of insect pests. He is the author of 65 peer-reviewed research manuscripts in scientific journals. He has mentored number of PhD and MPhil students. In addition, he is the member of Editorial boards of scientific journals, the Fellow of Royal Entomological Society (UK), TWAS-UNESCO Associate and a Fulbright Scholar (USA).

Jose Romeno Faleiro is from Goa (India) and has a PhD in Entomology from the Indian Agriculture Research Institute, New Delhi, India. His professional career as a Research Entomologist began at the Indian Council of Agricultural Research (ICAR), Goa, during 1985. He is specialized in tropical insect pest management and renowned for his work on red palm weevil (RPW). His work has been published in internationally renowned peer-reviewed journals. Besides writing book chapters, he has presented invited talks on RPW in several countries. Since 2008, he has completed consultancy assignments for FAO on RPW in several countries including Saudi Arabia, UAE, Yemen, Morocco, Libya and Tunisia. During 2015, Faleiro was selected for the prestigious "Khalifa International

Date Palm Award" in the distinguished figure category by the Government of the UAE in recognition of his work on RPW in date palm sector.

Thomas A. Miller is a Professor of Entomology at the University of California, Riverside, USA. He received his BA in physics in 1962 and his PhD in entomology in 1967, both from the University of California, Riverside. Since 2010, he has served as a Jefferson Science Fellow with the U.S. Department of State. He is a Fellow of the Entomological Society of America, AAAS and the Royal Entomological Society, UK. His discoveries have included the myogenicity of insect heart beat as a graduate student, the mode of action of pyrethroid insecticides, early diapause in pink bollworm and transposable element cause of unstable genome of Heliothine moths. Additionally, Miller developed transgenic pink bollworm, perfected symbiotic control of Pierce's disease of grapevines and patented the sterilization of green coffee beans. He has also been the editor of the Contemporary Topics in Entomology series.

Chapter 1
Date Palm Production and Pest Management Challenges

Waqas Wakil, Jose Romeno Faleiro, Thomas A. Miller, Geoffrey O. Bedford, and Robert R. Krueger

Abstract Date palm, *Phoenix dactylifera*, is a monocotyledonous species belonging to the palm family (Arecaceae or Palmae) which is perennial and dioecious and cultivated mostly in the arid regions of the world. Date palm is important to the agrarian economy of several countries, with the ability to withstand severe abiotic stresses prevalent in the world's arid regions, including hot and dry climatic conditions, water stress and salinity. A recent report on the arthropod fauna of date palm lists 112 species of insects and mites associated with it worldwide, including 22 species attacking stored dates. In several date producing countries, the monoculture type of date palm cultivation, climate change, unrestrained use of chemical insecticides and extensive international trade is likely to impact the pest complex and its natural enemies in the date agroecosystems. Considering the significance of date palm, we summarize the biology and sustainable management of major insect and mite pests addressing related challenges and future research areas. The emerging role of semiochemicals in date palm IPM is described including new strategies in

W. Wakil (✉)
Department of Entomology, University of Agriculture, Faisalabad, Punjab, Pakistan
e-mail: waqaswakeel@hotmail.com; arid1972@yahoo.com

J.R. Faleiro
Food and Agriculture Organization of the United Nations (UTF/SAU/043/SAU), Date Palm Research Center, Al-Ahsa, Saudi Arabia
e-mail: jrfaleiro@yahoo.co.in

T.A. Miller
Department of Entomology, University of California, Riverside, CA, USA
e-mail: thomas.miller@ucr.edu

G.O. Bedford
Department of Biological Sciences, Macquarie University, NSW 2109, Australia
e-mail: geoffrey.bedford@mq.edu.au

R.R. Krueger
USDA-ARS, National Clonal Germplasm Repository for Citrus and Dates, Riverside, CA, USA
e-mail: robert.krueger@ars.usda.gov

W. Wakil et al. (eds.), *Sustainable Pest Management in Date Palm: Current Status and Emerging Challenges*, Sustainability in Plant and Crop Protection,
DOI 10.1007/978-3-319-24397-9_1

mating disruption, "attract-and-kill" and "push-pull" technologies. Also phytoplasma diseases and their insect vectors are discussed, besides innovative methods for managing storage pests of dates.

1.1 Introduction

Date palm, *Phoenix dactylifera* L. (Arecaceae, or Palmae) is an important species cultivated mostly in arid areas of the world. It is a perennial and dioecious with female (fruit bearing) and male (pollen bearing) plants growing separately. Date palms were cultivated 6000 years ago in the Mesopotamian region (present day Iraq) (Wrigley 1995; Zohary and Hopf 2000; Johnson et al. 2013). The importance of date palm in the economy of several date producing countries is based on its ability to withstand severe climatic conditions, water stress and salinity ranging up to 2000 ppm. Among the 100 million date palms of the world, 60 % exist in North Africa and Middle East, where it has become important to the life and culture of the people in these regions.

According to Johnson (2011) propagation of superior palm types targeting increased crop productivity and fruit quality is among the early date palm-specific technologies applied. Crop and water management of palms, tree segregation on the basis of sex, artificial pollination, assigning names to cultivars, and characterization of different growth stages of fruit, fruiting seasonality and fruit flesh texture are all other old practices and characteristics that are still important today (Johnson 2011). Among the modern date palm production technologies, plantlets are developed commercially through tissue-culture techniques, a practice that has enabled rapid expansion of planting and replanting of date palm orchards, besides incorporating traits like resistance against insect pests and pathogens, and increasing fruit yields (Johnson et al. 2013).

1.2 Production, Area and Yield

With the major proportion of world's total date palm production in the Middle East and North Africa, date palms have also been introduced in Australia, India, Mexico, Southern Africa, South America, Pakistan and the USA during the last three decades. Dates are not only a staple food for local populations in many countries, but their production also contributes significantly to the economy, society and environment of those countries (Chao and Krueger 2007).

Date palm has a wide genetic diversity due to extensive out breeding (Popenoe 1992). There are around 3000 cultivars of date palm worldwide (Zaid 2002), but some of the varieties could represent synonyms (Johnson et al. 2013). Though mainly propagated by offshoots, seedling date palms are the original source of most of the existing established cultivars in several countries. Date palms are cultivated from seeds for two major reasons; (i) germplasm preservation and (ii) conservation of desirable traits (Johnson et al. 2013). Male palms are also an important genetic resource that provides valuable pollen to pollinate the female cultivars, which is carried out artificially either manually, by inserting pollen strands in individual female flowers, or mechanically by using pollen dusters so as to ensure fruit setting and to sustain yields (Al-Wusaibai et al. 2012).

The date fruit is classified as a drupe exhibiting high variation in color, texture, shape and chemical composition that mainly depends on prevailing environmental conditions, cultivar genotype and practices as well as growing season. Date varieties are usually characterized by the fruit attributes (El-Hadrami and Al-Khayri 2012). The fruit development on the palm is categorized mainly into four stages using Arabic terms viz. Kimri, Khalal, Rutab and Tamar stages, where the averages for moisture content is around 80 %, 60 %, 40 % and 20 %, respectively (Fayadh and Al-Showiman 1990; Al-Shahib and Marshall 2002). The most desirable character in dates after production is a large fruit size (Johnson et al. 2013). According to the Food and Agriculture Organization (FAO) of the United Nations, the global date production has increased from 1.8 million tons in 1962 to nearly 8 million tons in 2012. Egypt, Saudi Arabia and Iran are the current top three date producing countries (FAOSTAT 2012) (Fig. 1.1).

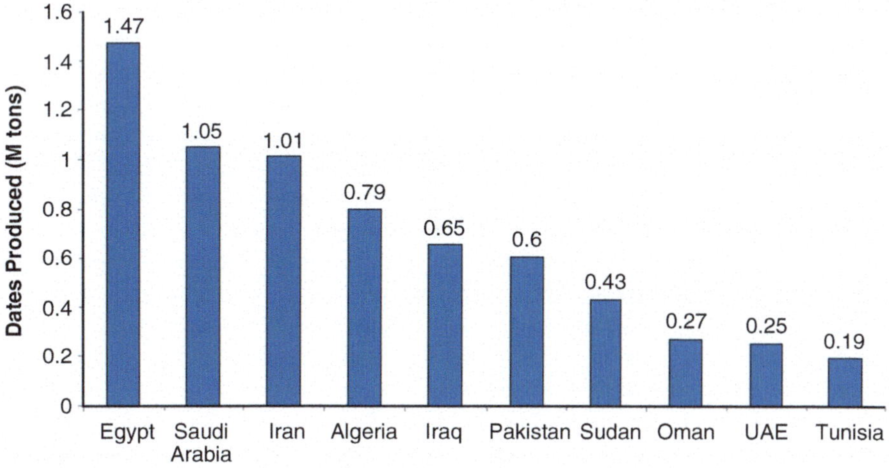

Fig. 1.1 Production of dates (million tons) in top ten producing countries during 2012 (Source: FAOSTAT 2012)

1.3 Date Palm and Global Food Security

Dates contain high amounts of essential nutrients: minerals (Mg, Ca, Fe, K), carbohydrates (total sugars 44–88 %, glucose and fructose 65–80 % of dry weight) (Al-Shahib and Marshall 2003; Chandrasekaran and Bahkali 2013), vitamins (niacin, B1, B2), dietary fibers (6.4–11.5 %), fatty acids and proteins. They could play an important role in emergency food relief programs. In certain cultures dates are considered to have several medicinal qualities viz. antifungal, antibacterial, antiulcer, immuno-modulatory and antitumor properties. Also, some date cultivars have antioxidant activity due to phenolic compounds (Vayalill 2002; Al-Farsi et al. 2005; Baloch et al. 2006; El-Hadrami and Al-Khayri 2012). Efficient utilization of raw dates and date seeds is expected to generate a number of new products by bioprocessing, especially through the exploration of these technologies on commercial scale by the pharmaceutical and food industries (Chandrasekaran and Bahkali 2013). Date seeds are also used in animal feed due to their high protein, fat and dietary fiber contents (Besbes et al. 2004). About 11–18 % of date fruit weight is from the seed, where a recent report suggests that the antioxidant content of date seed oil is comparable to that of olive oil (Abdul Afiq et al. 2013).

Increasing global production of dates will significantly improve the livelihood among rural communities of date farmers in several date producing countries, leading to increased internal and external trade of dates and date-based products. Trade of dates especially in the USA and EU is governed by international marketing norms. In North Africa, the international standards and CODEX norms are available for standardization of dates for the date varieties, like Medjool and Deglet Noor, and widely used by date importing and exporting countries. Besides the standards recommended by CODEX, standards set by the EU, USDA and date exporting countries of North Africa (Morocco and Tunisia) are also important. Dates with standardized color, size, texture, resistance against insect pests and less number of infested fruits are well accepted internationally (Anonymous 1985).

The CODEX limit for dates is 6 % of defective fruits, whether the damage is visual or contamination is due to the incidence of dead insects or presence of body parts of insects or mites, their exuvia or excreta. Currently, absence of internationally accepted standards for marketing dates is adversely impacting the date industry in several countries, especially in the Gulf Cooperation Council countries (Saudi Arabia, United Arab Emirates, Oman, Kuwait, Bahrain, Qatar) of the Middle East which accounts for around 30 % of the global date production. This important issue in the date palm industry was highlighted in the International Dates Council (IDC) establishment meeting, recently organized by Ministry of Agriculture, Saudi Arabia and FAO in Riyadh in December, 2013. Jaradat and Zaid (2004) studied quality parameters of dates in reference to consumers' preference and found that fruit size, ripening, shape and their associated interactions were primarily important, together with fruit color, softness and consumption stage accounting for most of the variability in the economic value.

1.4 Climate Change, Desertification and Date Palm

Climate change is adversely impacting human societies. The Intergovernmental Panel on Climate Change (IPCC) has estimated that by 2100 the global mean surface temperature will increase by 1–3.5 °C from the 1990 level (Cannon 2004). According to FAO over one third land area of the world is desert which mainly occurs due to prevailing weather patterns and over-exploitation by inappropriate land use. Lal (2001) indicated that dry lands of the world occupy 6.31 billion ha or nearly half of the earth's land area. This situation leads to overgrazing, deforestation and faulty irrigation practices that in turn could cause deterioration in the soil fertility and crop productivity.

The major threat to biodiversity conservation is given by desertification, particularly in the dry regions of the world. Desertification is a persistent land degradation characterized by reduced water bodies, vegetation and wildlife as a result of climate changes and human activities, thus converting a relatively dry region into arid land. Factors like climate change, land degradation, loss of biodiversity, desertification and water shortage are affecting over one billion people worldwide. Several parts of the world including the Indian sub-continent, North Africa, China, Saudi Arabia, Eastern Australia, Central Asia, Egypt, parts of Southern Africa, Northern Mexico and Southwestern USA are already experiencing water shortages. Water availability for agriculture is becoming limited with greater demands from other sectors (increased urbanization of farming regions, for example). The relationship between desertification and climate change, combined with biodiversity loss, degradation of land and water shortage makes farming in the dry lands of the world increasingly difficult and challenging.

In this context, the date palm is very useful, with its ability to withstand adverse climatic variations, in addition to providing an environment with a micro-climate for farmers to grow a variety of crops. It can also be used for bio-ethanol production as a bio-energy crop. In several countries date palm constitutes an important economic factor that provides income to a large number of farming communities particularly those residing in the arid regions. Moreover, according to FAO, the ability of date palm to tolerate harsh climatic conditions and salinity renders it a significant potential resource in holding back or combating desertification.

1.5 Insect Pests of Date Palm: Current Status, Challenges and Future Priorities

Integrated Pest Management (IPM) requires information on pest biology, ecology, sampling and monitoring, for developing action and identifying thresholds. Fully integrated pest management approaches combine elements of plant resistance, chemical, semiochemical, biological and microbial control. In this context an assessment of the pest complex and associated biological control agents is essential.

The earliest report on the insect pests of date palm goes almost one hundred years back when Buxton (1920) documented the insect pests of date palm in Mesopotamia (present day Iraq). Later, Carpenter and Elmer (1978) listed 54 species of insect pests and mites on dates and date palms. A comprehensive report from Israel on the arthropod pest complex of date palm and their management lists 16 major and 15 minor insect pests (Blumberg 2008). Recently, El-Shafie (2012), listed 112 species of mites and insects worldwide associated with date palm, including 22 species attacking stored dates. Of the arthropod pests listed in this report, only ten are considered major. This review also lists 45 predators and parasitoids associated with the insect pest complex of date palm.

Management of palm pests usually starts with the application of insecticides, but gradually progresses into the IPM mode, mainly due to the negative aspects of insecticide-based control programs. Scale insects (Hemiptera: Coccidea) cause heavy infestations in date palms, but the indiscriminate use of organophosphates against this and several other pests led farmers in Israel to gradually adopt IPM based programs. Preliminary IPM trials in date palm were aimed at controlling fruit pests and scale insects and also tissue borers like red palm weevil (RPW) in Israel (Soroker et al. 2005; Blumberg 2008).

In Iraq, Al-Jboory (2007) emphasized the importance of biological control for pests attacking date palm. Sustainable IPM programmes implemented by the Ministry of Agriculture in Iraq along with international collaborators in area-wide programs between 2009 and 2012, targeting stem borers, lesser date moth and dubas bug resulted in 90.5 %, 80 % and 96.7 % decrease in infestation levels, respectively. IPM interventions included solar light traps and hand collection of borers, application of neem based (azadirachtin) sprays against dubas and biological control of the lesser date moth employing treatments with *Bacillus thuringiensis* (http://www.icarda.org/bio-control-date-palm-pests).

Geographic Information System (GIS) based techniques and threshold based sampling plans could serve as vital tools to assess the performance of area-wide IPM programs in date palm. Massoud et al. (2012) used GIS tactics to determine the spatial-temporal distribution of RPW in food baited pheromone traps installed in vast stretches of date plantations in Al-Ahsa, Saudi Arabia, during 2009 and 2010. The fluctuation in spatial activity of RPW revealed lower weevil incidence in 2009 than in 2010, while maximum weevil activity was observed during early summer and late spring in 2009 and 2010. Furthermore, threshold based sequential sampling plans revealed that during 2011, the pheromone based IPM program against RPW in the same date palm oasis of Saudi Arabia was controlling the pest (infestations below 1 %) in the east of the oasis while needing minor adjustments (infestations approaching 1 %) in the centre of Al-Ahsa, and required major reinforcement (infestations above 1 %) in the north (Faleiro et al. 2010; Al-Shawaf et al. 2012). In Israel, the overall improvement in controlling insect and mite pests of date palm through the IPM approach involving mechanical and biological control, use of botanicals and semiochemicals together with crop management procedures enabled reduction in the use of synthetic insecticides with significant preservation of natural enemies (Blumberg 2008).

Monoculture of date palm in several growing countries, coupled with global warming, unrestrained use of chemical insecticides and extensive international trade affect the pest complex and its natural enemies. Changes are observed in the date palm agroecosystem in terms of species diversity and density of individual species per unit area. During the last two decades, there has been a significant increase in the area under date palm cultivation (FAOSTAT 2012) with several new plantations cultivated in vast monoculture stretches that offer an ideal ecological niche for biotic stresses, including insect pests and diseases. Between 1992 and 2012 the Maghreb region of North Africa (comprising of Algeria, Morocco, Tunisia, Libya and Mauritania) and the Gulf Cooperation Council (GCC) countries of the Middle East (comprising of Saudi Arabia, UAE, Sultanate of Oman, Bahrain, Kuwait and Qatar) showed a significant increase in new date plantations (Fig. 1.2).

At a plant density of 100 palms · Ha^{-1} (Zaid 2002) the new plantations in the Maghreb and GCC countries are estimated to account for around 25 million date palms (Faleiro et al. 2012). These young plantations are prone to attacks by several insect pests especially RPW, which prefers to infest date palms less than 20 years old (Abraham et al. 1998). Besides RPW, farmers will face new challenges arising either from new introductions or minor pests becoming major threats. Thus, they will have to rely on sustainable IPM techniques as proposed by Blumberg (2008) for date palm farmers in Israel, which integrates semiochemical mediated mating disruption or "attract-and-kill" methods, improved monitoring systems, deployment of biocontrol agents and inundating releases of egg parasitoids.

Recent genome sequencing in date palm indicates that stress resistance genes are present in the chromosomal regions where the density of single-nucleotide polymorphisms is comparatively low (Al-Mssallem et al. 2013). Advanced techniques like gene silencing or RNA interference (RNAi), aiming at delivering durable multiple resistance traits towards important date palm pests through the development of genetically improved resistant transgenic palms, represent potential strategies for future management approaches.

Novel IPM approaches using species-specific and environment-friendly management options of insect pests include: mating to control insects by rearing, sterilizing and releasing large numbers of insects (sterile insect technique: SIT) or release of insects with dominant lethal gene (RIDL). These strategies could be pursued in the context of area-wide sustainable IPM programs. The pheromone/semiochemical-based IPM strategy most widely adapted in date palm is the one used to manage RPW (Faleiro 2006). Use of semiochemicals for behavioral manipulation involves lower risks of resistance development in target pest species, a main disadvantage associated with the use of traditional insecticides (Miller et al. 2006). Furthermore, unlike conventional insecticides where the entire crop is sprayed posing a hazard to non-target organisms, semiochemicals are deployed selectively. As new pheromones and repellents are identified and deployed against more insect pests of date palm, semiochemical-based IPM strategies will have an increasing role to play in future IPM programs.

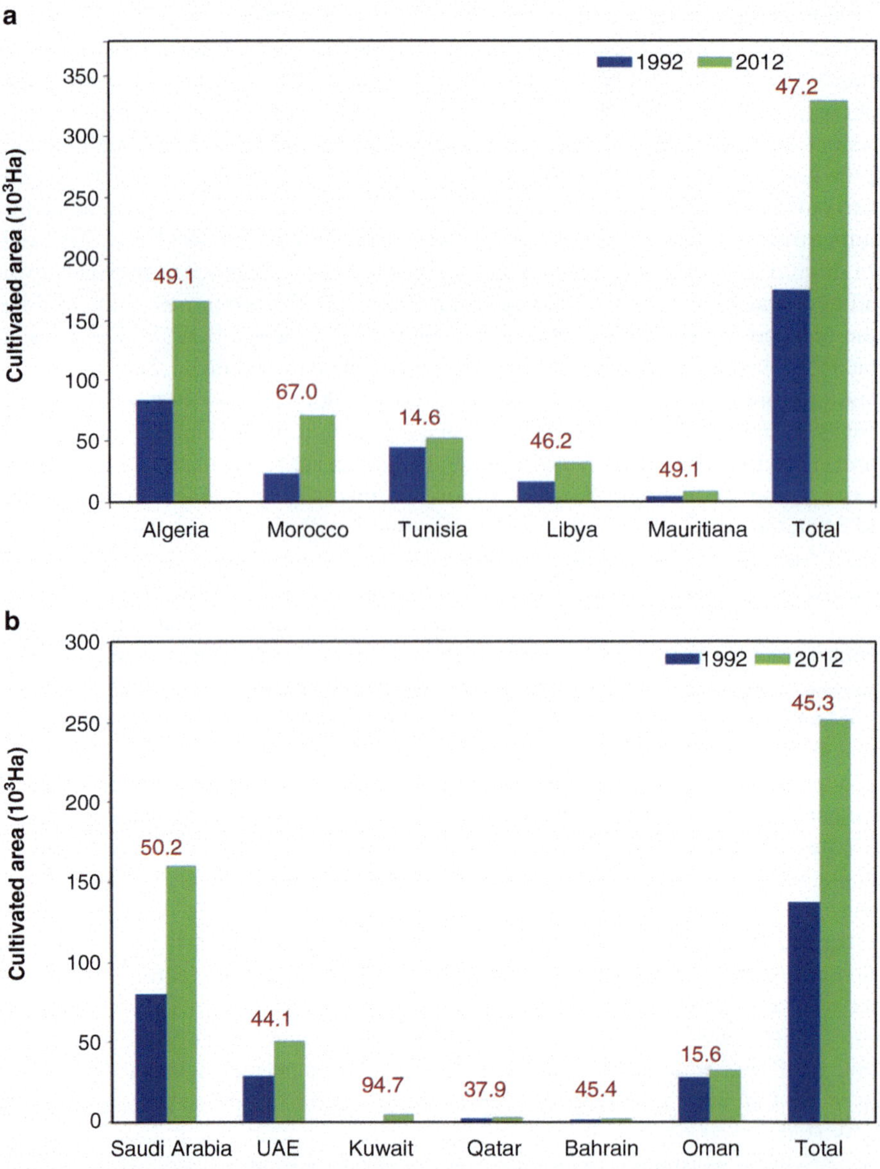

Fig. 1.2 Changes in the area under date palm cultivation (with % increase shown in *red*) in the Maghreb region of North Africa (**a**) and the GCC countries of the Middle East (**b**), between 1992 and 2012 (Source: FAOSTAT 2012)

According to the report of IPCC, the increase in mean temperature of earth surface could likely result in an increased abundance of several insects, with far reaching consequences on the way IPM strategies will evolve. According to Porter et al. (1991), climatic variations may bring changes in geographical distribution, crop-pest synchrony, inter-specific interactions, increased over-wintering, elongation of development times, increasing number of generations and increased chances of invasive pest species episodes. Furthermore, Cannon (2004) reported that as temperatures increase, insects become more abundant through various inter-related processes, such as phenological changes and range extensions, as well as population growth, development, over-wintering and migration. Consequently, the monitoring and evaluation of insect pests and their natural enemies is becoming increasingly an imperative.

The adaptation of RPW – a tropical weevil from its warm and humid speciation area in South and Southeast Asia where it is regarded as a key pest of coconut – to the arid and hot regions of the Middle East on date palm, provides a classical example of how some expected changes may affect distant agroecosystems. Large scale and rapid movement of infested planting material has resulted in the abundance of RPW population in Middle East on date palm, and also on its adaptation to the Canary Island palm, *Phoenix canariensis* Chabaud in the Mediterranean basin countries and the transfer of *Oryctes agamemnon arabicus* from the Middle East and its establishment in Tunisia.

In view of the importance of date palm as an emerging crop of the future, and of the need to develop and adopt ecologically sound and socially acceptable IPM techniques, this volume aims at anticipating the issues related to the sustainable management of major insect and mite pests on this crop, by reviewing the current IPM strategies available and addressing emerging challenges and future research priorities. The book provides an overview of the biology (life cycle, damage, losses, geographical distribution, host range) and management (monitoring, action thresholds, cultural practices, physical and mechanical measures, use of semiochemicals, biopesticides and botanicals, biological and chemical control) of major insect pests of date palm from the orders Coleoptera (RPW, long horn beetle, dynastid beetles, frond borer, sap beetles), Hemiptera (dubas date bug, issid bug, scale insects, mealy bug) and Lepidoptera (lesser date moth, carob moth, raisin moth), together with other pests like mites. Issues pertaining to the role of semiochemicals in date palm IPM involving new strategies revolving around "attract-and-kill" and "push-pull" technologies, phytoplasmas, their insect vectors and innovative methods for managing storage pests of dates are also addressed.

References

Abdul Afiq, M. J., Abdul Rahman, R., Che Man, Y. B., Al-Kahtani, H. A., & Mansor, T. S. T. (2013). Date seed and date seed oil. *International Food Research Journal, 20*(5), 2035–2043.

Abraham, V. A., Al Shuaibi, M., Faleiro, J. R., Abozuhairah, R. A., & Vidyasagar, P. S. P. V. (1998). An integrated approach for the management of red palm weevil *Rhynchophorus ferrugineus* Oliv. – a key pest of date palm in the Middle-East. *Sultan Qaboos University Journal of Scientific Research (Agricultural Science), 3*, 77–83.

Al-Farsi, M., Alasalvar, C., Morris, A., Barron, M., & Shahidi, F. (2005). Compositional and sensory characteristics of three native sun-dried date (*Phoenix dactylifera* L) varieties grown in Oman. *Journal of Agricultural and Food Chemistry, 53*, 7586–7591.

Al-Jboory, I. J. (2007). Survey and identification of the biotic factors in date palm environment and its application for designing IPM-program of date palm pests in Iraq. *University of Aden Journal of Natural and Applied Sciences, 11*, 423–457.

Al-Mssallem, I. S., et al. (2013). Genome sequence of the date palm *Phoenix dactylifera* L. *Nature Communications, 4*, 2274. doi:10.1038/ncomms3274. 1–9 pp.

Al-Shahib, W., & Marshall, R. J. (2002). Dietary fibre content of 13 varieties of date palm (*Phoenix dactylifera* L.). *Journal of Food Science and Technology, 37*, 719–721.

Al-Shahib, W., & Marshall, R. J. (2003). The fruit of the date palm: It's possible use as the best food for the future? *International Journal of Food Sciences and Nutrition, 24*, 247–259.

Al-Shawaf, A. M., Al-Abdan, S., Al-Abbad, A. H., Ben Abdallah, A., & Faleiro, J. R. (2012). Validating area-wide management of *Rhynchophorus ferrugineus* (Coleoptera: Curculionidae) in date plantation of Al-Hassa. *Indian Journal of Plant Protection, 40*(4), 255–259.

Al-Wusaibai, N. A., Ben Abdallah, A., Al-Husainai, M. S., Al-Salman, H., & Elballaj, M. (2012). A comparative study between mechanical and manual pollination in two premier Saudi Arabian date palm cultivars. *Indian Journal of Science and Technology, 5*(4), 2487–2490.

Anonymous. (1985). Codex standards for dates. CODEX STAN. 6 p.

Baloch, M. K., Saleem, S. A., Ahmad, K., Baloch, A. K., & Baloch, W. A. (2006). Impact of controlled atmosphere on the stability of Dhakki dates. *Swiss Society of Food Science and Technology, 39*, 671–676.

Besbes, S., Blecker, C., Deroanne, C., Bahloul, N., Drira, N. E., & Attia, H. (2004). Date seeds: Chemical composition and characteristic profiles of the lipid fraction. *Food Chemistry, 84*, 577–584.

Blumberg, D. (2008). Review: Date palm arthropod pests and their management in Israel. *Phytoparasitica, 36*(5), 411–448.

Buxton, B. A. (1920). Insect pests of date and date palm in Mesopotamia and elsewhere. *Bulletin of Entomological Research, 11*, 287–303.

Cannon, R. J. C. (2004). The implications of predicted climate change for insect pests in the UK, with emphasis on non-indigenous species. *Global Change Biology, 4*(7), 785–796.

Carpenter, J. B., & Elmer, H. S. (1978). *Pests and diseases of the date palm* (Agricultural Research Service Handbook. No. 527; 42 pp). Washington, DC: United States Department of Agriculture.

Chandrasekaran, M., & Bahkali, H. A. (2013). Valorization of date palm (*Phoenix dactylifera*) fruit processing by-products and wastes using bioprocess technology – Review. *Saudi Journal of Biological Sciences, 20*, 105–120.

Chao, T. C., & Krueger, R. R. (2007). The date palm (*Phoenix dactylifera* L.): Overview of biology, uses, and cultivation. *Horticulture Science, 42*(5), 1077–1082.

El-Hadrami, A., & Al-Khayri, J. M. (2012). Socioeconomic and traditional importance of date palm. *Emirates Journal of Food and Agriculture, 24*(5), 371–385.

El-Shafie, H. A. F. (2012). Review: List of arthropod pests and their natural enemies identified worldwide on date palm, *Phoenix dactylifera* L. *Agriculture and Biology Journal of North America, 3*(12), 516–524.

Faleiro, J. R. (2006). A review of the issues and management of the red palm weevil *Rhynchophorus ferrugineus* (Coleoptera: Rhynchophoridae) in coconut and date palm during the last one hundred years. *International Journal of Tropical Insect Science, 26*, 135–154.

Faleiro, J. R., Abdallah, B. A., & Ashok, K. J. (2010). Sequential sampling plan for area-wide management of *Rhynchophorus ferrugineus* (Olivier) in date palm plantations of Saudi Arabia. *International Journal of Tropical Insect Science, 30*(3), 145–153.

Faleiro, J. R., Ben Abdallah, A., El Bellaj, M., Al-Ajlan, A. M., & Oihabi, A. (2012). Threat of red palm weevil, *Rhynchophorus ferrugineus* (Olivier) to date plantations of the Maghreb region in North Africa. *Arab Journal of Plant Protection, 30*, 274–280.

FAOSTAT. (2012). Food and agricultural commodities production. Available at: http://faostat.fao.org/site/567/default.aspx#ancor. Accessed on 29 Apr 2015.

Fayadh, J. M., & Al-Showiman, S. S. (1990). Chemical composition of date palm (*Phoenix dactylifera* L). *Journal of the Chemical Society of Pakistan, 12*, 84–103.

Jaradat, A. A., & Zaid, A. (2004). Quality traits of date palm fruits in a center of origin and center of diversity. *Food, Agriculture and Environment, 2*(1), 208–217.

Johnson, D. V. (2011). Date palm biotechnology from theory to practice. In S. M. Jain, J. M. Al-Khayri, & D. V. Johnson (Eds.), *Date palm biotechnology* (pp. 1–11). Dordrecht: Springer.

Johnson, D. V., Al-Khayri, J. M., & Jain, S. M. (2013). Seedling date palms (*Phoenix dactylifera* L.) as genetic resources. *Emirates Journal of Food and Agriculture, 25*(11), 809–830.

Lal, R. (2001). Potential of desertification control to sequester carbon and mitigate the greenhouse effect. *Climatic Change, 15*, 35–72.

Massoud, M. A., Sallam, A. A., Faleiro, J. R., & Al-Abdan, S. (2012). Geographic information system-based study to ascertain the spatial and temporal spread of red palm weevil *Rhynchophorus ferrugineus* (Coleoptera: Curculionidae) in date plantations. *International Journal of Tropical Insect Science, 32*(2), 108–115.

Miller, J. R., Gut, L. J., de Lame, F. M., & Stelinski, L. L. (2006). Differentiation of competitive vs. non-competitive mechanisms mediating disruption of moth sexual communication by point sources of sex pheromone (Part II): Case studies. *Journal of Chemical Ecology, 32*, 2115–2143.

Popenoe, P. (1992). The pollination of the date palm. *Journal of the American Oriental Society, 42*, 343–354.

Porter, J. H., Parry, M. L., & Carter, T. R. (1991). The potential effects of climatic change on agricultural insect pests. *Agricultural and Forest Meteorology, 57*(1–3), 221–240.

Soroker, V., Blumberg, D., Haberman, A., Hamburger-Rishard, M., Reneh, S., Talebaev, S., Anshelevich, L., & Harari, A. R. (2005). Current status of RPW in date palm plantations in Israel. *Phytoparasitica, 33*, 97–106.

Vayalill, P. K. (2002). Antioxidant and antimutagenic properties of aqueous extract of date fruit (*Phoenix dactylifera* L. Arecaceae). *Journal of Agricultural and Food Chemistry, 50*, 610–617.

Wrigley, G. (1995). Date-palm (*Phoenix dactylifera* L.). In J. Smartt & N. W. Simmonds (Eds.), *The evolution of crop plants* (2nd ed., pp. 399–403). Essex: Essex Longman.

Zaid, A. (2002). *Date palm cultivation* (FAO Plant Production and Protection Paper No. 156, Revision 1). Rome: Food and Agriculture Organization of the United Nations.

Zohary, D., & Hopf, M. (2000). *Domestication of plants in the old world: The origin and spread of cultivated plants in west Asia, Europe, and the Nile Valley* (3rd ed.). Oxford University Press Inc., New York, USA. 328 pp.

Chapter 2
Biology and Management of Red Palm Weevil

Óscar Dembilio and Josep A. Jaques

Abstract The red palm weevil, *Rhynchophorus ferrugineus*, an indigenous species from South East Asia, has recently become one of the most dangerous pests of palms around the globe. The weevil is extensively dispersed in the Old World, and described on 26 palm species in 16 different genera. In date palm, *Phoenix dactylifera*, control methods revolve around the use of food baited pheromone traps, while in the Canary Island date palm, *Phoenix canariensis*, control of this pest is mostly based on the use of insecticides. Environmental and non-target side effects argue for development of more environment friendly and sustainable management practices. In this context, there is an urgent need to (a) develop and deploy tools for early detection of infested palms, (b) enhance the knowledge of the relationships between *R. ferrugineus* and its hosts, especially the preferred association with *P. canariensis* in the Mediterranean basin, and (c) explore alternatives to synthetic organic insecticides. The latter include biopesticides and new semiochemical mediated strategies involving "attract-and-kill", "attract-and-infect" and "push-pull" techniques for eco-friendly management of this insect pest.

2.1 Introduction

The red palm weevil (RPW), *Rhynchophorus ferrugineus* (Olivier) (Coleoptera: Curculionidae) is indigenous to South Asia, but has quickly spread during the last three decades mostly due to the movement of infested palms from the Middle East and Africa to the Mediterranean area. With an extensive geographical distribution in different climates and a broad host range, this pest has been reported in Asia, Oceania, Africa, Europe and South America. The RPW reportedly attacks over 26 species of palm belonging to 16 genera globally. It was originally reported on coconut (*Cocos nucifera* L.) as a pest in Southeast Asia, but now it is known as the major

Ó. Dembilio (✉) • J.A. Jaques
Department of Agricultural and Environmental Sciences, Campus del Riu Sec, Universitat Jaume I, 12071 Castelló de la Plana, Spain
e-mail: odembili@uji.es; josep.jaques@camn.uji.es

W. Wakil et al. (eds.), *Sustainable Pest Management in Date Palm: Current Status and Emerging Challenges*, Sustainability in Plant and Crop Protection,
DOI 10.1007/978-3-319-24397-9_2

pest of the date palms including *Phoenix canariensis* Chabaud and *P. dactylifera* L. in several regions of the world (Morici 1998).

Severe quarantine restrictions have been imposed by different countries to avoid the spread of this notorious pest. The early detection of RPW infestations is critical to prevent palm death and thus serves as an essential element of a successful Integrated Pest Management (IPM) program used to control RPW. As the first signs of RPW infestation are difficult to detect and only later stages of attack are conspicuous, efforts to develop better early detection methods are underway. Once palms are infested by RPW, they are very difficult to treat, especially *P. canariensis* that often dies from the extensive damage already caused by the pest till detection. However, if infested palms are detected and treated in the early stages of attack, there is a good chance of palm recovery. IPM programs, comprising agronomic practices (sanitation), biological and chemical controls and the use of semiochemicals for monitoring and mass trapping of the pest, have been developed and implemented around the world. The present chapter summarizes the research on different aspects of the biology of RPW, and all recent advances related to its management to mitigate the economic impact.

2.2 Distribution and Host Range

Rhynchophorus ferrugineus is one of the most devastating insect pests of palm trees. According to a survey conducted in 2008 (EPPO 2008a, 2009), there is a wide distribution of this pest in Asia, Oceania, Northern Africa, Europe and in the Caribbean (Aruba and Curaçao). This pest is now present on all continents of the world, except Antarctica. It was first described attacking *Cocos nucifera* in South and Southeast Asia, *P. dactylifera* in the Middle East and other regions wherever these two palm species overlap (Fiaboe et al. 2012).

RPW has been described on more than 20 palm species and almost on all Arecaceae including *Arecastrum romanzoffianum* (Cham.) Becc., *Areca catechu* (L.), *Borassus flabellifer* (L.), *Calamus merrillii* (Becc.), *Arenga pinnata* (Wurmb) Merr., *Caryota cumingii* Lodd. ex Mart., *C. nucifera*, *Chamaerops humilis* (L.), *Corypha utan* (Lam.), *Metroxylon sagu* (Rottb.), *Elaeis guineensis* (Jacq.), *Livistona decipiens* (W. Bull) Dowe, *L. chinensis* (Jacquin) R. Brown (ex Martius in C.F.P. von Martius et al.), *Oncosperma horridum* (Griff.) Scheff., *O. tigillarium* (Jack) Ridl., *Roystonea regia* (Kunth) O.F. Cook, *P. dactylifera*, *P. canariensis*, *Washingtonia robusta* H. Wendl. and *Trachycarpus fortunei* (Hook.) H. Wendl. All these species are more or less susceptible to the weevil (Esteban-Durán et al. 1998b; Malumphy and Moran 2007; Dembilio et al. 2009; OJEU 2008; EPPO 2008a, 2009). Although, RPW is native to Central, South and Southeast Asia and is reported primarily from *C. nucifera* (Wattanapongsiri 1966a), only 15 % of the coconut growing countries have reported this pest (Faleiro 2006; Faleiro et al. 2002, 2012, 2014). On *P. dactylifera*, the spread has been rapid for the last two decades, and now it is reported on *P. canariensis* from the entire Mediterranean basin and most of the date palm growing countries.

Chamaerops humilis is a native European species initially thought to exhibit resistance against *R. ferrugineus* (Barranco et al. 2000). But this palm along with *Washingtonia* spp. was included in the list of vulnerable plant species by the European Union (OJEU 2008). However, experiments conducted in 2007 showed that *R. ferrugineus* did not significantly infest *W. filifera* (Lindl.) H. Wendl., whereas *W. robusta* was highly susceptible (Dembilio et al. 2009). Moreover, *C. humilis* could not be naturally infested by RPW. Antixenosis was the major resistance mechanism occurring in this plant species. Interestingly, this mechanism could be by-passed in palms previously damaged, e.g. suffering an attack from the lepidopteran *Paysandisia archon* Burmeister (EPPO 2008b; Dembilio et al. 2009).

Another palm species, *Phoenix theophrasti* Greuter, is indigenous to the eastern Mediterranean area and only found in Crete and some Aegean Islands (Greece) (IUCN 2010). The susceptibility of *P. theophrasti* to RPW was studied using 4 year old plants which were not detected with any natural infestation, but first instar larvae introduced into artificially drilled holes in the trunks did cause infestation. This is in line with some previous findings for *C. humilis*, where infestation of formerly harmed palms (either mechanically by wind or by the activity of other palm pests) served as natural hosts for RPW (Barranco et al. 2000).

2.3 Economic Importance and Damage

The palm family Arecaceae contains 250 genera with about 2,900 species mainly tropical, with some representatives in warm temperate regions (Dransfield 1978). *Phoenix* is one of the largest genus in this family and most popular as containing commercially important palm species abundantly used in landscaping. These are very adaptable palm species growing in variable climates i.e. from the tropics to the deserts, and even in the cooler subtropical and temperate climates throughout the world. As a group, these palm species are probably the most commonly grown palms in the world, particularly in Canary Islands across northern and Central Africa, southeast of Europe and southern Asia including China and Malaysia. The fruit of *P. dactylifera* is high in energy (1 kg pulp of prepared dates provides 3,000 cal) and is a staple diet of people in several countries of the world (Al-Sahib and Marshall 2003). Other *Phoenix* species have only a thin layer of fruit pulp. Among the latter, *P. canariensis*, is an indigenous palm to the Canary Islands (Barrow 1998) where its sap is used to produce palm syrup.

Besides their value as food for humans and cattle, palms are largely used in interior and landscapes. Some monumental specimens have a high ornamental value, especially those at historic sites, like the palm grove of Elche (Valencia, Spain), which is included in the UNESCO World Heritage Site list (UNESCO 2014). In developed countries, like USA, Japan, Australia and Southern Europe, palms are grown by the ornamental industry for use as interior and field-grown plants in landscapes. In contrast, palm orchards are raised for the production of food, oil and several other commercial uses in northern Africa, the near East, India and Central and South America. In Spain, palms grown in urban areas and gardens account for almost half of the products of ornamental nurseries (Horticom 2008).

Larvae of RPW attack stems and growing tissues of the palms and the detection of this pest is very challenging at the start of its infestation. The RPW infestation in *P. dactylifera* usually occurs in young palms (<20 years old) on the trunk within a meter from the ground. Fresh wounds on palm tissue due to frond shaving and offshoot removal pre-dispose the palm to RPW. In *P. canariensis* where infestation mostly occurs in the crown of the palm, only thorough examination of damaged palms leads to the detection of this pest. Currently detection of infested palms in both date palm and the Canary Islands palm is done manually and largely depends on the detection skills and experience of the staff involved. The following symptoms indicate the presence of the pest: perforations in the trunk and/or crown where the digested fibers from larvae are ejected (secretion of brown sticky liquid with a characteristic bitter smell) (Fig. 2.1), and long after the infestation, crown or frond loss with dried offshoots is seen (Fig. 2.2). Secondary infections by opportunistic fungi and bacteria may occur within these injured tissues that contribute to the deterioration of the palm.

2.4 Biology

The complete life cycle of RPW is exhibited in Fig. 2.3. The eggs are laid individually by female weevils in separate holes excavated by its snout. Eggs are creamy white, shiny and oblong with an average size of 2.62×1.12 mm (Menon and Pandalai 1960). Prior to the hatching, larval mouthparts can be observed through the egg shell. Egg hatching occurs between three to several days depending on mean temperatures (Murphy and Briscoe 1999).

Larvae are creamy white, legless, pyriform, up to 50 mm long and 20 mm wide with a brown head and body comprised of 13 segments. The head capsule is brown russet-red to brilliant brown-black with strongly chitinized mouthparts. Depending on diet (Table 2.1) and prevailing temperature, the larval development may last from 24 to 128 days (Butani 1975; Salama et al. 2009). The number of larval instars varies depending on the diet or host plant. For instance, Martín and Cabello (2006) observed 17 instars of RPW, whereas Nirula (1956) described 3 instars when the pest was fed on a meridic diet. On the other hand, Dembilio and Jacas (2011) reported that RPW larvae completed a total of 13 instars in live *P. canariensis* palms. The newly emerged larvae seek their way into the palm by boring through the palm tissues and making tunnels, where damage increases with each molt. Young larvae have a tendency to feed on soft tissues present around apical meristems, then mature larvae move towards the periphery of the trunk or crown and form a cocoon enwrapped in palm fibers.

Fig. 2.1 Holes in the crown and trunk in palms produced by RPW larvae with ejected chewed-up fibers (Photos: Óscar Dembilio)

The larval stage converts to pre-pupae and this stage lasts for 3 days (Viado and Bigornia 1949). The pre-pupae then convert into pupae inside the cocoon (Murphy and Briscoe 1999). The pupae have an average size of 80×35 mm (Murphy and Briscoe 1999). Initially pupae are creamy, and then turn brown in color. They are reticulated, greatly furrowed with a shiny surface. The range of pupal development period varies from 11 to 45 days (Viado and Bigornia 1949; Esteban-Durán et al. 1998a).

Adults of RPW are large, rusty red in color (35×10 mm long) with a characteristic long and curved rostrum which comprises about one-third of the total length. On the dorsal side of the thorax the weevils exhibit dark spots. In males, the anterior dorsal half of the rostrum has short brownish setae (hairs). By contrast, the female rostrum lacks any hair, and is comparatively narrower, more curved and longer than the male rostrum. The adult weevils have well-developed wings and are capable to

Fig. 2.2 Infested palms with visible frond loss and dried offshoots (Photos: Josep A. Jaques)

undertake long flights (Lepesme 1947), and to cover long distances of 500–800 m (Wattanapongsiri 1966b). Computer-aided flight mill studies to analyze the flying ability of *R. ferrugineus* revealed that 54 % adult weevils are short distance flyers (covering <100 m), 36 were classified as medium (100–5,000 m), and 10 % were categorized as long distance (>5,000 m) flyers (Ávalos et al. 2014). The emerging adults can stay in the same palm until its meristem is completely consumed, which results in the death of either the palm (commonly in *P. canariensis*) and or its offshoots (*P. dactylifera*). Subsequently, adults disperse from the dead host to look for the new palm hosts.

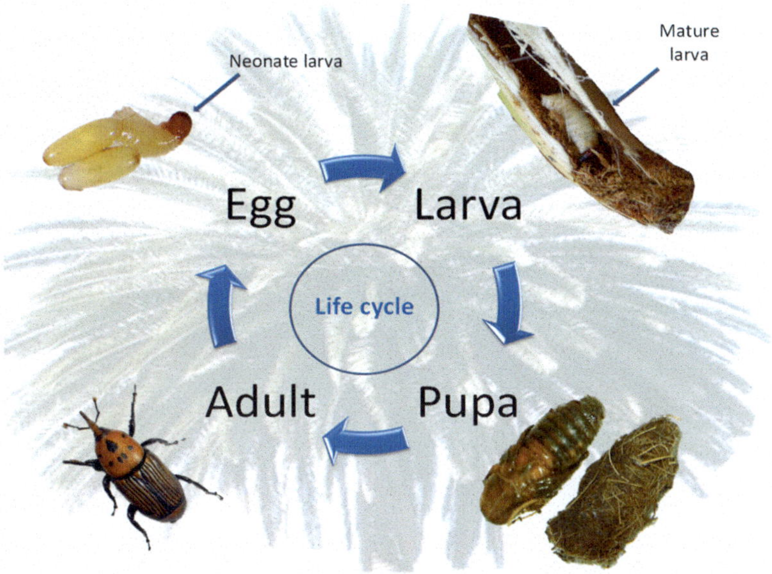

Fig. 2.3 Life cycle of *Rhynchophorus ferrugineus* (Photo: Óscar Dembilio)

Table 2.1 Effect of feeding substrate on life history parameters of *Rhynchophorus ferrugineus* (Dembilio and Jacas 2012)

Feeding substrate	Development time (days)				No. of instar	Reference
	Egg	Larva	Pupa	Total		
Honey in cotton	4–5	–	–	–	4	Shahina et al. (2009)
Sugarcane lumps	4–5	50–80	20–30	74–115	9	Shahina et al. (2009)
Apple slices	4–5	–	–	–	4	Shahina et al. (2009)
Apple slices	–	–	–	–	12	Abe et al. (2009)
Banana slices	5	90	16–20	111–115	5	Salama et al. (2009)
Sugarcane lumps	5	128	25–29	158–162	5	Salama et al. (2009)
Squash fruit	5	83	20–24	108–112	5	Salama et al. (2009)
Apple slices	5	103	16–18	124–126	5	Salama et al. (2009)
Palm crown lumps	5	69	16–19	90–93	5	Salama et al. (2009)
Sugarcane lumps	3–4	82	19	108	–	Kaakeh (2005)
Palm heart lumps	3–4	86	21	124	–	Kaakeh (2005)
Palm leaf base	3–4	84	18	119	–	Kaakeh (2005)
Artificial diet	3–4	70–102	16–23	93–131	–	Kaakeh (2005)
Sugarcane lumps	3–4	88	25	116	11–17	Martin-Molina (2004)
Artificial diet	3–4	93	30	128	7–12	Martin-Molina (2004)

(continued)

Table 2.1 (continued)

Feeding substrate	Development time (days)				No. of instar	Reference
	Egg	Larva	Pupa	Total		
Palm lumps	–	–	–	–	8–15	Martin-Molina (2004)
Banana slices	–	–	13–22	–	–	Salama et al. (2002)
Sugarcane lumps	–	81–89	–	–	7	Jaya et al. (2000)
Sugarcane lumps	–	76–102	19–45	139	–	Esteban-Duran et al. (1998)
Palm lumps	1–6	41–78	–	–	–	Avand Faghih (1996)
Not specified	2–3	60	14–21	76–84	–	Kranz et al. (1982)
Sago palm pith	–	–	–	105–210	–	Kalshoven (1981)
Sugarcane lumps	2–4	24–61	18–34	44–100	–	Butani (1975)
Sugarcane lumps	3–4	32–51	15–28	50–82	–	Rahalkar et al. (1972)
Coconut slices	2–5	36–67	12–21	54–120	3	Nirula (1956)
Coconut slices	3	35–38	11–19	49–70	9	Viado and Bigornia (1949)
Not specified	3	60	15	90–180	–	Lepesme (1947)
Not specified	3	60–120	14	74–134	–	Dammerman (1929)
Sago palm lumps	–	60	13–15	73–75	–	Leefmans (1920)
Palm lumps	3–4	25–61	18–33	48–82	–	Ghosh (1912), (1923)

2.5 Management

2.5.1 Cultural Control

Field and crop sanitation including the elimination of hidden breeding sites is vital for the successful management of RPW. Severely infested palms beyond recovery should be destroyed by shredding. Burning of palms is not recommended as palms do not burn easily when green and could result in pest stages embedded deep inside the palm, escaping the burning procedure and remaining alive. As a component of phytosanitation, it is also essential to treat freshly cut/injured palm surfaces after frond shaving and offshoot removal with insecticide so as to eliminate the palm volatiles that could attract invading female weevils for oviposition.

2.5.2 Trapping and Monitoring

Monitoring the activity of RPW is essential to protect palms against infestation (Faleiro 2006; Rochat 2006; Guarino et al. 2013; Vacas et al. 2013) and depending on mean temperatures, oviposition and egg hatching period may vary in different regions (Fig. 2.4). Mass trapping of adults using food baited pheromone traps plays an important role in the management of *R. ferrugineus* (Faleiro 2006).

Fig. 2.4 The estimated oviposition (OP) and egg hatching periods (EHP) of *Rhynchophorus ferrugineus* in different temperature regions from the Mediterranean basin (Dembilio and Jacas 2012)

Ferrugineol is the main aggregation pheromone of *R. ferrugineus* (Hallett et al. 1993; Vacas et al. 2014) and is complemented with 4-methyl-5-nonanone in mass trapping used in various countries facing the havoc of RPW (Abraham et al. 1998; Hallett et al. 1999; Vidyasagar et al. 2000). These pheromone traps pre-ferentially catch female over adult male weevils, usually in the ratio of two females for one male weevil captured (Oehlschlager 1998; Vidyasagar et al. 2000; Abraham et al. 2001; Faleiro 2005; Jayanth et al. 2007; Vacas et al. 2013, 2014). Jayanth et al. (2007) reported that 74 % of the pheromone trap-captured adults were female weevils. This result is important for the management of RPW as females are captured before they initiate oviposition triggering new infestations.

Adopting optimum trapping protocols (Hallett et al. 1999; Faleiro 2006; Giblin-Davis et al. 2013) is vital to ensure the benefits of pheromone trapping to manage RPW. The standard four window bucket trap (Faleiro 2006) is widely used in trapping this pest. Besides the pheromone lure, food bait viz. palm tissue, dates, sugarcane (Hallett et al. 1999; Faleiro 2006) mixed in 1 l water in the trap ensures bait-lure synergy and enhanced captures. Black colored RPW have been found to enhance weevil captures (Hallett et al. 1999; Abuagla and Al-Deeb 2012).

While the food bait in RPW traps needs to be replaced every 1–2 weeks, the field longevity of the commercial lure varies according to the product. Ethyl acetate, when incorporated in RPW pheromone traps along with fermenting food, also enhances captures by a factor of two to five times (Oehlschlager 1998; Sebay 2003; Al-Saoud 2013).

In date plantations, pheromone traps are set at ground level (buried half into the soil) or hung on the palm trunk, but the latter practice should be avoided as long as possible. To ensure maximum lure longevity, it is recommended to set traps under shade. In mass trapping programs, trap density depends on the density of the pest in the field and varies from one to ten traps per Ha (Faleiro et al. 2011). Additional traps would require extra work force for their periodical services (change food bait and water). In this context bait free trapping using 'attract-and-kill' technique could be a viable option to food baited traps (El-Shafie et al. 2011).

Several studies have highlighted the potential benefits of the use of food-baited pheromone traps (Faleiro et al. 2003) in mass trapping programs to keep the pest population below economic threshold. Faleiro et al. (2010) recommended that pheromone based area-wide IPM programs need to be implemented at a low infestation level of just 1 % infested date palms. Implementing area-wide management of RPW on the basis of trap captures or infestation reports can be erroneous, as they may either under- or over-estimated the pest density in the field and therefore decisions on infestation levels due to RPW in date palm based on sequential sampling or Geographic Information System (GIS) based models have been developed (Faleiro et al. 2010; Massoud et al. 2012).

2.5.3 Host Plant Resistance

The main mechanisms of host plant resistance to insect pests are categorized as antibiosis, antixenosis and tolerance (Kogan and Ortman 1978; Wiseman 1999; Ju et al. 2011). Antibiosis is a type of resistance in which the mutual interaction of host plant and insect causes physiological or developmental disorders in the insect. In antixenosis (also known as non-preference resistance) the insect is either repelled or even not attracted to its host plant. During a study carried out in Valencia (Spain), these mechanisms of resistance were studied in *C. humilis* and *W. filifera*. The results showed that *W. filifera* had both antixenotic and antibiotic mechanisms of resistance to *R. ferrugineus*, whereas *C. humilis* had antixenosis only (Dembilio et al. 2009). *Chamaerops humilis* fronds are fibrous and only the tender tissues inside the palm apex are accessible for oviposition, this could be the reason why natural infestation is unsuccessful.

Recently, Ju et al. (2011) carried out different experiments based on population growth parameters, where it was revealed that *W. filifera* and *P. canariensis* were more appropriate hosts than *Pinus sylvestris* L. for RPW. In coconut, the cultivar Chowghat Green Dwarf was highly preferred for oviposition by *R. ferrugineus* while very few eggs were laid on Malayan Yellow Dwarf (Faleiro and Rangnekar 2001).

Al-Ayedh (2008) established that the date palm cultivars containing high sugar content promote the growth of RPW. Faleiro et al. (2014) studied the mechanisms of resistance against the weevil in seven major date palm cultivars of the Al-Ahsa Oasis in Saudi Arabia. They determined the extent of attraction of female adults to fresh palm volatiles emitted from frond tissues through a four arm-choice olfactometer® (Analytical Research Systems, Inc., Florida, USA; http://www.ars-fla.com/mainpages/Bio-Assay/4&6-Choice/4&6-Choice.htm). Besides determining oviposition, the antixenosis and feeding antibiotic effects by these cultivars were also assessed. The results revealed that the popular date palm cultivar Khalas was highly susceptible to *R. ferrugineus*, and none of the cultivars tested exhibited any antibiotic effect against the pest. Further studies are needed to understand palm defense mechanisms against this pest.

2.5.4 *Plant Quarantine Treatments*

The main issue associated with infestations by RPW is the difficulty to detect early symptoms of its attack (Nakash et al. 2000; Al-Shawaf et al. 2013; Giblin-Davis et al. 2013). Red palm weevil has been unwittingly dispersed by deliberate exportation of infested plants and this situation gave rise to a rapid spread of *R. ferrugineus* in the Mediterranean basin (EPPO 2008a, b). During 2007, the EU imposed emergency preventive measures to avoid new invasions of this pest within the European Community (OJEU 2007, 2008, 2010). However, plant quarantine procedures are not included in these measures.

Chemical and physical treatments are the key protocols for managing arthropod pests from a produce. Among chemical treatments, fumigants are widely used to deal with the insect infestation problems. For instance phosphine (PH_3) and methyl bromide (CH_3Br) penetrate the commodity and are highly toxic to target pests. Methyl bromide, a commonly used fumigant globally, has been banned as a post-harvest treatment due to its role as an ozone-depleting substance, according to the UN Montreal Protocol. However, the PH_3 fumigation continues as an effective and economically feasible fumigant against a wide array of insect infestations (MARM 2014).

Llácer and Jacas (2010) studied aluminum phosphide as a safe quarantine treatment against RPW. The results of theses studies recommended that a 1.14 g/m^3 dose rate of aluminum phosphide applied for 3 days in laboratory conditions is quite sufficient to kill all stages of the weevil. Moreover, fumigation of infested *P. canariensis* palm crowns with phosphine gave a complete control of *R. ferrugineus*. Recently, Dembilio and Jaques (2015) confirmed that a dose of 1.14 g/m^3 for 2 days (1 day less than previous work) was enough to kill all stages of RPW in live palms and demonstrated that there are no phytotoxic effects for up to 1 year after the treatment. This procedure, which could be easily applied in sealed containers used for palm trade, could efficiently reduce risks associated to palm movement worldwide. These findings may provide the basis for the establishment of an excellent quarantine protocol against RPW. Researchers in Saudi Arabia developed an insecticide based quarantine protocol to treat date palm offshoots, wherein dipping of offshoot in 0.004 % fipronil for 30 min before transporting is recommended against RPW (Al-Shawaf et al. 2013).

2.5.5 Biological Control

Biological control of RPW or other *Rhynchophorus* species has been studied in different countries (Faleiro 2006; Mazza et al. 2014). In India, trials conducted in laboratory and field using a predacious black earwig, *Chelisoches morio* (F.) a natural enemy of RPW (Abraham and Kurian 1975) did not show a noticeable effect. A cytoplasmic polyhedrosis virus capable of infecting all developmental stages of RPW has been reported in Kerala, India (Gopinadhan et al. 1990). Infected larvae emerged as deformed adults and led to significant suppression of the weevil populations. Several parasitic mites of RPW have also been reported from India (Nirula 1956; Peter 1989). Recent reports from the Middle East indicate that three species of phoretic mites are associated with adult RPW collected from date plantations of the United Arab Emirates (Al-Deeb et al. 2011).

Entomopathogenic nematodes (EPNs) are also effective control measure alternative to chemicals against *R. ferrugineus* (Saleh and Alheji 2003; Elawad et al. 2007; Llácer et al. 2009; Dembilio et al. 2010a). These nematodes are harmless to non-target vertebrates and the environment. Previous field trials in date palms conducted several years ago yielded conflicting results (Abbas et al. 2001a). However, recent experiments performed by Dembilio et al. (2010a) using *Steinernema carpocapsae* in a formulation of chitosan showed good efficacies. The trials conducted under semi-natural field conditions, including both preventative and curative assays confirmed the efficacy of this nematode against RPW. The treatments were 80 % effective as a curative, and up to 98 % in preventing RPW infestation to Canary Island date palms (Llácer et al. 2009). The high efficacy of this treatment was also proven in *P. theophrasti* (Dembilio et al. 2011). The effects of *S. carpocapsae* and imidacloprid alone or as a combined treatment under field conditions were not significantly different from each other, and efficacies range was 73–95 % (Dembilio et al. 2010a). On the basis of these findings, it is suggested that EPNs should be considered while developing new strategies to control RPW.

Recently in laboratory bioassays, the populations of *Bacillus* (Family: Bacillaceae), entomopathogenic bacteria isolated from RPW cadavers were tested for their ovicidal (at four dose rates ranging 10^3–10^6 CFU/mL), and larvicidal (at a dose of 10^6 CFU/mL) effect against RPW (Francesca et al. 2015). The egg hatchability was significantly affected by nine isolates belonging to seven species, viz. *Bacillus amyloliquefaciens*, *B. cereus*, *B. licheniformis*, *B. megaterium*, *B. pumilus*, *B. subtilis* and *Lysinibacillus sphaericus*. However, *B. licheniformis* (CG62) was the only strain that gave a satisfactory control of RPW larvae, rendering these bacterial isolates more useful as a preventive treatments rather than as curative measures (Francesca et al. 2015).

Entomopathogenic fungi (EPF) constitute another possible biological means to control RPW. On contact with the host insect cuticle, EPF spores germinate and grow through the insect body. The infection spreads in two ways, firstly by direct treatment/contact, and secondly by spread from infected cadavers (horizontal transmission) (Quesada-Moraga et al. 2004). These exclusive qualities make EPF potential candidates in the management of cryptic insects like RPW. The effectiveness of various strains of *Beauveria bassiana* (Balsamo) Vuillemin and *Metarhizium anisopliae* (Metchnikoff) Sorokin has been evaluated against this weevil (Gindin et al. 2006; Dembilio et al. 2010b; Ricaño et al. 2013; Lo Verde et al. 2015). Different strains of these fungi have also been isolated from field collected specimens of RPW.

In 2005–2006, *B. bassiana* isolated from *R. ferrugineus* was used by El-Sufty et al. (2009) in field trials. The fungus generated mortality ranging from 12.8 to 47.1 % in the adult population of RPW. More recently, an indigenous isolate of *B. bassiana* obtained from a cadaver of RPW was successfully used to reduce the field occurrence of RPW in Egypt (Sewify et al. 2009). In 2007, pupae of RPW infected with *B. bassiana* collected from date palms in Spain, successfully infected the egg, larval and adult stages of RPW (Dembilio et al. 2010b). Furthermore, the adult lifespan of infected weevils was reduced from one half to approximately one tenth compared with control treatment. The adult males and females treated with the fungal strain spread the infection to healthy adults at transmission rates ranging from 55 to 60 %. This fungal treatment also significantly affected the fecundity and egg hatchability, which were reduced to 62.6 and 32.8 %, respectively. Moreover, 30–35 % mortality of larvae emerged from the eggs obtained from fungus treated couples led to progeny suppression of 78 % compared to the control treatment.

In preventive field assays *B. bassiana* caused mortality up to 85.7 %, suggesting that infection to exposed healthy adults occurred on contact with the infected ones and confirmed this strain to be an excellent biocontrol agent against RPW. Additionally, the males irradiated by gamma-radiation proved to be an efficient vector for the same strain of *B. bassiana* (Llácer et al. 2013). A recent report from Saudi Arabia found the commercial formulation of *B. bassiana* (Broadband) to have high efficacy against RPW in semi-field trials, with the potential of infecting adult weevils by deploying the fungus in the field through pheromone traps involving the "attract-and-infect" method (Hajjar et al. 2015).

In another study, the successful use of acoustic methods for determining the effect of *B. bassiana* against RPW demonstrated the usefulness of this technique for the evaluation of post-treatment fungal efficacy against internally feeding insect pests (Jalinas et al. 2015). Therefore, EPF should be seriously considered as a new tactic in developing effective biological control of RPW.

2.5.6 Reproductive Control

Initial studies carried out using irradiated sterile RPW adult male weevils during the 1970s in India did not appear promising, due to the high frequency of fertile female weevils collected from the field, after mass release of sterile males (Rahalkar et al. 1974, 1977). In the Middle East, as a first step to develop the Sterile Insect Technique (SIT) against RPW, studies on the impact of γ radiation on the mating behavior, and the effect of different relative humidity levels on the efficacy of SIT, were performed by Al-Aydeh and Rasool (2010). The γ radiation did not affect the mating behavior of RPW and the weevils were sexually stimulated during aggregation. However, the usefulness of this technique remains quite uncertain due to the high cost of mass rearing. Furthermore, the RPW adult mates several times during its life span and a single mating would be sufficient for female weevils to lay fertile eggs (Llácer et al. 2013).

2.5.7 Chemical Control

Since symptoms of infestations in palms by RPW are difficult to identify at early stages, the preventive tactics are, therefore, critically important. Chemical control has been the most common practice and consists of repeated applications of pesticides as preventative and curative measures to suppress the spread of infestation. In India during 1970s, the use of carbamate and organophosphate insecticides was established as the foundation of the chemical approach (Murphy and Briscoe 1999). In recent years, however, besides the above chemical groups, phenylpyrazole and neonicotinoid-based insecticides are used as preventive and curative applications to control this pest (Hernandez-Marante et al. 2003; Dembilio et al. 2010a; Llácer et al. 2012; Al-Shawaf et al. 2013).

In Spain, eight preventative insecticide treatments per season (March–November) were proposed (Valencian Department of Agriculture) with six active substances (neonicotinoids, imidacloprid and thiamethoxam, avermectin, abamectin and organophosphates chlorpyrifos and phosmet) authorized by the Spanish Ministry of Agriculture against RPW (MARM 2014). All these recommended pesticides could be applied in three different ways, as spray on the crown, injected into the trunk, or as a soil drench.

In laboratory and semi-field conditions, imidacloprid SL formulation successfully controlled RPW (Martín et al. 1998; Kaakeh 2006). Reports from Israel recommend preventive treatments against RPW with imidacloprid applied to soil at 5 ml per palm through irrigation water during March, April, and May and again after the harvest of dates in September (Soroker et al. 2005). An oil dispersion (OD) formulation of this insecticide was applied through irrigation in preventative and curative semi-field assays to *P. canariensis* in Spain (Dembilio et al. 2010a; Llácer et al. 2012), and proved highly effective (Llácer et al. 2012). Additionally, preventative efficacy values of more than 95 % were also observed for at least 45 days post-treatment (Llácer et al. 2012). In field trials, just two annual applications of this formulation, successfully decreased *P. canariensis* palms infestation to less than 27 % compared to >84 % infestation observed in palms receiving no treatment (Dembilio et al. 2010a). This insecticide was also tested as trunk injection and compared with abamectin (Dembilio et al. 2015) which revealed that the imidacloprid applied as trunk injection not only showed a better distribution within the palm but also yielded greater persistence than its application as crown spray. When abamectin

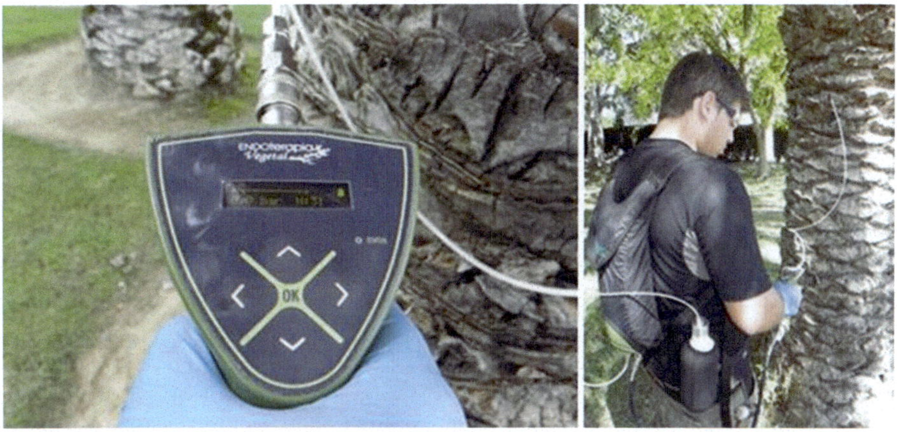

Fig. 2.5 New injection device using a pressure control (Photos: Óscar Dembilio)

was injected to infested palms, the accumulated concentration of abamectin in the fronds resulted in 50–90 % mortality in young larvae up to 1 month post-treatment. On the other hand, injected imidacloprid was capable to cause more than 90 % mortality in young grubs for more than 2 months after treatment (Dembilio et al. 2015). On the basis of its performance against RPW under field conditions, imidacloprid was found as a better choice over abamectin. Currently, in date palms stem injection using pressure machines to control RPW infestation is widely practiced technique (Fig. 2.5). It is important to ensure that this approach should be carried out under the supervision of technical staff so that treatment pressure does not exceed 1 bar, to avoid palm tissue damage.

2.6 Future Research

Many scientific documents covering different perspectives of RPW bioecology and control have been published, and research has provided important tools to improve its management. Some of the areas studied and aspects that require further research are as follows.

2.6.1 Early Detection

Arguably, the most vital issue concerning infestations by RPW is early detection. Since new infestations are hardly detected, involuntary movement of infested plants is the common and main cause for the wider distribution of this weevil. Some research groups have focused on the development of acoustic sensors for early detection (Levsky et al. 2007; Potamitis et al. 2009; Gutiérrez et al. 2010). However, sensors developed so far have been limited use in practice. Other detection techniques, including detection of chemical signatures of infested palms, thermal imaging and molecular tools, should be explored.

2.6.2 *Plant Quarantine*

In addition to current pre-departure and post-entry quarantine protocols (OJEU 2007), the development of either physical or chemical quarantine treatment to disinfest palms (Llácer and Jacas 2010), would greatly reduce the high risk that palm exportations currently impose world-wide. Coordination among enforcement agencies at the national, regional and international level is vital to check the spread of this cryptic pest through planting material either for farming of landscape gardening.

2.6.3 *Semiochemicals and Trapping*

Semiochemicals are key factors to the detection and management of *R. ferrugineus* in palm groves, and will probably have a central role for its management in non-agricultural areas. However, there are many questions related to their use in monitoring and mass trapping (trap density, maintenance and servicing, location) as well as using new possibilities ("push-pull", "attract-and-kill", "attract-and-infect", "attract-and-sterilize" strategies) which are currently being investigated. Recently, Guarino et al. (2013) identified repellents against RPW that could be used in a push-pull strategy in conjunction with pheromones, to repel adult weevils from the palms.

2.6.4 *Biological Control*

There is an incomplete knowledge of the natural enemies of RPW in its native habitat (Murphy and Briscoe 1999; Faleiro 2006; Mazza et al. 2014), and this precludes the implementation of any classical biological control program against this insect pest. On the other hand, different entomopathogenic nematodes (Abbas et al. 2001a, b; Llácer et al. 2009; Dembilio et al. 2011) and fungi (Gindin et al. 2006; Sewify et al. 2009; Dembilio et al. 2010b; Hajjar et al. 2015) have been identified. This opens new possibilities of inundative releases of these biocontrol agents, and of their combined use with semiochemicals in "attract-and-infect" strategies.

2.6.5 Chemical Control

Even though highly effective pesticides exist (Barranco et al. 1998; Llácer et al. 2010; Dembilio et al. 2010a), there are many problems related both to the delivery of the product to the target (tunneling larvae within the palm) and also regarding the ecotoxicological profile of these chemicals. New environment friendly products are urgently needed, and alternative application methods (such as trunk injection) or uptake mechanisms (such as systemic products) have already been demonstrated (Dembilio et al. 2015).

2.6.6 Resistance and Induced Plant Defenses

Both antibiotic and antixenotic mechanisms of defense have been identified in some palm species (Barranco et al. 2000; Dembilio et al. 2009). Further research is needed to clarify the basis for such mechanisms, and studies on induced defenses could result in novel approaches for the management of weevil in two major *Phoenix* palms viz. date palm and the Canary Island palm, where infestations due to RPW are known to be severe.

2.7 Conclusion

The invasive red palm weevil, one of the most destructive pests of palms in the world is widely distributed in all continents. In the Mediterranean basin, RPW has become the major pest of palms. A better understanding of the biology and the ecology of RPW and their natural enemies and early detection of infested palms is critical in order to avoid death of palms and is the fundamental to the success of any IPM strategy implemented to reduce this pest. However, in the early stages of attack, palms can recover after treatment with insecticides and/or sanitation procedures. IPM strategies, including field sanitation, agronomic practices, chemical and biological controls and the use of semiochemicals both for adult monitoring and mass trapping, have been developed and implemented in several countries. Further research is needed to improve the management of this pest and to protect the palms against this havoc and thereby continue enjoying *Phoenix* and other palms in our farms, gardens, natural landscapes and parks.

References

Abbas, M. S. T., Hanounik, S. B., Mousa, S. A., & Mansour, M. I. (2001a). On the pathogenicity of *Steinernema abbasi* and *Heterorhabditis indicus* isolated from adult *Rhynchophorus ferrugineus* (Coleoptera). *International Journal of Nematology, 11*, 69–72.

Abbas, M. S. T., Saleh, M. M. E., & Akil, A. M. (2001b). Laboratory and field evaluation of the pathogenicity of entomopathogenic nematodes to the red palm weevil, *Rhynchophorus ferrugineus* (Oliv.) (Col.: Curculionidae). *Journal of Pest Science, 74*, 167–168.

Abe, F., Hata, K., & Sone, K. (2009). Life history of the red palm weevil, *Rhynchophorus ferrugineus* (Coleoptera: Dryophthoridae), in Southern Japan. *Florida Entomologist, 92*, 421–425.

Abraham, V.A., & Kurian, C. (1975). An integrated approach to the control Rhynchophorus ferrugineus F. the red weevil of coconut palm. In: Proceedings of the 4th Session of the FAO Technical Working Party on Coconut Production, Protection and Processing, September 14–25, Kingston.

Abraham, V. A., Al Shuaibi, M., Faleiro, J. R., Abozuhairah, R. A., & Vidyasagar, P. S. P. V. (1998). An integrated approach for the management of red palm weevil *Rhynchophorus ferrugineus* Oliv. – A key pest of date palm in the middle-East. *Sultan Qaboos University Journal of Scientific Research (Agricultural Science), 3*, 77–83.

Abraham, V. A., Faleiro, J. R., Al Shuaibi, M. A., & Al-Abdan, S. (2001). Status of pheromone trap captured female red palm weevils from date gardens in Saudi Arabia. *Journal of Tropical Agriculture, 39*, 197–199.

Abuagla, A. M., & Al-Deeb, M. A. (2012). Effect of bait quantity and trap color on the trapping efficacy of the pheromone trap for the red palm weevil, *Rhynchophorus ferrugineus. Journal of Insect Science, 12*, 120. doi:10.1673/031.012.12002.

Al-Aydeh, H. Y., & Rasool, K. G. (2010). Determination of the optimum sterilizing radiation dose for control of the red date palm weevil *Rhynchophorus ferrugineus* Oliv. (Coleoptera: Curculionidae). *Crop Protection, 29*, 1377–1380.

Al-Ayedh, H. (2008). Evaluation of date palm cultivars for rearing the red date palm weevil, *Rhynchophorus ferrugineus* (Coleoptera: Curculionidae). *Florida Entomologist, 91*, 353–358.

Al-Deeb, M. A., Muzaffar, S. B., Abuagla, A. M., & Sharif, E. M. (2011). Distribution and abundance of phoretic mites (Astigmata, Mesostigmata) on *Rhynchophorus ferrugineus* (Coleoptera: Curculionidae). *Florida Entomologist, 94*, 748–755.

Al-Sahib, W., & Marshall, R. J. (2003). The fruit of the date palm: Its possible use as the best food for the future? *Journal of Food Science and Nutrition, 54*, 247–259.

Al-Saoud, A. H. (2013). Effect of ethyl acetate and trap colour on weevil captures in red palm weevil *Rhynchophorus ferrugineus* (Coleoptera: Curculionidae) pheromone traps. *International Journal of Tropical Insect Science, 33*, 202–206.

Al-Shawaf, A. M., Al-Shagag, A., Al-Bagshi, M., Al-Saroj, S., Al-Bather, S., Al-Dandan, A. M., Abdallah, A. B., & Faleiro, J. R. (2013). A quarantine protocol against red palm weevil *Rhynchophorus ferrugineus* (Olivier) (Coleoptera: Curculionidae) in date palm. *Journal of Plant Protection Research, 53*, 409–415.

Ávalos, J. A., Martí-Campoy, A., & Soto, A. (2014). Study of the flying ability of *Rhynchophorus ferrugineus* (Coleoptera: Dryophthoridae) adults using a computer-monitored flight mill. *Bulletin of Entomological Research, 104*, 462–470.

Avand Faghih, A. (1996). The biology of red palm weevil, *Rhynchophorus ferrugineus* Oliv. (Coleoptera: Curculionidae) in Saravan region (Sistan and Balouchistan Province, Iran). *Applied Entomology and Phytopathology, 63*, 16–18.

Barranco, P., Peña, J. A., De la Martin, M. M., & Cabello, T. (1998). Efficacy of chemical control of the new palm pest *Rhynchophorus ferrugineus* (Olivier 1790) (Curculionidae: Coleoptera). *Boletin de Sanidad Vegetal Plagas, 24*(2), 301–306.

Barranco, P., De La Pena, J. A., Martin, M. M., & Cabello, T. (2000). Rango de hospedantes de *Rhynchophorus ferrugineus* (Olivier, 1790) y diametro de la palmera hospedante. (Coleoptera, Curculionidae). *Boletín de Sanidad Vegetal de Plagas, 26*, 73–78.

Barrow, S. (1998). A revision of *Phoenix*. *Kew Bulletin, 53*, 513–575.
Butani, D. K. (1975). Insect pests of fruit crops and their control, sapota-11. *Pesticides Research Journal, 9*, 40–42.
Dammerman, K. W. (1929). *The agricultural zoology of the Malay archipelago*. Amsterdam: J.H. Bussy Limited. 493 pp.
Dembilio, Ó., & Jacas, J. A. (2011). Basic bio-ecological parameters of the invasive red palm weevil, *Rhynchophorus ferrugineus* (Coleoptera: Curculionidae), in *Phoenix canariensis* under Mediterranean climate. *Bulletin of Entomological Research, 101*, 153–163.
Dembilio, Ó., & Jacas, J. A. (2012). Bio-ecology and integrated management of the red palm weevil, *Rhynchophorus ferrugineus* (Coleoptera: Curculionidae), in the region of Valencia (Spain). *Hellenic Plant Protection Journal, 5*, 1–12.
Dembilio, Ó., & Jaques, J.A. (2015). Short and long-term efficacy and phytotoxicity of phosphine against *Rhynchophorus ferrugineus* in live *Phoenix canariensis* palms. *Spanish Journal of Agricultural Research, 13*(4), e10SC01, 1–4.
Dembilio, Ó., Jacas, J. A., & Llácer, E. (2009). Are the palms *Washingtonia filifera* and *Chamaerops humilis* suitable hosts for the red palm weevil, *Rhynchophorus ferrugineus* (Col. Curculionidae)? *Journal of Applied Entomology, 133*(7), 565–567.
Dembilio, Ó., Llácer, E., Martinez de Altube, M. M., & Jacas, J. A. (2010a). Field efficacy of imidacloprid and *Steinernema carpocapsae* in a chitosan formulation against the red palm weevil *Rhynchophorus ferrugineus* (Coleoptera: Curculionidae) in *Phoenix canariensis. Pest Management Science, 66*, 365–370.
Dembilio, Ó., Quesada-Moraga, E., Santiago- Álvarez, C., & Jacas, J. A. (2010b). Potential of an indigenous strain of the entomopathogenic fungus *Beauveria bassiana* as a biological control agent against the red palm weevil, *Rhynchophorus ferrugineus. Journal of Invertebrate Pathology, 104*, 214–221.
Dembilio, Ó., Karamaouna, F., Kontodimas, D. C., Nomikou, M., & Jacas, J. A. (2011). Susceptibility of *Phoenix theophrasti* to *Rhynchophorus ferrugineus* (Col.: Curculionidae) and its control using *Steinernema carpocapsae. Spanish Journal of Agricultural Research, 9*, 623–626.
Dembilio, Ó., Riba, J. M., Gamón, M., & Jacas, J. A. (2015). Mobility and efficacy of abamectin and imidacloprid against *Rhynchophorus ferrugineus* in *Phoenix canariensis* by different application methods. *Pest Management Science, 71*, 1091–1098.
Dransfield, J. (1978). Growth forms of rain forest palms. In P. B. Tomlinson & M. H. Zimmermann (Eds.), *Tropical trees as living systems* (pp. 247–268). Cambridge: Cambridge University Press.
Elawad, S. A., Mousa, S. A., Shahdad, A. S., Alawaash, S. A., & Alamiri, A. M. A. (2007). Efficacy of entomopathogenic nematodes against red palm weevil in UAE. *ISHS Acta Horticulturae, 736*, 415–420.
El-Shafie, H. A. F., Faleiro, J. R., Al-Abbad, A. H., Stoltman, L., & Mafra-Neto, A. (2011). Bait-free attract and kill technology (Hook™ RPW) to suppress red palm weevil, *Rhynchophorus ferrugineus* (Coleoptera: Curculionidae) in date palm. *Florida Entomologist, 94*(4), 774–778.
El-Sufty, R., Al-Awash, S. A., Al Bgham, S., Shahdad, A. S., & Al Bathra, A. H. (2009). Pathogenicity of the fungus *Beauveria bassiana* (Bals.) Vuill to the red palm weevil, *Rhynchophorus ferrugineus* (Oliv.) (Col.: Curculionidae) under laboratory and field conditions. *Egyptian Journal of Biological Pest Control, 19*, 81.
EPPO (European and Mediterranean Plant Protection Organization). (2008a). Data sheets on quarantine pests *Rhynchophorus ferrugineus. EPPO Bulletin, 38*, 55–59.
EPPO (European and Mediterranean Plant Protection Organization). (2008b). Data sheets on quarantine pests *Paysandisia archon. EPPO Bulletin, 3*, 163–166.
EPPO. (2009). *Rhynchophorus ferrugineus found* on *Howea forsteriana* in Sicilia, Italy. No. 3 2009/051. European and Mediterranean Plant protection Organization, Paris, France. Available at: https://archives.eppo.int/EPPOReporting/2009/Rse-0912.pdf. Accessed on: 12 June 2015.

Esteban-Durán, J., Yela, J. L., Beitia Crespo, F., & Jiménez Álvarez, A. (1998a). Exotic curculionids liable to be introduced into Spain and other EU countries through imported vegetables. *Boletín de Sanidad Vegetal Plagas, 24*, 23–40.

Esteban-Durán, J., Yela, J. L., Beitia-Crespo, F., & Jimenez-Alvarez, A. (1998b). Biologia del curculionido ferruginoso de las palmeras *Rhynchophorus ferrugineus* (Olivier) en laboratorio y campo: Ciclo en cautividad, peculiaridades biologicas en su zona de introduccion en Espana y metodos biologicos de deteccion y posible control (Coleoptera: Curculionidae: Rhynchophorinae). *Boletín de Sanidad Vegetal Plagas, 24*, 737–748.

Esteban-Durán, J., Yela, J. L., Crespo, F. B., & Alvarez, A. J. (1998c). Biology of red palm weevil, *Rhynchophorus ferrugineus* (Olivier) (Coleoptera: Curculionidae: Rhynchophorinae), in the laboratory and field life cycle, biological characteristics in its zone of introduction in Spain, biological method of detection and possible control. *Boletin Sanidad Vegetal Plagas, 24*, 737–748.

Faleiro, J. R. (2005). *Pheromone technology for the management of red palm weevil, Rhynchophorus ferrugineus (Olivier) (Coleoptera: Rhynchophoridae) – A key pest of coconut* (Technical Bulletin No. 4). ICAR Research Complex, Goa, 40 pp.

Faleiro, J. R. (2006). A review of the issues and management of the red palm weevil *Rhynchophorus ferrugineus* (Coleoptera: Rhynchophoridae) in coconut and date palm during the last one hundred years. *International Journal of Tropical Insect Science, 26*, 135–154.

Faleiro, J. R., & Rangnekar, P. A. (2001). Location specific seasonal activity of red palm weevil, *Rhynchophorus ferrugineus* Oliv. coconut plantations of Goa. *Indian Journal of Applied Entomology, 15*, 7–15.

Faleiro, J. R., Ashok Kumar, J., & Rangnekar, P. A. (2002). Spatial distribution of red palm weevil *Rhynchophorus ferrugineus* Oliv. (Coleoptera: Cuculionidae) in coconut plantations. *Crop Protection, 21*, 171–176.

Faleiro, J. R., Rangnekar, P. A., & Satarkar, V. R. (2003). Age and fecundity of female red palm weevils *Rhynchophorus ferrugineus* (Olivier) (Coleoptera: Rhynchophoridae) captured by pheromone traps in coconut plantations of India. *Crop Protection, 22*, 999–1002.

Faleiro, J. R., Abdallah, A. B., Kumar, J. A., Shagagh, A., & Abdan, S. A. (2010). Sequential sampling plan for area-wide management of red palm weevil, *Rhynchophorus ferrugineus* (Olivier) in date plantations of Saudi Arabia. *International Journal of Tropical Insect Science, 30*, 145–153.

Faleiro, J. R., Abo-El-Saad, M., & Al-Abbad, A. H. (2011). Pheromone trap density to mass trap *Rhynchophorus ferrugineus* (Coleoptera: Curculionidae/Rhynchophoridae/Dryophthoridae) in date plantations of Saudi Arabia. *International Journal of Tropical Insect Science, 31*(1–2), 75–77.

Faleiro, J. R., Ben Abdullah, A., El-Bellaj, M., Al Ajlan, A. M., & Oihabi, A. (2012). Threat of red palm weevil, *Rhynchophorus ferrugineus* (Olivier) to date palm plantations in North Africa. *Arab Journal of Plant Protection, 30*, 274–280.

Faleiro, J. R., El-Shafie, H. A. F., Ajlan, A. M., & Sallam, A. A. (2014). Screening date palm cultivars for resistance to red palm weevil *Rhynchophorus ferrugineus* (Coleoptera: Curculionidae). *Florida Entomologist, 97*, 1529–1536.

Fiaboe, K. K. M., Peterson, A. T., Kairo, M. T. K., & Roda, A. L. (2012). Predicting the potential worldwide distribution of the red palm weevil *Rhynchophorus ferrugineus* (Olivier) (Coleoptera: Curculionidae) using ecological niche modeling. *Florida Entomologist, 95*(3), 659–673.

Francesca, N., Alfonzo, A., Verde, G. L., Settanni, L., Sinacori, M., Lucido, P., & Moschetti, G. (2015). Biological activity of *Bacillus* spp. evaluated on eggs and larvae of red palm weevil *Rhynchophorus ferrugineus*. *Annals of Microbiology, 65*, 477–485.

Ghosh, C. C. (1912). Life-histories of Indian insects, III. The rhinoceros beetle (*Oryctes rhinoceros*) and the red palm weevil (*Rhynchophorus ferrugineus*). *Memoires of the Department of Agriculture of India, 2*, 193–217.

Ghosh, C. C. (1923). The palm beetles in Burma with notes on other pests. *Bulletin of the Department of Agriculture Burma, 19*, 1–40.

Giblin-Davis, R. M., Faleiro, J. R., Jacas, J. A., Peña, J. E., & Vidyasagar, P. S. P. V. (2013). Biology and management of the red palm weevil, *Rhynchophorus ferrugineus*. In J. E. Peña (Ed.), *Potential invasive pests of agricultural crop species* (pp. 1–34). Oxfordshire: CAB International, CABI Wallingford.

Gindin, G., Levski, S., Glazer, I., & Soroker, V. (2006). Evaluation of the entomopathogenic fungi *Metarhizium anisopliae* and *Beauveria bassiana* against the red palm weevil *Rhynchophorus ferrugineus*. *Phytoparasitica, 34*, 370–379.

Gopinadhan, P. B., Mohandas, N., & Nari, K. P. V. (1990). Cytoplasmic polyhedrosis virus infecting red palm weevil of coconut. *Current Science, 59*, 577–580.

Guarino, S., Peri, E., Bue, P. L., Germanà, M. P., Colazza, S., Anshelevich, L., Ravid, U., & Soroker, V. (2013). Assessment of synthetic chemicals for disruption of *Rhynchophorus ferrugineus* response to attractant-baited traps in an urban environment. *Phytoparasitica, 41*, 79–88.

Gutiérrez, A., Ruiz, V., Moltó, E., Tapia, G., & del Mar Téllez, M. (2010). Development of a bioacoustic sensor for the early detection of red palm weevil (*Rhynchophorus ferrugineus* Olivier). *Crop Protection, 29*, 671–676.

Hajjar, M. J., Ajlan, A. M., & Al-Ahmad, M. H. (2015). New approach of *Beauveria bassiana* to control the red palm weevil (Coleoptera: Curculionidae) by trapping technique. *Journal of Economic Entomology, 108*(2), 425–432.

Hallett, R. H., Gries, G., Gries, R., Borden, J. H., Czyzewska, E., Oehlschlager, A. C., Pierce, H. D., Jr., Angerilli, N. P. D., & Rauf, A. (1993). Aggregation pheromones of two Asian palm weevils *Rhynchophorus ferrugineus* and *R. vulneratus*. *Naturwissenschaften, 80*, 328–331.

Hallett, R. H., Oehlschlager, A. C., & Borden, J. H. (1999). Pheromone trapping protocols for the Asian palm weevil, *Rhynchophorus ferrugineus* (Coleoptera: Curculionidae). *International Journal of Pest Management, 45*, 231–237.

Hernandez-Marante, D., Folk, F., Chez, A., & Fernadez-Escobar, R. (2003). Control of red palm weevil (*Rhynchophorus ferrugineus* Olivier) using trunk injections and foliar sprays. *Boletin de Sanidad Vegetal Plagas, 29*(4), 563–573.

Horticom. (2008). *El semanal de horticultura*. Edición Internacional August 22, 2008. Available at: www.horticom.com. Accessed on 30 Jul 2014.

IUCN. (2010). *IUCN red list of threatened species, Version 2010*. Available at www.iucnredlist.org. Accessed on 30 Jul 2014.

Jalinas, J., Güerri-Agulló, B., Mankin, R. W., López-follana, R., & Lopez-Llorca, L. V. (2015). Acoustic assessment of *Beauveria bassiana* (Hypocreales: Clavicipitaceae) effects on *Rhynchophorus ferrugineus* (Coleoptera: Dryophthoridae) larval activity and mortality. *Journal of Economic Entomology, 108*(2), 444–453.

Jaya, S., Suresh, T., Sobhitha-Rani, R. S., & Sreekumar, S. (2000). Evidence of seven larval instars in the red palm weevil, *Rhynchophorus ferrugineus* Olivier reared on sugarcane. *Journal of Entomological Research, 24*, 27–31.

Jayanth, K. P., Mathew, M. T., Narabenchi, G. B., & Bhanu, K. R. M. (2007). Impact of large scale mass trapping of red palm weevil, *Rhynchophorus ferrugineus* Olivier in coconut plantations in Kerala using indigenously synthesized aggregation pheromone lures. *Indian Coconut Journal, 38*, 2–9.

Ju, R. T., Wang, F., Wan, F. H., & Li, B. (2011). Effect of host plants on development and reproduction of *Rhynchophorus ferrugineus* (Olivier) (Coleoptera: Curculionidae). *Journal of Pest Science, 84*, 33–39.

Kaakeh, W. (2005). Longevity, fecundity, and fertility of the red palm weevil, *Rhynchophorus ferrugineus* Olivier (Coleoptera: Curculionidae) on natural and artificial diets. *Emirates Journal of Agricultural Sciences, 17*, 23–33.

Kaakeh, W. (2006). Toxicity of imidacloprid to developmental stages of *Rhynchophorus ferrugineus* (Curculionidae: Coleoptera): Laboratory and field tests. *Crop Protection, 25*, 432–439.

Kalshoven, L. G. E. (1981). *Pests of crops in Indonesia*. Jakarta: P.T. Ichtiar Baru-Vanhoeve. 720 pp.

Kogan, M., & Ortman, E. E. (1978). Antixenosis, a new term proposed to replace Painter's 'Nonpreference? Modality of resistance. *Bulletin of Entomological Society of America, 24*, 175–176.

Kranz, J. H., Schmutterer, H., & Koc, W. (1982). *Enfermedades, plagas y malezas de los cultivos subtropicales*. Berlin: Verlag Paul Parey. 722 pp.

Leefmans, S. (1920). De palmsnuitkever (*Rhynchophorus ferrugineus* Olivier). *Mededelingen van Institute voor Plantenziekten Buitenzorg, 43*, 1–90.

Lepesme, P. (1947). *Les insectes des palmiers*. Paris: P. Lechevalier. 904 pp.

Levsky, S., Kostyukovsky, M., Pinhas, J., Mizrach, A., Hetzroni, A., Nakache, Y., & Soroker, V. (2007). Detection and control methods for the red palm weevil in date palm offshoots. In *The XXVI annual meeting of the entomology Society of Israel*, Haifa.

Llácer, E., & Jacas, J. A. (2010). Efficacy of phosphine as a fumigant against *Rhynchophorus ferrugineus* (Coleoptera: Curculionidae) in palms. *Spanish Journal of Agricultural Research, 8*(3), 775–779.

Llácer, E., Martinez, J., & Jacas, J. A. (2009). Evaluation of the efficacy of *Steinernema carpocapsae* in a chitosan formulation against the red palm weevil, *Rhynchophorus ferrugineus*, in *Phoenix canariensis*. *BioControl, 54*, 559–565.

Llácer, E., Dembilio, Ó., & Jacas, J. A. (2010). Evaluation of the efficacy of an insecticidal paint based on chlorpyrifos and pyriproxyfen in a microencapsulated formulation against *Rhynchophorus ferrugineus* (Coleoptera: Curculionidae). *Journal of Economic Entomology, 103*, 402–408.

Llácer, E., Negre, M., & Jacas, J. A. (2012). Evaluation of an oil dispersion formulation of imidacloprid as a drench against *Rhynchophorus ferrugineus* (Coleoptera, Curculionidae) in young palm trees. *Pest Management Science, 68*(6), 878–882.

Llácer, E., Santiago-Álvarez, C., & Jacas, J. A. (2013). Could sterile males be used to vector a microbiological control agent? The case of *Rhynchophorus ferrugineus* and *Beauveria bassiana*. *Bulletin of Entomological Research, 103*(02), 241–250.

Lo Verde, G., Torta, L., Mondello, V., Caldarella, C. G., Burruano, S., & Caleca, V. (2015). Pathogenicity bioassays of isolates of *Beauveria bassiana* on *Rhynchophorus ferrugineus*. *Pest Management Science, 71*(2), 323–328.

Malumphy, C., & Moran, H. (2007). Red palm weevil *Rhynchophorus ferrugineus*. Central Science Laboratory Plant Pest Notice 5. Available at: http://faculty.ksu.edu.sa/10439/Documents/fifty.pdf. Accessed on 23 Jul 2012.

MARM. (2014). *Registro de productos fitosanitarios*. Available at: http://www.marm.es/es/agricultura/temas/medios-de-produccion/productos-fitosanitarios/registro/menu.asp. Accessed on 30 Jul 2014.

Martín, M. M., & Cabello, T. (2006). Manejo de la cría del picudo rojo de la palmera, *Rhynchophorus ferrugineus* (Olivier, 1790) (Coleoptera, Dryophthoridae), en dieta artificial y efectos en su biometría y biología. *Boletín de Sanidad Vegetal de Plagas, 32*, 631–641.

Martín, M. M., García, T. C., de la Peña Peña, J., & Vega, P. B. (1998). Eficacia del control químico de la nueva plaga de las palmeras *Rhynchophorus ferrugineus* (Oliver, 1790) (Col: Cuculionidae). *Boletín de sanidad vegetal Plagas, 24*(2), 301–306.

Martin-Molina, M. M. (2004). Biologia y ecologia del curculionido rojo de las palmeras, Rhynchophorus ferrugineus (Olivier, 1790) (Coleoptera: Dryophthoridae). PhD Dissertation, Universidad de Almeria, Spain, 203 pp.

Massoud, M. A., Sallam, A. A., Faleiro, J. R., & Al-Abdan, S. (2012). Geographic information system-based study to ascertain the spatial and temporal spread of the red palm weevil *Rhynchophorus ferrugineus* (Coleoptera: Curculionidae) in date plantations. *International Journal of Tropical Insect Science, 32*(2), 108–115.

Mazza, G., Francardi, V., Simoni, S., Benvenuti, C., Cervo, R., Faleiro, J. R., & Roversi, P. F. (2014). An overview on the natural enemies of *Rhynchophorus palm* weevils, with focus on *R. ferrugineus*. *Biological Control, 77*, 83–92.
Menon, K. P. V., & Pandalai, K. M. (1960). *A monograph of the coconut palm*. Ernakulum: Central Coconut Committee. 384 p.
Morici, C. (1998). *Phoenix canariensis* in the wild. *Principles, 42*, 85–93.
Murphy, S. T., & Briscoe, B. R. (1999). The red palm weevil as an alien invasive: Biology and prospects for biological control as a component of IPM. *BioControl, 20*, 35–45.
Nakash, J., Osem, Y., & Kehat, M. (2000). A suggestion to use dogs for detecting red palm weevil (*Rhynchophorus ferrugineus*) infestation in date palms in Israel. *Phytoparasitica, 28*, 153–155.
Nirula, K. K. (1956). Investigations on the pests of coconut palm. Part. *Rhynchophorus ferrugineus*. *Indian Coconut Journal, 9*, 229–247.
Oehlschlager, A. C. (1998, December 15–17). Trapping of the date palm weevil. In *FAO workshop on date palm weevil (Rhynchophorus ferrugineus) and its control*, Cairo.
OJEU. (2007). Commission Decision 2007/365/EC on emergency measures against the introduction and spread within the EU of *R. ferrugineus* (Olivier) [notified under document number C (2007) 2161]. *Official Journal of the European Union L, 139*, 24–27.
OJEU. (2008). Commission Decision of October 06, 2008 amending Decision 2007/365/EC on emergency measures to prevent the introduction into and the spread within the Community of *R. ferrugineus* (Olivier) [notified under document number C (2008) 5550]. *Official Journal of the European Union L, 266*, 51–54.
OJEU. (2010). Commission decision of 17 August 2010 amending Decision 2007/365/EC on emergency measures to prevent the introduction into and the spread within the Community of *R. ferrugineus* (Olivier) [notified under document number C (2010) 5640]. *Official Journal of the European Union L, 226*, 42–44.
Peter, C. (1989). A note on the mites associated with *Rhynchophorus ferrugineus* in Tamil Nadu. *Journal of Insect Science, 2*, 160–161.
Potamitis, I., Ganchev, T., & Kontodimas, D. (2009). On automatic bioacoustic detection of pests: The cases of *Rhynchophorus ferrugineus* and *Sitophilus oryzae*. *Journal of Economic Entomology, 102*(4), 1681–1690.
Quesada-Moraga, E., Santos-Quiros, R., Valverde-Garcia, P., & Santiago-Álvarez, C. (2004). Virulence, horizontal transmission, and sublethal reproductive effects of *Metarhizium anisopliae* (anamorphic fungi) on the German cockroach (Blattodea: Blattellidae). *Journal of Invertebrate Pathology, 87*, 51–58.
Rahalkar, G. W., Harwalkar, M. R., & Rananavare, H. O. (1972). Development of red palm weevil, *Rhynchophorus ferrugineus* Oliv. on sugarcane. *Indian Journal Entomology, 34*, 213–215.
Rahalkar, G. W., Harwalkar, M. R., Rananvare, H. D., Shantaram, K., & Goplayengar, A. R. (1974). Laboratory studies on radiation sterilization of the red palm weevil, (*Rhynchophorus ferrugineus* Oliv.) males. *Journal of Plantation Crops, 1*, 141–146.
Rahalkar, G. W., Harwalkar, M. R., Rananvare, H. D., Kurian, C., Abrham, V. A., & Koya, A. K. M. (1977). Preliminary field studies on the control of the red palm weevil, *Rhynchophorus ferrugineus* using radio sterilized males. *Journal of Nuclear Agriculture and Biology, 6*, 65–68.
Ricaño, J., Güerri-Agulló, B., Serna-Sarriás, M. J., Rubio-Llorca, G., Asensio, L., Barranco, P., & Lopez-Llorca, L. V. (2013). Evaluation of the pathogenicity of multiple isolates of *Beauveria bassiana* (Hypocreales: Clavicipitaceae) on *Rhynchophorus ferrugineus* (Coleoptera: Dryophthoridae) for the assessment of a solid formulation under simulated field conditions. *Florida Entomologist, 96*(4), 1311–1324.
Rochat, D. (2006). Trapping: Drawbacks and prospects: need for more research. In *Proceedings of the first international workshop on red palm weevil*, November 28–29, 2005, Valencia, Spain. Fundacion Agroalimed, Valencia, Spain, 99–104.

Salama, H., Hamdy, M., & Magd El-Din, M. (2002). The thermal constant for timing the emergence of the red palm weevil, *Rhynchophorus ferrugineus* (Oliv.), (Coleoptera, Curculionidae). *Journal of Pest Science, 75*, 26–29.

Salama, H. S., Zaki, F. N., & Abdel-Razek, A. S. (2009). Ecological and biological studies on the red palm weevil Rhynchophorus ferrugineus (Olivier). *Archives of Phytopathology and Plant Protection, 42*, 392–399.

Saleh, M. M. E., & Alheji, M. (2003). Biological control of red palm weevil with entomopathogenic nematodes in the eastern province of Saudi Arabia. *Egyptian Journal of Biological Pest Control, 13*, 55–59.

Sebay, Y. (2003). Ecological studies on the red palm weevil, *Rhynchophorus ferrugineus* Oliv (Coleoptera: Curculionidae) in Egypt. *Egyptian Journal of Agricultural Research, 81*, 523–529.

Sewify, G. H., Belal, M. H., & Al-Awash, S. A. (2009). Use of the entomopathogenic fungus, *Beauveria bassiana* for the biological control of the red palm weevil, *Rhynchophorus ferrugineus* Olivier. *Egyptian Journal of Agricultural Research, 19*, 157.

Shahina, F., Salma, J., Mehreen, G., Bhatti, M. I., & Tabassum, K. A. (2009). Rearing of *Rhynchophorus ferrugineus* in laboratory and field conditions for carrying out various efficacy studies using EPNs. *Pakistan Journal of Nematology, 27*, 219–228.

Soroker, V., Blumberg, D., Haberman, A., Hamburger-Rishad, M., Reneh, S., Talebaev, S., Anshelevich, L., & Harari, A. R. (2005). Current status of red palm weevil infestation in date palm plantations in Israel. *Phytoparasitica, 33*(1), 97–106.

UNESCO (United Nations Educational, Scientific and Cultural Organization). (2014). World heritage. February 19, 2010. Available at: http://whc.unesco.org/en/list/930. Accessed on 30 Jul 2014.

Vacas, S., Primo, J., & Navarro-Llopis, V. (2013). Advances in the use of trapping systems for *Rhynchophorus ferrugineus* (Coleoptera: Curculionidae): Traps and attractants. *Journal of Economic Entomology, 106*(4), 1739–1746.

Vacas, S., Abad-Payá, M., Primo, J., & Navarro-Llopis, V. (2014). Identification of pheromone synergists for *Rhynchophorus ferrugineus* trapping systems from *Phoenix canariensis* palm volatiles. *Journal of Agricultural and Food Chemistry, 62*, 6053–6064.

Viado, G. B., & Bigornia, A. E. (1949). A biological study of the Asiatic palm weevil, *Rhynchophorus ferrugineus* (Olivier), (Curculionidae, Coleoptera). *Philippine Agriculturist, 33*, 1–27.

Vidyasagar, P. S. P. V., Hagi, M., Abozuhairah, R. A., Al-Mohanna, O. E., & Al-Saihati, A. A. (2000). Impact of mass pheromone trapping on red palm weevil adult population and infestation level in date palm gardens of Saudi Arabia. *Planter, 76*, 347–355.

Wattanapongsiri, A. (1966a). A revision of the genera *Rhynchophorus* and *Dynamism* (Coleoptera: Curculionidae). PhD Dissertation, Department of Entomology, Oregon State University, USA, 418 pp.

Wattanapongsiri, A. (1966b). A revision of the genera *Rhynchophorus* and *Dynamism* (Coleoptera: Curculionidae), Bangkok, Thailand. *Department of Agriculture Science Bulletin, 1*, 328.

Wiseman, B. R. (1999). Mechanisms of plant resistance against arthropod pests. In J. R. Rubertson (Ed.), *Handbook of pest management* (pp. 155–174). New York: Marcel Dekker.

Chapter 3
Bridging the Knowledge Gaps for Development of Basic Components of Red Palm Weevil IPM

Ali M. Idris, Thomas A. Miller, Ravi Durvasula, and Nina Fedoroff

Abstract A meeting was held at King Abdullah University of Science and Technology (KAUST), Thuwal (Saudi Arabia) in March 16–18, 2013 on research and development needs for the control of red palm weevil (RPW), *Rhynchophorus ferrugineus*. Participants included both RPW researchers of long standing and a select group of experts associated with the development of new biotechnology approaches for pest and plant disease control. Participants made presentations on several aspects of RPW management including monitoring, sterile insect technique (SIT), semiochemical, biological and chemical control methods. Relatively little is known about RPW biology, behavior, and resistance to chemical pesticides. The resistance of date palm cultivars has not been investigated and it is not known whether there is a species complex or cryptic subspecies of this serious insect pest of palm tree in the Middle East. The meeting participants also discussed new technologies being used for other pest/crop complexes and how these might be applied to address the RPW problem of palm trees, as well as how genetic modification should be regulated and whether it would be accepted socially. Participants further proposed the creation of a collaborative organization to coordinate and encourage pursuit of innovative research strategies for the monitoring and control of RPW through the development of a strategic plan and establishment of a research program funded through national and international organizations.

A.M. Idris (✉) • N. Fedoroff
Center for Desert Agriculture, King Abdullah University of Science and Technology, Thuwal, Saudi Arabia
e-mail: idris@email.arizona.edu; nvf1@psu.edu

T.A. Miller
Department of Entomology, University of California, Riverside, CA, USA
e-mail: thomas.miller@ucr.edu

R. Durvasula
Center for Global Health, Department of Internal Medicine, UNM School of Medicine, Albuquerque, NM, USA
e-mail: ravi.durvasula@med.va.gov

W. Wakil et al. (eds.), *Sustainable Pest Management in Date Palm: Current Status and Emerging Challenges*, Sustainability in Plant and Crop Protection,
DOI 10.1007/978-3-319-24397-9_3

3.1 Introduction

The red palm weevil (RPW), *Rhynchophorus ferrugineus* (Olivier), belonging to the order Coleoptera, family Curculionidae (true weevils or snout beetles), is native to Southeast Asia, where it has been considered a major pest of palm plantations including oil, coconut and sago palms (Murphy and Briscoe 1999a). In its native range in the Indian subcontinent and Southeast Asia, the RPW negatively impacts the local economy of several Asian countries, including India, Indonesia, and Philippines. While this insect was a minor insect pest of palms in its native range, it developed into a major insect pest after coconut and oil palm became widespread (Nirula 1956). This insect pest appeared to have been restricted to its native range until the mid-1980s, when it was detected on date palm (*Phoenix dactylifera* L.) in the United Arab Emirates (UAE) in 1986 and in eastern region of Saudi Arabia in 1987. The insect was inadvertently introduced into UAE, Saudi Arabia, and other Gulf countries in the 1980s (Murphy and Briscoe 1999b) and is now established in all date palm growing areas. The first record of date palm infestation in Saudi Arabia was in 1987 in Al-Qatif in the eastern region (Abraham et al. 1998). Since then, it has slowly spread to adjacent date growing areas including Al-Ahsa, Al-Kharj, Wadi Al-Dawasir, Mecca, Najran, Qasim, and Tabuk. Numerous reports have cited the spread of RPW to all of the countries of the Mediterranean basin and Aruba, Australia, California, China, Laos, Japan, Malaysia, Madagascar, Papua New Guinea, Philippines and countries in North Africa. The magnitude of this problem in Saudi Arabia, the third-largest date producer globally (FAOSTAT 2014), and the widespread infestation of RPW around the world demand the development of better integrated pest management (IPM) strategies, including earlier and more sensitive detection methods (Faleiro 2006).

In areas where the insect is not yet present, the best management strategy is "RPW exclusion", although this has proved to be elusive and hard to implement in the most affected Gulf countries. Exclusion includes barring the introduction of RPW-contaminated palm trees and seedlings to an area that is not infested yet. Exotic wood-boring beetles were excluded from entering the US by adoption of an approach to detect, monitor, control and eradicate them (USDA-APHIS 2011). This would have been the best approach to manage RPW when it was discovered in the 1980s because most of the date palm growing areas are isolated from each other by vast inhospitable desert, a natural barrier to the movement and spreading of RPW. However, RPW is now present in all date palm growing areas and only an area-wide approach can be effective. A good example of such an approach is provided by the eradication of the cattle ticks, *Boophilus* spp., vectors of cattle fever in the USA. The strategy was based on rigorous restriction cattle movement, combined with cattle-dipping. The cattle ticks were eliminated in 1943 at a cost equivalent to the annual losses prior to the implementation of this program (Vreysen and Robinson 2011).

3.2 Biology of RPW

The destructive stage of the RPW is the larva, which excavates the trunk of the palm tree, eventually killing it (Dembilio and Jacas 2010). The transport of mature trees and seedlings from infested farms to new territories accounts for most of the spread of RPW to and within the Kingdom of Saudi Arabia (Faleiro 2006; Mukhtar et al. 2011). The larvae are inaccessible, hard to detect and difficult to treat with chemical pesticides. The commonly used detection methods rely on external symptoms of infestation that often become visible only when it is already too late to save the tree by chemical treatment (Abraham et al. 1998). The cryptic nature of the RPW and lack of reliable detection methods, especially at early infestation stages, encourages the indiscriminate use of chemical pesticides by date palm growers (Ferry and Gomez 2002). Such indiscriminate use of chemical pesticides results in pesticide-contamination of dates that reaches hazardous levels (El-Saeid and Al-Dosari 2010). Thus the development of an early, reliable detection method is of paramount importance to progress toward a sound strategy for the sustainable management of RPW infestation (Ferry and Gomez 2002; Faleiro 2006).

The RPW undergoes complete metamorphosis, progressing through a four-stage life cycle: egg, larva, pupa and adult. Adult females live for several weeks during which they lay as many as 500 eggs in the crown of a palm. The eggs hatch in about 6 days and the neonate larvae enter the tree through wounds, caused by pruning green growing fronds, feeding of other insects, and other types of physical damage. Entry can be made at the trunk/soil interface or the crown of the tree. Irrigation too close to the base of a tree creates a humid environment that facilitates entry of RPW. The larvae (grubs), characterized by dark prominent heads, cause serious damage to the palm tree by tunneling the tissue often leading to the death of the palm.

The larvae pass through three to seven instars, growing into mature grubs that migrate to the periphery of the trunk where they build cocoons, entering the pupal stage. Adults (Fig. 3.1) emerge from the cocoons a few weeks later. The entire life cycle takes about 3 months (Dembilio et al. 2012). The weevil often completes multiple generations in the same tree host before the tree collapses and the adult weevils abandon it to seek uninfested palm trees. Because of the concealed nature of the larval and pupal stages, infestation of the date palm stem is often hard to detect until the palm tree is dying. Early clues are provided by the drying of the youngest fronds due to larvae feeding, which damages the meristem, resulting in larval excrement accumulation at the base of fronds or at a point of injury.

3.3 Invasive Species

Modern transportation systems facilitate the rapid movement of humans and goods between countries and continents in the current era of globalization. Along with goods come uninvited guests including weeds, rodents, fish, diseases and insect

Fig. 3.1 Adult red palm weevil on date palm (Photo: Ali M. Idris)

pests. These alien invaders often rapidly became established as major problems in their new ecosystems. Another contributory factor to the increasing number of invasive species is climate change (Pyšek and Richardson 2010; van der Putten et al. 2010). Most invasive species were introduced into new territories without their natural enemies, pathogens and predators, which normally keep these species under the economic threshold in their original habitat. Natural enemies are important; but pest outbreaks happen (Nair 1998). Abundant food, including diverse host species, and lack of competition from other related species make organisms more successful in their new ecosystems.

It is likely that the vast majority of invasive species are not noticed because they do not immediately cause crop damage or health problems. The size of the response to invasive species is directly related to the economic, observable environmental or health impact. The cost of invasive species to economies globally is very high. It has been estimated that the cost of pest management and loss in production in the USA alone due to invasive species is as much as $130 billion per year (Pimentel et al. 2001).

The oil boom in the Gulf countries has changed the life style of the people through wealth accumulation and the influence of foreign nationals attracted by employment opportunities. The local populations became more interested in ornamental plants, especially drought-tolerant plant species, to change the inhospitable desert into modern green landscapes. Ornamental palm trees from South and Southeast Asia were imported into the Gulf countries to alter the landscape. Because of the concealed nature of the RPW, infested ornamental palm trees went undetected, resulting in the initial inoculum of RPW and subsequent spread to date palm throughout the Middle East. The oil boom and cheap labor contributed to the expansion of small-holdings and monocultures of date palm plantations. This rapid expansion of date palm monoculture took place through transportation of offshoots and

Fig. 3.2 Wild palm (*Hyphaene thebaica*) in the southwest desert of Saudi Arabia (Photo: Ali M. Idris)

adult date palm trees within and between countries in the region. Combined with concealed nature of the RPW, the monocultures and the transportation of date palm trees contributed to the spread and successful establishment of the RPW throughout the Middle East and into North Africa and Southern Europe.

Records indicate that RPW is now present in all date palm growing regions of Saudi Arabia, especially the Central region, the highest producer of dates in Saudi Arabia, Qassim Region, Eastern Region, and Asir Region. Date palm plantations in these regions are separated by vast stretches of desert, which should make quarantine a relatively easy task. However, now that the RPW is widely established in the country, there is a chance that it will also become established in the Doum palm tree (Fig. 3.2), also known as gingerbread tree (*Hyphaene thebaica* Mart.), a wild palm tree growing in the wadis in the west coastal area and the southern region of Saudi Arabia.

3.4 Detection and Available Control Methods

Because the movement of mature trees is a major means of RPW spread, detection and treatment before trees are shipped is critical. However, it is also difficult to assess the efficacy of the treatments applied. Hence there is a critical need for sensitive means of detecting the persistence of live adults and larvae after treatment to avoid introduction of the insect into new areas. Monitoring devices currently being used have significant limitations (Faleiro 2006; Mankin et al. 2011). Although, monitoring traps are used, these detect the pest only after adults emerge and infestation is well-established. Nevertheless, monitoring traps alert pest managers on the presence of RPW in an area and help in planning area-wide operations such as inspection of palms to detect infestations, mass trapping of adult weevils, preventive and curative chemical treatment, identification of hidden breeding sites, and eradication of severely infested palms (Abraham et al. 1998). Food-baited aggregation pheromone traps that attract male and female weevils are widely used in both monitoring and mass trapping of RPW (Abraham et al. 2001). Since fermentation of the food bait in RPW pheromone traps is necessary, there is the possibility that bacteria or fungi growing in the traps create volatiles that contribute to weevil attraction.

Detection at early stages of infestation and monitoring of RPW are essential components of IPM approach. There is, at present, no reliable protocol for detecting infestation of a given palm tree. As well, there are currently no methods that interfere with RPW mating. Thus, the major hurdles in dealing with RPW infestation are (1) the difficulty of early infestation detection (Abraham et al. 1998), (2) the failure to control the transportation of infested trees and seedlings (Al-Shawaf et al. 2013) rendering "exclusion" ineffective, and (3) the absence of effective treatments other than, mass trapping of adult weevils, drenching and injecting of broad spectrum insecticides (Abbas et al. 2001; Prabhu et al. 2010).

The national and international transportation of RPW infested offshoots is the likely culprit in weevil dissemination, probably as concealed immature larvae (Faleiro 2006). Several countries have implemented preventative and curative measures, as well as cultural practices to control the insect. These include quarantine and certifications (Salama and Abd-Elgawad 2003), inspections coupled with pre- and post-entry pesticide treatments (Faleiro 2006), stem injections (Abdallah and Al-Khatri 2000), and soaking of offshoot bases with pesticides (Abraham et al. 1998; Vidyasagar et al. 2000). Moreover, both pheromone traps and chemical injection are routinely used for area-wide management (Oehlschlager 2005; Faleiro 2006; Abraham et al. 2001). However, early detection of insect infestation remains a significant challenge in the management of this pest.

A number of different techniques for early detection have been explored, including acoustic detection of the noise made by feeding larvae (Rach et al. 2013). Current state-of-the-art techniques for detection rely on using a simple conditioning circuit consisting of an analog-to-digital converter (ADC) that is attached directly to the acoustic transducer (Rach et al. 2013). This procedure results in a simple system in which the data can be further processed using digital filters (Mankin 2011). However, this comes at the expense of significant power consumption, since the ADC will effectively digitize both the signal and the interfering noise. Hence, the selected ADC must be characterized with at least 13–16 bits of resolution, a power-hungry electrical component of the system that limits its portability due to battery size and the operational time of the system. In addition, once the signal is digitized, the common software signal processing technique will not be sufficient to remove the noise that may exist in the same frequency band as the signal. These two drawbacks are the main limitations of the systems currently under development (Mankin 2011; Rach et al. 2013).

Better management strategies, including reductions in pesticide use, require the development of earlier and more sensitive methods to detect the presence of the insect at an early stage of infestation before visible symptoms of infestation develop (Abraham et al. 1998). In this chapter, the conventional management methods employed to reduce the impact of introduced insect pest like RPW are described and the information gaps that need to be bridged for sound management techniques are addressed. In addition, the importance of recent advances in biotechnologies and microbiology is underscored as potential components of integrated RPW management and concludes with regulation and public acceptance of genetically modified organisms.

3.4.1 *Cultural Control*

Sound cultural practices contribute to the protection of the date palm trees from RPW infestation and should be used when implementing other components of IPM (Abraham et al. 1998). These agronomic practices include the protection of wounds on the palm surface used by RPW for egg laying. As well, a proper irrigation system is important for the management of RPW to avoid wetting the base of the trunk, which softens the stem and facilitates the entry of the larvae. Wetting can be avoided by raising the soil around trunk to keep the irrigation water away (Fig. 3.3). Abiotic stresses, such as drought or excess salt, weaken the trees and render them vulnerable

Fig. 3.3 Irrigation system to avoid wetting of the trunk, Sarah Farm, Al-Kharj, Riyadh, Saudi Arabia (Photo: Saleh A.A. Aldosari, King Saud University, Riyadh, Saudi Arabia; with permission)

to RPW infestation. Removing dead leaves eliminates RPW hiding places. Removing and destroying palm trees in abandoned groves eliminates a breeding ground for insect pests and a major source for re-infestation (Mukhtar et al. 2011). Pruning the leaves and separating the offshoots creates openings ideal for the entry of RPW larvae and kairomones released by wounded trees attract weevils (Gunawardena et al. 1998). These wounds should be sealed to avoid infestation of the tree. The removal of infested trees should be followed by destroying them to kill the concealed insects. That can be done by either burning the trees or chopping the trunk into manageable log sizes, wrapping the logs in insect-proof black plastic, and exposing them to the sun. The logs can be stored away from date palm farms and checked periodically to assure that no RPW stages survive.

3.4.2 Biological Control

Biological control classically meant importation of natural enemies, both parasites and predators, from the native ranges of invasive species. This was later modified to include biopesticides such as insecticidal bacteria, viruses and fungi. In the modern era it has to also come to include such newer methods as sterile insect technique (SIT), symbiotic control and now the use of *Wolbachia* transplantation to create pest species that out-compete conspecifics in targeted areas (Hedges et al. 2008). With the advent of genetic engineering, genes from *Bacillus thuringiensis* responsible for order-specific toxins have been transferred to crop plants, removing them from the category of biological control and place them in the plant resistance component of pest management (Hellmich and Hellmich 2012).

Examples of innovative methods given as models of IPM upon which to draw for palm tree are described below:

3.4.3 Sterile Insect Technique

The sterile insect technique (SIT) has been an important component of IPM for certain insect pests. SIT involves the production and mass release of sterile males to mate with wild female to produce non-viable eggs. This has been used successfully to eradicate or suppress a number of insect species of animal including human as well as insect pests of crops in many countries (Hendrichs et al. 1995). SIT is species-specific insect management measure suitable for area-wide application and an environment friendly approach that does not affect other species and not pollute the environment. The requirement for a successful SIT is establishment of mass-rearing facilities to produce sterile males by irradiation, creating chromosomal aberrations and dominant lethal mutations. Frequent release of large numbers of sterile males reduces the wild population in subsequent generations and the ratios of released sterile males to wild males increases until the target species is eliminated (Knipling 1955). The success of SIT depends on the ability of the irradiated sterile males to be sexually active and compete effectively with wild males.

The paradigm for this technique was the successful eradication of the screw-worm (*Cochliomyia hominivorax* Coquerel) in North America and much later in Libya. Screwworm is a lethal insect pest of livestock and humans in the Americas (Bushland et al. 1955; Lindquist et al. 1992). One of the problems facing the implementation of SIT is the manual sorting of males to reduce female contamination. Sorting is a hard and laborious process and frequently results in release of unacceptable levels of females (Delprat et al. 2002). To reduce the amount of females released in the SIT system, a transgenic line of target insects was created in which the males express Green Fluorescent Protein (Condon et al. 2007) that allows for separation by high-throughput instruments. Modern versions of SIT include inserting lethal genes into target species (Morrison et al. 2010). This revolutionary advance allows insects that could not be produced successfully by radiation based methods to be considered as SIT targets. One of the classic examples of this is the cotton boll weevil (Mayer and Brazzel 1966; Cross 1973; Miller 2013).

Godfray (2013) recently summarized advances in mosquito control efforts to combat malaria. These include new attempts to produce strains to facilitate sorting of males and females (Yamada et al. 2012). Because of the concealed nature of RPW, it seems a little pre-mature to apply these approaches to manage this insect.

Very little attention has been given to the potential use of SIT as a possible measure for the management of RPW (Al-Ayedh and Rasool 2010). The main reason for lack of serious SIT research is the concealed nature of the RPW, where the insect spends several generations in a single trunk before adults leave for a new host palm. This makes SIT control of RPW possible only when adults can be enticed to leave their host palm before laying eggs so that sterile males will have a chance to complete for mating with native males. In addition to the concealed nature of the RPW, the SIT option faces several unresolved issues, including mass rearing of RPW on artificial diets to attain native adult size, production of sterile males that are capable of competing with native males for mating, and the feasibility of separating males from females. Recent advances in biotechnology may resolve the issue of separating males from females. Biotechnology can also be used to create markers to distinguish released insects from natives. The other issues facing SIT are the lack of knowledge regarding the number of RPW subspecies/cryptic species evolved in the region and whether there is a mating barrier between populations. This is important for selecting the number of populations to be reared for insect release (Alphey et al. 2008).

3.4.4 Symbiotic Control Methods

An alternative to the SIT method is *Wolbachia* infection now being pursued by a number of researchers globally (Serbus et al. 2008). *Wolbachia* is known to create two reproductive conditions that favor its own survival. The first is Cytoplasmic Incompatibility (CI), which renders the mating of infected males with uninfected sterile females. The infected male in essence functions as if it has a sperm defect that can only be rescued by mating with an infected female. When populations of infected individuals are mixed with uninfected counterparts, mating is biased in favor of an infected population. The second major difference conferred by *Wolbachia* infection is parthenogenesis. This is very common in parasitic Hymenoptera where only females are known or the sex ratio is biased heavily in favor of females (Werren et al. 2008; Bian et al. 2013; McGraw and O'Neill 2013). This method transmits the endosymbiotic *Wolbachia* bacteria from one species into another. The target host can be malaria mosquito, *Anopheles gambiae* Giles (Jin et al. 2009) or the dengue mosquito *Aedes aegypti* L. (Hoffmann et al. 2011; Iturbe-Ormaetxe et al. 2011; Rasgon 2011; Walker and Moreira 2011). Dengue mosquito was a recipient of a *Wolbachia* strain from *Drosophila* (Hoffmann et al. 2011); while the malaria mosquito *Anopheles stephensi* Liston (Bian et al. 2013) was a recipient of a *Wolbachia* from *Anopheles albopictus* Skuse. The first surprise from these manipulations was that transplanted *Wolbachia* not only induced CI, as expected, but made the host insect vector incompetent, which was unexpected. The clear advantage of *Wolbachia* is the reproductive strategy of CI it confers on the host (Brelsfoard and Dobson 2009). This favors the production of *Wolbachia* infected offspring and theoretically an infected area need only to be seeded with the infected individuals a relative small number of times (five separate releases in 40 day for population replacement in the Cairns, Australia project to displace the local vectoring *Ae. aegypti* population with a vector incompetent *Wolbachia* infected strain (Rasgon 2011).

The strategy for using the *Wolbachia* method is to determine the nature of *Wolbachia* infection in native populations and then to determine the reproductive strategy that this might induce in the population. Since little is known about *Wolbachia* associated with palm tree pests, a significant amount of work will be needed to determine whether it can be used as an effective control measure.

Another approach to symbiotic control is to consider a number of other endosymbionts (intracellular symbionts) that are very likely associated with palm tree pests. In certain insects, the symbiotic gut bacteria and other endosymbionts determine the host plant choice and a number of other traits (Hosokawa et al. 2007; Little and Currie 2007; Hosokawa et al. 2010; Tsuchida et al. 2010; Uematsu et al. 2010; Hamdi et al. 2011; Himler et al. 2011; Kikuchi et al. 2012). In some cases the female employs strategies that ensure all offspring obtain the correct gut symbiotic bacteria. The same is true for certain beetles.

Yet another method fitting under symbiotic control is that being pioneered at Louisiana State University of Baton Rouge, Louisiana, USA where gut microbes of termites are employed to deliver anti-protozoan strategies that are spread through the colonies by social behavior and are selectively lethal to the only mechanism available to digest wood (Husseneder and Collier 2009; Husseneder et al. 2010a). Termites use protozoa and cellulolytic bacteria to digest wood, the termite's only nutrient source (Husseneder 2010; Husseneder et al. 2010b). In view of the formation of a fermentation column in the trunks of palm trees by red palm weevil, this approach should be examined seriously. It offers the possible advantage of delivery by a special bait concoction and can be extremely selective.

3.4.5 Chemical Control

Chemical pesticides constitute an important component of IPM that must be compatible with other environmentally safer, and economically cheaper options. These options include cultural practices, pheromone trapping, used both for both surveillance and reducing the weevil population, and biological control (Mukhtar et al. 2011). Since the discovery of RPW in the Gulf countries, chemical pesticides have played a significant role in the management of RPW. The relatively low prices of insecticides coupled with weak enforcement of regulations led to the growing use of chemical pesticides in the region. Generally, chemical pesticides are used on date palm trees for the management of insect pests. These include spraying the canopy and the trunk, tree injection, treatment of the soil and dipping the offshoots before transplantation (Al-Shawaf et al. 2013). Application of chemicals to the soil as a preventive measure before transplanting date palm offshoots requires increasing the pesticide amount (5× to 10×, the recommended dose) to compensate for the losses due to leaching, volatilization and fixation by soil particles (Doccola and Wild 2012). The sandy soil, its high pH and salinity in date palm growing regions in Saudi Arabia have limited the use of soil treatment to the planting hole for date palm transplantation. However, dipping the bole of date palm offshoots in pesticide solution (0.004 % Fipronil for 30 min) before planting has proved effective in killing larvae and adults without phytotoxicity (Al-Shawaf et al. 2013). Pesticide soil application is restricted in the urban landscape due to concerns of inadvertent human exposure and contamination of underground water due to leaching in sandy soil regions. Canopy and stem sprays can be very effective against insect pests that feed on leaves of dwarf/young date palms, but is not used for tall trees because of drift, inadequate coverage and the requirement for special equipment (Doccola and Wild 2012). Moreover, canopy spraying is not used in the management of RPW in date palm, although periodic preventive sprays of the date palm trunk is done in areas with high weevil activity (Abraham et al. 1998).

The third and widely used chemical method for the management of RPW is tree injection (Mukhtar et al. 2011). Two kinds of tree injections are commonly used on date palm trees: trunk infusion and pressurized trunk injection. Tree injection has been used successfully for the management of Dutch elm disease (DED), a fungal vascular wilt, when fungicide spraying failed to save elm trees in the USA (Shigo et al. 1980). This method also has been used to control lethal yellowing (LY), a bacterial disease of coconut, caused by a Mycoplasma-Like Organism (MLO) (McCoy 1975). Tree injection is gaining popularity because less chemical pesticides are used with less non-target contamination (Doccola and Wild 2012). However, certain concerns need to be addressed including pesticide residue levels in the fruits and the mortality rate of adults, larvae and the non-feeding pupae. There is a need for improved injection technology and pesticide formulations designed especially for date palm injection. Any new formulations for trunk injection should consider residual chemical pesticide level in the fruits and the mortality rate of adults, larvae and the non-feeding pupae.

3.5 Advances in Biotechnologies: Delivery Systems for Biopesticides

A significant improvement on SIT has been proposed by in which a repressible dominant lethal gene is introduced into the target insect under the control of a female-specific promoter (Thomas et al. 2000). The female specific promoter is repressed by addition of the antibiotic tetracycline to the diet. This SIT system, which is termed as release of insects carrying a dominant lethal (RIDL) gene, eliminates the need for genetic sexing system because the females die when the repressor is removed from the diet. This is a very important advancement because all the released insects are male and there is no need for laborious manual separation of the sexes. Female survival is conditional, permitting the population to be increased and maintained under laboratory conditions. The released males compete with native males for native females only in the absence of their sterile females. In relatively recent study, emanation of females was found to increase the efficiency of released males by three to fivefolds (Rendón et al. 2004). The technology was improved further by addition of genes encoding fluorescent marker proteins to facilitate identification of transgenic insects both in the laboratory and for monitoring of released insects in the field (Marinotti et al. 2013). The RIDL system has been used successfully to rear and release Medfly (*Ceratitis capitata* Wiedemann) in Guatemala and the Mexican fruitfully (*Anastrepha ludens* Loew) (Alphey et al. 2008). Currently, there are several research projects aiming at developing RIDL mosquitoes for the management of vector-borne diseases like malaria and dengue fever. Some of these studies are in advanced field trials and RIDL mosquitoes have recently been approved for commercial release in Brazil (Alfred 2014).

Recent advances in biotechnology made easy to identify and introduce these endosymbionts into uninfected arthropods through embryonic cytoplasmic microinjections (McGraw et al. 2002). The symbiotic microbes associated with RPW are only just beginning to be examined. The RPW has an internal structure called a symbiotic organ and is full of bacteria that have recently been identified including bacteria of the genus, *Nardonella* (Murray and Stackebrandt 1995). At a reduced genome size of around 0.2 Mega bases, this is one of the smallest genomes known, and suggests a very highly specialized function. *Nardonella* has lost the genetic pathways to produce most amino acids, with the exception of tyrosine (Toju et al. 2010). Since the RPW is a very large insect with a very thick cuticle made largely of cross-linked protein and since it is known that tyrosine is the most important source of the quinones used in cross-linking, it is hypothesized that *Nardonella* in the symbiotic organ is essential to the growth and development of RPW. This represents a critical point that could be targeted in a new control strategy. When *Nardonella* is eliminated, the resulting cuticle of RPW is very thin and weak, showing that this microorganism is an important source of tyrosine (Kuriwada et al. 2010).

The reproductive strategy of RPW is not well understood. It is likely that females attract males by pheromone release and there is evidence that mating occurs inside of the tree. However, females appear to re-mate readily with available fit male (Faleiro 2006). Reported sex ratios of pheromone-trapped females vary from 1.2 to 2.5 females per male. Nothing is known about the presence and role of *Wolbachia* bacteria in the RPW. However, in other insects *Wolbachia* bacteria determine reproductive strategies designed to ensure continued infection by vertical transmission in eggs (Presgraves 2010). These strategies can be reversed by antibiotic treatment. *Wolbachia* bacteria are purely parasitic and cured insects may completely change the reproductive strategy.

Once inside a palm tree, a female weevil lays eggs and the eggs hatch to begin excavating galleries or tunnels through the palm tree material (Dembilio et al. 2012). It is known that infestations are accompanied by microbial attack on the substance of the palm tree associated with weevil feeding including fermentation. The source of these microbes is unknown. It was suggested that this microbial action amounts to pre-digestion of the palm tree material to aid the developing larvae in feeding. There is precedent for this strategy in female fruit flies of the genus *Tephritidae*, which deliberately defecate at oviposition sites on fruit, thereby depositing bacteria (Behar et al. 2008a). The deposited bacteria accelerate the rotting process and allow larvae to gain access to nutrients. It is also known that Tephritids acquire gut complements of bacteria that are essential to metabolism (Behar et al. 2008b). Bacteria are not the only microorganisms that develop symbiotic relationship with insects; there are several instances of fungi associated in mutualistic ectosymbiotic relationships with tree-boring insects. The bark beetles (*Xyleborus dispar* F.) excavate tunnels in their host plants to form galleries, where there cultivate ambrosia fungi (*Ambrosiella hartigii* Batra) occur as their sole food (Six 2003). Radiation used on Tephritid pupae in SIT programs is known to kill most of the gut bacteria. The resulting adults starve to death in 3 days. For this reason the International Atomic Energy Agency (IAEA) in Vienna, Austria initiated a research program to identify probiotic strategies to restore essential gut microflora and thus boost viability and longevity of SIT adults. Whether a similar vital symbiotic association occurs in RPW larvae and adults is not known.

Insect pests such as the RPW inflict billions of dollars in damage to agricultural crops and greatly threaten global food security (van der Valk 2007). Control strategies that depend on chemical pesticides are complicated by prohibitive costs, increasing target-insect resistance, environmental contamination of soil systems and ground water and lethal effects on humans (Showler 2002; Miller 2013). Biological pesticides such as entomopathogenic fungi, nematodes and bacteria are gaining importance worldwide as alternatives to chemical insecticides (Alves et al. 1998). Microbial pesticides such as *Bacillus thuringiensis*, *Metarhizium anisopliae* (Metchnikoff) Sorokin, *Beauveria bassiana* (Balsamo) Vuillemin and *Lecanicillium* Zare & W. Gams species have proven to be effective against ticks, termites, coffee berry borer, diamondback moths, and whiteflies, and for control of desert and Asian locust populations (Inglis et al. 1996; Kramm and West 1982; Betz et al. 2000). Fungal and bacterial biopesticides exhibit ideal killing characteristics and are comparable to chemical pesticides in laboratory trials. However, several major hurdles exist when these organisms are deployed in the field, especially in arid and desert regions of the world (Shah et al. 1998; Lacey and Siegel 2000; Côté et al. 2001). Bacterial and fungal spores/conidia are rapidly inactivated by ultra-violet (UV) radiation, (Hedimbi et al. 2008) extreme temperatures, aridity (Fargues et al. 1996) and rain wash-off (Elçin 1995; Elçin and Öktemer 1995; Betz et al. 2000). Many of the most vulnerable target areas of the world are in desert zones such as North and Sub-Saharan Africa and the Middle East. Indeed, the use of entomopathogenic fungi such as *B. bassiana* and *M. anisopliae* against RPW had been described and offers several advantages over chemical pesticide based eradication strategies. Given the susceptibility of these fungal isolates to heat, UV radiation and aridity, new biotechnological approaches to enhance the efficiency and longevity of these biopesticide formulations in the field are needed.

Attempts to attenuate UV-mediated deactivation have largely involved oil-based formulations for pesticide delivery (Alves et al. 1998; Hedimbi et al. 2008). Oil emulsions (10 % v/v) can increase conidial viability of *M. anisopliae* by 21 % after 6 h of simulated solar-UV exposure; other attempts have yielded similarly low results (Alves et al. 1998; Hedimbi et al. 2008). Sunscreens that have been incorporated into oil emulsions have enhanced UV protection by nominal amounts (4–34 % survival after 1 h simulated sun exposure), but desiccation under arid conditions remained problematic (Hedimbi et al. 2008). Several UV blockers [congo red, benzaldehyde and para-aminobenzoic acid (PABA)] have also been added to oil formulations with varying fungal survival advantages (McGuire et al. 1999; Hadapad et al. 2008).

Perhaps, one of the most promising approaches to protection of entomopathogenic fungi in harsh environmental conditions involves use of physical UV-barriers such as starch and calcium-cross-linked sodium alginate. These materials have demonstrated higher success rates at prolonging encapsulated bacterial or fungal survival (50 % spore survival at 6 h UVB and 7 h UVC, respectively) under conditions of UV radiation (Elçin 1995; Zohar-Perez et al. 2003).

The Durvasula Laboratory, with funding support from The United States Department of Agriculture and The Bill and Melinda Gates Foundation, is developing novel bioencapsulation strategies for the protection of biopesticides in the environment. Initial efforts have been focused on the glassy winged sharpshooter, *Homalodisca vitripennis* Germar, and the desert locust, *Schistocerca gregaria* Forsskål. Subsequent efforts will involve encapsulation of entomopathogenic fungi to target RPW.

A calcium-alginate microcapsule for containment of *M. anisopliae* has been developed. For related applications, similar particles for a bacterial biopesticide, *Pantoea agglomerans*, have also been developed. To assure that the particles would coat plants evenly and be ingested by feeding locusts, a modified aerosolization-coacervation process to generate particles pre-dominantly in the 20–60 μm diameter range was used (Fig. 3.4). The encapsulation process protected *P. agglomerans* and *M. anisopliae* from desiccating conditions (Fig. 3.5). Addition of 0.5–1.0 % India ink to the microparticle conferred high-level UV-resistance, even under grossly exaggerated conditions of UVC exposure (Figs. 3.6 and 3.7). The calcium-alginate-10 % glycerol hydrogel particle should be well suited to protect and deliver

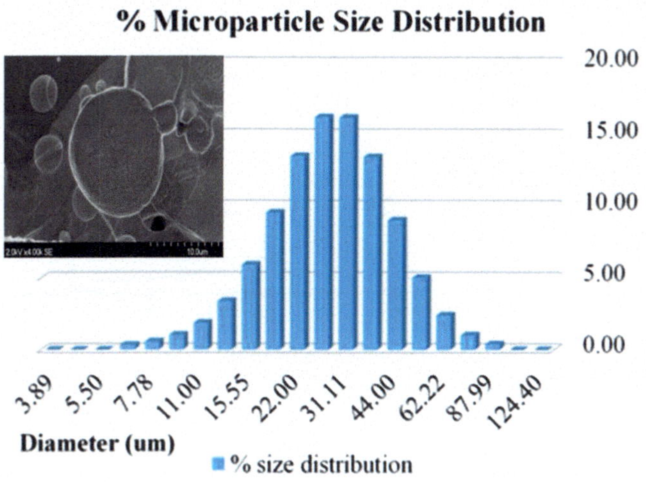

Fig. 3.4 Calcium-alginate microparticle size distribution. Insert: electron micrograph of particles containing GFP-producing *Pantoea agglomerans* (Photo: Ravi Durvasula – unpublished)

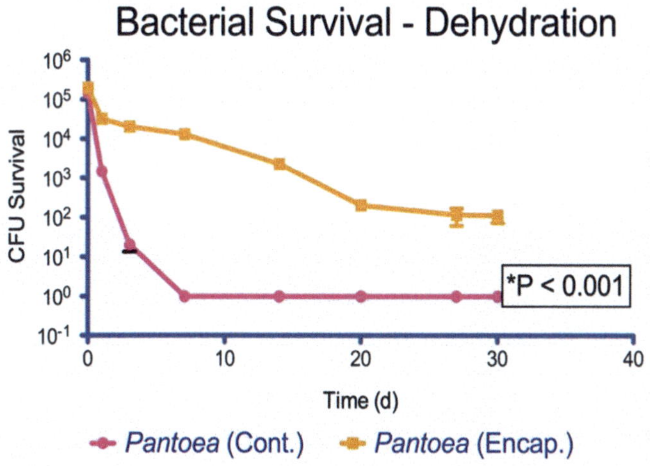

Fig. 3.5 Protection of microbial payload under highly arid conditions. This graph demonstrates survival of *Pantoea agglomerans*; similar results were obtained with *Metarhizium anisopliae*. Survival of encapsulated *Pantoea agglomerans* was significantly greater at 30 day using Student *T*-test (P<0.001) (Photo: Ravi Durvasula – unpublished)

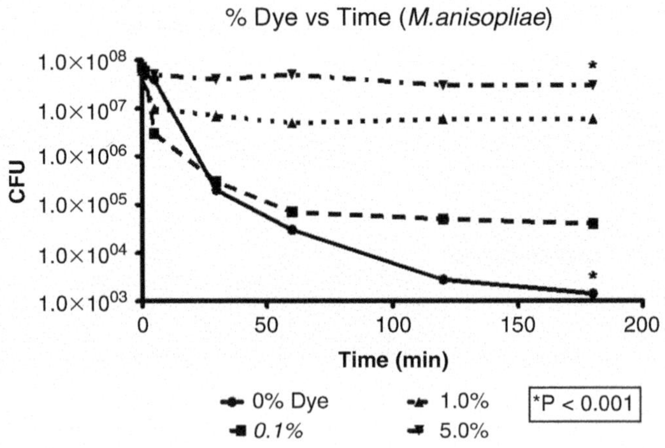

Fig. 3.6 Increasing concentration of India ink (*dye*) in microparticles conferred high-level protection of *Metarhizium anisopliae* against UVC radiation. A concentration of 5 % dye resulted in survival of fungal spores at 120 min; absence of dye resulted in fungal death (P<0.001) (Photo: Ravi Durvasula – unpublished)

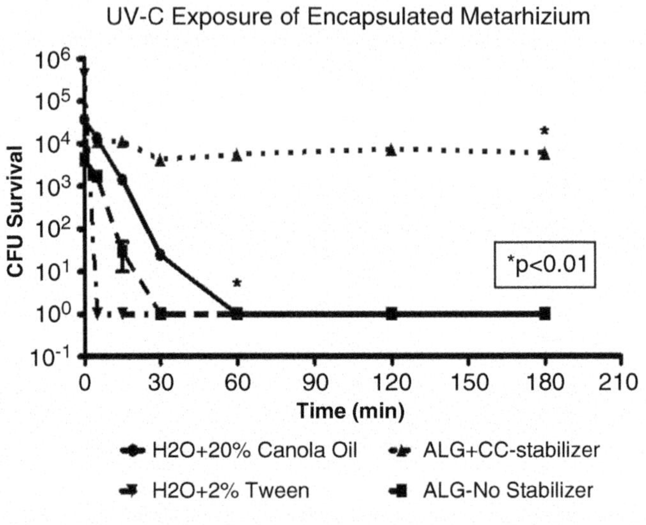

Fig. 3.7 Absence of dye led to death of fungal spores within 60 min of UVC exposure, whereas microparticles containing India ink dye protected *Metarhizium anisopliae* to at least 180 min (P<0.001) (Photo: Ravi Durvasula – unpublished)

M. anisopliae under desert field conditions. Enhanced encapsulation strategies are at the field validation stage against the American grasshopper in Montana, USA and the desert locust in Tunisia. Similar approaches may hold promise for delivery of entomopathogenic fungi that kill RPW under desert conditions.

3.6 Regulation and Public Acceptance of Genetic Modification

For most of their evolutionary history, humans were hunter-gatherers, spending their time finding and capturing food. Roughly 10–20,000 years ago, humans began to shape plants and animals to be more useful as food and for other domestic purposes. As well, they became less mobile and settled down in favorable places to grow crops and tend animals. People domesticated dogs, cattle and many other animals, as well as such important crop plants as wheat, rice and corn long before they knew they were practicing genetic modification. Commencing with Mendel's pioneering studies in the late nineteenth century, the contribution of genetic modification to agriculture began to accelerate, particularly with the adoption of chemical and radiation mutagenesis in crop improvement. The late twentieth century's molecular revolution began with the explication of the structure and function of the genetic material, deoxyribonucleic acid (DNA). The ability to clone genes followed in the 1960s and 1970s on the discovery of restriction enzymes and the molecular characterization of bacterial plasmids and viruses, leading to the development of what came to be called "recombinant DNA (rDNA)" technology.

When rDNA technology was first being developed in the U.S., scientists became concerned about the possibility that combining genes across species, genera and kingdoms might create organisms with unexpected new traits. A group of prominent scientists published a note in Science magazine requesting that the USA National Institutes of Health (NIH) establish an advisory committee to oversee an experimental program to evaluate the hazards of the new techniques and devise guidelines for investigators working with the techniques (Berg et al. 1974). The requested advisory body, subsequently known as the NIH Recombinant DNA Advisory Committee (RAC), was duly constituted by the NIH director and developed guidelines for containment procedures.

As experience with recombinant DNA-containing organisms accumulated, it became increasingly clear that the initial estimates of risks had been exaggerated and the guidelines were gradually relaxed. A very important point is that the NIH guidelines could and did evolve with accumulating evidence that rDNA containing organisms were not hazardous. The U.S. regulatory picture began to change when the first requests came in to the RAC in the early 1980s to field test transgenic plants. The White House Office of Science and Technology Policy (OSTP) organized an interagency committee to consider how transgenic plants should be regulated. The committee wrote a document titled "Coordinated Framework for Regulation of Biotechnology" (OSTP 1986), implemented in 1986 after public comment. The committee concluded that the use of rDNA techniques was not inherently risky and therefore did not require new legislation, but could be regulated under existing statues. The relevant regulatory agencies, which were the Environmental Protection agency (EPA), the United States Department of Agriculture (USDA) and the Food and Drug Administration (FDA) were directed to find existing statutes that could provide the basis of regulating the release of rDNA-modified organisms into the environment. The EPA decided to use the Toxic Substances Control Act (TSCA) and the Federal Insecticide, Fungicide and Rodenticide Act (FIFRA). The USDA used the Plant Quarantine Act of 1912 and the Federal Plant Pest act of 1957, under which it has the power to decide if a new plant variety or microorganism is likely to become a pest. And the FDA used the 1938 Federal Food, Drug, and Cosmetic Act (FFDCA).

The decision to use laws originally designed to regulate toxic chemicals, plant diseases and pests, and food additives shaped early public opinion about what is now called "genetically modified organisms" or GMOs. The same statutes are still used to regulate agricultural biotechnology in the USA today. By contrast to the rapid evolution of the original guidelines under the RAC, however, the regulatory requirements have evolved very little, despite the decades of biosafety experimentation and experience that have accumulated since then.

Alarmed by the emerging regulatory environment for GMOs, the Council of the National Academy of Sciences issued a white paper in 1987 titled "Introduction of Recombinant DNA-Engineered Organisms into the Environment: Key Issues" (Kelman 1987). Its basic conclusions remain consistent with all of the assessments that have been done in the intervening quarter century. These are: (1) there is no evidence that unique hazards exist either in the use of R-DNA techniques or in the movement of genes between unrelated organisms, (2) the risks associated with the introduction of R-DNA-engineered organisms are the same in kind as those associated with the introduction of unmodified organisms and organisms modified by other methods, and (3) assessment of the risks of introducing R-DNA-engineered organisms into the environment should be based on the nature of the organism and the environment into which it is introduced, not on the method by which it was produced. Indeed, the 2010 EC report on GMO biosafety research is well-summarized in the following (Economidis et al. 2010): "The main conclusion to be drawn from the efforts of more than 130 research projects, covering a period of more than 25 years of research and involving more than 500 independent research groups, is that biotechnology, and in particular GMOs, are not per se more risky than e.g. conventional plant breeding technologies". Every credible scientific body that has examined the evidence has come to the same conclusion.

Nonetheless, despite near unanimity on the part of the scientific community, significant regulatory barriers have arisen to the use in agriculture of organisms modified by molecular techniques in response to public rejection, which has intensified in the last decade. The origins of the contemporary situation can be found in the very early regulatory directions taken and in the development of the organic food industry. Paradoxically, although the intent of the Coordinated Framework was to avoid creating the perception of hazard where none was known to exist, the very process of using statutes created to regulate toxic substances, plant pests and food contaminants created precisely such a perception.

The contemporaneous development that proved profoundly formative of public opinion in the U.S. is the formulation of the national Organic Rule and the explosive growth of the organic food industry. In 1990, the U.S. Congress passed the Organic Food Production Act with the objective of bringing some uniformity to the organic certification process, then done by dozens of different organizations. The act established the National Organic Standards Board and charged it with developing a uniform set of national standards for growing food organically. The Board was appointed in 1992 and it issued its first recommendations, popularly known as the "Organic Rule", in 1997 (http://www.ams.usda.gov/AMSv1.0/nop).

As is customary with U.S. government recommendations, these were first published in the Federal Register. They provoked an unprecedented volume of comments and the most vociferously protested issue was the USDA's proposal to allow GMOs to be used in organic farming. The ensuing controversy catalyzed significant modifications of the Organic Rule and the final version, published in 2000, specifically prohibited the use of GMOs, both plants and animals. Hence from the outset, the burgeoning organic food industry and its supporters rejected the use of modern molecular techniques in organic agriculture. It is worth emphasizing that this was not based on evidence of any kind of danger inherent in GM technology, but on public perception and perhaps as much as anything on the organic food industry's vision of what it sought to sell their consumers, the view that organically grown food was more natural and more healthful than conventionally grown food. The past several years have seen the emergence of a movement in the USA demanding that all food containing GM ingredients be labeled. Although recent state labeling initiatives in California and Washington, USA were defeated, demands for mandatory labeling continue with proposed legislation in a number of states.

Despite the anomalies of regulating genetically modified organisms as toxic chemicals and plant pests, GM crops began to move into the market in the US, with the first commercial plantings, primarily corn and cotton, commencing in 1996. Things took a rather different course in Europe, where organisms modified by rDNA techniques were viewed and regulated as a separate category from the outset. In 1990, the European Union (E.U.) established a process for approving the "deliberate release" of GMOs that required the approval of either all member states or a majority of a committee made up of representatives of member states. In spite of the complexity of this process, the E.U. approved 18 GM crops for commercial marketing between 1992 and 1998. Regulatory approvals reached a political impasse in 1998 and in 1999, the E.U. Council formalized a moratorium on approvals to market GM crops until it could be demonstrated that there was neither a human health nor an environmental impact (the pre-cautionary principle) and that they satisfied labeling and traceability requirements. Moreover, Denmark, France, Greece, Italy and Luxembourg declared they would block future GM crop approvals altogether, effectively imposing a total moratorium on them.

The first GM crop approved by the EU Commission since 1998 was the BASF Amflora potato cultivar, an approval subsequently challenged in court and overturned, prompting BASF to move its biotech headquarters to the US and stop commercialization of GM products in Europe. This was followed in 2013 by Monsanto's decision to withdraw most of its pending GM crop applications with the EU. Today, the U.S. regulations are still in large part governed by the OSTP's Coordinated Framework. GM crops are reaching the market, albeit at a very high cost. In Europe, GM crop approvals are essentially at a standstill, as they are in Japan and a number of African countries, but there are also many countries that have established regulatory regimes that have allowed GMO commercialization. However, the recent move to allow member states to decide whether they permit GMOs may break the impasse, although it is likely to create additional regulatory dilemmas.

While Europe's anti-GMO attitudes and regulatory policies have exerted a severe inhibitory influence on the acceptance of GM crops in most African countries (Paarlberg 2009), they have flourished elsewhere. GM crop acreage has increased rapidly worldwide, driven primarily by cotton, corn, canola and soybeans. In 2013, GM crops were grown in 27 countries on 175.2 million hectares (ISAAA 2014). That represents a remarkable 100-fold increase over the 1.7 million hectares planted in the first year that biotech crops became commercially available in 1996. Importantly, 90 % of the 18 million farmers growing biotech crops worldwide are small-holder farmers. Half of the biotech acreage today is in developing countries. Farmers migrate to GM crops because their yields increase and their costs decrease.

According to the recent comprehensive assessment of their cumulative economic impact over 17 years, GM crops have added more than a hundred billion dollars to farmer income globally, about half of the increase accruing to the pre-dominantly small-holder farmers of the developing world (Brookes and Barfoot 2014). And despite persistent claims to the contrary, GM traits have increased yield by 10.4 % for corn and 16.1 % for cotton over that period. Adoption of insect-resistant GM crops has reduced pesticide spraying by 503 million kg of active ingredient between 1996 and 2012. While the amount of herbicide used decreased only slightly, the Environmental Impact Quotient (EIQ) improved by 15.5 % because of the use of more benign herbicides. Most of the reduction in pesticide use, however, was attributable to the use of insect resistant varieties, primarily of corn and cotton. The adoption of herbicide tolerant and insect-resistant crops has also resulted in fuel savings from reduced tillage and spraying. Over the entire period from 1996 to 2012, the reduction in fuel use has reduced the amount of CO_2 added to the atmosphere by about 16.7 billion kilograms. The greater adoption of reduced till and no till farming has resulted in additional soil carbon sequestration of as much as another order of magnitude. The decreased input of CO_2 for just 2012 was equivalent to taking ten million cars off the road.

Regulatory approval of GM animals and insects has been and remains difficult. AquaBounty Technologies (http://aquabounty.com) developed a transgenic salmon that grows faster on less feed than its wild counterparts and a biological containment system. Another example is the above described RIDL system developed by the British company Oxitec to produce male sterile mosquitoes (http://www.oxitec.com). In trials, these have proved remarkably effective in decreasing the population of dengue bearing mosquitoes. Brazil has just approved their commercial release (http://www.oxitec.com/press-release-high-tech-solution-for-controlling-the-dengue-mosquito-is-approved-by-ctnbio/), but it remains in the regulatory approval process in the U.S. While the possibility of using RIDL technology for the control of the RPW is very attractive, the covert lifestyle of the insect provides a significant barrier to its use and will require extensive research both to develop an appropriate genetic system and to identify a strategy to render the transgenic males competitive.

3.7 Summary

RPW is a serious insect pest of date palm, an important tree especially in the desert areas of the Middle East and North Africa. The destructive stage of the RPW is the larva, which excavates the trunk of the palm tree, eventually killing it. RPW has invaded the entire Mediterranean area and threatens ornamental palms as well as those providing oil and fruit crops. The national and international transportation of RPW infested offshoots is the likely culprit in weevil dissemination, probably as concealed immature larvae. Attention to quarantine to prevent spread is urgently needed in the infested regions. An integrated pest management strategy is being studied for the management of this invasive insect pest including the use of semiochemicals, biotechnology and the development of a new biopesticide delivery systems. The concealed nature of RPW and inability to readily detect new infestations makes control challenging. Information on the biology and life cycle of RPW are urgently needed, as is the development new methods of biological control.

Acknowledgments We thank Dr. Saleh A.A. Aldosari, King Saud University, Riyadh, Saudi Arabia for providing the picture of irrigation system at Sarah Farm, Al-Kharj, Riyadh, Saudi Arabia.

References

Abbas, M. S. T., Hanounik, S. B., Mousa, S. A., & Mansour, M. I. (2001). On the pathogenicity of *Steinernema abbasi* and *Heterorhabditis indicus* isolated from adult *Rhynchophorus ferrugineus* (Coleoptera). *International Journal of Nematology, 11*, 69–72.

Abdallah, F. F., & Al-Khatri, S. A. (2000). The effectiveness of trunk injection and fumigation for the control of red palm weevil *Rhynchophorus ferrugineus* Olivier in date palm. *Journal of Plant Protection in the Tropics, 13*, 17–21.

Abraham, V. A., Al Shuaibi, M. A., Faleiro, J. R., Abozuhairah, R. A., & Vidyasagar, P. S. P. V. (1998). An integrated management approach for red palm weevil *Rhynchophorus ferrugineus* Oliv. – A key pest of date palm in the Middle East. *Agricultural Sciences, 3*, 77–83.

Abraham, V. A., Faleiro, J. R., Al Shuaibi, M. A., & Al Abdan, S. (2001). Status of pheromone trap captured female red palm weevils from date gardens in Saudi Arabia. *Journal of Tropical Agriculture, 39*, 197–199.

Alfred, J. (2014). GM mosquitoes set to be released in Brazil to combat dengue. Available at: http://www.iflscience.com/health-and-medicine/gm-mosquitoes-set-be-released-brazil-combat-dengue-0. Accessed on 13 Apr 2015.

Alphey, L., Nimmo, D., O'Connell, S., & Alphey, N. (2008). Insect population suppression using engineered insects. In S. Aksoy (Ed.), *Transgenesis and the management of vector-borne disease* (pp. 93–103). New York: Springer.
Al-Ayedh, H. Y., & Rasool, K. G. (2010). Sex ratio and the role of mild relative humidity in mating behaviour of red palm weevil Rhynchophorus ferrugineus Oliv. (Coleoptera: Curculionidae) gamma-irradiated adults. *Journal of Applied Entomology, 134*, 157–162.
Al-Shawaf, A. M., Al-Shagag, A., Al-Bagshi, M., Al-Saroj, S., Al-Bather, S., Al-Dandan, A. M., Abdallah, A. B., & Faleiro, J. R. (2013). A quarantine protocol against red palm weevil *Rhynchophorus ferrugineus* (Olivier) (Coleoptera: Curculionidae) in date palm. *Journal of Plant Protection Research, 53*, 409–415.
Alves, R. T., Bateman, R. P., Prior, C., & Leather, S. R. (1998). Effects of simulated solar radiation on conidial germination of *Metarhizium anisopliae* in different formulations. *Crop Protection, 17*, 675–679.
Behar, A., Jurkevitch, E., & Yuval, B. (2008a). Bringing back the fruit into fruit fly–bacteria interactions. *Molecular Ecology, 17*, 1375–1386.
Behar, A., Yuval, B., & Jurkevitch, E. (2008b). Gut bacterial communities in the Mediterranean fruit fly (*Ceratitis capitata*) and their impact on host longevity. *Journal of Insect Physiology, 54*, 1377–1383.
Berg, P., Baltimore, D., Boyer, H. W., Cohen, S. N., Davis, R. W., Hogness, D. S., Nathans, D., Roblin, R., Watson, J. D., & Weissman, S. (1974). Potential biohazards of recombinant DNA molecules. *Science, 185*, 303.
Betz, F. S., Hammond, B. G., & Fuchs, R. L. (2000). Safety and advantages of *Bacillus thuringiensis* protected plants to control insect pests. *Regulatory Toxicology and Pharmacology, 32*, 156–173.
Bian, G., Joshi, D., Dong, Y., Lu, P., Zhou, G., Pan, X., Xu, Y., Dimopoulos, G., & Xi, Z. (2013). Wolbachia invades Anopheles stephensi populations and induces refractoriness to Plasmodium infection. *Science, 340*, 748–751.
Brelsfoard, C. L., & Dobson, S. L. (2009). *Wolbachia*-based strategies to control insect pests and disease vectors. *Asia Pacific Journal Molecular Biology Biotechnology, 17*, 55–63.
Brookes, G., & Barfoot, P. (2014). GM crops: Global socio-economic and environmental impacts 1996–2012. PG Economics Ltd. Available at: http://www.pgeconomics.co.uk/page/36/-gm-crop-use-continues-to-benefit-the-environment-and-farmers. Accessed on: 07 Apr 2015.
Bushland, R. C., Lindquist, A. W., & Knipling, E. F. (1955). Eradication of screw-worms through release of sterilized males. *Science, 122*, 287–288.
Condon, K. C., Condon, G. C., Dafa'Alla, T. H., Fu, G., Phillips, C. E., Jin, L., Gong, P., & Alphey, L. (2007). Genetic sexing through the use of Y-linked transgenes. *Insect Biochemistry and Molecular Biology, 37*, 1168–1176.
Côté, J.-C., Vincent, C., Son, K.-H., & Bok, S. H. (2001). Persistence of insecticidal activity of novel bio-encapsulated formulations of *Bacillus thuringiensis* var. *kurstaki* against Choristoneura rosaceana [Lepidoptera: Tortricidae]. *Phytoprotection, 82*, 73–82.
Cross, W. H. (1973). Biology, control, and eradication of the boll weevil. *Annual Review of Entomology, 18*, 17–46.
Delprat, M. A., Stolar, C. E., Manso, F. C., & Cladera, J. L. (2002). Genetic stability of sexing strains based on the locus *sw* of *Ceratitis capitata. Genetica, 116*, 85–95.
Dembilio, O., & Jacas, J. A. (2010). Basic bio-ecological parameters of the invasive red palm weevil, *Rhynchophorus ferrugineus* (Coleoptera: Curculionidae), in *Phoenix canariensis* under Mediterranean climate. *Bulletin of Entomological Research, 101*, 153–163.
Dembilio, Ó., Tapia, G. V., Téllez, M. M., & Jacas, J. A. (2012). Lower temperature thresholds for oviposition and egg hatching of the Red Palm Weevil, *Rhynchophorus ferrugineus* (Coleoptera: Curculionidae), in a Mediterranean climate. *Bulletin of Entomological Research, 102*, 97–102.
Doccola, J. J., & Wild, P. M. (2012). Tree injection as an alternative method of insecticide application. In S. Soloneski & M. Larramendy (Eds.), *Insecticides-basic and other applications* (pp. 61–78). Rijeka: InTech.

Economidis, I., Cichocka, D., & Högel, J. (2010). A decade of EU-funded GMO research (2001–2010). Available at: http://ec.europa.eu/research/biosociety/pdf/a_decade_of_eu-funded_gmo_research.pdf. Accessed on 13 Apr 2015.

Elçin, Y. M. (1995). *Bacillus sphaericus* 2362-calcium alginate microcapsules for mosquito control. *Enzyme and Microbial Technology, 17*, 587–591.

Elçin, M. Y., & Öktemer, A. (1995). Larvicidal and sporal behaviour of *Bacillus sphaericus* 2362 in carrageenan microcapsules. *Journal of Controlled Release, 33*, 245–251.

El-Saeid, M. H., & Al-Dosari, S. A. (2010). Monitoring of pesticide residues in Riyadh dates by SFE, MSE, SFC, and GC techniques. *Arabian Journal of Chemistry, 3*, 179–186.

Faleiro, J. R. (2006). A review of the issues and management of red palm weevil *Rhynchophorus ferrugineus* (Coleoptera: Rhynchophoridae) in coconut and date palm during the last one hundred years. *International Journal of Tropical Insect Science, 26*, 135–154.

FAOSTAT. (2014). Food and agricultural commodities production. Available at: http://faostat.fao.org/site/339/default.aspx. Accessed on 16 Apr 2015.

Fargues, J., Goettel, M. S., Smits, N., Ouedraogo, A., Vidal, C., Lacey, L. A., Lomer, C. J., & Rougier, M. (1996). Variability in susceptibility to simulated sunlight of conidia among isolates of entomopathogenic hyphomycetes. *Mycopathologia, 135*, 171–181.

Ferry, M., & Gomez, S. (2002). The red palm weevil in the Mediterranean area. *Palms, 46*, 172–178.

Godfray, H. C. J. (2013). Mosquito ecology and control of malaria. *Journal of Animal Ecology, 82*, 15–25.

Gunawardena, N. E., Kern, F., Janssen, E., Meegoda, C., Schäfer, D., Vostrowsky, O., & Bestmann, H. J. (1998). Host attractants for red weevil, *Rhynchophorus ferrugineus*: Identification, electrophysiological activity, and laboratory bioassay. *Journal of Chemical Ecology, 24*, 425–437.

Hadapad, A. B., Vijayalakshmi, N., Hire, R. S., & Dongre, T. K. (2008). Effect of ultraviolet radiation on spore viability and mosquitocidal activity of an indigenous ISPC-8 *Bacillus sphaericus* Neide strain. *Acta Tropica, 107*, 113–116.

Hamdi, C., Balloi, A., Essanaa, J., Crotti, E., Gonella, E., Raddadi, N., Ricci, I., Boudabous, A., Borin, S., & Manino, A. (2011). Gut microbiome dysbiosis and honeybee health. *Journal of Applied Entomology, 135*, 524–533.

Hedges, L. M., Brownlie, J. C., O'Neill, S. L., & Johnson, K. N. (2008). *Wolbachia* and virus protection in insects. *Science, 322*, 702.

Hedimbi, M., Kaaya, G. P., Singh, S., Chimwamurombe, P. M., Gindin, G., Glazer, I., & Samish, M. (2008). Protection of *Metarhizium anisopliae* conidia from ultra-violet radiation and their pathogenicity to *Rhipicephalus evertsi evertsi* ticks. *Experimental and Applied Acarology, 46*, 149–156.

Hellmich, R. L., & Hellmich, K. A. (2012). Use and impact of *Bt* maize. *Nature Education Knowledge, 3*, 4.

Hendrichs, J., Franz, G., & Rendon, P. (1995). Increased effectiveness and applicability of the sterile insect technique through male only releases for control of Mediterranean fruit flies during fruiting seasons. *Journal of Applied Entomology, 119*, 371–377.

Himler, A. G., Adachi-Hagimori, T., Bergen, J. E., Kozuch, A., Kelly, S. E., Tabashnik, B. E., Chiel, E., Duckworth, V. E., Dennehy, T. J., & Zchori-Fein, E. (2011). Rapid spread of a bacterial symbiont in an invasive whitefly is driven by fitness benefits and female bias. *Science, 332*, 254–256.

Hoffmann, A. A., Montgomery, B. L., Popovici, J., Iturbe-Ormaetxe, I., Johnson, P. H., Muzzi, F., Greenfield, M., Durkan, M., Leong, Y. S., & Dong, Y. (2011). Successful establishment of *Wolbachia* in *Aedes* populations to suppress dengue transmission. *Nature, 476*, 454–457.

Hosokawa, T., Kikuchi, Y., Shimada, M., & Fukatsu, T. (2007). Obligate symbiont involved in pest status of host insect. *Proceedings of the Royal Society B: Biological Sciences, 274*, 1979–1984.

Hosokawa, T., Koga, R., Kikuchi, Y., Meng, X.-Y., & Fukatsu, T. (2010). *Wolbachia* as a bacteriocyte-associated nutritional mutualist. *Proceedings of the National Academy of Sciences, 107*, 769–774.

Husseneder, C. (2010). Symbiosis in subterranean termites: A review of insights from molecular studies. *Environmental Entomology, 39*, 378–388.

Husseneder, C., & Collier, R. E. (2009). Paratransgenesis in termites. In K. Bourtzis & T. A. Miller (Eds.), *Insect symbiosis* (Vol. 3, pp. 361–376). Boca Raton: CRC Press/Taylor & Francis.

Husseneder, C., Ho, H.-Y., & Blackwell, M. (2010a). Comparison of the bacterial symbiont composition of the Formosan subterranean termite from its native and introduced range. *Open Microbiology Journal, 4*, 53–66.

Husseneder, C., Sethi, A., Foil, L., & Delatte, J. (2010b). Testing protozoacidal activity of ligand-lytic peptides against termite gut protozoa in vitro (protozoa culture) and in vivo (microinjection into termite hindgut). *Journal of Visualized Experiments, 46*, e2190. doi:10.3791/2190.

Inglis, D. G., Johnson, D. L., & Goettel, M. S. (1996). Effect of bait substrate and formulation on infection of grasshopper nymphs by *Beauveria bassiana. Biocontrol Science and Technology, 6*, 35–50.

ISAAA. (2014). ISAAA Brief 46-2013: Executive summary, global status of commercialized biotech/GM crops: 2013. [Electronic resource]/International Service for the Acquisition of Agribiotech Applications, 2014. Available at: http://www.isaaa.org/resources/publications/briefs/46/executivesummary/. Accessed on 12 Apr 2015.

Iturbe-Ormaetxe, I., Walker, T., & O'Neill, S. L. (2011). *Wolbachia* and the biological control of mosquito-borne disease. *EMBO Reports, 12*, 508–518.

Jin, C., Ren, X., & Rasgon, J. L. (2009). The virulent *Wolbachia* strain wMelPop efficiently establishes somatic infections in the malaria vector *Anopheles gambiae. Applied and Environmental Microbiology, 75*, 3373–3376.

Kelman, A. (1987). *Introduction of recombinant DNA-engineered organisms into the environment: Key issues*. Washington, DC: National Academies Press. 24 p.

Kikuchi, Y., Hayatsu, M., Hosokawa, T., Nagayama, A., Tago, K., & Fukatsu, T. (2012). Symbiont-mediated insecticide resistance. *Proceedings of the National Academy of Sciences, 109*, 8618–8622.

Knipling, E. F. (1955). Possibilities of insect control or eradication through the use of sexually sterile males. *Journal of Economic Entomology, 48*, 459–462.

Kramm, K. R., & West, D. F. (1982). Termite pathogens: Effects of ingested *Metarhizium, beauveria*, and *Gliocladium* conidia on worker termites (*Reticulitermes* sp.). *Journal of Invertebrate Pathology, 40*, 7–11.

Kuriwada, T., Hosokawa, T., Kumano, N., Shiromoto, K., Haraguchi, D., & Fukatsu, T. (2010). Biological role of Nardonella endosymbiont in its weevil host. *PloS One, 5*, e13101.

Lacey, L. A., & Siegel, J. P. (2000). Safety and ecotoxicology of entomopathogenic bacteria. In J. F. Charles, A. Delécluse, & C. Nielsen-le Roux (Eds.), *Entomopathogenic bacteria: From laboratory to field application* (pp. 253–273). Dordrecht: Springer.

Lindquist, D. A., Abusowa, M., & Hall, M. J. R. (1992). The new world screwworm fly in Libya: A review of its introduction and eradication. *Medical and Veterinary Entomology, 6*, 2–8.

Little, A. E. F., & Currie, C. R. (2007). Symbiotic complexity: Discovery of a fifth symbiont in the attine ant-microbe symbiosis. *Biology Letters, 3*, 501–504.

Mankin, R. (2011). Recent developments in the use of acoustic sensors and signal processing tools to target early infestations of red palm weevil in agricultural environments. *Florida Entomologist, 94*, 761–765.

Mankin, R. W., Hagstrum, D. W., Smith, M. T., Roda, A. L., & Kairo, M. T. K. (2011). Perspective and promise: A century of insect acoustic detection and monitoring. *American Entomologist, 57*, 30–44.

Marinotti, O., Jasinskiene, N., Fazekas, A., Scaife, S., Fu, G., Mattingly, S. T., Chow, K., Brown, D. M., Alphey, L., & James, A. A. (2013). Development of a population suppression strain of the human malaria vector mosquito, Anopheles stephensi. *Malaria Journal, 12*, 142.

Mayer, M. S., & Brazzel, J. R. (1966). Laboratory studies to sterilize the boll weevil with radiation. *Annals of the Entomological Society of America, 59*, 284–290.

McCoy, R. E. (1975). Effect of oxytetracycline dose and stage of disease development on remission of lethal yellowing in coconut palm. *Plant Disease Report, 59*, 717–720.
McGraw, E. A., & O'Neill, S. L. (2013). Beyond insecticides: New thinking on an ancient problem. *Nature Reviews Microbiology, 11*, 181–193.
McGraw, E. A., Merritt, D. J., Droller, J. N., & O'Neill, S. L. (2002). Wolbachia density and virulence attenuation after transfer into a novel host. *Proceedings of the National Academy of Sciences, 99*, 2918–2923.
McGuire, M. R., Connick, W. J., Jr., & Quimby, P. C., Jr. (1999). Formulation of microbial pesticides. In H. B. Scher (Ed.), *Controlled-release delivery systems for pesticides* (pp. 173–193). New York: Marcel Dekker.
Miller, T. A. (2013). Delivery. *Pest Management Science, 69*, 1199–1204.
Morrison, N., Franz, G., Koukidou, M., Miller, T., Saccone, G., Alphey, L., Beech, C., Nagaraju, J., Simmons, G., & Polito, L. (2010). Genetic improvements to the sterile insect technique for agricultural pests. *Asia-Pacific Journal of Molecular Biology Biotechnology, 18*, 275–295.
Mukhtar, M., Rasool, K. M., Parrella, M. P., Sheikh, Q. I., Pain, A., Lopez-llorca, L. V., Aldryhim, Y. N., Mankin, R. W., & Aldawood, A. S. (2011). New initiatives for management of red palm weevil threats to historical Arabian date palms. *Florida Entomologist, 94*, 733–736.
Murphy, S. T., & Briscoe, B. R. (1999). The red palm weevil as an alien invasive: Biology and the prospects for biological control as a component of PM. *Biocontrol News Information, 20*, 35–45.
Murray, R. G. E., & Stackebrandt, E. (1995). Taxonomic note: Implementation of the provisional status Candidatus for incompletely described procaryotes. *International Journal of Systematic Bacteriology, 45*, 186–187.
Nair, K. S. S. (1998). KFRI's tryst with the teak defoliator. *Evergreen (Trichur), 40*, 1–7.
Nirula, K. K. (1956). Investigations on the pests of coconut palm. Part IV. *Rhynchophorus ferrugineus* F. *Indian Coconut Journal, 9*, 229–247.
Oehlschlager, A. C. (2005). Current status of trapping palm weevils and beetles. *Planter, 81*, 123–143.
Ostp, U. (1986). Coordinated framework for regulation of biotechnology. *Federal Register, 51*, 23302–23350.
Paarlberg, R. L. (2009). *Starved for science: How biotechnology is being kept out of Africa*. Cambridge, MA: Harvard University Press. 256 p.
Pimentel, D., McNair, S., Janecka, J., Wightman, J., Simmonds, C., O'connell, C., Wong, E., Russel, L., Zern, J., & Aquino, T. (2001). Economic and environmental threats of alien plant, animal, and microbe invasions. *Agriculture Ecosystems and Environment, 84*, 1–20.
Prabhu, S. T., Dongre, T. K., & Patil, R. S. (2010). Effect of irradiation on the biological activities of red palm weevil, *Rhynchophorus ferrugineus* Olivier. *Karnataka Journal of Agricultural Sciences, 23*(1), 186–188.
Presgraves, D. C. (2010). The molecular evolutionary basis of species formation. *Nature Reviews Genetics, 11*, 175–180.
Pyšek, P., & Richardson, D. M. (2010). Invasive species, environmental change and management, and health. *Annual Review of Environment and Resources, 35*, 25–55.
Rach, M. M., Gomis, H. M., Granado, O. L., Malumbres, M. P., Campoy, A. M., & Martín, J. J. S. (2013). On the design of a bioacoustic sensor for the early detection of the red palm weevil. *Sensors, 13*, 1706–1729.
Rasgon, J. L. (2011). Dengue fever: Mosquitoes attacked from within. *Nature, 476*, 407–408.
Rendón, P., McInnis, D., Lance, D., & Stewart, J. (2004). Medfly (Diptera: Tephritidae) genetic sexing: Large-scale field comparison of males-only and bisexual sterile fly releases in Guatemala. *Journal of Economic Entomology, 97*, 1547–1553.
Salama, H. S., & Abd-Elgawad, M. M. (2003). Quarantine problems: An analytical approach with special reference to palm weevils and Phytonematodes. *Archives of Phytopathology and Plant Protection, 36*, 41–46.

Serbus, L. R., Casper-Lindley, C., Landmann, F., & Sullivan, W. (2008). The genetics and cell biology of *Wolbachia*-host interactions. *Annual Review of Genetics, 42*, 683–707.

Shah, P. A., Aebi, M., & Tuor, U. (1998). Method to immobilize the aphid-pathogenic fungus Erynia neoaphidis in an alginate matrix for biocontrol. *Applied and Environmental Microbiology, 64*, 4260–4263.

Shigo, A. L., Campana, R., Hyland, F., & Anderson, J. (1980). Anatomy of elms injected to control Dutch elm disease. *Journal of Arboriculture, 6*, 96–100.

Showler, A. T. (2002). Effects of kaolin-based particle film application on boll weevil (Coleoptera: Curculionidae) injury to cotton. *Journal of Economic Entomology, 95*, 754–762.

Six, D. L. (2003). Bark beetle-fungus symbioses. *Insect Symbiosis, 1*, 97–114.

Thomas, D. D., Donnelly, C. A., Wood, R. J., & Alphey, L. S. (2000). Insect population control using a dominant, repressible, lethal genetic system. *Science, 287*, 2474–2476.

Toju, H., Hosokawa, T., Koga, R., Nikoh, N., Meng, X. Y., Kimura, N., & Fukatsu, T. (2010). "Candidatus Curculioniphilus buchneri", a novel clade of bacterial endocellular symbionts from weevils of the genus Curculio. *Applied and Environmental Microbiology, 76*, 275–282.

Tsuchida, T., Koga, R., Horikawa, M., Tsunoda, T., Maoka, T., Matsumoto, S., Simon, J.-C., & Fukatsu, T. (2010). Symbiotic bacterium modifies aphid body color. *Science, 330*, 1102–1104.

Uematsu, K., Kutsukake, M., Fukatsu, T., Shimada, M., & Shibao, H. (2010). Altruistic colony defense by menopausal female insects. *Current Biology, 20*, 1182–1186.

USDA-APHIS. (2011). *New pest response guidelines: Exotic wood-boring and bark beetles.* Riverdale: USDA-APHIS-PPQ-EDP-Emergency Management. 256 p.

van der Putten, W. H., Macel, M., & Visser, M. E. (2010). Predicting species distribution and abundance responses to climate change: Why it is essential to include biotic interactions across trophic levels. *Philosophical Transactions of the Royal Society, B: Biological Sciences, 365*, 2025–2034.

van der Valk, H. (2007). Review of the efficacy of *Metarhizium anisopliae* var. *acridum* against the desert locust. Food and Agriculture Organization, Plant Production and Protection Division, Locusts and other Migratory Pests Groups, No. AGP/DL/TS/34, Rome, Italy, 81 p.

Vidyasagar, P. S. P. V., Al-Saihati, A. A., Al-Mohanna, O. E., Subbei, A. I., & Mohsin, A. A. M. (2000). Management of red palm weevil *Rhynchophorus ferrugineus* Olivier, a serious pest of date palm in AlQatif, Kingdom of Saudi Arabia. *Journal of Plantation Crops, 28*, 35–43.

Vreysen, M. J. B., & Robinson, A. S. (2011). Ionising radiation and area-wide management of insect pests to promote sustainable agriculture-a review. *Agronomy for Sustainable Development, 31*, 233–250.

Walker, T., & Moreira, L. A. (2011). Can *Wolbachia* be used to control malaria? *Memórias do Instituto Oswaldo Cruz, 106*, 212–217.

Werren, J. H., Baldo, L., & Clark, M. E. (2008). *Wolbachia*: Master manipulators of invertebrate biology. *Nature Reviews Microbiology, 6*, 741–751.

Yamada, H., Benedict, M. Q., Malcolm, C. A., Oliva, C. F., Soliban, S. M., & Gilles, J. R. L. (2012). Genetic sex separation of the malaria vector, Anopheles arabiensis, by exposing eggs to dieldrin. *Malaria Journal, 11*, 208.

Zohar-Perez, C., Chernin, L., Chet, I., & Nussinovitch, A. (2003). Structure of dried cellular alginate matrix containing fillers provides extra protection for microorganisms against UVC radiation. *Radiation Research, 160*, 198–204.

Chapter 4
Longhorn Stem Borer and Frond Borer of Date Palm

Mohammad Ali Al-Deeb and Mohammed Zaidan Khalaf

Abstract Date palm trees are attacked by the longhorn date palm stem borer, *Jebusaea hammerschmidtii*, and the frond borer, *Phonapate frontalis* in several countries. The frond borer is not a major economic pest of date palm trees, however, the larvae of *J. hammerschmidtii* cause severe damage by boring and feeding on plant tissues, which leads to infection by pathogens and also breaking of fronds and trunks. Both the larvae and adults of *P. frontalis* cause damage by feeding on green fronds and the infestation by this pest results in either the breaking or the gradual drying of the frond. Management of *J. hammerschmidtii* relies mainly on the use of light traps, which attract the male and female adults at night, and can be used successfully for monitoring and mass trapping. Moreover, balanced and adequate irrigation as well as proper fertilization of date palm trees appear to be important factors in reducing the infestation levels. Similarly, light traps were found to be highly efficient in attracting the adults of *P. frontalis*. Other methods such as biological and chemical control are also effective against *J. hammerschmidtii* and *P. frontalis*.

4.1 Longhorn Stem Borer (*Jebusaea hammerschmidtii*)

4.1.1 Distribution and Host Range

Although the geographic distribution of *Jebusaea hammerschmidtii* (Reiche) (Coleoptera: Cerambycidae), is not as large as that of the red palm weevil *Rhynchophorus ferrugineus* (Olivier), the situation can be worrying in certain

M.A. Al-Deeb (✉)
Department of Biology, College of Science, United Arab Emirates University, P.O. Box 15551, Al-Ain, United Arab Emirates
e-mail: m_aldeeb@uaeu.ac.ae

M.Z. Khalaf
Integrated Pest Control Research Center, Directorate of Agricultural Research, Ministry of Science and Technology, Karrada, Jadyria, Baghdad, Iraq
e-mail: mkhalaf34@yahoo.co.uk

W. Wakil et al. (eds.), *Sustainable Pest Management in Date Palm: Current Status and Emerging Challenges*, Sustainability in Plant and Crop Protection, DOI 10.1007/978-3-319-24397-9_4

infested areas. This pest has been reported in Iraq, Bahrain, Kuwait, Oman, Qatar, Saudi Arabia, UAE, Egypt, Algeria, Jordan, Iran, and India (Al-Haideri and Al-Hafidh 1986; Gassouma 1991; Elwan 2000; Al-Deeb 2012). The infestation rate is higher in old date palm trees, with variations depending on the variety of the palm trees (Al-Turaihi and Al-Khulaifi 2012). In Iraq, the distribution of this pest varies from one region to another; sometimes overlapping with that of other borers such as *Oryctes* spp., another major coleopteran palm borer species of the region. The date palm tree is the major host of *J. hammerschmidtii* and is spread by the movement of trees or offshoots infested with its larvae. Also, the adults are strong fliers and can spread the infestation to new areas.

4.1.2 Biology and Seasonal Incidence

There is a great shortcoming in our knowledge on the fundamental aspects of the biology and ecology of *Jebusaea hammerschmidtii*, which negatively influences the success of its management programs (El-Shafie 2015). *Jebusaea hammerschmidtii* has one generation per year (Hussein 1974). Mature adults (Fig. 4.1) are mostly observed early in May, when mating occurs. The female deposits single eggs between frond bases or in small cracks on the tree trunk in June. Eggs hatch after 2 weeks giving rise to small larvae. Newly emerged larvae bore into the trunks of the trees. The duration of the larval stage is about 3 months (Al-Azawi 1980; Khalaf and Al-Taweel 2014), however, it can reach up to 10 months. Larvae are about 45 mm long and legless, and have a creamy to white color (Fig. 4.2c). Usually there is one larva per tunnel. The first two segments of the thorax are broader than the rest of the body, which tapers towards the tail end. This is a major difference between this pest and the other legless larva like that of the red palm weevil, which is stouter and more spindle shaped.

Fig. 4.1 Adult male of *Jebusaea hammerschmidtii* (Photo: Mohammad A. Al-Deeb)

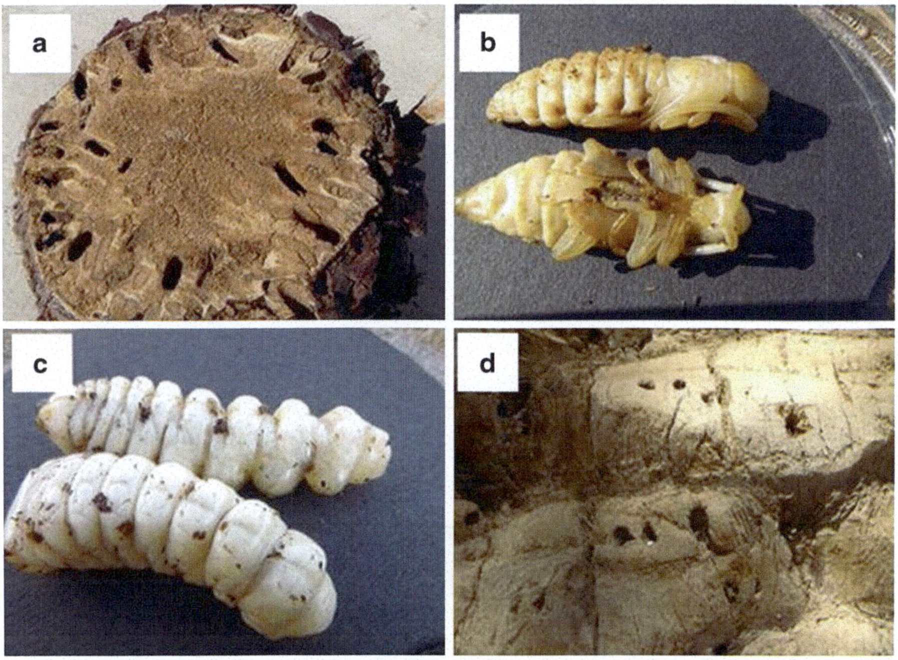

Fig. 4.2 *Jebusaea hammerschmidtii*; larval damage inside date palm tree trunk (**a**), pupae (**b**), larvae (**c**), larval holes on the trunk from outside (**d**) (Photos: Mohammed Z. Khalaf)

In the literature, *J. hammerschmidtii* was reported attacking the date palm tree top (Hussein 1974). Larvae were found boring into green frond bases for about 3 months, and then they bored into the trunk to spend the winter. The larvae stopped feeding and remained dormant at the end of the tunnel during the winter time. Pupae are exarate and vary from 36 to 45 mm in length (Fig. 4.2b). Pupation lasts for 3 weeks and occurs in a chamber located at the end of the tunnel during March and April. Adult emergence occurs from May to August (Al-Haideri and Al-Hafidh 1986). Adults chew round exit holes to leave the trunk (Hussein 1974).

The female is bigger than the male and it is about 30–45 mm long, while the male is 21–31 mm. The body is light-to-dark brown in color and antennae are as long as the length of the body.

4.1.3 Damage, Economic Threshold and Losses

Longhorn stem borer is one of the most destructive pests of date palm trees. It causes damage by boring into and feeding on plant tissues, which leads fronds and trunks to breakage and infection by pathogens (Dhiab et al. 1979; Al-Jboory 2007). The holes bored by *J. hammerschmidtii* larvae inside and outside of date palm tree trunk are exhibited in Fig. 4.2a, d. In addition, the damage is accompanied by plant weakness and low productivity. The main damage is caused by the larvae, which are able to burrow into the base of fronds at the top of date palm trees and the trunk (Fig. 4.3) (Al-Azawi 1980; Khalaf and Al-Taweel 2014).

The infestation shortens tree longevity and reduces tree yields. It also decreases the market value of the wood (Al-Haideri and Al-Hafidh 1986). Although, *J. hammerschmidtii* is a trunk borer like the red palm weevil, there are some differences: (1) *J. hammerschmidtii* tends to affect old and/or neglected date palm trees, while the red palm weevil prefers younger palms <20 years old (Abraham et al. 1998); (2) it has one generation per year, whereas the red palm weevil has several overlapping generations per year; and (3) the progress of *J. hammerschmidtii* damage is slower than that of the red palm weevil, and in some cases it does not kill the tree. The main identifying character of infestation by *J. hammerschmidtii* is the presence of brown sticky materials oozing from holes, made by the pest in different regions of the

Fig. 4.3 *Jebusaea hammerschmidtii* larval damage on date palm tree frond bases and trunk (Photos: Mohammad A. Al-Deeb)

trunk. During a field survey in some regions of Iraq, a total of 265 holes were recorded in a 1 m length of tree trunk (Khalaf and Al-Taweel 2014). Damage can be severe in more neglected orchards. Larvae bore inside frond bases, and usually there is one larva per frond. However, in some cases two to three larvae can be found in one frond base. Larval tunnels can reach the center of the trunk. In some countries like Iraq, the number of holes per tree in an orchard is considered a major factor in determining the market price of the trees (Hussein 1974). The economic threshold for this insect is undetermined, which shows that more research needs to be conducted on this pest.

4.1.4 Management

4.1.4.1 Cultural Control

There is no single method to prevent *J. hammerschmidtii* attack. However, balanced and adequate irrigation as well as proper fertilization of trees appear to be important factors for reducing the infestation of date palm plantations. Phytosanitation through the removal of the badly infested trees, which can serve as reservoirs of infestation, is a significant control measure. The pest is known to flourish in neglected farms where irrigation is inadequate. Thus, farm practices that keep the trees strong, healthy and in a vigorous condition are the best defense against *J. hammerschmidtii* and other trunk borers. It is also important to inspect trees regularly for borer symptoms.

4.1.4.2 Physical Control

Damaged frond bases and trunk tissues should be removed, and the resulting cavities in the tree trunks need to be treated with insecticides and filled with cement to avoid attracting other pests, especially the females of the red palm weevil, which are known to exploit tree wounds.

4.1.4.3 Trapping and Monitoring

Light traps attract *J. hammerschmidtii* adults at night, and they can be used for monitoring and mass trapping. In the United Arab Emirates, light traps were effective for capturing the adults in infested date palm plantations (Al-Deeb – unpublished data). In such traps, mercury vapor light bulbs producing white light were used. Besides capturing *J. hammerschmidtii* adults, light traps can also help in the monitoring and mass trapping of the *Oryctes* sp. beetles on the farms where both pests exist. On remote farms where electricity is not available, light traps equipped with solar panels were used successfully.

4.1.4.4 Host Plant Resistance

Host plant resistance currently is not an option against *J. hammerschmidtii*, since high market value date palm varieties are not resistant to the pest. Even if some low quality date varieties appear to be less affected by *J. hammerschmidtii*, date palm growers will not give up their good date varieties for the sake of insect resistance. Hence, the other control options such as trapping, biological control and chemical control remain the available choices in Integrated Pest Management (IPM) programs against this insect pest.

4.1.4.5 Biological Control

It was reported that the fungus *Beauveria bassiana* (Balsamo) Vuillemin, can control the larvae (Al-Haideri and Al-Hafidh 1986). For instance, *B. bassiana* caused 72.7 % mortality among the tested *J. hammerschmidtii* larvae (Fayyadh et al. 2013). Al-Haideri and Al-Hafidh (1986) mentioned another fungus, *Cordyceps* sp., as well as two mite species, *Hypoaspis* sp. and *Ameroseius* sp., which can also be used as biological control agents. Larvae of *J. hammerschmidtii* were found to be infected by a pathogenic poxviridae virus (Al-Jboory and Salih 2002). A pathogenic nematode of the genus *Steinernema*, a parasitic fly of the genus *Megaselia*, and a poxviridae virus were reported affecting the long horn stem borer adults and larvae in Iraq (Al-Jboory 2007). More studies need to be carried out to effectively deploy and incorporate biological control as an IPM tool against *J. hammerschmidtii*.

4.1.4.6 Chemical Control

Despite the efficacy of light traps against the adult *J. hammerschmidtii*, insecticides can also be used as a control measure, however, a thorough coverage of the trunk is essential. Systemic insecticides can be applied to infested trees in a way similar to that used in the chemical control of the red palm weevil. This involves spraying insecticides so as to thoroughly soak the trunk with insecticides solution or injecting the insecticides into the trunk. However, the effectiveness of the chemical control can be, at times, limited or unreliable, because the damaging larval stage of *J. hammerschmidtii* lives inside the tree trunk where it is protected from direct exposure to the insecticides. Chemical treatment should not be used unless borer activity is known to be occurring in the tree, because there are cases where the adult exit holes are visible on the trunk and frond bases but there are no living larvae inside the tunnels. The timing of applications is very important, especially for topical treatments, when using insecticides with limited residual toxicity. The insecticide residues must be on the trunk during the period between the egg's hatching and the borer's entry into the tree. In this way, newly hatched larvae are exposed to a lethal dose before they bore inside the trunk.

4.2 Frond Borer (*Phonapate frontalis*)

4.2.1 Distribution and Host Range

This insect, *Phonapate frontalis* (Fåhraeus) (Coleoptera: Bostrichidae), is not a major economic pest of date palm trees. It has been reported to occur in Iraq, Saudi Arabia, Egypt, Bahrain, Yemen, Libya, Tunisia, Algeria and Oman. This insect is reported to attack pomegranates and grapes in addition to date palm trees. In Iraq, it is mainly found in the middle region of the country (Al-Haideri and Al-Hafidh 1986; Atia et al. 2009).

4.2.2 Biology and Seasonal Incidence

The mature adult is a small beetle (15–22 mm), dark brown to black in color (Fig. 4.4). Its body is almost rectangular in shape, and the first segment of the thorax covers the head. The anterior dorsal part of the first thoracic segment is serrated, while the posterior dorsal part is smooth and glossy. The larva is creamy white in color, legless, and more or less C-shaped (about 20 mm). Both the larva and the adult live inside the tunnels, which they make in the fronds (Khalaf and Al-Taweel 2014).

The seasonal activity of *P. frontalis* varies according to the region and environmental conditions. A field study conducted on three orchards in Iraq during 2010 showed that the highest flight activity of *P. frontalis* occurred during the months of May, June, and July (Khalaf et al. 2012). In Saudi Arabia, the adult activity started in March (El-Zayat et al. 2002).

Fig. 4.4 Adult of *Phonapate frontalis* (Photo: Mohammed Z. Khalaf)

Fig. 4.5 Damage by *Phonapate frontalis* (Photo: Mohammed Z. Khalaf)

4.2.3 Damage, Economic Threshold and Losses

Both the larvae and adults of *P. frontalis* cause damage by feeding on green fronds (Fig. 4.5). The larvae usually feed inside the trunk by boring tunnels, causing the secretion of sticky materials from the entrance holes. Infestation due to this pest causes either the breaking or the gradual drying of the frond. The infestation usually continues in the dry fronds, which makes the fronds unsuitable for use in making farmhouse roofs or traditional handcrafts. Feeding by this pest can convert the inner frond tissues into a powder-like material. This pest is also capable of boring inside fruit stalks, resulting in dryness; consequently, the fruits become unsuitable for human consumption. The main characteristic of infestation due to this pest is the appearance of sticky spots (tree sap secretions) and holes, which are used for additional infestation by adults and for movement into and out of the frond. Field observations showed differences in infestation among different date palm tree varieties (Al-Azawi 1980; Khalaf and Al-Taweel 2014).

4.2.4 Management

4.2.4.1 Cultural Control

Infestation can be significantly reduced through balanced and adequate irrigation and by keeping the trees healthy. Also, phytosanitation can help in removing infested plant materials that could be present on the date palm farms. Regular inspection of trees for signs of infestation is very important in achieving successful management.

4.2.4.2 Trapping and Monitoring

Solar light traps (Magna Traps with lamps of 320–420 nm) were found to be highly efficient in attracting the adults. These traps can be used in IPM programs for monitoring and controlling this pest (Khalaf et al. 2012). A light trap can also catch *J. hammerschmidtii* and *Oryctes* sp. beetles in date palm plantations. This makes the non-chemical monitoring and mass trapping device a very important tool in the management programs of the three pests collectively.

4.2.4.3 Chemical Control

In case there is a need for chemical control, systemic insecticides can be used by farmers to kill the newly emerged larvae boring inside green fronds during peak beetle activity. However, chemical control needs to be justified before being launched. This is very important because with the presence of light traps, which work effectively, chemical control becomes a less preferred option in the management programs of this pest.

4.3 Future Research

More research should be directed towards *J. hammerschmidtii*. So far, a few studies have been conducted *J. hammerschmidtii*, and very limited literature is available on the biology and management of this insect pest. Therefore, this pest should receive the necessary research attention, particularly with regard to (1) studying the population dynamics and bioecology in the affected countries, (2) identifying and deploying potential biological control agents (predators, parasites, parasitoids and microorganisms) and (3) designing pheromone traps for monitoring and mass trapping.

4.4 Summary

The *J. hammerschmidtii* and *P. frontalis* are two pests of the date palm trees. The former is more important because the larvae can cause severe damage by boring and feeding on plant tissues, while the latter is considered as a minor pest. Management of both pests can be done using a variety of methods. Light traps were successfully used in the monitoring and mass trapping. Additionally, proper irrigation and balanced fertilization of date palm trees appear to be important factors for reducing the infestation. Integrated pest management programs against *J. hammerschmidtii* and *P. frontalis* should select the most effective management methods. Also, pesticide use should be reduced to minimize the possible harm to humans and the environment.

References

Abraham, V. A., Al-Shuaibi, M. A., Faleiro, J. R., Abozuhairah, R. A., & Vidyasagar, P. S. P. V. (1998). An integrated approach for the management of red palm weevil *Rhynchophorus ferrugineus* (Oliv.) – A key pest of date palm in Middle East. *Sultan Qaboos University Journal of Scientific Research (Agricultural Science), 3*, 77–84.

Al-Azawi, A. F. (1980). *General and applied entomology* (540 pp). Iraq: Alzahraa, Ministry of High Education and Scientific Research.

Al-Deeb, M. A. (2012). Date palm insect and mite pests and their management. In A. Manickavasagan, M. Mohamed Essa, & E. Sukumar (Eds.), *Dates production, processing, food, and medicinal values* (pp. 113–128). Boca Raton: Taylor & Francis.

Al-Haideri, H. S., & Al-Hafidh, E. N. T. (1986). *Seasonal pests of palm and dates in the near east and north Africa. Regional project for Palm and Dates Research in the near east and north Africa* (126 pp). Baghdad: Food and Agriculture Organization.

Al-Jboory, I. J. (2007). Survey and identification of the biotic factors in the date palm environment and its application for designing IPM-program of date palm pests in Iraq. *Aden University Journal of Natural and Applied Sciences, 11*, 423–457.

Al-Jboory, I. J., & Salih, H. A. A. M. (2002). New record of pathogenic virus of the longhorn date palm stem borer *Jebusaea hammerschmidtii* Reich (Coleoptera: Cerambycidae). *Basra Date Palm Research Journal, 2*(1–2), 1–3.

Al-Turaihi, E., & Al-Khulaifi, Y. (2012). *Pests and diseases of date palm in Qatar* (229 pp). Qatar: Ministry of Environment.

Atia, Z. M., Kara, H., Al- Dankali, A., & Kafo, A. A. (2009). Ecology and biology studies on palm frond borer, *Phonapate frontalis* in the western coastal regions of Libya. *Arab Journal of Plant Protection (Special Issue, Supplementary), 27*, 21.

Dhiab, M., Swayir, I. A., & Abdul-Hadi, I. (1979). Investigation on palm-stem borer *Pseudophilus testasceus* Gah. (Coleoptera: Cerambycidae) Yearbook of Plant Protection, Ministry of Agriculture. *Iraq, 2*, 103–112.

Elwan, A. A. (2000). Survey of the insect and mite pests associated with date palm trees in Al-Dakhliya region, Sultanate of Oman. *Egyptian Journal of Agricultural Research, 78*(2), 653–664.

El-Shafie, H. A. F. (2015). Biology, ecology and management of the longhorn date palm stem borer *Jebusaea hammerschmidtii* (Coleoptera: Cerambycidae). *Outlooks on Pest Management, 26*(1), 20–23.

El-Zayat, M. M., Al-Kaeet, S. I., Lukmah, H. M., Dhafran, H. A., & Abdulsalam, K. S. (2002). *The most important diseases and pests of date palm trees in the Kingdom of Saudi Arabia and their integrated control methods* (369 pp). Kingdom of Saudi Arabia: Ministry of Agriculture and Water. (In Arabic)

Fayyadh, M. A., Abdalqader, A. Z., Fadel, A. A., Muhammed, E. A., Manea, A. A., Jaffer, M. A., & Fadel, A. A. (2013). Formulation of biopesticide from *Beauveria bassiana* as part of biological control of date palm stem borer *(Jebusaea hammerschmidtii). Advances in Agriculture and Botanics International Journal of the Bioflux Society, 5*(2), 84–90.

Gassouma, M. S. S. (1991). *Agricultural pests in the United Arab Emirates. Part 1. Fruit trees* (148 pp). United Arab Emirates: Ministry of Agriculture and Fisheries.

Hussein, A. A. (1974). *Date palm and dates with their pest in Iraq* (166 pp). Iraq: University of Baghdad, Ministry of High Education and Scientific Research.

Khalaf, M. Z., & Al-Taweel, A. A. (2014). *Palm borers in Iraqi environment: Species, damages, methods of control.* Republic of Iraq: Ministry of Science and Technology, Agricultural Research Directorate. Available at http://www.iraqi-datepalms.net/Uploaded/file/Manual%20of%20Palm%20Borers.pdf. Accessed on 20 May 2014.

Khalaf, M. Z., Shabar, A. K., Naher, F. H., Jabo, N. F., Abdulhamza, B. H., & Sami, R. A. (2012). Activity of insect fauna during the night in palm orchards of central Iraq. *Journal of Food Science and Engineering, 2*, 277–282.

Chapter 5
Dynastid Beetle Pests

Geoffrey O. Bedford, Mohammad Ali Al-Deeb, Mohammed Zaidan Khalaf, Kazem Mohammadpour, and Rasmi Soltani

Abstract Two main species of dynastid or rhinoceros beetle (Coleoptera: Scarabaeoidea: Scarabaeidae: Dynastinae: Tribe Oryctini) *Oryctes elegans*, and subspecies of *Oryctes agamemnon*, attack date palms causing significant and documented damage. Adults of *O. elegans* bore into the stalks of inflorescences and fruit bunches to feed, and oviposit in leaf axils where the larvae develop and may invade the trunk. *Oryctes agamemnon* larvae bore into frond bases, the trunk, and respiratory roots where their tunnelling may cause the palm to fall. It is difficult to distinguish the larvae of the two species in regions where both coexist. For control, annual servicing of palms includes cutting off old fronds at their bases using the correct technique which enables removal of larvae and their breeding places, and this may be integrated with light trapping for catching adults. Quarantine measures may hinder the spread of these pests to uninfested areas. In India, adults of a third species, *O. rhinoceros*, have been noted boring into the soft tissue of the growing point, and this species has also been reported from Yemen. Pheromone trapping is available for *O. elegans* but for effectiveness it requires the addition of fresh date palm tissue to

G.O. Bedford (✉)
Department of Biological Sciences, Macquarie University, NSW 2109, Australia
e-mail: geoffbedford@bigpond.com

M.A. Al-Deeb
Department of Biology, College of Science, United Arab Emirates University,
P.O. Box 15551, Al-Ain, United Arab Emirates
e-mail: m_aldeeb@uaeu.ac.ae

M.Z. Khalaf
Integrated Pest Control Research Center, Directorate of Agricultural Research,
Ministry of Science and Technology, Karrada, Jadyria, Baghdad, Iraq
e-mail: mkhalaf34@yahoo.co.uk

K. Mohammadpour
Plant Protection Research Department, South Khorasan Agricultural and Natural Resources
Research and Education Centre, AREEO, Birjand, Iran
e-mail: mohammadpour_k@yahoo.com

R. Soltani
Laboratory of Plant Protection (INRA Tunis), Regional Center of Agriculture Research,
B.P. 357, Sidi Bouzid 9100, Tunisia
e-mail: soltani_rasmi@yahoo.com; rasmi.soltani@iresa.agrinet.tn

W. Wakil et al. (eds.), *Sustainable Pest Management in Date Palm: Current Status and Emerging Challenges*, Sustainability in Plant and Crop Protection,
DOI 10.1007/978-3-319-24397-9_5

the traps as a synergist. The entomopathogenic fungus *Metarhizium anisopliae* and the nematode *Rhabditis* sp. may have potential in integrated pest management but their possible natural occurrence in an area should be determined, prior to propagation and release. The pathogenic *Oryctes* Nudivirus, was successful against *Oryctes rhinoceros* in lowering its populations and damage to coconut palms when introduced into areas where this virus did not previously exist, and should be tested against date palm dynastids. *Oryctes agamemnon* and *O. elegans* adults attacking date palms do not appear to bore into the heart or meristem causing V-cuts to unfurling fronds or the death of the palm, in contrast to attacks by *O. rhinoceros* on coconut and young oil palms. Also *O. rhinoceros* larvae are found only in dead decomposing wood or other organic material, whereas larvae of date palm pests *O. elegans* and *O. agamemnon* may tunnel in living tissues.

5.1 Introduction

Dynastid or rhinoceros beetles may cause serious damage to date palms. Adults attack fruit bunches or frond bases. Larvae may tunnel into frond bases, the trunk or the root mass causing the death or collapse of the palm. The main pests are two species of *Oryctes*, namely *Oryctes elegans* (Prell), and subspecies of *Oryctes agamemnon* (Burmeister) (El-Shafie 2012). In Iraq, there is evidence that adults of *Oryctes* spp. act as vectors of the fungal pathogen *Fusarium proliferatum* (Matsushima) Nirenberg ex Gerlach & Nirenberg which causes wilt disease symptoms in date palms (Khudhair et al. 2014b). Adults of the well-known pest of coconut and oil palms *Oryctes rhinoceros* L. have been noted as attacking date palms in India (Butani 1974, 1975) and United Arab Emirates (UAE) (Gassouma 1991) and have also been reported from Yemen (Al-Habshi et al. 2006).

Taxonomic studies, with illustrations and a key to adults of the species *Oryctes* Illiger (Tribe Oryctini) have been provided by Endrödi (1985) and Dechambre and Lachaume (2001). In the order Coleoptera, species of *Oryctes* are members of the superfamily Scarabaeoidea, family Scarabaeidae, subfamily Dynastinae, and tribe Oryctini. In the field the three species mentioned above (Figs. 5.1, 5.2, and 5.3) can

Fig. 5.1 *Oryctes elegans* adult male (Dechambre and Lachaume 2001; with permission from Hillside Books, Canterbury, UK)

Fig. 5.2 *Oryctes rhinoceros*: male (*left* and *middle*), female (*right*) (Dechambre and Lachaume 2001; with permission from Hillside Books, Canterbury, UK)

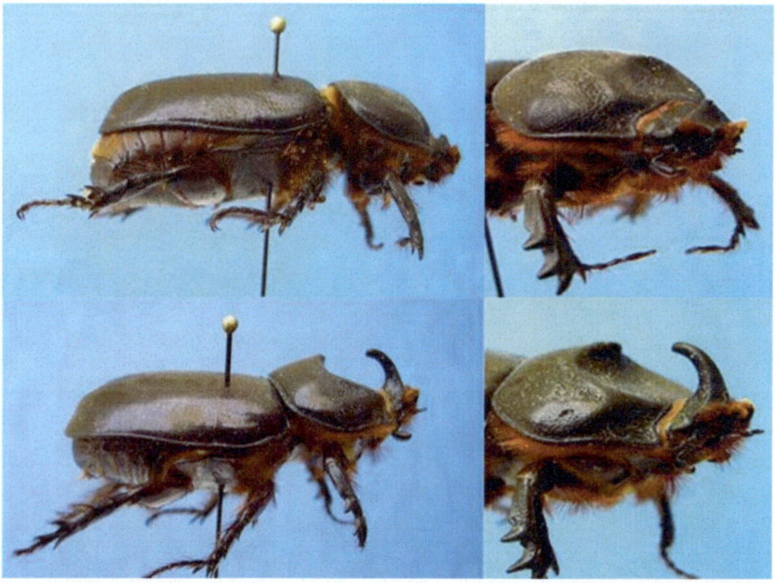

Fig. 5.3 *Oryctes agamemnon arabicus* adult. Female (*top left*), female head with small tuber-like horn (*top right*), male (*bottom left*), and male head with big horn (*bottom right*) (Photos: Mohammad A. Al-Deeb)

be separated from each other based on leg morphology (Al-Deeb 2012). The apex of the hind tibia has two large fixed teeth in *O. rhinoceros*, while it has three fixed teeth in *O. agamemnon* and *O. elegans*. In turn these two species can be separated based on the presence of a fixed tooth on the underside of the front tibia in *O. elegans* and its absence in *O. agamemnon*. A description and illustrations of these features has been provided for *O. elegans* (Hurpin and Fresneau 1969).

Data on the larvae of *O. elegans* with illustrations were given by Hurpin and Fresneau (1969), and the larva together, with eggs and pupa, are shown in Fig. 5.4. The larva of *O. agamemnon arabicus* and eggs are shown in Fig. 5.5.

Fig. 5.4 *Oryctes elegans*: egg; larva; pupa (*left* to *right*) (Photos: Kazem Mohammadpour)

Fig. 5.5 *Oryctes agamemnon arabicus* larva (*left*), larval head showing big mandibles (*top right*), and eggs in dissected abdomen of female in dorsal view (*bottom right*) (Photos: Mohammad A. Al-Deeb)

However, a full description of the larvae of *O. elegans* and *O. agamemnon* using the method of Ritcher (1966), and a key for differentiating them are not available, causing difficulty in distinguishing larvae of each in regions where the two species coexist i.e. are sympatric. A full description of the larva of *O. rhinoceros* is available (Bedford 1974).

5.2 Distribution

Oryctes elegans is endemic in Iran and Iraq (Buxton 1920; Endrödi 1985; Rochat et al. 2004; Payandeh and Dehghan 2010); Northern Pakistan (Ratcliffe and Ahmed 2010); Saudi Arabia (Al-Deghairi 2007), UAE, Bahrain and Qatar (El-Haidari and Al-Hafidh 1986; Gassouma 1991). In Iran, locations where *O. elegans* has been reported (Gharib 1970; Endrödi and Petrovitz 1974; Mohammadpour unpublished) are shown in Fig. 5.6. The distribution of *Oryctes* species in Iraq is shown in Fig. 5.7.

Various subspecies of *O. agamemnon* are reported as attacking the date palm. In the Arabian Peninsula, *O. agamemnon arabicus* Fairmaire is endemic in Saudi Arabia (Endrödi 1985), Oman (Al-Sayed and Al-Tamiemi 1999). In the eastern part of UAE in the Al-Ain Region, it is the only species of *Oryctes* found (Gassouma 1991; Al-Deeb et al. 2012a). In Iraq, it coexists with *O. elegans* (Buxton 1920; Khalaf et al. 2013a, b) (Fig. 5.7). Subsequent research has clarified that the material referred to as *O. elegans* in Khalaf et al. (2010, 2011, 2012) is now identified as *O. agamemnon arabicus* (Khalaf, pers. comm. 2015), and the adjustment applies throughout this chapter.

In Tunisia, *O. agamemnon arabicus* was accidentally introduced into the oasis of Mrah Lahouar in the Djerid zone (Fig. 5.8) from the UAE in the late 1970s. The pest

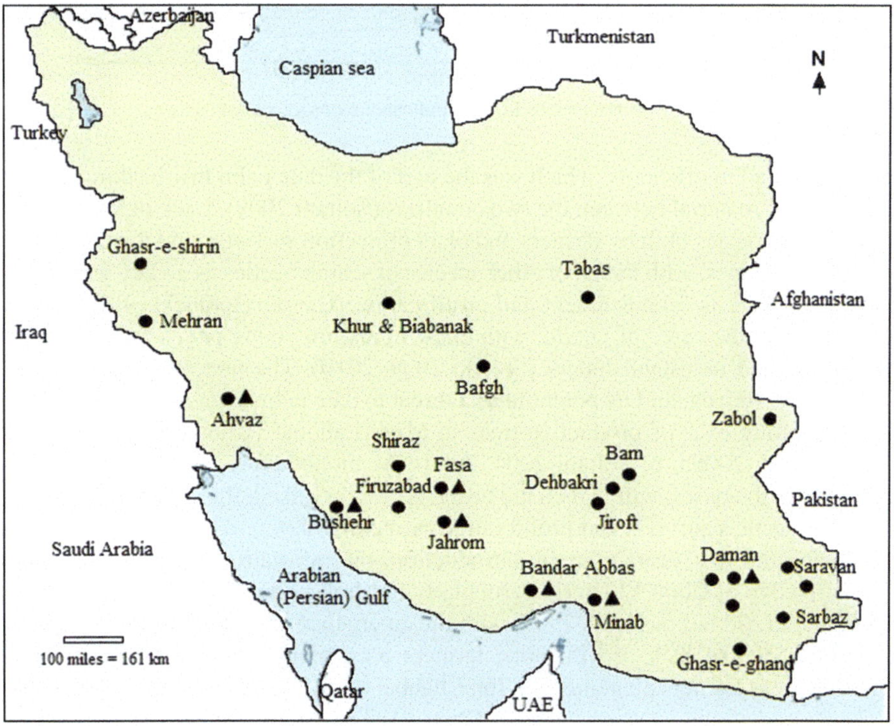

Fig. 5.6 Distribution of *Oryctes elegans* (●) and *Oryctes agamemnon* (▲) in Iran

Fig. 5.7 Distribution of *Oryctes* species in Iraq (indicated by *black spots*)

was concealed in offshoots, which was the part of the date palm involved in varietal exchange of material between the two countries (Soltani 2009). Lack of knowledge of this new species both by farmers and plant protection services, and the confusion of its larval stages, with larvae of other species of scarab beetles as no key was available, permitted its establishment and proliferation. The development of date palm cultivation in the oases of Djerid, with many immature young palms and offshoots also facilitated its establishment (Soltani 2009, 2010). The first recognition of its presence in Tunisia, and its potential as a threat to date palms, was in 1995 after the sudden falling over of productive trees in Mrah Lahouar (Khoualdia et al. 1997; Soltani et al. 2008a, b; Soltani 2004, 2009). As monoculture of the date palm is common in the oases, with which the beetle is closely linked, this association facilitates the beetle's survival and proliferation (Soltani 2009).

During the first years after its introduction, the infestation was limited to the northern edge of Chott El Djerid particularly in Mrah Lahouar (387 ha) and Dhraa El Janoubi (200 ha) oases. In 1987, with the emergence of date palm cultivation in the zone and the lack of offshoots, farmers used plants originating from these infested areas for new plantations in Ibn Chabbat oasis, located to the north of Mrah Lahouar and in the southern edge of Chott El Gharsa. This allowed the insect to spread and develop there, as the importance of sanitary and quarantine measures had been overlooked (Soltani 2010).

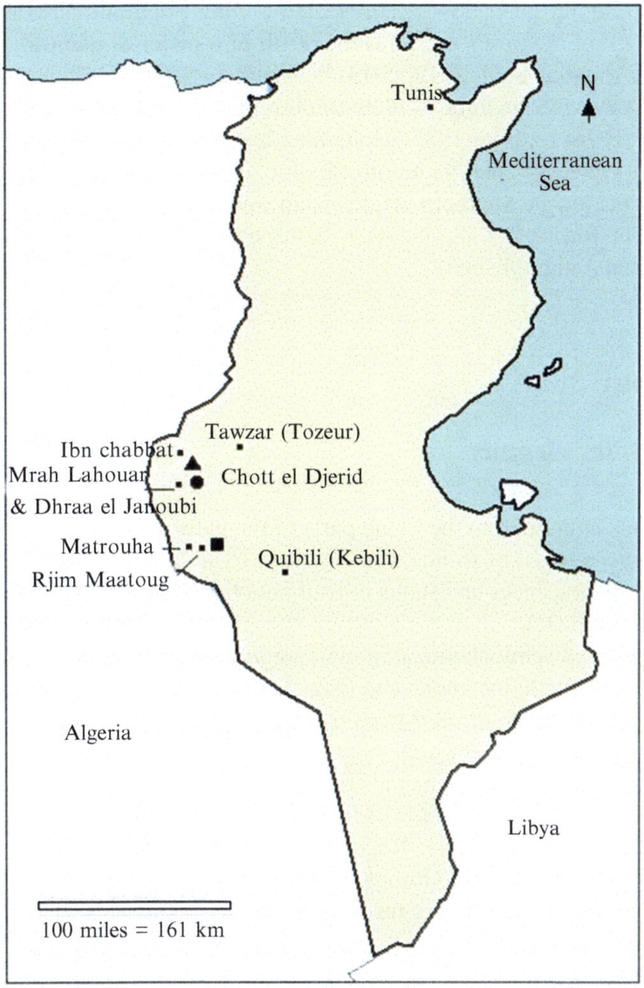

Fig. 5.8 Geographical distribution of *Oryctes agamemnon arabicus* in Southwest Tunisia. The main foci of infestation are the oases of Mrah Lahouar (●), Ibn Chabbat (▲) and Rjim Maatoug (■)

At the beginning of the 1990s, in offshoots originating from Ibn Chabbat, the pest reached the oases of Rjim Maatoug on the southern edge of Chott El Djerid and presently covers the entire border zone except eastern Matrouha oases (Soltani 2009, 2010). As the habits of this pest were not fully understood at that time, *O. agamemnon arabicus* was able to spread mainly through offshoots, as well as by flying adult beetles (Soltani 2010).

Analysis of DNA markers in *O. agamemnon arabicus* populations in Tunisia showed no significant differences in relation to location or host plant varieties, suggesting a limited number of genotypes would have been present at the start of the invasion. Substantial gene flow was revealed among populations, confirming the belief that much of the pest's spread is the result of transfer of planting material for propagation (Abdallah et al. 2012, 2013).

Oryctes agamemnon matthiesseni Reitter is reported from Southern Iran (Endrödi and Petrovitz 1974; Endrödi 1985; Mohammadpour unpublished) and Iraq (Khalaf unpublished). However, there is no information about its biology from these countries. It also occurs in Southern Afghanistan and North West Pakistan (Endrödi 1985). Another form of *O. agamemnon* (subspecies unknown) has been reported from Iraq (Khalaf unpublished).

5.3 Biology

5.3.1 *Oryctes elegans*

The adult bores a tunnel into the living parts of the palm such as the midrib (rachis) and leaf bases (petioles) of fronds, causing them to break off during wind, and into the stems of inflorescences and stalks of fruit bunches (Buxton 1920; Hussain 1963; Mohammadpour 2002).

Males produce a semiochemical, a male aggregation pheromone which attracts both sexes. It is 4-methyloctanoic acid (Fig. 5.9a), and its attractiveness is strongly enhanced by the presence of odor from fresh date palm tissue (Rochat et al. 2004). The laboratory protocols for the synthesis of these pheromones are reported (Ragoussis et al. 2007; Tabrizian et al. 2009; Sultanov et al. 2013).

The larvae are found in the axils of fronds and at the junction of dead and living tissue in crowns (Hussain 1963; Rochat et al. 2004) and may tunnel toward the growing point (Buxton 1920). Groups of larvae may tunnel in moist tissues inside dying or newly dead palm trunks forming a large hole (Gharib 1970), which may topple the palm (Hussain 1963).

In laboratory studies in which the larvae were reared on a food mixture of decayed wood or compost and cow dung at 28–30 °C, the average duration of development was: egg, 10 days; larva, 91 days (three instars); prepupa, 7 days and pupa, 31 days (Hurpin and Fresneau 1969). Larvae were also able to feed on living plant

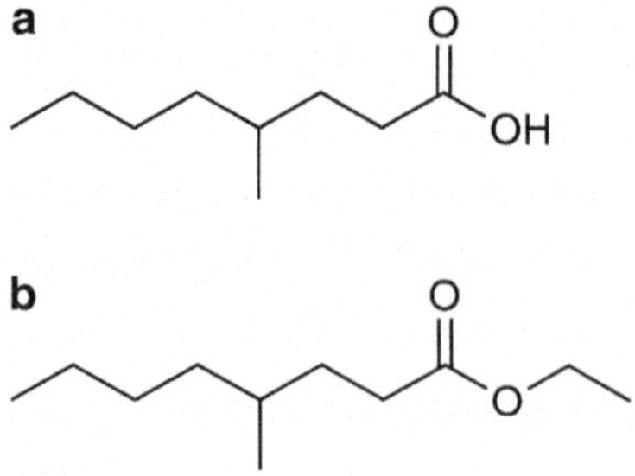

Fig. 5.9 4-methyloctanoic acid (4-MO) pheromone of *Oryctes elegans* (**a**), ethyl-4 methyloctanoate (E4-MO) (**b**) (Bedford 2013; with permission from Annual Review of Entomology Volume 58. © 2013 by Annual Reviews)

material e.g. apple, but development time was longer and the resulting adults were smaller sized (Hurpin 1970). In another laboratory study, at 27 °C, somewhat longer development times were observed, with mortalities of 5, 23 and 3 % recorded for the egg, larval and pupal stages, respectively (Payandeh and Dehghan 2010).

Adult life span was about 4 months, and egg-laying began 2–3 weeks after individuals were brought together as couples, with an average of 60 eggs laid by a single female per life cycle (Hurpin and Fresneau 1969).

5.3.2 Oryctes agamemnon arabicus

In Tunisia, no attacks by adults on the inflorescence and green fronds were registered and little sign of feeding was found under laboratory rearing conditions (Soltani et al. 2008a, b; Soltani 2009, 2010). However, in Saudi Arabia, adults attacked fruit bunch stalks and the rachis of fronds (Al-Sayed and Al-Tamiemi 1999).

In Tunisia, the larvae are found feeding in the dead respiratory roots around the base of date palm trunks (Soltani et al. 2008b), but no attacks on the inflorescences and green fronds by larvae were reported (Soltani et al. 2008a, b; Soltani 2009, 2010). In Tunisia, Saudi Arabia and Oman, it breeds in dead dry bark, dry frond petioles, fiber masses in axils, and offshots (Khoualdia et al. 1997; Al-Sayed and Al-Tamiemi 1999; Soltani et al. 2008a, b), while in Iraq they occur in debris in the bases of lower fronds and in tunnels in dead frond bases, where there may be up to 12–13 larvae per tree (Khalaf et al. 2010).

The life cycle of *O. agamemnon arabicus* in Iraq is illustrated in Fig. 5.10. Here at 25 °C, the average durations of the stages are; eggs: 13 days; total larval stages: 196 days and pupae 29 days (Khalaf et al. 2014). In Tunisia, laboratory rearing at 23 °C resulted in average durations of egg, first-, second-, third-instar larva, total larval stages, pupa, and total immature stages of 14, 35, 51, 141, 227, 24 and 266 days, respectively; at 27 °C, 12, 35, 44, 118, 197, 22 and 231 days, respectively; and at 30 °C, 9.5, 36, 45, 142, 223, 21 and 254 days, respectively (Soltani et al. 2008a; Soltani 2012).

Males of *O. agamemnon arabicus* emitted a blend of four compounds: (1) ethyl 4-methyloctanoate (Fig. 5.9b), (2) 4-methyloctanoic acid (Fig. 5.9(a), (3) 4-methyloctanyl acetate, and (4) 4-methyloctanol. Laboratory observations using an olfactometer showed that compounds 1 and 3 attracted both sexes but females were more attracted by compound 1, virgin females by compound 2, and males by compound 3. Field studies revealed that a mixture of compounds 1 and 2 with date palm core odor was the most effective in attracting both sexes, with females pre-dominating (Saïd et al. 2015). Ethyl 4-methyloctanoate (E4-MO, Fig. 5.9b) is the pheromone of the coconut and oil palm pests *O. rhinoceros* and *O. monoceros* (Olivier), and is commercially available and widely used in trapping and IPM (Integrated Pest Management) programs directed against them (Bedford 2013).

Fig. 5.10 Life cycle of Arabian rhinoceros beetle, *Oryctes agamemnon arabicus*; *1*: eggs, *2*: larva, *3–4*: pupae, *5*: emerging adult, *6*: adult female, *7*: adult male (Photos: Mohammed Z. Khalaf)

5.4 Seasonal Incidence

In Iraq, adults of *O. elegans* are attracted to light from April to September, but especially during April and May (Buxton 1920; Hussain 1963). Light trapping in three orchards showed that flying activity of *O. agamemnon arabicus* started during April (Khalaf et al. 2013b) and peaked during July (82 individuals caught per trap and month). Population densities as measured by light trap captures then decreased dramatically in October (Khalaf et al. 2011, 2012).

In Iran, *O. elegans* is univoltine, having one generation per year. Here this species and *O. agamemnon matthiesseni* are sympatric; *O. elegans* adults are nocturnal and attracted towards light. Light trapping revealed that the adults emerged from March and April (spring) to late September, and were most abundant in June and July (summer) (Fasihi 2011), and were not found in late autumn. The female of *O. elegans* makes a small cavity and lays eggs individually in palm fiber in the axils of fronds and inflorescences, or living tissue where petioles join the trunk, or in damp dead half-rotten trunks. The larvae overwinter in their feeding sites in the palm, and each larva weaves a cocoon of fiber (6×2 cm). The pupal period in March and early April lasts for 3–4 weeks (Gharib 1970). Pheromone trap catches also showed the population of *O. elegans* reached to a maximum during summer season (mid-July to mid-August) (Rochat et al. 2004).

In Oman, light trapping showed that *O. agamemnon* appeared from mid-April to early May, with maximum catches recorded from early June to early July, the warmest and driest season, and then catches fall and cease from late August to early October (Al-Sayed and Al-Tamiemi 1999).

In the Al-Ain region of the UAE, *O. agamemnon arabicus* adults were attracted to street lights and found crawling on the grass of gardens and parks under lights after sunset (Al-Deeb unpublished). In a 2-year study using light-trapping in UAE, Al-Deeb et al. (2012a) found that *O. agamemnon arabicus* males and females appeared in the field around April and May and the population continued to build until maximum numbers were reached in mid-June. However, at the end of July and early August, the numbers declined sharply and no adults were found after the end of September. *Oryctes agamemnon arabicus* has one generation per year (Al-Deeb 2012; Al-Deeb et al. 2012a) unlike the red palm weevil *Rhynchophorus ferrugineus* (Olivier), which has several overlapping generations per year. In Saudi Arabia, Al-Deghairi (2007) reported that the highest activity of *O. elegans* occurred from April to July and the peak monthly activity was observed during June, and no beetle activity during January and February. In date orchards in Riyadh Province (Saudi Arabia) 58 % of scarabs light-trapped were *O. elegans* and reached highest abundance in August (Al Dhafer and Alayeid 2014).

In Tunisia, light trapping showed a single peak – indicating that *O. agamemnon arabicus* has one generation per year (univoltine). The adults emerged from the date palm breeding sites in late May/mid-June and flights continued to mid-October/early November, with the peak flight activity between mid-July and mid-August, the warmest part of the year. Females were the most numerous in the catches overall. The trap captures decreased afterwards, becoming rare in late September/early October, and then ceasing from mid-October/early November until June of the following year (Soltani 2009; Ehsine et al. 2014). Nocturnal activity started about 40 min after sundown and continued until 1 h before sunrise. Most beetles were caught in the first 2 h of flight activity, females dominating in the first hour, and males the second hour (Ehsine et al. 2014). Adults spend most of their time within the green parts of the palm, from where they emerge to seek breeding sites and disperse. In oases, both adult and immature stages also occur in compost heaps (Soltani 2009).

In Yemen, when light traps were used to monitor the *O. rhinoceros* population, it appeared in March, increased in April, peaked in June, decreased to low numbers by September, with very low numbers caught from October to February. The single peak of adult captures (sex ratio male: female 1:1.3) indicated there was one generation per year (Al-Habshi et al. 2006). In coconut-growing areas of the South Pacific, it is considered that *O. rhinoceros* is only occasionally attracted by light (Gressitt 1953).

5.5 Damage and Losses

5.5.1 Oryctes elegans

When the adult bores into the stalk of the fruit bunch (Fig. 5.11), it makes a tunnel on one side of the stalk and the dates on that side of the bunch remain undersized, while on the other side of the bunch, fruit is of normal size. A heavily attacked stalk can cause all the dates on a bunch to shrivel, or the stalk breaks and the whole bunch falls. In Iraq, the stalks were attacked straight after fruit set in April.

In the early 1960s in Iraq the infestation of bunches was less than 2 %, so it was not regarded as a serious pest at that time (Hussain 1963) whereas heavy damage to bunches was reported in 1918 (Buxton 1920). Similar damage to fruit stalks occurs in Saudi Arabia (Al-Deghairi 2007).

Fig. 5.11 Damage from *Oryctes elegans* adults on base of fruit bunch (Photo: Kazem Mohammadpour)

In Iran, *O. elegans* damage is more significant in young date palms (aged around 10–20 years) than in old plantations, and was more severe where planting distance between palms was less than 5 m (Latifian et al. 2012). The long feeding tunnels in the stalks of inflorescences (especially during the pollination period) and bunches, weaken them and cause them to break. They also feed on other living parts of the palm such as the base (petiole) and midrib (rachis) of fronds, causing them to break off during wind. In some areas the damage caused by *O. elegans* represents 5–20 % of the total harvest (Gharib 1970).

The larval feeding sites of *O. elegans* may also serve as the egg deposition sites by *R. ferrugineus*, however only a weak correlation between the abundance of both species was observed in 16 date palm orchards of Saudi Arabia monitored from June 2007 to May 2008 (Al-Ayedh and Al Dhafer 2015). In areas where both *O. elegans* and *O. agamemnon* (regardless of subspecies) coexist e.g. Iraq, at times it is not only difficult to attribute damage to the larval or the adult stages, but also to ascribe it to one or other of the species. Moreover, the morphological differentiation of the larvae of both species is also difficult as no key is available. In Iran, *Oryctes* larvae live in the crown and trunk, feeding on petioles and rachis of fronds, and this injury may allow entry of fungi and secondary insect pests. When tunnelling in the trunk they may reach the central bud and kill the palm. They also attack the base of the palms to feed on sap.

5.5.2 *Oryctes agamemnon arabicus*

In Oman, adults were reported as attacking fruit bunch stalks and the rachis of fronds and, in Saudi Arabia, fruit stalks and the rachis and bases of fronds (Al-Sayed and Al-Tamiemi 1999).

In Iraq, all ages of date palm are infested by *O. agamemnon arabicus* larvae, but the trees aged over 30 years scored the highest level of infestation in comparison with middle age and young trees in all experimental date palm orchards, in different provinces. However, infestation level varied among provinces as Wasit (Al-Numaniyah) (Fig. 5.7) had the highest infestation level leading to breakdown of trunks, especially during wind storms (Khalaf et al. 2013a, b, 2014; Khalaf and Al-Taweel 2014).

In situations where *O. agamemnon arabicus* is clearly involved, it can infest most parts of the tree. In the young date palm it infests aerial roots at the base of the trunk, due to the presence of frond bases that are close to the soil surface, leading to yellowing and drying of the crown (Khalaf et al. 2014; Khalaf and Al-Taweel 2014). However, this type of infestation does not occur in middle aged and old trees. The lower part of the trunk (about 1 m above the soil surface) is infested in palms of all ages. The insect infests the middle of the trunk in the middle-age and old palms (Khalaf et al. 2013a, b, 2014). No infestation has been recorded in the upper stem part, close to the crown region in the case of young trees, but severe infestation was recorded here on middle aged and old date palm trees. The upper crown, stalk and fronds have not been infested by this species on palms of all ages. The date palm parts that are attacked, in relation to the age of the palm, are summarized in Table 5.1.

In Tunisia, outbreaks of *O. agamemnon arabicus* develop inside plantations from those that are newly planted up to about 35 years old, with regular spacing between trees of 9×9 m. This pest does not occur in plantations cultivated in the traditional system where palms are irregularly spaced and intercropped with other crops (Soltani 2009). The feeding activity of larvae causes damage to offshoots, and to aerial roots and dry petioles of young palms beyond the offshoot stage (Soltani 2009, 2010).

In offshoots eggs are laid in the mass of fibers in the axils of fronds, and here young larvae feed unnoticed until the third instar larvae bore into the heart. Most often, the attacked palms are killed in the first year after planting, and losses can reach up to 100 % (Fig. 5.12).

In palms that are past the offshoot stage, adults bore holes into the aerial root mass to oviposit, where damage is mostly due to the subsequent tunnelling by feeding larvae which produces cavities. These cavities provide access to subsequent generations of adults for entry and oviposition. Groups of adults, mainly ovipositing females, can coexist in these cavities, along with larvae, and the continuous feeding and tunnelling activity by larvae (Fig. 5.13) can lead to collapse of the palm after a few years (Fig. 5.14) (Ehsine et al. 2009, Soltani 2009).

When eggs are laid in the mass of fiber in frond petioles, the early larval stages feed here, and the third instar larvae tunnel into the dry petioles, feeding on the soft wood and pupate at the end of the tunnels (Fig. 5.15). Any attack to the trunk itself is less important, as only external dead wood is affected (Soltani et al. 2008b; Soltani 2009) - it is unlikely to cause the palm to fall, but provides an ongoing breeding site (Ehsine et al. 2009) (Fig. 5.16).

Table 5.1 Parts of the date palm infested by *Oryctes agamemnon arabicus* in Iraq

Date palm tree part	Young trees (4–10 years)	Middle age trees (10–20 years)	Old trees (over 30 years)
Stem base close to the root i.e. aerial roots at base of trunk	+	–	–
Lower stem or trunk, about 1 m above soil	+	+	+
Middle of the stem (Trunk)	–	+	+
Upper stem (when the old fronds are present)	–	+	+
Lower crown	–	+	+
Upper crown	–	–	–
Fronds	–	–	–
Stalks (frond bases i.e. petioles)	–	–	–

(–) No infestation, (+) Infestation

Fig. 5.12 Damage caused to offshoot by a third instar *Oryctes agamemnon arabicus* larva (*left*), and offshoot death after larval attack (*right*) (Photos: Rasmi Soltani)

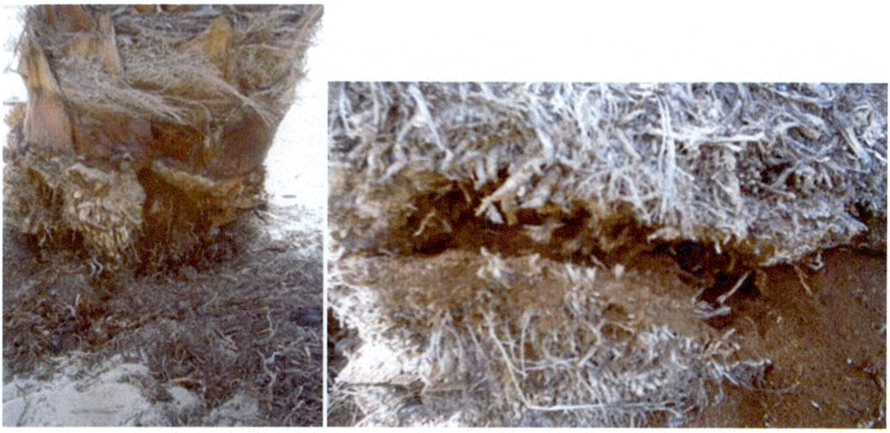

Fig. 5.13 Aerial roots attacked (*left*) and damage due to *Oryctes agamemnon arabicus* larvae (*right*) (Photos: Rasmi Soltani)

Fig. 5.14 Date palm highly liable to collapse following severe attack by *Oryctes agamemnon arabicus* larvae at the collar region (Photo: Rasmi Soltani)

Fig. 5.15 Damage due to *Oryctes agamemnon arabicus* larvae and adults on dry petioles (Photo: Rasmi Soltani)

Fig. 5.16 Attack by larvae of *Oryctes agamemnon arabicus* on fiber masses in axils and on bark of the trunk (Photo: Rasmi Soltani)

5.6 Management

5.6.1 Physical Control

This approach is very important in all date growing regions. Old and almost dried frond bases are pruned and removed from the palm trunk in an annual servicing by cutting them at a downward sloping 45° angle, the cut ends allowing the farmer to climb the tree. Layers of fiber between fronds, frond thorns, and old dried bunch stalks are also removed (Figs. 5.17, 5.18, and 5.19). This eliminates sites for beetles to hide and oviposit, and for larvae to develop, and allows hand picking of any larvae found (Figs. 5.20 and 5.21). The number of larvae found later on serviced i.e. pruned palms was significantly less than the number present on unpruned control palms (Khalaf and Al-Abid 2013). Pruning also reduces humidity around the trunk and frond bases so discouraging oviposition. Disinfecting any wounds on trees with approved pesticides may deter adults (Mohammadpour 2002). Plantation sanitation is recommended to remove potential breeding sites such as dead palms or rotting trunks along with clippings and fronds after pruning, and to try to reduce excess humidity due to irrigation in palm groves (Gharib 1970).

Fig. 5.17 A climbing harness being used in annual servicing of a middle-age date palm (Photo: Mohammed Z. Khalaf)

Fig. 5.18 Young date palm showing a stage in annual servicing (Photo: Mohammed Z. Khalaf)

Fig. 5.19 Previously pruned trunk showing frond bases cut off at the recommended 45° angle (Photo: Mohammed Z. Khalaf)

5.6.2 Trapping and Monitoring

5.6.2.1 Light Trapping

This technique has been used for monitoring populations of *O. elegans* and *O. agamemnon* in Iran (Fasihi 2011), *O. agamemnon arabicus* in Iraq (Khalaf et al. 2010, 2011, 2012), UAE (Al-Deeb et al. 2012a) and Oman (Al-Sayed and Al-Tamiemi 1999), *O. elegans* in Saudi Arabia (Al-Deghairi 2007) and *O. rhinoceros* in Yemen (Al-Habshi et al. 2006). Where electricity supply is not available light traps with solar panels can be used (Fig. 5.22). In UAE and Saudi Arabia, light traps emitting white light from a mercury vapour bulb may be used for mass trapping (Al-Deghairi 2007; Al-Deeb et al. 2012a).

In Iraq, the light traps, in conjunction with hand removal of larvae, have proven very efficient as a control method against *Oryctes attacks*, reducing beetle numbers by 90 % (Khalaf et al. 2011). In fields where hand picking of immature stages from old frond bases is neglected, the subsequent catch of adults by light traps was found to be high.

Moonlight affected the light trap catches of adult beetles. During the nights of bright moon light, adult captures (considered to be *O. agamemnon arabicus)* decreased compared to nights with less moon light (Khalaf et al. 2011). A similar trend was also suggested by the data collected using coconut wood traps (i.e. non-light traps) for monitoring *O. rhinoceros* in New Britain (Bedford 1975).

Fig. 5.20 Larvae of *Oryctes agamemnon* detected in their tunnels in frond bases during important annual servicing in Iraq. *Top right* picture shows black tunnels (Photos: Mohammed Z. Khalaf)

Fig. 5.21 *Oryctes* spp. larvae collected during annual servicing of date palms in Iraq (Photo: Mohammed Z. Khalaf)

Fig. 5.22 Solar light trap (Magna Trap with lamp of 320–420 nm wavelength, Russell IPM Limited, UK) used for monitoring and controlling *Oryctes* spp. in date palm orchards in Iraq (Photo: Mohammed Z. Khalaf)

5.6.2.2 Pheromone Trapping

In Iran, monitoring of population fluctuations of *O. elegans* was conducted for 25 weeks in the Saravan region between April and September, 2002. Ten traps baited with a piece of date palm core and the male aggregation pheromone 4-methyloctanoic acid were placed about 5 m above the ground on trunks of date palms in groves. The traps consisted of 24 l plastic buckets with lids perforated by eight radial 8 × 5 cm openings (Fig. 5.23).

Weekly captures of *O. elegans* between April and September increased gradually from mid-April until the beginning of June (5.3 ± 1.4 beetles per trap), reaching a peak in mid-July to early August (11.5 ± 1.9 beetles per trap), then the catch fell (Fig. 5.24).

Considering the mean captures per trap calculated on monthly basis, the population reached a maximum in summer (mid-July to mid-August: 5.9 ± 0.8 beetles). Both sexes seemed to emerge simultaneously in spring and were reported until the temperature dropped in late autumn. The sex ratio of catches fluctuated through the trial, being approximately balanced in spring and then showing an excess of females through summer, especially in August. The female excess recorded in summer may reflect a shorter life span of the males or a greater mobility of females (Rochat et al. 2004). Furthermore, a trial with traps using 4-methyloctanoic acid and the odor from fresh date palm tissue as synergist, caught in Saravan more than 4000 *O. elegans* over the two trapping seasons during which the synthetic pheromone was evaluated. The captures averaged 6.3 beetles per trap and week (Rochat et al. 2004).

Fig. 5.23 Plastic bucket pheromone trap for trapping *Oryctes elegans* (Photo: Kazem Mohammadpour)

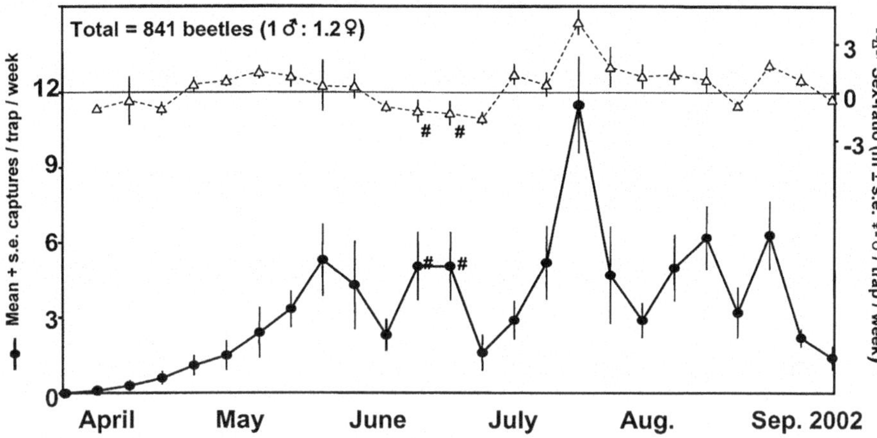

Fig. 5.24 Weekly captures of *Oryctes elegans* per trap (*bottom, left* scale) with corresponding sex-ratios (*top, right* scale) from ten traps baited with a piece of date palm core and 4-methyloctanoic acid, the major component of the male aggregation pheromone, emitted at a rate of 2.2±0.1 mg/day (mean ± SE) for 25 weeks between April and September 2002 in Eastern Iran (Rochat et al. 2004)

Monitoring by traps could be integrated with insecticide application at the time of leaf-pruning around fruit bunches, and could reduce the beetle population and hence the need for insecticide use.

Experiments on the possibility of co-mass trapping of red palm weevil, *R. ferrugineus* and *O. elegans*, using pheromone traps, were carried out in infested date palm groves of the Saravan Region in the Sistan and Baluchistan Province of Iran, during the years 2004–2005. Results showed that the traps baited with separate dispensers of aggregation pheromones of the two insects, when compared with traps that were baited with pheromone mix (50:50) in one dispenser, attracted significantly more *O. elegans*. But all forms of the baits were equally attractive for *R. ferrugineus*. Also in this trial there was no significant difference in the number of *O. elegans* caught by traps placed at different heights i.e. at ground level, or 1.5 or 4 m above, but the traps placed on the ground attracted significantly more *R. ferrugineus* weevils in comparison with traps placed about 1.5 and 4 m above ground. The placement of traps on the ground is hence convenient for catching both species (Mohammadpour and Avand-Faghih 2008). However, placing traps at a higher level (4 m) gave a higher catch for *O. elegans* and, depending on the main target, the technique has to be adapted consequently (Rochat et al. 2004). A similar benefit, i.e. improved catch by elevating the trap, has been noted in the pheromone trapping of *O. rhinoceros* in young oil palm plantations (Bedford 2014, Oehlschlager 2007). The effect of date palm core tissue ageing on the catch of *O. elegans* and *R. ferrugineus* was similar. Captures in traps where replacement of date palm core tissue was done every week, were significantly greater than in traps where replacement of date palm core was done every 2–3 weeks. Increasing age of the date palm core, hence, decreased captures in traps. These results indicated that pheromone traps with separate dispensers of pheromone of *R. ferrugineus* and *O. elegans*, and placed on the ground, can be used for co-mass trapping of these beetles (Mohammadpour and Avand-Faghih 2008). Pheromone traps may also be used against *O. elegans* in Iraq.

To date, neither *O. agamemnon matthiesseni* in Iran nor *O. agamemnon arabicus* in Iraq were reported to be attracted to traps using the *O. elegans* pheromone (Khalaf and Mohammadpour unpublished).

5.6.3 Host Plant Resistance

Different varieties of date palm have varying susceptibility to *Oryctes* attack. In Iran, *O. elegans* damage is more prominent in young date palms (10–20 years) and on short varieties (Mozafati) than in old plantations and tall varieties (i.e. Halili, Krout and Mordarsang) (Gharib 1970). No other plant species are reported as hosts (Mohammadpour 2002).

In Iraq the date varieties Brem and Ustaomran (Umrani) were the most susceptible to attack and infestation by *O. agamemnon arabicus*, and 10 and 9 larvae per tree, respectively, were found in the parts exposed during the annual frond pruning of palms (Khalaf et al. 2011, 2014). Date palm varieties show much difference in their morphological characters e.g. length and orientation of leaves, type of growth and shape of frond etc. In Iraq, a study on the characteristics of the dried base of fronds for different commercial date palm varieties in relation to borer infestation showed that varieties having fragile textured fronds are preferred by *Oryctes*, compared to the varieties with solid and hard textured fronds (Khalaf et al. 2010).

5.6.4 Quarantine Measures

Females of *Oryctes* spp. are attracted to the odors of animal manure or organic compost for oviposition (Ehsine et al. 2009). In the UAE organic fertilizer in bags or loose piles on the farm can serve as good breeding sites. They were found to chew holes in the fertilizer bags to enter and lay eggs. Therefore, proper storage and handling of the organic fertilizer on the date palm farm is very important to reduce infestation. In addition, importing organic fertilizer from countries infested with *Oryctes* beetles can serve as a new source of infestation. Thus, it is not recommended to bring in organic fertilizers from infested farms or countries. In this regard, internal and national quarantine procedures are a legislative method to prevent the spread of *Oryctes* infestation by helping in examination, detention, treatment or disposal of infested organic fertilizers.

In Tunisia, after defining infested areas, quarantine protocols were applied by plant protection services and law enforcement authorities at checkpoints to prevent the spread of infested offshoots (Soltani 2010).

5.6.5 *Biological Control*

5.6.5.1 Entomopathogenic Fungus

In Iran, laboratory studies using *Metarhizium anisopliae* (Metchnikoff) Sorokin showed that it killed *O. elegans* adults 6–11 days post-infection (LT50 values). Additionally, the adults fed less and females laid fewer eggs compared to untreated insects (Latifian and Rad 2012). It is the second most frequently found entomopathogenic fungus in Iraq, occurring in 18 % of soil samples collected from date and date/citrus orchard sites (Khudhair et al. 2014a). The fungus infects *Oryctes* under laboratory conditions (Fig. 5.25).

Spraying of spore suspensions against larvae, or light traps allowing automatic infection by spores followed by release of the contaminated adults, may have potential for incorporating the fungus in future IPM programs against *Oryctes* spp. in the date palm environment.

5.6.5.2 Nematodes

In the UAE infective juveniles of the entomopathogenic nematode *Steinernema riobrave* (Cabanillas, Poinar and Raulston) at various concentrations caused a range of 44–100 % mortality in third instar larvae of *O. agamemnon* under laboratory conditions, and 33–78 % mortality when applied to soil in a fig orchard (Abbas and Mahmoud 2009).

Rhabditis sp. has been isolated in Southern Iraq from *Oryctes* larvae which had developed an abnormal brown color. It is being propagated and investigated for its possible potential in pest control (Khalaf unpublished). *Rhabditis* spp. and *Heterorhabditis* spp. Poinar in Iran (Mohammadpour unpublished) are potential candidates as biological control agents. However, there are no data on their natural incidence in the field.

Fig. 5.25 Adult of *Oryctes* infected by *Metarhizium anisopliae* fungus under laboratory conditions (Photo: Mohammed Z. Khalaf)

5.6.5.3 *Oryctes* Nudivirus

This non-endemic virus was introduced into Oman and significantly lowered the damage caused by *O. rhinoceros* to fronds of coconut palms (Kinawy 2004; Bedford 2013). It is not known if *O. elegans* or *O. agamemnon* are susceptible to this pathogen, but its introduction into Oman means it is now present in the Arabian Peninsula, and free to disseminate.

5.6.5.4 Mites

In the UAE, *O. elegans* adults often carry a large load of the phoretic mites *Sancassania* sp. and *Hypoaspis rhinocerotis* Oudemans on their body and under the elytra (Fig. 5.26) (Al-Deeb and Enan 2010; Al-Deeb et al. 2012b). However, it has

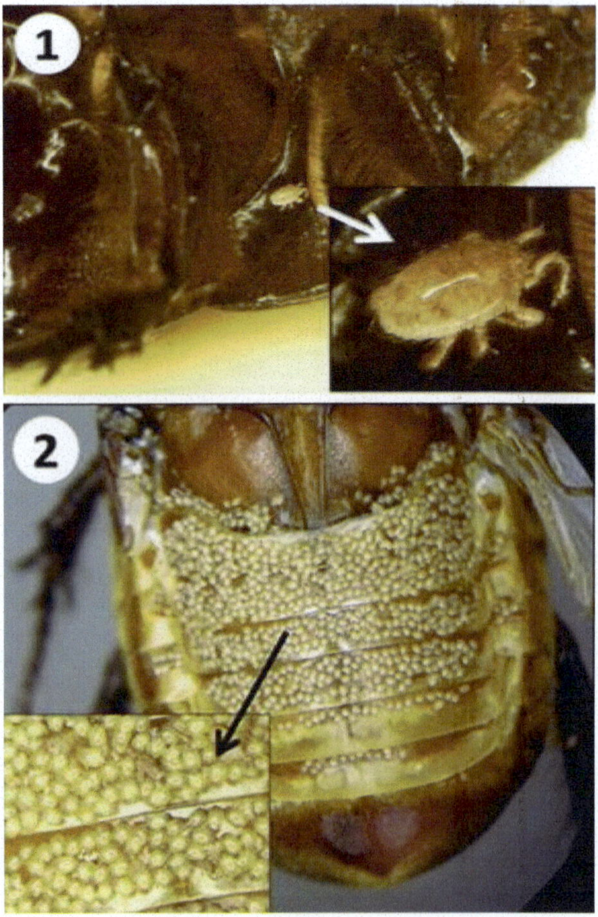

Fig. 5.26 Mites on *Oryctes agamemnon* adult in UAE; (*1*) *Hypoaspis* sp. Adult, (*2*) *Sancassania* sp. deutonymphs on abdominal tergites (Photos: Mohammad A. Al-Deeb)

not yet been established if the mites have a harmful effect on the pest, or are only phoretic. The same holds for *Hypoaspis* sp. in Iraq (Figs. 5.27 and 5.28).

5.6.5.5 Vertebrates

In Iran a species of squirrel, *Funambulus palmarum* L. inhabits the palm groves of Baluchistan Region (its local name is Herdak) and feeds on larvae of *O. elegans*. The average body length is 142 mm, general body color is gray with three strips of

Fig. 5.27 *Hypoaspis* sp. mite (Photo: Mohammed Z. Khalaf)

Fig. 5.28 *Hypoaspis* mites on adult *Oryctes* (to *left* of base of front legs) (Photo: Mohammed Z. Khalaf)

bright yellow color on the back, while the abdomen is yellowish to white (Gharib 1970). However, its effect on the pest population is unknown.

5.6.6 *Chemical Control*

In Iran, as the adult beetles were usually active in the inflorescence of palm trees, chemical control of dynastid pests was difficult and expensive. In areas where control measures consisted of insecticides against the lesser date moth, *Batrachedra amydraula* Meyrick and the dubas bug, *Ommatissus lybicus* De Bergevin, Asche & Wilson, adult beetles of *O. elegans* that were active and flying in this season, may also be eliminated.

Placing a bran bait with insecticide in the crowns was recommended in the past (Gharib 1970). In Tunisia, chemical control was difficult due to the concealed behavior of the pest, the cost and high quantities of pesticide that would be needed, and the risk of pollution. However, treatment by dipping of offshoots in insecticide solution for 5 min is recommended to kill larvae, prior to the transfer of the offshoots to new areas (Soltani 2010). In Iraq, a trial comparing three insecticides and a control showed spraying was not effective against *Oryctes* larvae in their concealed habitat (ICARDA 2011). In the UAE, chemical control was not recommended as a major control method however, the use of chemical control can be applicable in certain situations: (1) soil insecticides can be used on infested farms where light traps are not available, (2) animal manure or organic compost could be treated with soil insecticides as a pre- or post-application treatment, and (3) insecticides can be applied to treat the root systems of infested trees.

5.6.7 *Integrated Pest Management*

Combining hand picking of larvae and light trapping of adults in orchards south of Baghdad (Iraq) from 2009 to 2012 compared to control orchards (without integrated control) reduced adult numbers caught in the location by about 90 % and increased yield by about 28 and 31 % in 2011 and 2012, respectively (Table 5.2) (Khalaf and Al-Abid 2013; Khalaf et al. 2011, 2013a, b, 2014; Khalaf and Al-Taweel 2014).

5.6.8 *Comparison of Oryctes spp. on Date Palms with Oryctes rhinoceros on Coconut and Oil Palms*

It is of interest to compare the biology and damage caused by the two species of *Oryctes* on date palms, with that caused by the well-known pest *O. rhinoceros* on coconut (*Cocos nucifera* L.) and oil palms (*Elaeis guineensis* Jacquin). Adults of *O.*

Table 5.2 Effect of IPM program against *Oryctes* on yield of Barhee date palms

Treatment	No. of bunches per tree		Yield (kg per tree)	
	2011	2012	2011	2012
Control (no hand collection of larvae)	10.91	10.7	104.00	106.00
Almost complete control, with hand collection of larvae and light traps	11.27	11.1	133.36	139.00
Yield/bunch increase (quantity)	0.36	0.40	29.36	33.00
Yield increase (%)	3.3	3.74	28.20	31.13

Fig. 5.29 Coconut palm in Fiji showing damaged V-cut fronds resulting from *Oryctes rhinoceros* adults boring into the immature fronds (Bedford 2013; with permission from Annual Review of Entomology, © 2013 by Annual Reviews)

elegans bore into fruit bunch stalks of the date palm to feed, while *O. agamemnon* adults apparently feed little, or not at all. On coconut or oil palms *O. rhinoceros* adults crawl down frond axils to bore into the heart or meristem feeding on sap as they go. This tunnelling damages the still immature fronds which as a result display V-shaped cuts when they unfurl (Fig. 5.29).

Such attacks by adult *O. rhinoceros* if continued and repeated, kill the growing meristem and the coconut palm dies (Fig. 5.30).

Coconut palms of all ages may be attacked and killed, while young oil palms in particular may be thus killed while older oil palms seem less affected. This type of damage is not reported in the case of date palms attacked by *Oryctes* spp. However, *O. rhinoceros* adults boring into young oil palms can cause the rachis of fronds to

Fig. 5.30 Coconut palms killed and reduced to poles due to repeated heavy attacks by *Oryctes rhinoceros* at Drauniivi, Fiji. Breeding now occurs in the tops of the dead poles (Bedford 2013; with permission from Annual Review of Entomology, © 2013 by Annual Reviews)

break and the upper section of the frond to hang down, as has been noted with *Oryctes* spp. damage in date palms.

In date palms *Oryctes* spp. oviposit in leaf axils, or debris or fiber therein, and from here the larvae may penetrate into the leaf bases and trunk. Oviposition also occurs in the masses of respiratory roots at the base of the trunk. These attacks occur in the living date palm.

Oryctes rhinoceros only oviposits, and the larvae only develop, in dead wood e.g. the tops of dead standing coconut poles (often those killed by adults' attacks, Fig. 5.30), decaying coconut logs and stumps, shredded pulverised oil palm trunk material, and a variety of other breeding sites consisting of various types of decaying organic matter (Bedford 2013). However, it has been reported as breeding in debris in coconut palm leaf axils in Guam (University of Guam, 2010; http://www.biologynews.net/archives/2010/06/21/unusual_rhino_beetle_behavior_discovered.html), perhaps similar to *O. elegans*.

Adults of subspecies of *Scapanes australis* Sternberg in Melanesia attack only young low coconut and oil palms, but in a manner similar to *O. rhinoceros*. Its breeding sites are at soil level under decaying bush logs or in decaying tree roots. Larvae are not found in living coconut or oil palms (Bedford 1980, 2013).

5.7 Future Research

A full description of the larvae of *O. elegans* and *O. agamemnon* and its subspecies, with a key to identify them, would be very useful so they can be reliably distinguished in locations where both species coexist.

The virus *Oryctes* Nudivirus (OrNV) has been successful in lowering populations and damage by *O. rhinoceros* to coconut palms in a number of countries where this virus did not occur prior to its introduction (summarised in Bedford 1980, 2013). It kills larvae and adults, and infected adults act as vectors spreading it before they succumb. It would be of interest to test whether larvae or adults of *O. elegans* and *O. agamemnon* are susceptible to this virus. If found susceptible, prior to any release it would be important to observe samples of larvae and adults taken from the field to see if any other endemic viruses already exist in their natural habitat. As regards the Arabian Peninsula, it would be important to note that this virus has now been introduced and established in Oman (Kinawy 2004) to lower damage by *O. rhinoceros* on coconut palms. Starting from there it is free to self-disseminate elsewhere, and would be available for introduction into locations where this species is attacking date palms.

The fungus *M. anisopliae* is ubiquitous so it would be important to observe samples of field-collected date dynastid larvae and adults for its presence to obtain data on its natural incidence in an area, prior to wide scale release against date palm dynastids. This would provide a base line from which the effect of releases may be assessed. Possibly, light traps may be adapted to auto-infect adults with the fungus allowing their release to disseminate it. A pheromone trap (utilising the already widely-used commercially available pheromone ethyl 4-methyloctanoate) has been adapted for trial use in oil palm plantations to capture and auto-infect *O. rhinoceros* adults with *M. anisopliae*, followed by theirrelease in the field (Ramle et al. 2011). Perhaps pheromone traps for *O. elegans*, or light traps for *O. elegans* and *O. agamemnon*, could be adapted for capture, infection of adults with *M. anisopliae*, then allow their release to disseminate the entomopathogen.

As the components of the male aggregation pheromone of *O. agamemnon arabicus* have now been identified (Saïd et al. 2015) as primarily a mix of 2 already known *Oryctes* spp. semiochemicals, ethyl 4-methyloctanoate and 4-methyloctanoic acid, can this discovery be developed for trapping and future use in an IPM program (Vuts et al. 2014)? For *O. elegans*, the odor or volatiles from fresh date palm tissue, usually from shoots, is synergistic with the pheromone 4-methyloctanoic acid in traps, but as offshoots are sold for propagation, a cheap substitute for date palm offshoots is needed. So extraction and identification of date palm volatiles is needed, which might then be synthesized and added to the pheromone so trapping can be more fully developed, enhanced and utilised. In locations where *O. rhinoceros* attacks date palms, use of the commercially available male aggregation pheromone ethyl 4-methyloctanoate for trapping may be applicable (Bedford 2013).

Are there volatiles common to date, coconut and oil palms which attract *Oryctes* species e.g. *O. elegans* and *O. rhinoceros* to feed, or to oviposit in the case of *O. agamemnon*, and thus these *Oryctes* species are attracted to palms and not to other plants ? Is it a similar question to that which applies to *O. rhinoceros* – do date palm volatiles attract date palm *Oryctes* spp. pests, and then the release of the male produced aggregation pheromone promotes intraspecific competition thence mating, and possibly subsequent oviposition?

While males of *O. elegans* confined in traps, or 4-methyloctanoic acid lures in traps, attract both sexes to traps (Rochat et al. 2004), there is lacking information or observations on where or when during their adult life span, males in the wild emit the pheromone, or both sexes respond (Vuts et al. 2014). Does this occur, with subsequent mating, in feeding sites, or breeding sites, or both? This question also applies to *O. agamemnon arabicus* where more than one key pheromone molecule are produced (Saïd et al. 2015).

A prior assessment of the natural incidence of entomopathogenic nematodes in an area would be advisable if their propagation and widespread release is envisaged.

5.8 Summary

Adults of *O. elegans* cause damage to date palms by boring into the stalks of inflorescences, fruit bunches and frond bases. Their larvae bore into frond bases and trunks. Larvae of subspecies of *O. agamemnon* bore into frond bases thence trunks and into aerial root masses and may thus cause the palm to fall subsequently. Physical control measures such as pruning off old frond bases removes oviposition sites and hence ensuing damage by larvae.

Light trapping enables monitoring of both species and may be used as a component of IPM. Pheromone trapping is also an important tool available in IPM for control of *O. elegans*. Various entomopathogens and other biological control agents may well be investigated as potential or possible components in IPM against these date palm pests.

The damage caused by adults of *O. rhinoceros* which bore into the heart of coconut and oil palms causing the emergence of typical V-cut fronds, and the death of the palm if repeated attacks destroy the meristem, does not seem to occur with the two species of *Oryctes* attacking date palms. *Oryctes rhinoceros* only breeds in dead decomposing woody or organic material whereas immature stages of date palm dynastids may tunnel into living date palm tissue.

References

Abbas, M. S. T., & Mahmoud, B. (2009). On the pathogenicity of the entomopathogenic nematode, *Steinernema riobravis* on some economic insect pests. *Egyptian Journal of Biological Pest Control, 19*, 49–53.

Abdallah, Z., Mezghani-Kkhemkhem, M., Bouktila, D., Makni, H., & Makni, M. (2012). Genetic diversity of an invasive pest (*Oryctes agamemnon* Burmeister, Coleoptera: Scarabaeidae) of date palm in Tunisia, inferred from random amplified polymorphic DNA (RAPD) markers. *African Journal of Agricultural Research, 7*, 1170–1176.

Abdallah, Z., Mezghani-Khemakhem, M., Bouktila, D., Makni, H., & Makni, M. (2013). Genetic variation and invasion pattern of the Arabian rhinoceros beetle, *Oryctes agamemnon arabicus* (Burmeister) (Coleoptera: Scarabaeidae), in Tunisia, deduced from mitochondrial DNA sequences. *African Entomology, 21*, 362–367.

Al-Ayedh, H. Y., & Al Dhafer, H. M. (2015). Does *Oryctes elegans* (Coleoptera: Scarabaeidae) abundance determine future abundance of *Rhynchophorus ferrugineus* (Coleoptera: Rhynchophoridae) in the date palms of Saudi Arabia? *African Entomology, 23*, 43–47.

Al Dhafer, H. M., & Alayeid, H. Y. (2014). Survey and relative abundance of insects (Insecta) excluding Lepidoptera from sixteen commercial date palm orchards using light traps at Riyadh Province, Saudi Arabia. *African Entomology, 22*, 93–103.

Al-Deeb, M. A. (2012). Date palm insect and mite pests and their management. In A. Manickavasagan, M. Mohamed Essa, & E. Sukumar (Eds.), *Dates production, processing, food, and medicinal values* (pp. 113–128). Boca Raton: Taylor and Francis Group.

Al-Deeb, M. A., & Enan, M. R. (2010). First record of a phoretic astigmatid mite *Sancassania* sp. (Acaridae: Astigmata) on *Oryctes agamemnon* (Coleoptera: Scarabaeidae) in UAE. *International Journal of Agriculture and Biology, 12*, 157–160.

Al-Deeb, M. A., Muzaffar, S., & Sharif, E. M. (2012a). Interactions between phoretic mites and the Arabian rhinoceros beetle, *Oryctes agamemnon arabicus*. *Journal of Insect Science, 12*, 128. doi:10.1673/031.012.12801.

Al-Deeb, M. A., Mahmoud, S. T., & Sharif, E. M. (2012b). Use of light traps and differing light colour to investigate seasonal abundance of the date palm pest, *Oryctes agamemnon arabicus* (Coleoptera: Scarabaeidae). *Journal of Economic Entomology, 105*, 2062–2067.

Al-Deghairi, M. (2007). Seasonal fluctuation of the date palm fruit stalk borer, *Oryctes elegans* Prell (Coleoptera: Scarabaeidae), in date palm plantations in Al-Qassim region, Saudi Arabia. *Agricultural and Marine Sciences, 12*, 67–70.

Al-Habshi, K. A., Ba-Angood, S. A., & Al-Baiti, S. O. (2006). The occurrence of the date palm borer *Oryctes rhinoceros* (Linnaeus) in light traps in Wadi Hadramout in 2002. *University of Aden Journal of Natural and Applied Sciences, 10*, 419–426.

Al-Sayed, A. E., & Al-Tamiemi, S. S. (1999). Seasonal activity of the fruit-stalk borer, *Oryctes agamemnon* (Burm.) (Coleoptera Scarabaeidae) in Sultanate of Oman. *Egyptian Journal of Agricultural Research, 77*, 1597–1605.

Bedford, G. O. (1974). Descriptions of the larvae of some rhinoceros beetles (Col.: Scarabaeidae, Dynastinae) associated with coconut palms in New Guinea. *Bulletin of Entomological Research, 63*, 445–472.

Bedford, G. O. (1975). Trap catches of the coconut rhinoceros beetle *Oryctes rhinoceros* (L.) (Coleoptera, Scarabaeidae, Dynastinae) in New Britain. *Bulletin of Entomological Research, 65*, 443–451.

Bedford, G. O. (1980). Biology, ecology and control of palm rhinoceros beetles. *Annual Review of Entomology, 25*, 309–339.

Bedford, G. O. (2013). Biology and management of palm dynastid beetles: Recent advances. *Annual Review of Entomology, 58*, 353–372.

Bedford, G. O. (2014). Advances in the control of rhinoceros beetle, *Oryctes rhinoceros* in oil palm. *Journal of Oil Palm Research, 26*, 183–194.

Butani, D. K. (1974). Les insectes parasites du palmier-dattier en Inde et leur controle. *Fruits, 29*, 689–691.

Butani, D. K. (1975). Insect pests of fruit crops and their control – 15. Date palm. *Pesticides, 9*, 40–42.

Buxton, P. A. (1920). Insect pests of dates and the date palm in Mesopotamia and elsewhere. *Bulletin of Entomological Research, 11*, 287–303.

Dechambre, R.-P., & Lachaume, G. (2001). *Les Coléoptères du monde* [Beetles of the world] – Dynastidae – Volume 27, *Oryctes* (72 pp). Canterbury: Hillside Books.

Ehsine, M., Belkadhi, M. S., & Chaieb, M. (2009). Bio-ecologic observations on rhinoceros beetle *Oryctes agamemnon* (Burmeister 1847) on the palm dates oasis of Rjim Maatoug in south-western Tunisia. *Journal of Arid Land Studies, 19*, 379–382.

Ehsine, M., Belkadhi, M. S., & Chaieb, M. (2014). Seasonal and nocturnal activities of the rhinoceros borer (Coleoptera: Scarabaeidae) in the North Saharan oases ecosystems. *Journal of Insect Science, 14*, 1–5. doi:10.1093/jisesa/ieu118.

El-Haidari, H. S., & Al-Hafidh, E. M. (1986). *Palm and date arthropod pests in the Near East and North Africa* (126 pp). Regional Project for Palm and Dates Research Center in the Near East and North Africa, Food and Agriculture Organization. Baghdad: Al-Watan Press.

El-Shafie, H. A. F. (2012). Review: List of arthropod pests and their natural enemies identified worldwide on date palm, *Phoenix dactylifera* L. *Agriculture and Biology Journal of North America, 3*, 516–524.

Endrödi S. (1985). *The Dynastinae of the World* (Series Entomologica 28, 800 pp). The Hague: Junk.

Endrödi, S., & Petrovitz, R. (1974). Die Arten und Rassen der Gattung *Oryctes* Illiger in Iran. *Applied Entomology and Phytopathology, 36*, 4–19.

Fasihi, M. T. (2011). *A study on the population dynamism of sympatric species of Oryctes agamemnon and O. elegans* (Annual report of project, 11 pp). Bushehr: Agricultural and Natural Resources Center.

Gassouma, M. S. S. (1991). *Agricultural pests in the United Arab Emirates. Part 1. Fruit trees* (In Arabic) (148 pp). Dubai: Ministry of Agriculture and Fisheries.

Gharib, A. (1970). *Oryctes elegans* Prell. (Coleoptera-Dynastidae). *Applied Entomology and Phytopathology, 29*, 20–29.

Gressitt, J. L. (1953). The coconut rhinoceros beetle (*Oryctes rhinoceros*) with particular reference to the Palau Islands. *Bulletin of the Bernice P Bishop Museum, 212*, 157.

Hurpin, B. (1970). Essais sur l'alimentation des larves d' *Oryctes* aux dépens de végétaux vivants. *Oleagineux, 25*, 657–660.

Hurpin, B., & Fresneau, M. (1969). Contribution a l'étude de *Oryctes elegans* (Col. Dynastidae). *Annales de la Societé Entomologique de France (N.S.), 5*, 595–612.

Hussain, A. A. (1963). Notes on borers of date palms in Iraq. *Bulletin of Entomological Research, 54*, 345–348.

ICARDA. (2011). Integrated control for stem borers of date palm in Iraq. In *Improved livelihoods of small farmers in Iraq through integrated pest management and organic fertilization* (4 p). Available at: https://apps.icarda.org/wsInternet/wsInternet.asmx/DownloadFileToLocal?filePath=Research _publications_archive/Plant_protection_IPM/Stem_Borers.pdf&fileName=Stem_Borers.pdf. Accessed on 17 Apr 2015.

Khalaf, M. Z., & Al-Abid, M. (2013). *Photographical manual for annual regular practices for date palm trees and its efficiency on palm borers* (23 pp). Baghdad: Ministry of Science and Technology.

Khalaf, M. Z., & Al-Taweel, A. A. (2014). *Palm borers in Iraqi environment: species, damages, methods of control* (31 pp). Republic of Iraq: Ministry of Science and Technology, Agricultural Research Directorate.

Khalaf, M. Z., Naher, F. H., & Abdul-Sattar, A. A. (2010). Population density of *Oryctes elegans* Prell (Coleoptera: Scarabaeidae) on some date palm varieties in South Baghdad orchards. *Agriculture and Biology Journal of North America, 1*, 238–242.

Khalaf, M. Z., Shbar, A. K., Al-Seria, M. H., Sami, R. A., & Naher, F. H. (2011). Some aspects of biology and control methods of Fruit Stalk Borer, *Oryctes elegans* Prell (Coleoptera: Scarabaeidae). *Journal of Agricultural Science and Technology, A1*, 142–147.

Khalaf, M. Z., Shbar, A. K., Naher, F. H., Jabo, N. F., Abdulalhamza, B. H., & Sami, R. A. (2012). Activity of insect fauna during the night in palm orchards of central Iraq. *Journal of Food Science and Engineering, 2*, 277–282.

Khalaf, M. Z., Al-Derawie, M. M., Naher, F. H., Al-Rubeae, H. F., & Al-Zaidi, S. (2013a). Palm borers in the Iraqi environment. In: *Abstract book of 6th national conference of environment and natural resources* (26 pp). , Iraq: University of Basrah.

Khalaf, M. Z., Alrubeae, H. F., Al-Taweel, A. A., & Naher, F. H. (2013b). First record of Arabian rhinoceros beetle, *Oryctes agamemnon arabicus* Fairmaire on date palm trees in Iraq. *Agriculture and Biology Journal of North America, 4*, 349–351.

Khalaf, M. Z., Naher, F., Khudair, M. W., Hamood, J. B., & Khalaf, H. S. (2014). Some biological and behavioural aspects of Arabian rhinoceros beetle, *Oryctes agamemnon arabicus* Fairmaire

(Coleoptera: Scarabaeidae: Dynastinae) under Iraqi conditions. *Iraqi Journal of Agricultural Research (Special Issue), 19*, 122–133.

Khoualdia, O., Rhouma, A., Marro, J. P., & Brun, J. (1997). First observations on the date palm pest *Oryctes agamemnon* in Tunisia. *Fruits, 52*, 111–115.

Khudhair, M. W., Aboud, H. M., Khalaf, M. Z., & Shbar, A. K. (2014a). *Oryctes elegans*, a vector of *Fusarium proliferatum* causing wilt disease symptoms of date palm. *International Journal of Phytopathology, 3*(2), 67–72.

Khudhair, M. W., Al Rubeai, H. F., Khalaf, M. Z., Shbar, A. K., Hamad, B. S., & Khalaf, H. S. (2014b). Occurrence and distribution of entomopathogenic fungi in Iraqi agro-ecosystems. *International Journal of Entomological Research, 2*, 117–124.

Kinawy, M. M. (2004). Biological control of the coconut palm rhinoceros beetle (*Oryctes rhinoceros* L. Coleoptera: Scarabaeidae) by using *Rhabdionvirus oryctes* Hüger in Sultanate of Oman. *Egyptian Journal of Biological Pest Control, 14*, 113–118.

Latifian, M., & Rad, B. (2012). Pathogenicity of the entomopathogenic fungi *Beauveria bassiana* (Balsamo) Vuillemin, *Beauveria brongniartii* Saccardo and *Metarhizium anisopliae* Metsch. to adult *Oryctes elegans* Prell, and effects on feeding and mortality. *International Journal of Agriculture and Crop Sciences, 4*, 1026–1032.

Latifian, M., Rahnama, A. A., & Sharifnezhad, H. (2012). Effects of planting pattern on major date palm pests and diseases injury severity. *International Journal of Agriculture and Crop Sciences, 4*, 1443–1451.

Mohammadpour, K. (2002). *Investigation on development of control methods of Date palm fruit stalk borer, Oryctes elegans Prell., by semiochemicals* (81 pp). MSc thesis, Faculty of Agriculture, Ahvaz Shahid Chamran University, Iran. (In Persian with English Summary)

Mohammadpour, K., & Avand-Faghih, A. (2008). Investigation on the possibility of co-mass trapping of the populations of red palm weevil, *Rhynchophorus ferrugineus* and date palm fruit stalk borer *Oryctes elegans* using pheromone traps. *Applied Entomology and Phytopathology, 75*, 39–53.

Oehlschlager, C. (2007). Optimizing trapping of palm weevils and beetles. *Acta Horticulturae, 736*, 347–368.

Payandeh, A., & Dehghan, A. (2010). Demography of date palm fruit stalk borer, *Oryctes elegans* (Col.: Scarabaeidae), on date palm under laboratory conditions. *Plant Protection Journal, 2*, 255–263.

Ragoussis, V., Giannikopoulos, A., Skoka, E., & Grivas, P. (2007). Efficient synthesis of (+/-)-4-methyloctanoic acid, aggregation pheromone of rhinoceros beetles of the genus *Oryctes* (Coleoptera: Dynastidae: Scarabaeidae). *Journal of Agricultural and Food Chemistry, 55*, 5050–5052.

Ramle, M., Norman, K., & Mohd, B. W. (2011). Trap for the auto dissemination of *Metarhizium anisopliae* in the management of rhinoceros beetle *Oryctes rhinoceros*. *Journal of Oil Palm Research, 23*, 1011–1017.

Ratcliffe, B. C., & Ahmed, Z. (2010). Additions to the distribution of Scarabaeidae (Insecta: Coleoptera) in northern Pakistan. *Pakistan Journal of Zoology, 42*, 827–830.

Ritcher, P. O. (1966). *White grubs and their allies* (219 pp). Corvallis: Oregon State University Press.

Rochat, D., Mohammadpour, K., Malosse, C., Avand-Faghih, A., Lettere, M., Beauhaire, J., Morin, J. P., Pezier, A., Renou, M., & Abdollahi, G. A. (2004). Male aggregation pheromone of date palm fruit stalk borer *Oryctes elegans*. *Journal of Chemical Ecology, 30*, 387–407.

Saïd, I., Hasni, N., Abdallah, Z., Couzi, P., Ouhichi, M., Renou, M., & Rochat, D. (2015). Identification of the aggregation pheromone of the date palm root borer *Oryctes agamemnon*. *Journal of Chemical Ecology, 41*, 446–457.

Soltani, R. (2004). *Oryctes agamemnon Burmeister (Coleoptera: Scarabaeidae): Contribution à l'étude de sa biologie et des dégâts occasionnés dans les oasis du Djérid au Sud Ouest Tunisien* (81 pp). Diplôme d'Etudes Approfondies. Ecole Supérieure d'Horticulture et d'Elevage de Chott Marien, Tunisie.

Soltani, R. (2009). *Oryctes agamemnon arabicus Fairmaire (1896): Etude bioécologique et éthologique dans les oasis de Rjim Maâtoug au sud ouest Tunisien* (181 pp). Thèse de Doctorat, Institut des Sciences Agronomique, Chott Mariem, Tunisie.

Soltani, R. (2010). The rhinoceros beetle *Oryctes agamemnon arabicus* in Tunisia: Current challenge and future management perspectives. *Tunisian Journal of Plant Protection, 5*, 179–194.

Soltani, R. (2012). Laboratory rearing of immature stages of *Oryctes agamemnon arabicus* under three constant temperatures. *Tunisian Journal of Plant Protection, 7*, 35–43.

Soltani, R., Chaieb, I., & Ben Hamouda, M. (2008a). The life cycle of the root borer, *Oryctes agamemnon*, under laboratory conditions. *Journal of Insect Science, 8*, 61.

Soltani, R., Ikbel, C., & Ben Hamouda, M. (2008b). Descriptive study of damage caused by the rhinoceros beetle, *Oryctes agamemnon*, and its influence on date palm oases of Rjim Maatoug, Tunisia. *Journal of Insect Science, 8*, 57.

Sultanov, R. M., Ismagilov, R. R., & Dzhemilev, U. M. (2013). One-step synthesis of racemic 4-methyloctanoic acid, a component of the aggregation pheromone of *Oryctes rhinoceros. Chemistry of Natural Compounds, 48*, 1122–1123.

Tabrizian, M., Mohammadpour, K., & Tabak, S. N. (2009). Synthesis and field evaluation of aggregation pheromone of date palm fruit stalk borer *Oryctes elegans. Applied Entomology and Phytopathology* (Pesticide Special Issue No. 87), 51–60.

Vuts, J., Imrei, Z., Birkett, M. A., Pickett, J. A., Woodcock, C. M., & Tóth, M. (2014). Semiochemistry of the Scarabaeoidea. *Journal of Chemical Ecology, 40*, 190–210.

Chapter 6
Carob Moth, Lesser Date Moth, and Raisin Moth

Thomas M. Perring, Hamadttu A.F. El-Shafie, and Waqas Wakil

Abstract Lepidopteran pest species attacking date palms and fruits are problematic in stored dates or in the date garden. Three such species, the carob moth, lesser date moth, and raisin moth, have biologies that are timed with the seasonal occurrence of date fruit. Thus they have become primary pests for the date producer and have been the focus of scientific research toward the development of sustainable management. This chapter provides an in-depth coverage of these three insects, with particular emphasis on their distribution, natural history, and management in date gardens.

6.1 Introduction

The date palm, *Phoenix dactylifera* L., has a well-defined process of fruit development from the inflorescence located deep within the crown through five fruit stages (Reuveni 1986). These stages are known as the "Hababouk," the small green fruit stage that appears after pollination; "Kimri," the period during which the fruit expands in size, retaining a hard green morphology; "Khalal," when the fruit grows to maximum size and weight, with the color gradually changing to yellow, purplish-pink, or red; "Rutab," a period when the fruit loses water, becoming soft, sweet, light brown in color and less astringent; and "Tamar," when the fruit gains maximum sweetness and becomes wrinkled in shape and dark brown in color (Hussain 1974; Nixon and Carpenter 1978; Barrevald 1993; Zaid and De Wet 1999; Ashraf and Hamidi-Esfahani 2011).

T.M. Perring (✉)
Department of Entomology, University of California, Riverside, CA 92521, USA
e-mail: thomas.perring@ucr.edu

H.A.F. El-Shafie
Date Palm Research Center of Excellence, King Faisal University,
Hofuf, Al-Ahsa, Saudi Arabia
e-mail: elshafie62@yahoo.com

W. Wakil
Department of Entomology, University of Agriculture, Faisalabad, Punjab, Pakistan
e-mail: waqaswakeel@hotmail.com; arid1972@yahoo.com

W. Wakil et al. (eds.), *Sustainable Pest Management in Date Palm: Current Status and Emerging Challenges*, Sustainability in Plant and Crop Protection,
DOI 10.1007/978-3-319-24397-9_6

Depending on variety, dates are harvested at the last three stages of development. It is also during these last stages that they are infested with lepidopteran insects. Blumberg (2008) listed four lepidopteran pest species that attack dates in Israel, and Carpenter and Elmer (1978) and El-Shafie (2012) listed nine moth species that are pests of dates. Of these species, three are major pests in the date garden, the carob moth, raisin moth, and lesser date moth.

6.2 Carob Moth (*Ectomyelois* = *Apomyelois ceratoniae*)

The carob moth, *Ectomyelois* (*Apomyelois*) *ceratoniae* (Zeller) (family: Pyralidae) (Fig. 6.1) is a highly polyphagous insect with a broad distribution throughout the world. It infests many agronomic as well as non economic hosts, reaching severe pest status in many commodities. Often it infests the crop in the field during the ripening stages and remains in infested fruit through the post harvest processing, becoming a pest of the stored product. This is the nature of its damage in dates, particularly in varieties that are allowed to remain in the field to dry prior to harvest. While *E. ceratoniae* will infest dates that are harvested during the soft-ripe stage, the most serious problems are found in dates harvested later in the season. This is the case in the United States where the two major varieties that are grown are Medjool and Deglet Noor. Medjool dates typically are harvested as soft-ripe dates in August and September, while Deglet Noor dates typically are harvested in October and November, although some fields may be harvested in December depending on fruit ripening, labor, and market conditions. The seasonal biology of *E. ceratoniae* in this region and the environmental conditions synchronize to produce the largest density of adult moths from late August through mid-October. This peak occurrence of carob moth timed with maximum host abundance in date gardens makes it a pest of dates, in many parts of the world.

Fig. 6.1 Adult female carob moth, *Ectomyelois ceratoniae* (Photo: Justin E. Nay)

6.2.1 Systematics and Taxonomy

The carob moth also is known as the locust bean moth (Goater 1986), the carob bean moth (Gonzalez and Cepeda 1999), the pomegranate fruit moth (Krasil'Nikova 1964; Moawad 1979), and the pomegranate fruit worm (Al-Jamali 2006).

Ectomyelois ceratoniae originally was described by Zeller (1839) as *Myelois ceratoniae* Zeller. Since that time there have been a number of species described that have been synonymized. These include *Phycis ceratoniella* Fischer von Röslerstamm (1839), *Trachonitis pryerella* Vaughan (1870), *Euzophera zellerella* Sorhagen (1881), *Myelois tuerckheimiella* Sorhagen (1881), *Phycita dentilinella* Hampson (1896), *Hypsipyla psarella* Hampson (1903), *Heterographis rivulalis* Warren and Rothschild (1905), *Myelois oporedestella* Dyar (1911), *Myelois phoenicis* Durrant (1915) (Pintureau and Daumal 1979), *Spectrobates artonoma* Meyrick (1935), and *Laodamia duradi* Lucas (1950).

Corbet and Tams (1943) and Real (1948) also list *Myelois ceratoniella* Fischer von Röslerstamm (1839), *Myelois pryerella* Vaughan (1870), and *Myelois zellerella* Sorhagen (1881), and Aitken (1963) lists *Myelois pryerella* (Vaughan 1870). These three species were not confirmed in the original literature, so it is possible that these authors were referring to *Phycis ceratoniella*, *Trachonitis pryerella*, and *Euzophera zellerella*, respectively.

Following the synonymization of the various species under *Myelois ceratoniae*, this species has been placed in several genera. Heinrich (1956) described two new genera *Apomyelois* and *Ectomyelois*, and he placed *M. ceratoniae* in the genus *Ectomyelois*. Roesler (1968) made *Ectomyelois* and *Apomyelois* junior synonyms to *Spectrobates*, a genus proposed by Meyrick in 1935 as "probably allied with *Myelois*" (Meyrick 1935). Klimesch (1968) referred to the species as *Spectrobates* (*Myelois*) *ceratoniae*. Neunzig (1979) described the larvae of *S. ceratoniae* and discussed the biology, distribution, hosts, and parasitoids of this species. *Spectrobates ceratoniae* existed for a little over a decade, when Roesler and Küppers (1981) treated *Ectomyelois* as a junior synonym of *Apomyelois*.

Several prominent systematists did not adopt the placement of *ceratoniae* in either *Spectrobates* or *Apomyelois* (Munroe 1983; Goater 1986; Palm 1986; Sinev 1986), and Neunzig (1990) provided evidence for separating the taxa *Ectomyelois*, *Spectrobates*, and *Apomyelois*. Furthermore, he placed *ceratoniae* back in the genus *Ectomyelois*. Meyer et al. (1997) followed the conventions of Roesler and Küppers (1981) and listed *Apomyelois ceratoniae* in their collections from the Azores but noted that Shaffer (1995) continued to hold the placement of *ceratoniae* in *Ectomyelois*. Leraut (2002) also upheld the synonymization of *Ectomyelois* with *Apomyelois*, and *Apomyelois ceratoniae* Zeller is listed as the "provisionally accepted name" by The Species 2000, ITIS Catalogue of Life (2011), and the LepIndex (Beccaloni et al. 2011). This is also the name found in the London Natural History Museum Global Lepidoptera Names Index (2014) current website and the LepIntercept website (Gilligan and Passoa 2014). The Entomological Society of America Common Names of Insects Database (2014) refers to the carob moth as

Ectomyelois ceratoniae (Zeller), and this binomen will be used for the rest of this chapter. To be thorough, researchers interested in studies on this insect would be wise to search the literature for *Spectrobates ceratoniae*, *Ectomyelois ceratoniae*, and *Apomyelois ceratoniae*.

6.2.2 Distribution and Host Range

The carob moth has a cosmopolitan distribution. Heinrich (1956) states that this moth is "apparently of Mediterranean origin," and a review of the literature shows the carob moth has become established in Afghanistan (Harvey 2013), Albania (de Jong 2013), Algeria (Durrant 1915; Trabut 1923; Widiez 1932; Jacobs 1933), Antilles (De Stefani 1919), Argentina (Lange 1991; Gonzalez and Cepeda 1999), Armenia (Arutyunyan 1988), Australia (Michael 1968), Austria (Zeller 1839), Azores (Meyer 1996; Meyer et al. 1997), Belgium (de Jong 2013), Bulgaria (de Jong 2013), Canary Islands (Baez 1998), Corsica (de Jong 2013), Crete (de Jong 2013), Croatia (de Jong 2013), the Czech Republic (Sumpich and Skyva 2008), Chile (Gonzalez and Cepeda 1999; Gonzalez 2002), Cyprus (Wilkinson 1925; Jacobs 1933; Anonymous 1938), Egypt (Shafik 1938), England (Vaughan 1870; Jacobs 1933), France (Real 1948), Greece (Georgevits 1981), Hungary (de Jong 2013), India (Hampson 1903), Iran (Farzaneh 1984; Shakeri 1993; Ehteshami et al. 2013; Taki et al. 2014), Iraq (Al-Maliky and Al-Izzi 1986), Israel (Avidov and Gothilf 1960; Wysoki 1977, 1986; Gothilf 1984; Halperin 1986b; Navarro et al. 1986), Italy (Real 1948; Deseo et al. 1981), Jamaica (McFarlane 1962), Japan (Sonda 1963; Yoshiyasu and Kitatsuji 2008), Libya (Bitaw et al. 1988), Luxembourg (de Jong 2013), Macedonia (de Jong 2013), Madagascar (De Stefani 1919), Malta (de Jong 2013), Morocco (Trabut 1923; Bouka et al. 2001), the Netherlands (Huisman 1974), Poland (de Jong 2013), Portugal (Alfken 1928), Puerto Rico (Leonard 1933), Romania (de Jong 2013), Russia (Krasil'Nikova 1966), Sardinia (de Jong 2013), Saudi Arabia (Adham 1965), Sicily (De Stefani 1919; Mineo 1967; Schiliro and Bellini 1978), South Africa (De Stefani 1919; Lounsbury 1919; Anonymous 1926; Jacobs 1933; Catling 1970), South America (Real 1948), Spain (Sorhagen 1881; Agenjo 1959), Switzerland (de Jong 2013), Syria (Real 1948), Taiwan (Hung et al. 2003), Trinidad (CDFA 1983), Tunisia (Trabut 1923), Turkey (Tokmakogiu et al. 1967; Gençer et al. 2005; Ozturk et al. 2005), Turkmenistan (Krasil'Nikova 1965), Ukraine (de Jong 2013), and the United States, (Dyar 1911; CDFA 1983), including Hawaii (Swezey 1923).

The carob moth also was picked up in a shipment of cork sent from Portugal to Germany, migrating to the cork from adjoining broken cases of figs that also originated in Portugal (Alfken 1928). Similarly it was found in shelled almonds that were shipped from Spain to Canada (Sheppard 1926). There are no further reports from Germany or Canada; thus we assume it to be absent in those countries.

The carob moth has been reported on 43 hosts from 18 plant families (Table 6.1). Twenty-one of these plants are of economic importance as agricultural commodities

Table 6.1 Plants reported to be hosts of *Ectomyelois ceratoniae*

Family name	Scientific name	Common name	Classification	Reference
Anacardiaceae				
	Pistacia vera L.	Pistachio nut	Agricultural	Halperin (1986b), Mehrnejad (1993, 1995, 2001)
Apocynaceae				
	Carissa macrocarpa (Ecklon) de Candolle. Synonym of *Carissa grandiflora*	Natal plum	Agricultural	Heinrich (1956)
Arecaceae				
	Livistona chinensis (Jacquin) Brown ex Martius	Chinese Fan Palm	Ornamental	Heinrich (1956)
	Phoenix dactylifera L.	Date palm	Agricultural	De Stefani (1919), Bodenheimer (1930), Jacobs (1933), Heinrich (1956), CDFA (1983)
Euphorbiaceae				
	Ricinus communis L.	Castor bean	Agricultural	Zimmerman (1958)
Fabaceae				
	Acacia pinetorum Hermann. Synonym of *Vachellia (= Acacia) farnesiana* (L.) Wight and Arnott *and Vachellia insularis* Small	Needle bush (weed), pineland Acacia, pineland wattle	Ornamental	Bodenheimer (1930), Real (1948), Heinrich (1956), Zimmerman (1958), Palmony (1969)
	Albizia saman Mueller. Synonym of *Samanea saman*	Saman, rain tree, monkeypod Peanut Flame of the forest	Ornamental	Zimmerman (1958)
	Arachis hypogaea L.	Peanut	Agricultural	Sheppard (1926)
	Butea monosperma (Lamarck) Taubert. Synonym of *Erythrina monosperma* Lamarck	Flame of the Forest	Ornamental	Swezey (1923), Zimmerman (1958)

(continued)

Table 6.1 (continued)

Family name	Scientific name	Common name	Classification	Reference
	Caesalpinia sappan L.	Sappanwood	Ornamental, Medicinal	Zimmerman (1958)
	Ceratonia siliqua L.	Carob Tree, locust bean	Agricultural, Ornamental	Wilkinson (1925), Bodenheimer (1930), Jacobs (1933), Anonymous (1938), Real (1948), Heinrich (1956), Zimmerman (1958), Gothilf (1970)
	Erythrina variegata L. Synonym of *Erythrina indica* Lamarck	Coral Tree	Ornamental	Warner (1988)
	Gleditsia triacanthos L.	Honey Locust	Ornamental	Doumandji (1981)
	Haematoxylon campechianum L.	Logwood	Ornamental, Dyes	Zimmerman (1958)
	Parkinsonia florida (Benth. ex Grey) S. Wats.	Blue Palo Verde	Ornamental	Nay and Perring (2008b)
	Prosopis juliflora (Swartz) de Candolle	Mesquite	Ornamental	Zimmerman (1958)
	Retama raetam (Forsk) Webb	Juniper bush	Ornamental	Doumandji (1981)
	Robinia sp.	Species not stated	Ornamental	Lounsbury (1919), Heinrich (1956)
	Robinia pseudoacacia L.	Black Locust	Ornamental	Georgevits (1981)
	Senna bicapsularis (L.) Roxburgh. Synonym of *Cassia bicapsularis*	Cassia, rambling Senna, Christmas Bush	Ornamental, Spice	Heinrich (1956), Zimmerman (1958)
	Tamarindus indica L.	Tamarind	Agricultural	Leonard (1933), Heinrich (1956)
	Vachellia (Acacia) caven (Molina) Seigler and Ebinger	Roman Cassie	Ornamental	Warner (1988)
Fagaceae				
	Castanea sativa Miller	Chestnut	Agricultural, Ornamental	Lounsbury (1919)
Juglandacea				
	Juglans regia L.	Walnut	Agricultural	Lounsbury (1919), Heinrich (1956), Zaviezo et al. (2007)

Lythraceae				
	Punica granatum L.	Pomegranate	Agricultural	Lounsbury (1919), Bodenheimer (1930), Adham (1965), Krasil'Nikova (1965), Al-Izzi et al. (1985), Ozturk et al. (2005)
Malvaceae				
	Brachychiton acerifolius (Cunningham ex. Don) Macarthur and Moore. Synonym of *Sterculia acerifolia* Cunningham	Illawarra Flame Tree	Ornamental	Doumandji (1981)
	Hildegardia populifolia (Roxburgh) Schott and Endlicher. Synonym of *Sterculia populifolia* Cunningham	Common name unknown	Ornamental	Doumandji (1981)
Moraceae				
	Ficus carica L.	Fig	Agricultural	De Stefani (1919), Trabut (1923), Alfken (1928), Widiez (1932), Heinrich (1956), Shakeri (1993), Gencer et al. (2005), Mozaffarian et al. (2008)
Oleaceae				
	Olea europaea L.	Olive	Agricultural	Mineo (1967)
Proteaceae				
	Macadamia integrifolia	Macadamia nut	Agricultural	Wysoki (1977), DeVilliers and Wolmarans (1980), Wysoki (1986)
Rhamnaceae				
	Ziziphus spina-christi	Jujube	Ornamental	Bodenheimer (1930)
Rosaceae				
	Cydonia oblonga Miller	Quince	Agricultural, Ornamental	De Stefani (1919), Jacobs (1933), Krasil'Nikova (1965)
	Eriobotrya japonica (Thunberg) Lindley	Loquat	Agricultural	Sheppard (1926), Heinrich (1956)

(continued)

Table 6.1 (continued)

Family name	Scientific name	Common name	Classification	Reference
	Malus domestica Borkhausen	Apple	Agricultural	Krasil'Nikova (1965), Palmony (1969), Gonzalez (2002), Zaviezo et al. (2007)
	Prunus armeniaca L.	Apricot	Agricultural	Warner (1988)
	Prunus dulcis (Miller) Webb	Almond	Agricultural	Sheppard (1926), Bodenheimer (1930), Jacobs (1933), Michael (1968), Calderon et al. (1969), Gothilf (1984), Navarro et al. 1986
	Prunus persica (L.) Stokes	Peach	Agricultural	Warner (1988), Yoshiyasu and Kitatsuji (2008)
	Pyrus sp.	Pear	Agricultural	Krasil'Nikova (1965)
Rutaceae				
	Varied species	Citrus	Agricultural	Anonymous (1926), Agenjo (1959), Avidov and Gothilf (1960), Tokmakogiu et al. (1967), Catling (1970), Sternlicht (1971), Schiliro and Bellini (1978), Ozturk et al. (2011)
Solanaceae				
	Solanum tuberosum L.	Potato	Agricultural	Kashkuli and Eghtedar (1976)
Vitaceae				
	Vitis sp.	Raisin	Agricultural	De Stefani (1919)
	Vitis vinifera L.	Grape	Agricultural	Martinez-Sanudo et al. (2013)
Zygophyllaceae				
	Balanites aegyptiaca	Soap berry tree, soap berry bush, desert date	Agricultural, Ornamental	Singh (1991)

(pistachio, natal plum, date palm, castor bean, peanut, tamarind, walnut, pomegranate, fig, olive, macadamia nut, loquat, apple, apricot, almond, peach, pear, citrus, potato, raisin grapes, and wine grapes), 15 are ornamental plants (Chinese fan palm, needle bush, saman, flame of the forest, coral tree, honey locust, blue palo verde, mesquite, juniper bush, *Robinia*, black locust, cassia, Illawarra flame tree, *Hildegardia populifolia*, and jujube), and 7 can be considered having multiple uses (sappanwood, carob, logwood, cassia, chestnut, quince, and soap berry tree). The vast majority of hosts are tree species, and the families with the most species attacked are the Fabaceae and the Rosaceae. The majority of damage caused by the *E. ceratoniae* is due to infestation of the fruit of these trees.

The broad host range of the carob moth suggests that it is an opportunistic feeder, able to obtain sufficient nutrients for growth and reproduction from a variety of food sources. Indeed, it has even been found feeding on non-plant hosts. De Stefani (1919), in his report of insects infesting carob pods, noted that "Professor De Joannis of Paris states that the larva has been found feeding on dried insects." Similarly, Cock (1987) reported on "unusual" observation of *E. ceratoniae* larvae feeding upon the pupae within cocoons of the white slug caterpillar, *Parasa philepida* Holloway. These cocoons are made of silk, but they are "toughened until it has the texture of fibre-board by a brown varnish-like liquid produced by the larva" (Cock 1987). There was no doubt that *E. ceratoniae* was feeding on the pupae, because frass had accumulated within the cocoon. The author could not confirm if they were acting as primary predators or scavengers. However, he noted that the polyphagous nature of *E. ceratoniae*, and the fact that it feeds on plant seeds with tough outer casings (like carob bean pods and almonds), make, it less surprising that it would feed on tough *P. philepida* cocoons.

Ectomyelois ceratoniae has been observed feeding on the carcass of apache cicadas, *Diceroprocta apache* Davis, collected in the Coachella Valley (Riverside County), California, USA (Perring, unpublished data). Similar to the *P. philepida* finding, it could not be confirmed whether this was a primary predator or if the carob moth was scavenging from the already dead carcass of the cicada. The omnivorous habit of *E. ceratoniae*, and the fact that it feeds on a number of economically important plant species, has generated a wealth of research studies on its biology, ecology, and management. There also has been considerable work on rearing this species on artificial diets (Levinson and Gothilf 1966; Gothilf 1968; Al-Izzi et al. 1987, 1988; Al-Izzi and Al-Maliky 1996; Hung et al. 2003; Ghavami 2006).

6.2.3 Biology, Ecology, and Natural History

Carob moth eggs (Fig. 6.2a) are ovoid in shape, measuring 0.75 mm long and 0.5 mm wide (Blumberg 2008). When the eggs are laid, they are opaque white in color and the eggs that are fertilized turn pinkish red as the embryos develop (Gothilf 1968; Navarro et al. 1986; Alrubeai 1987). Gothilf (1969c), Cox (1976), Alrubeai (1987), Al-Izzi et al. (1987), and Warner (1988) reported that the egg stage lasts for

Fig. 6.2 Carob moth, *Ectomyelois ceratoniae* life stages; (**a**) egg and first instar larva, (**b**) late instar larva in date fruit with copious amounts of frass, (**c**) silk "cap" at the calyx end of date fruit, (**d**) pupa displayed by opening silken chamber (Photos: Justin E. Nay)

an average of 3–4 days at 30 °C, with a lower temperature threshold for development of 15 °C (Warner 1988). On dates, the eggs are laid individually on the outer surface, and Warner (1988) noted that they are laid in wrinkles, openings, and tears, as well as on and under the calyx. Nay (2006) found eggs deep within a fruit that had abscised from the strand, an indication of the ovipositional behavior of females to "hide" their eggs.

The larvae of carob moth (Fig. 6.2a, b) typically are pink in color and can reach 15 mm in length (Blumberg 2008). They have a brown head and small, dark brown bumps on the dorsum (Abo-El-Saad and El-Shafie 2013). Upon hatching, larvae enter the date fruit under the calyx, and Nay (2006) observed that larvae crawl into the space between the seed (pit) and the mesocarp. He further noted that most often there was only a single larvae in a fruit, supporting the suggestion by Carpenter and Elmer (1978) and Al-Izzi et al. (1987) that the larvae are cannibalistic. There are five larval instars and the larva spin copious amounts of silk inside the date. Nay (2006) noted that they use this webbing to attach the fruit of abscised dates to the date strand at the calyx end (Fig. 6.2c). The silk also is used to construct a pupation chamber and a tube that leads from the chamber to the outside of the date. The length of time required for larval development is quite variable, lasting from 1 to 8 months (Carpenter and Elmer 1978).

Carob moths pupate within a silken pupation chamber (Fig. 6.2d). The obtect pupae[1] are light to dark brown in color, measuring about 10 mm in length. Duration of pupation likely depends more on temperature than the larval stages, since pupae do not feed. Thus host quality and moisture content should not directly impact pupal developmental time. However, the food resources obtained during the larval stage would have an indirect impact on pupal development.

Blumberg (2008) reported that the adult body is 8–10 mm in length and Carpenter and Elmer (1978) reported a wingspan of 22–24 mm. The adults vary in color from whitish to gray to brown (Carpenter and Elmer 1978), with two stripes across the wing (Abo-El-Saad and El-Shafie 2013). However, there is substantial morphological variation in the adult, depending on geographic origin (Mozaffarian et al. 2007c) and host plant. Mozaffarian et al. (2007a) found that the wings were smaller on moths from pomegranate compared to moths from fig, pistachio, and walnut. Regardless of host, Mozaffarian et al. (2007b) found the wings of the females were larger than those of the males.

Upon eclosion, adults crawl through the silken emergence tube to the outer surface of the dates, where their wings expand and dry. Gothilf (1969c) noted that adults emerged during the early hours of scotophase,[2] with males preceding females. Similar data were presented by Hung et al. (2003) and Aldini et al. (2010) who reported that females and males began to eclose just before the scotophase with peak emergence 1 h after scotophase began. The adult life span is from 2 to 15 days (Moawad 1979; Gothilf 1969c, 1984; Navarro et al. 1986; Al-Izzi et al. 1987; Alrubeai 1987). Flight, calling behavior, and ovipositional biology are directly linked to light conditions and temperature. Most studies have determined that female moths call during the last half of scotophase (Vetter et al. 1997; Hung et al. 2003; Soofbat et al. 2007; Aldini et al. 2010), while Sarjami et al. (2009) and Aldini et al. (2010) reported that in one population, most females called in the first half of scotophase. Interestingly, Sarjami et al. (2009) reported that moths reared at 30 °C started calling at a significantly later time than those reared at 20 or 25 °C. More eggs were laid during the first hour than any succeeding hour (Vetter et al. 1997; Hung et al. 2003). Carob moths raised in constant light do not oviposit viable eggs, confirming that mating takes place in the dark (Cox 1976; Nay 2006).

Gonzalez (2003) found that adults did not fly when the temperature was below 14 °C. Nay (2006) observed that at temperatures below 14.9 °C, adults did not fly, but they did mate successfully at temperatures as low as 13.7 °C. This was also the lowest temperature at which females were observed in a calling posture. Nay (2006) postulated that since the lower temperature threshold for mating was below the lower flight temperature, that males could walk to females and mating could still occur at this low temperature. Females have been reported to lay from 100 to 300 eggs during their life span, but most reported values lie within 100 and 200 eggs per female (Gothilf 1968, 1969c; Navarro et al. 1986; Al-Izzi et al. 1987).

[1] These are immobile chrysalis with all appendages tightly fastened to the body wall.

[2] Scotophase is the dark phase of the light-dark daily cycle.

It is thought that carob moth adults do not lay eggs on healthy, non-damaged, non-abscised dates. This belief comes from reports on other crops like carob pods (Gothilf 1970; Dobie 1978) and pistachio (Halperin 1986b; Mehrnejad and Panahi 2006). Halperin (1986b) noted that moths oviposited through emergence holes of the pistachio seed chalcid, *Megastigmus pistaciae* (Walker), Gothilf et al. (1975) found that *E. ceratoniae* females preferred to oviposit on carobs infested with fungi, and Serghiou (1983) reported that carob moth females were attracted to citrus fruit by the honeydew of feeding citrus mealybugs, *Planococcus citri* (Risso).

The biology and life cycle of the carob moth has been studied extensively (see review in Nay and Perring 2006). Other pertinent studies not contained in Nay and Perring (2006) give additional information (Hung et al. 2003; Yousefi et al. 2004; Basirat and Mehrnejad 2005; Norouzi et al. 2008; Aldini et al. 2010). It is clear from these studies that tremendous variability exists in the developmental biology and fitness of *E. ceratoniae* that depends on abiotic (temperature, humidity, and photoperiod) and biotic (host type and quality of the host) conditions in which the insect develops. For example, the lower temperature threshold for development varies from 9.4 °C (Basirat and Mehrnejad 2005) to 10.8 °C (Mart and Kilincer 1993b), to 11.8 °C (Yousefi et al. 2004) and 12.5 °C (Warner 1988).

Similarly, there is variation in the literature for the number of degree days required for *E. ceratoniae* to complete development. Basirat and Mehrnejad (2005) reported 769° days (DD) on pistachio, and 624 DD (Mart and Kilincer 1993b) and 707.2 DD were reported on pomegranate (Yousefi et al. 2004). On dates, the DD varies with fruit stage. Warner (1988) calculated 463, 569, and 749 DD for June drop, Rutab, and Tamar dates, respectively, and Nay and Perring (2008a) reported DD of 636, 649, and 658 for insects reared on Kimri, Khalal, and Tamar dates, respectively. It stands to reason, then, that the number of generations per year also depends on host and environmental conditions. As few as two generations per year to as many as five have been reported. Ashman (1968) found two generations for moths developing on carob pods in Cyprus, Blumberg (2008) reported three to four generations per year on dates in Israel, Wertheimer (1958) and Lepigre (1963) noted four generations on dates in North Africa, Mart and Kilincer (1993a) reported four generations on pomegranate in Turkey, and five generations were reported on pomegranates in Iraq by Al-Izzi et al. (1985) and by Warner (1988) and Nay and Perring (2008a) on dates in the United States. Regardless of location, the density of moths increases through the season as the availability of suitable hosts increases, and Calcat (1959) observed that the last two of four generations developed on date fruit in the bunches.

Nay and Perring (2006) concluded that under certain environmental conditions, the carob moth did not have typical temperature-dependent development, and host quality was known to be extremely important (Gothilf 1969c, 1984; Moawad 1979; Navarro et al. 1986; Al-Izzi et al. 1987; Alrubeai 1987). Nay and Perring (2006) found that the fruit moisture content had a significant impact on larval mortality and that moisture below 3.2 % was the lower threshold for larval development. Developmental rate also was influenced by host moisture content and development ranged from 49.3 days at 21.8 % fruit moisture content to 81 days at 5 % fruit moisture content. The maximum development occurred at 19.6 and 20.4 % fruit moisture content for females and males, respectively. In addition, the average number of eggs laid by females was linearly related to the fruit moisture at which they were reared as larvae. Additional studies by Nay and Perring (2008a) found that carob moths had the highest adult weight when reared on Kimri fruit and the lowest when reared on Tamar fruit, and more eggs were laid by females reared on Kimri fruit than by those reared on Tamar fruit.

The seasonal biology of *E. ceratoniae* is closely linked with the developmental stages of the date fruit (Nay and Perring 2008a). Carob moth does not feed on nonfruiting structures of dates, but they will infest the fruit of other hosts, if available (see host range earlier in this chapter). The larvae overwinter (November to February) in all instars, in date fruit on the ground, in the axils of the date palms, or in bunches that were left on the palms after the previous season (Wertheimer 1958; Gothilf 1970, 1984; Al-Izzi et al. 1985; Warner 1988; Nay 2006). These "waste" dates can be heavily infested with carob moth larvae from the previous season, and estimates as high as 50 % have been reported (Park and Perring 2010). Some authors believe that overwintering larvae are in true diapause, and Gothilf (1970) and Cox (1979) noted that this diapause was in response to low temperatures and short day length. However, others have found that larvae are active during the winter months (Warner 1988; Nay and Perring 2008a). As the season progresses and overwinter dates become dryer and suboptimal as a food resource, larval development can take up to 60 % (females) and 63 % (males) longer than when the dates are moist (Nay and Perring 2006).

Moths that emerge from the overwinter hosts will oviposit on waste dates, but they prefer more optimal hosts. New fruits become available in April and May, but undamaged, healthy dates that are attached to the strands are not infested (Warner et al. 1990b). Thus, the first stage that is available for infestation in the new crop is the green Kimri fruit. In production date gardens, where flowers are pollinated leading to large fruit sets, there is a natural abscission process, where 20–40 % of Kimri fruit abscised from the bunch (Kehat et al. 1976; Nixon and Carpenter 1978; Warner 1988). Some of these fall to the ground, a process called "June drop," because this occurs during June (Warner et al. 1990a; Nay et al. 2006). Many of the dates are stuck in the bunch. Nay et al. (2007) estimated this number between 34,000 and 205,000 dates per 0.4 ha. These abscised dates get stuck in bunches and they are heavily utilized by ovipositing carob moths. Larvae develop in these Kimri dates and Nay and Perring (2008a) determined that carob moths weighed more when reared on Kimri varieties. Tamar fruit and that females laid the most eggs when they developed as larvae on Kimri fruit. Adults emerge from Kimri fruit and oviposited on abscised Khalal fruit. There is another abscission of Khalal fruits during late July and August, and again fruit fall to the ground (termed August drop) or get stuck in the bunch. Larvae then develop in the Khalal fruit and development is rapid during the warmest months of the year.

In August and September, fruit in the Rutab stage loses moisture and there can be a slight loosening of the fruit from the strand (Warner et al. 1990b). Adult carob moths lay eggs on the wrinkled Rutab fruit, and larvae enter the fruit at the calyx end.

As the final fruit stage, the Tamar fruits become infested during September and they usually remain on the palm through November, depending on when harvest crews pick the crop. During harvest, the infested Tamar dates fall to the ground or in the axils of fronds; larvae in these dates overwinter and emerge to start the seasonal cycle the following year.

6.2.4 *Damage and Losses*

Carob moth damage to the fruit is caused by the developing larvae. First and second instar larvae are small, and fruits that are infested in the field can be fumigated during post harvest processing; these small larvae are difficult to detect in the finished date products. However, post-second instar larvae consume large portions of the date fruit. The feeding is accompanied by large quantities of frass (Fig. 6.2b).

The carob moth is known to cause significant economic losses to date producers. In the United States, dates are produced in the Coachella Valley (Riverside County, California, USA), and recent plantings of dates have taken place in the Bard/Winterhaven area of Imperial County, California, and near Yuma, Arizona, USA. Dates were harvested from 2468 ha in 2013, producing a crop worth US $38.4 million (USDA 2014). In this geographic region, the carob moth is the most important pest and fruit infested with larvae can cause damage reaching 10–40 % to the harvestable crop each year (Warner et al. 1990a; Nay et al. 2006). Bouka et al. (2001) reported 30 % yield losses in Morocco.

6.2.5 *Sampling and Economic Threshold*

Despite the fact that carob moth has been a pest in dates for a long time, there are no published economic thresholds. This has been a challenging problem since there are few studies that have demonstrated reliable sampling techniques that estimate pest density upon which mitigation measures can be implemented. Park and Perring (2010) developed a binomial sampling plan which inspects 150 abscised brown Khalal date fruits, collected from 15 trees after the June drop begins. Yet the relationship between the immature infestation rate from the sampled Khalal dates and the final carob moth infestation at the end of the season has yet to be determined, and future work in this area is essential before an economic threshold can be developed.

Important for this research is a method for estimating infestation. Studies conducted by Perring et al. (2005) evaluated the relationship between the number of fruit with visible webbing at the calyx end (Fig. 6.2c) and the total number of fruit with carob moth larvae. For this research, 100 dates were collected each from 18 bunches, and the presence of webbing was determined for each date. Fruits then were cut open to assess the actual infestation. The number of webbed fruits was regressed on the total number of infested fruits, and 96.6 % of the variation in

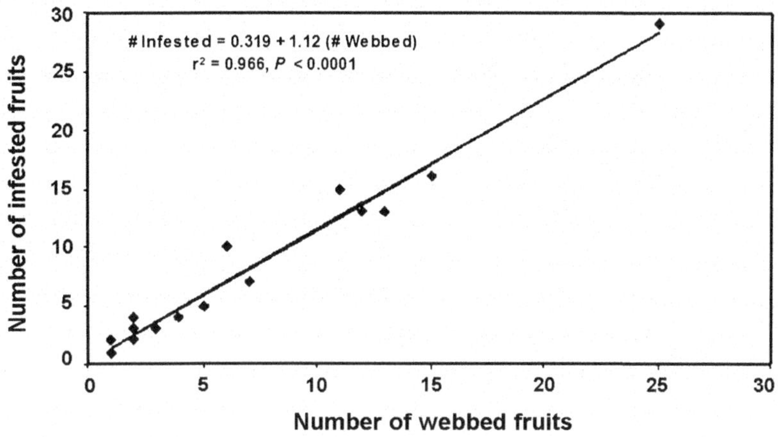

Fig. 6.3 Relationship between the number of date fruits with webbing at the calyx end and the number of fruits with a carob moth larvae (Perring et al. 2005)

infested fruits was explained by the webbed fruits (Fig. 6.3). This excellent relationship can be used to rapidly assess the infestation of dates in lieu of cutting the dates open. Another area of promise is sampling carob moth adult males with traps containing lures of female pheromone mimics (www.iscatech.com). This would require research to establish the relationship between male catches, at various times of the year, and larval infestation at harvest.

6.2.6 *Management*

While there are no true integrated pest management (IPM) programs for carob moth on dates, scientists have studied a variety of control strategies, some of which have been adopted by producers. These fall under the broad categories of cultural or physical control, mass trapping, host plant resistance, biological or chemical control, mating disruption, and sterile insect techniques.

6.2.6.1 Cultural Control

Cultural control strategies can be highly effective in reducing the overall densities of carob moths in dates. Some of these techniques are applied during off season, when there are no date fruits on the palms. Since carob moths overwinter in dates that fall to the ground in the previous harvest, producers can remove these waste dates on the floor of the garden (Carpenter and Elmer 1978). In the United States, this is done by discing dates between tree rows and cross discing between trees, when possible. This strategy also works for removing June and August drop dates

(Warner et al. 1990a, b). Kimri and Khalal dates that are stuck in bunches during this time are preferred oviposition sites for carob moth (Nay and Perring 2005). Thus any strategy to remove these dates from the bunch results in fewer insects. Nay et al. (2006) showed that up to 94 % of the abscised dates were dislodged from bunches with a cleaning tool, resulting in 99 % reduction in carob moth-infested fruit. When the dates dropped to the garden floor, they were exposed to high temperatures. Larvae quickly exited these dates and became prey to ants foraging the garden (Nay and Perring 2005). Over the course of the season, the one-time use of the bunch cleaning tool resulted in a 69.9 % reduction in infested fruit (Nay et al. 2007). Another method of facilitating the drop of abscised dates is to cut the center strands from the bunch (Nay and Perring 2009). This process reduced carob moth densities at harvest by up to 81 %.

6.2.6.2 Physical Control

Methods for excluding carob moth from the date bunches can be effective in reducing infestations. In the field, this is accomplished by placing mesh bags around the date bunches (Fig. 6.4). Bouka et al. (2001) determined that this practice not only reduced insect attacks, but it also prevented damage by birds, limited fruit loss due to winds, and facilitated harvest. Zirari and Laaziza Ichir (2010) found that bags with a mesh size of 1.7 mm and smaller resulted in less than 0.5 % infestation of *E. ceratoniae*.

Fig. 6.4 Mesh bags placed around date bunches to exclude *Ectomyelois* ceratoniae. Coachella Valley, California, USA (Photo: Thomas M. Perring)

Compared to non-treated controls, these bags had just 11 % of the infestation. Recent work has determined an 89 % decrease in carob moth densities when closed mesh bags were compared to open paper wraps (Perring and Nay 2015). In addition, there was a 46 % net increase in the weight of marketable fruit at the end of the season.

6.2.6.3 Mass Trapping

During the formulation and development stages of the Specialized Pheromone and Lure Application Technology (SPLAT EC-O®, Mafra-Neto et al. 2013), there were attempts to attract male moths to traps for mass trapping or attract-and-kill. These trials were not effective at reducing the density of infested fruits, and the density at which traps needed to be placed was cost prohibitive. These strategies were abandoned as the effectiveness of SPLAT EC-O® became more apparent. However, earlier studies were conducted for this purpose and they are worth mentioning here. Mansour (1984) mixed various substances (egg albumen, ammonium carbonate, milk, and certain petroleum oils) with pomegranate fruit, finding that the egg albumen was most attractive to *E. ceratoniae*. Cosse et al. (1994) identified the volatile compounds from fungus-infested dates as potential attractants. They found ethyl hexanoate as the dominant stimulant for attracting mated female carob moths. More recently, Al-Jamali (2006) evaluated the use of virgin females in traps. While attractive, this would be labor intensive to implement in a management program.

6.2.6.4 Host Plant Resistance

There has been very little research done on host plant resistance to *E. ceratoniae*, and the majority comes from pomegranate (Zand et al. 2013; Zare et al. 2013; Mamay et al. 2014). However, several studies in dates found infestation rates varying from 57% to 2 % (Idder et al. 2009) and 4.6 % to 1.5 % (Zirari and Laaziza Ichir 2010) depending on date variety. Interestingly, these authors noted that the size of the date fruits was directly correlated with the size of carob moth adults that emerged from those fruit. Another date study by Bouka et al. (2001) determined that one variety with a thin epidermis was most heavily infested.

It makes sense that different nutritive qualities and physical properties of the fruit of individual date varieties would have an impact on carob moth fitness, offering a rich area of research for carob moth control. But carob moth genetics may confound the success of plant resistance and tolerance. Mozaffarian et al. (2008) determined that *E. ceratoniae* maintained high within population genetic variation and gene flow between populations from different geographic locations and different host plants (pomegranate, fig, pistachio, and walnut). Thus, while certain varieties of dates could hold some promise in an IPM program, the carob moth may be able to adapt readily to any new varieties that are developed.

6.2.6.5 Biological Control

Parasitoids and Predators

There have been numerous reports of natural enemies associated with *E. ceratoniae* (Table 6.2). Although these parasitoids and predators attack pest eggs, larvae, and pupae, there has been little success in the implementation of these natural enemies in an integrated management program. This likely is due to the prophylactic use of insecticides, coupled with the limited effectiveness of the natural enemies. Of the parasitoids listed in Table 6.2, the highest rates of parasitism reported were for *Trichogramma cordubensis* Vargas and Cabello (64 %) (Idder et al. 2013), *Phanerotoma flavitestacea* Fischer (50 %) (Doumandji-Mitiche and Doumandji 1982), and *Trichogramma embryophagum* Hartig (19 %) (Doumandji-Mitiche and Doumandji 1982).

It is also clear that the host plant is important in the success of biological control agents against *E. ceratoniae*. Gothilf (1978) released *Pentalitomastix plethoricus* Caltagirone into almonds and carob plantations, and after several years, the parasites were recovered in the almonds but not in carobs. *Trichogramma embryophagum* (Hartig) reared on carob moth eggs collected from apples, Malus pumila, were more than twice as effective (45 % compared to 19 %) as parasitoids reared on moth eggs collected from other plants (Doumandji-Mitiche and Idder 1986). In addition, different parasitoid species may be more effective in different areas of the date garden. Bouka et al. (2001) found higher parasitism in date bunches with

Table 6.2 Natural enemies associated with *Ectomyelois ceratoniae*

Order and family	Scientific name	Life stage attacked	References
Acari: Pyemotidae	*Pyemotes ventricosus* Newport	Larvae	Gothilf (1969a)
Acari: Ascidae	*Blattisocius tarsalis* (Berlese)	Egg	Bitaw et al. (1988)
Diptera: Tachinidae	*Clausicella suturata* Rondani	Larvae	Gothilf (1969a), Kugler and Nitzan (1977), and Farahani et al. (2009)
	Fischeria bicolor Robineau-Desvoidy	Larvae	Farahani et al. (2009; 2010c)
Hymenoptera: Bethylidae	*Goniozus gallicola* Fouts	Egg	Neunzig (1979)
	Goniozus legneri Gordh	Egg	Gothilf and Mazor (1987), Zaviezo et al. (2007), and Ehteshami et al. (2013)
	Perisierola gallicola Kieffer	Larvae	Gothilf (1969a)
	Perisierola emigrata Rohwer	Larvae	Bridwell (1919a), Thompson (1946), and Zimmerman (1958)

(continued)

Table 6.2 (continued)

Order and family	Scientific name	Life stage attacked	References
Hymenoptera: Braconidae	*Apanteles angaleti* (Muesebeck)	Larvae	Al-Maliky and Al-Izzi (1990) and Al-Izzi et al. (1992)
	Apanteles lacteus (Nees)	Larvae	Aubert (1966), Gothilf (1969a), and Halperin (1986a)
	Apanteles laspeyresiellus Papp	Larvae	Norouzi et al. (2009)
	Apanteles myeloenta Wilkinson	Larvae	Wilkinson (1937), Gothilf (1969a), Nixon (1976), and Farahani et al. (2010c, 2012a, 2012c, 2013)
	Apanteles sp. *group ultor*	Larvae-pupae	Al-Maliky and Al-Izzi (1986) and Al-Maliky et al. (1988)
	Ascogaster sp.	Pupae	Al-Maliky and Al-Izzi (1986)
	Bracon mellitor Say	Larvae	Bridwell (1919b) and Thompson (1946)
	Chelenus sp.	Larvae	Farahani et al. (2010c)
	Habrobracon brevicornis Wesmael	Larvae	Thompson (1946), Lepigre (1963) and Gothilf (1969a)
	Habrobracon (Bracon) hebetor Say	Larvae	Gothilf (1969a), Al-Maliky and Al-Izzi (1986), Dhouibi and Jemmazi (1993), Bouka et al. (2001), Hameed et al. (2011) and Saadat et al. (2014)
	Microbracon pembertoni Bridwell	Larvae	Bridwell (1919b) and Zimmerman (1958)
	Phanerotoma dentata Panzer	Larvae	Thompson (1946), Lepigre (1963), and Aubert (1966)
	Phanerotoma flavitestacae Fischer	Egg-larvae	Gothilf (1969a, 1969b), Madkouri (1978), Biliotti and Daumal (1970), and Daumal et al. (1973)
	Phanerotoma leucobasis Kriechbaumer	Larva	Gothilf (1969b), Mesbah et al. (1998), and Bouka et al. (2001)
	Phanerotoma ocuralis Kohl	Larvae	Khoualdia et al. (1996), and Bouka et al. (2001)
	Phanerotoma sp.		Thompson (1946), and Al-Maliky and Al-Izzi (1986)
	Rhogas testaceus (Spinola)	Larvae	Gothilf (1969a)

(continued)

Table 6.2 (continued)

Order and family	Scientific name	Life stage attacked	References
Hymenoptera: Chalcididae	*Antrocephalus mitys* (Walker)	Larvae	Gothilf (1969a)
	Brachymeris aegyptiaca Masi	Pupae	Gothilf (1969a) and Al-Maliky and Al-Izzi (1986)
	Brachymeria ceratoniae Delvare	Pupae	Delvare et al. (2011)
	Brachymeris sp.	Pupae	Al-Maliky and Al-Izzi (1986)
	Proconura persica Delvare	Pupae	Delvare et al. (2011)
	Psilochalcis ceratoniae Delvare	Pupae	Delvare et al. (2011)
Hymenoptera: Encyrtidae	*Copidosomopsis* (*Pentalitomastix*) *plethoricus* Caltagirone	Egg-larvae	Gothilf (1978) and Gothilf and Mazor (1987)
Hymenoptera: Formicidae	*Pogonomyrmex californicus* (Buckley)	Larvae	Nay and Perring (2005)
	Solenopsis aurea Wheeler	Larvae	Nay and Perring (2005)
Hymenoptera: Ichneumonidae	*Campoplex tumidulus* Gravenhorst	Larvae	Farahani et al. (2010b)
	Diadegma (*Horogenes*) sp.	Larvae	Gothilf (1969a) and Gothilf and Mazor (1987)
	Diadegma oranginator Aubert		Aubert (1964) and Aubert et al. (1984)
	Gelis sp. (probably a hyperparasite)	Hyperparasitoid	Gothilf (1969a)
	Herpestomus arridens Gravenhorst	Larvae	Aubert (1966), Gothilf (1969a) and Aubert et al. (1984)
	Pristomerus vulnerator Panzer	Larvae	Gothilf (1969a)
	Nemeritis canescens Gravenhorst	Larvae	Lepigre (1963) and Al-Maliky and Al-Izzi (1986)
	Temelucha decorata Gravenhorst	Larvae	Farahani et al. (2010b)
	Venturia canescens Gravenhorst	Larvae	Aubert et al. (1984) and Farahani et al. (2012b)
Hymenoptera: Perilampidae	*Perilampus tristis* Mayr	Hyperparasitoid	Gothilf (1969a) and Al-Maliky and Al-Izzi (1986)
	Perilampus sp.	Larvae	Bitaw et al. (1988) and Bitaw and Saad (1990)
Hymenoptera: Pteromalidae	*Anisopteromalus mollis* Ruschka	Larvae	Gothilf (1969a)
	Pachycrepoideus vindemmiae Rondani	Hyperparasitoid	Farahani et al. (2010a)

(continued)

Table 6.2 (continued)

Order and family	Scientific name	Life stage attacked	References
Hymenoptera: Tricho-grammatidae	*Trichogramma bourarachae* Pintureau & Babault	Egg	Ksentini et al. (2010b, 2013)
	Trichogramma brassicae Bezdenko	Egg	Moezipour and Shojaei (2008), Moezipour et al. (2008), and Moezzipour et al. 2009)
	Trichogramma cacoeciae Marchal	Egg	Ksentini et al. (2010b, 2013)
	Trichogramma cordubensis Vargas & Cabello	Egg	Idder et al. (2013)
	Trichogramma embryophagum Hartig	Egg	Mirkarimi (2000), Doumandji-Mitiche and Doumandji (1982), Doumandji-Mitiche and Idder (1986), and Karami et al. (2011)
	Trichogramma evanescens Westwood	Egg	Ksentini et al. (2010b, 2013)
	Trichogramma oleae (Voegele & Pointel)	Egg	Ksentini et al. (2010b, 2013)
	Trichogramma sp.	Egg	Gothilf (1969a) and Poorjavad et al. (2011)

Phanerotoma ocularis Kohl, while *Habrobracon*(= *Bracon*) *hebetor* Say was more effective in dates on the garden floor.

Some work has been done using natural enemies in conjunction with other strategies. For example, Hameed et al. (2011) found that dates in storage were better protected from carob moth when *Bracon hebetor* Say was used in conjunction with pheromone traps. Ksentini et al. (2010a) determined that *Bacillus thuringiensis* (Berliner) application was harmless to *Trichogramma cacoeciae* Marchal, *Trichogramma bourarachae* Pintureau & Babault, and *Trichogramma evanescens* Westwood collected from carob moths on pomegranates in Tunisia, so it could be used along with the parasitoids.

Another example involves ant predation on carob moth larvae. Nay and Perring (2005) observed that two California, USA, native ant species, the fire ant *Solenopsis aurea* Wheeler and the California harvester ant *Pogonomyrmex californicus* (Buckley), foraged the date garden in search for prey. They found that larvae inside abscised date fruit that landed on the hot desert floor quickly emerged from the fruit and were preyed upon by the ants. They further developed a bunch cleaning tool which effectively dislodged abscised fruit from date bunches, resulting in up to 99 % reduction in carob moth-infested fruit (Nay et al. 2006).

With the diversity of natural enemies that have been found associated with *E. ceratoniae*, there is a reason to believe that biological control could play an important part in an integrated control program. Since the carob moth has a large number of host plants, many of which are of agricultural importance and thus under insecticide-heavy management programs, the likelihood of a classical biological strategy is slim. However, conservation of existing natural enemies, combined with augmentative releases properly timed with the carob moth life cycle and other management options, may be useful in reducing overall moth densities.

Bacillus thuringiensis and Other Pathogens

A few studies evaluated the effectiveness of *B. thuringiensis* (Bt) for *E. ceratoniae* control. This work shows high mortality in laboratory conditions that often fails to translate to efficacy in the field. In the laboratory, Alrubeai (1988) found that newly enclosed larvae were the most susceptible life stage, but more recent work found that the fourth instar larvae were more sensitive to the bacteria than second instars (Elsayed and Bazaid 2011). Harpaz and Wysok (1984) documented high mortality (95 %) of fourth instar larvae in lab tests using a high rate of bacterial cells, but found that mortality was much lower at rates that would be practical for field use. The best field efficacy was 82 % control, achieved after four applications in pomegranates (Alrubeai 1988).

Aytas et al. (1996) demonstrated 53 % control with *B. thuringiensis* subspecies *kurstaki* (Delfin) in navel oranges. On dates, Dhouibi (1992) evaluated *B. thuringiensis* subsp. *kurstaki* (Bactospeine) in the field and found little impact on the densities of *E. ceratoniae* larvae. Related work evaluated the insecticidal activity of the biosurfactant of *Bacillus subtilis* (Ehrenberg) Cohn and found this to be effective for *E. ceratoniae* control (Mnif et al. 2013). Further work found the optimal medium for growing the biosurfactant, which may show promise for carob moth control.

Finally, there is one report of a microsporidian (microsporidia) found in the Malpighian tubules of *E. ceratoniae* (Lange 1991). These authors found a 48 % infection in larvae, pupae, and adults in moths collected from walnuts. However, there was no further work to develop this pathogen as a biological control agent.

6.2.6.6 Reproductive Control

Mating Disruption

One of the most promising strategies for *E. ceratoniae* management is the use of mating disruption using a pheromone mimic. The female sex pheromone was identified by Baker et al. (1989, 1991), and the major component responsible for its attractiveness to males was found to be (Z,E)-9,11,13-tetradecatrienal, which was synthesized by Millar (1990). It was found to be highly labile, and field studies determined that synthetic blends using this component were not attractive to males.

Fig. 6.5 Applying SPLAT EC-O® with a hand applicator to the base of date palm (left); close-up of a 2.5 g dollop of SPLAT EC-O® (right), Coachella Valley, California, USA (Photos: Thomas M. Perring)

Thus, Todd et al. (1992) synthesized the more stable analogue (Z,E)-7,9,11-dodecatrienyl formate and demonstrated its effectiveness in field trapping studies. This formate was formulated for controlled release by ISCA Technologies, Inc., Riverside, California, USA (www.iscatech.com), using Specialized Pheromone and Lure Application Technology (SPLAT®), an approach that has been evaluated in numerous field trials (Vetter et al. 2006; Perring unpublished data). Results from these studies have consistently shown the inability of males to locate traps with pheromone-mimic lures, i.e., trap shutdown (Mafra-Neto et al. 2013). Furthermore, the carob moth infestation rates at harvest are similar to those obtained after using synthetic pesticides. With the single application of SPLAT® (Fig. 6.5) compared to multiple applications of pesticides, the cost of using SPLAT® is equal to the cost of other pesticides. The pheromone mimic has been registered as an organic use product in dates by the US Environmental Protection Agency (EPA) and is currently marketed as SPLAT EC-O® (Mafra-Neto et al. 2013; Perring et al. 2014).

Sterile Insect Technique

The use of radiation to sterilize males has been successful in several pest systems. For a number of years, this has been an interesting area of research for the carob moth as a pest in pomegranate. Early work determined the impact of various levels of gamma radiation on males of different ages (Al-Izzi et al. 1990). Subsequent research showed that gamma radiation could be used to successfully sterilize males, which when mated to normal females resulted in reduced progeny (Al-Izzi et al.

1993). This impact of lower egg viability was passed on to successive generations, suggesting that continual release of sterile males would result in declining densities through time. More recently, Zolfagharieh et al. (2011) determined that fewer eggs with lower survival rates were produced in crossings between irradiated males and females. In addition, they determined the best radiation dosage was 120 Gy and 160 Gy for 1–2-day-old pupae and 3–4-day-old pupae, respectively. Other studies indicate that irradiated males responded well to female pheromones and dispersed at the same level as non irradiated ones (Mediouni and Dhouibi 2007).

6.2.6.7 Chemical Control

Carob moth infests dates in the field, during post harvest processing and during storage. Most storage facilities rely on temperature-controlled chambers for heating or cooling dates that may be infested with larvae, or fumigants can be used to kill eggs, larvae, and pupae in infested dates. Early work showed that a constant temperature of 60 °C for 1.5 h killed all stages of the carob moth and did not harm the dates (Shafik 1938). Most recently the use of fumigants was studied. The most widely used fumigant is methyl bromide (Lepigre 1963; Nixon and Carpenter 1978; Abo-El-Saad and El-Shafie 2013), but this compound was phased out in 2005 under the "Montreal Protocol on Substances that Deplete the Ozone Layer" and the United States Clean Air Act. A search for alternatives is underway with several promising results.

One of the substitute materials is phosphine. In fumigation chambers at different temperatures, Mills et al. (2003a) found that 10, 7, 5, and 3 days at 15, 20, 25, and 30 °C, respectively, were effective exposure periods. There was no side effect on the quality of the dates in any of those trials. Another study evaluated carbon dioxide as a methyl bromide alternative (Mills et al. 2003b). At 30 °C, an exposure period of 4 days killed carob moth eggs and larvae. They also reported that using a vacuum did not improve control of the moths and at atmospheric pressure the dates did not absorb any CO_2.

Bessi et al. (2013) noted that treatment by ethyl formate fumigant resulted in 91.6 % mortality and had no negative impact on date fruit quality; the fumigant residue dissipated in 7 days. Plant essential oils also were evaluated as control agents. Ben Jemaa et al. (2012) studied the essential oils of five *Eucalyptus* species and found seasonal variations in the effectiveness of oils collected throughout the year. They found that for all species, the plant-based fumigant activity was greatest when leaves were collected in the summer, and the most effective oils were from *Eucalyptus rudis* (Endlicher). Further work by these authors found that the effectiveness of *Eucalyptus* oils depended on the availability of empty spaces surrounding the dates in the storage facility (Ben Jemaa et al. 2013). They found that the more dates in the facility, the less effective the fumigation. A recent study evaluated the impact of NaClO, UV-C germicidal radiation, ozonated water (O_3), alkaline electrolyzed water (AEW), and neutral electrolyzed water (NEW) to sanitize dates which then were stored for 30 days at 20 °C (Jemni et al. 2014). All these materials decreased moth densities in storage units, with the most effective being UV-C radiation and NEW with a pH 7.2, oxidation reduction potential of 814 mV, and 300 mg L^{-1} of free chlorine.

While post harvest control strategies tend to be very good, growers still must contend with insect parts and bodies in the fruit. Therefore it is important to reduce the density of carob moth in harvested dates by controlling them in the field. The primary method of controlling carob moth remains the use of synthetic chemicals, applied largely on a prophylactic basis, but insecticides are not particularly efficient due to the fact that larvae feed and develop inside the date fruit (Mediouni and Dhouibi 2007). Nevertheless, a large number of materials from most chemical classifications have been evaluated in the past (e.g., Lepigre 1963; Vincent and Lindgren 1954; Warner et al. 1990b; Blumberg 2008). In the United States, there are three materials registered for carob moth on dates. These are Delegate (spirodiclofen), Intrepid (methoxyfenozide), and malathion applied in a dust formulation (Fig. 6.6). Warner et al. (1990b) determined that treatments of malathion dust should begin just before the Rutab stage (soft and juicy dates), and earlier treatments did not improve control. This is an important result applicable to other chemicals, since the carob moth life cycle is aligned with infesting the Rutab stage. Malathion dust was the industry standard until 2010, when environmental concerns and the emergence of

Fig. 6.6 Applying malathion dust to dates for *Ectomyelois ceratoniae* control. Coachella Valley, California, USA circa 2009 (Photo: Thomas M. Perring)

new chemicals and technologies resulted in its phase out. Efficacy of insecticides may be improved by modifying sprayers to deliver ultra low volumes of atomized spray (Johnstone et al. 1972). Also, producers and researchers should be thoughtful about overuse of chemicals, since resistance has been documented in E. ceratoniae (Blumberg 2008; Ould et al. 2007).

There has been some work done with biorational materials, most notably spinosad (Khoualdia et al. 2002; Perring et al. 2014), azadirachtin (Khoualdia et al. 2002; Mehaoua et al. 2013), and various essential oils of *Pistacia lentiscus* L. (Bachrouch et al. 2010) and *Ferula assafoetida* L. (Peyrovi et al. 2011; Goldansaz et al. 2012) which act as toxicants, repellents, and mating disruptants. Growers must be careful to use low concentrations of plant oils to prevent burning the fruit.

6.2.6.8 Other Tactics

Often new ideas arise out of research conducted on other commodities. This may be the case with *E. ceratoniae* on pomegranate, a serious pest of this crop in Iran. Several studies have been done to alter the site of female oviposition by removing the stamens of the pomegranate fruit. These studies have shown a reduction in infestation of the fruit (Karami et al. 2011; Taki et al. 2014). Another idea was putting various materials in the "neck" of pomegranates (Mirkarimi 2002). This study, which evaluated sawdust, garden clay, and cotton lint, determined that all of the materials reduced carob moth incidence. However, using clay was the most effective, reducing the density of larvae to one third of the non-treated control. This resulted in higher fruit weight as well (Mirkarimi 2002).

6.3 Lesser Date Moth (*Batrachedra amydraula*)

The lesser date moth, *Batrachedra amydraula* Meyrick (family: Batrachedridae), is a serious pest of the date palm. Its biology and life history is such that it causes severe yield losses, particularly in young developing date fruit. The occurrence of the lesser date moth coincides with the presence of another lepidopteran insect, *Aphomia* (= *Arenipses*) *sabella* Hampson. This pyralid is larger (20 mm wingspan) than the lesser date moth (15 mm wingspan), and because of its relative size, it is commonly called the greater date moth, while *B. amydraula* is referred to as the lesser date moth. Interestingly, even though *A. sabella* is the larger of the two species and its damage is similar, it is not as severe a pest as the lesser date moth. Another common name for *B. amydraula* is the lesser fruit borer (Michael 1970; Elwan 2000). There are no satisfactory control tactics available for *B. amydraula*, and it is a limiting factor in date production in some countries.

6.3.1 Systematics and Taxonomy

The lesser date moth has many vernacular names such as "Hashaf" (or empty fruit), a name that derives from feeding activities on dates. However, the most commonly used vernacular name is "Humeira," which means "red or red-brown color." Humeira literally means "red" in Arabic, and the name comes from the distinctive color of infested fruits (Fig. 6.7b). Sohn et al. (2009) reviewed the systematic rank of the Batrachedridae, which comprises a group of small slender gelechioid moths, the adults of which usually are less than 15 mm in wingspan.

The Batrachedridae include around 150 species mostly occurring in the tropics, America, and tropical regions of Africa (Hodges 1998). They show morphological affinity to three other families, Momphidae, Cosmopterygidae, and Coleophoridae, that collectively are called microlepidoptera (Priesner 1989). Both forewings and hind wings are narrow with long fringes of hairy scales. Adults of Batrachedridae exhibit a peculiar resting posture in which they are raised anteriorly, while the legs are directed posteriorly. The larvae are without secondary setae, yellowish in color, and very mobile. They live in silken galleries between or inside the reproductive

Fig. 6.7 Damage on immature date fruit by *Batrachedra amydraula*. Larva penetrating young fruit at the calyx end (**a**). Infested fruit (**b**) turned red-brown and fastened with silken thread to the bunch strand. Empty bunch (**c**) with few fruits in strand and "Hashaf"(**d**) or evacuated fruits showing only the reddish-brown skins that have dropped to the ground (Photos: Hassan Al-Fardan, with permission)

organs of various plants including ferns. In addition to B. amydraula on date palm, some species of Batrachedridae are known pests of cultivated plants, e.g., *Batrachedra pinicolella* on Norway spruce (*Picea abies*) (Priesner 1989) and *Batrachedra nuciferae* on coconuts (Arnal et al. 1998).

6.3.2 Distribution and Host Range

The lesser date moth is found throughout the Middle East and North Africa where date palms are cultivated (Blumberg 2008). It is reported from Bahrain (Abdul-Jabar et al. 1982), UAE, Kuwait (Zaid et al. 1999), Sultanate of Oman (Elwan 2000), Qatar (Al-Khatri et al. 2009), Saudi Arabia (Talhouk 1983), Yemen (El-Bashir and El-Makaleh 1983), Iraq (Hussain 1974), Iran (Gharib 1968), Pakistan (Kakar et al. 2010), Israel (Blumberg 1975), Libya (Martin 1958), and Egypt (Michael 1970). The lesser date moth can infest and go through multiple generations in stored dates (Carpenter and Elmer 1978). Thus, it is possible for this pest to be introduced into new regions through anthropogenic activity. The lesser date moth is specific to the date palm and has not been reported on other hosts (Blumberg 1975, 2008).

6.3.3 Biology, Ecology, and Natural History

The adult lesser date moth is a gray-brown insect, with a body length of 8–10 mm. It has a wingspan of 11–14 mm and the forewings are narrow with brown scales and a marked longitudinal gray stripe in the center. The hind wings are narrow light gray in color with long fringes. Females display a yellowish lower segment on the abdomen, while males are distinguished by hairs, and sometimes two pinchers on the tip of their abdomen (Levi-Zada et al. 2013). The eggs are small with a diameter of 0.7 mm, whitish-yellow in color, and they usually are laid singly on newly set fruits and bunch strands. The head and the prothorax of the larva are light brown, while the rest of the body is translucent white. Mature larvae are creamy white to pinkish in color and they can reach a length of 12 mm. Lesser date moth larvae pass through five instars before pupation. Pupae are slender, long, and light brown in color and they are well protected by a white silken cocoon (Dowson 1982; Eitam 2001).

The biology of *B. amydraula* has been studied under laboratory conditions on the main varieties of dates in Iran as well as on a semi-artificial diet consisting of 400 g of dried date powder (variety Ghasb), 400 g of whole-wheat flour, 150 g of honey, 25 g of yeast, and 120 ml of glycerin. For this research, the authors evaluated survival, developmental time, fecundity, oviposition, post-ovipositional period, and longevity of adults (Shayesteh et al. 2010). The mean incubation period ranged from 6.3 to 7.8 days on dates, while it was 5.8 days on the semi-artificial diet. Total developmental time from egg to adult on the semi-artificial diet was 49.09 days, and the longest developmental time was 69 days for males and 65 days for females on the Ghasb date variety. The highest fecundity (45 eggs per female) occurred on the Deyri variety (Shayesteh et al. 2010). Adult males and females lived on the semi-artificial diet for 7.04 and 8.71 days, respectively (Rahmani et al. 2008).

Under field conditions, ecological studies have shown that the lesser date moth has three overlapping generations annually (Blumberg 1975; Venezian and Blumberg 1982; Talhouk 1991). The emergence of adults from the overwintering population coincides with the beginning of flower pollination, and the adult moths feed on the pollen, while females oviposit on the developing fruit clusters and petioles (Kakar et al. 2010).

The first-generation larvae feed on the new Hababouk dates from late March to late April. The second and third generations are present in May and June, respectively. Larval numbers decline during August. After August no larvae were found in the date garden (Blumberg 2008). Thus, lesser date moth larval damage occurs over a short time period (Venezian and Blumberg 1982). Third-generation larvae spin silk cocoons and they overwinter in diapause as full-grown larvae. In the spring of the next season, after diapause, the larvae require no additional feeding and they pupate without causing damage. Pupation takes place in well-hidden sites inside fibers at the bases of fronds, the crowns of the trees, and under dead tissues around the trunk (Dowson 1982; Talhouk 1991). Pupation gives rise to adults of the first generation, the larvae of which are responsible for the damage of newly formed fruits (Blumberg 1975). Initiation of larval diapause in late summer and its termination in early spring is triggered by certain day lengths (Blumberg 1975).

6.3.4 Damage and Losses

First-generation lesser date moth larvae feed on the newly formed inflorescences, causing injury to the Hababouk fruit. Approximately 80 % of the fruit damage occurs during this stage, when the fruits are between 0.6 and 1.0 cm in diameter (Eitam 2001; Blumberg 2008). The larva penetrates the fruit near the calyx and feeds on the internal fruit contents and seed (Fig. 6.7a). The feces of the larvae become attached to the penetration site, providing a visible diagnostic for damaged fruit. Each larva damages three or four fruits during its lifetime, feeding on a portion of each fruit. This feeding behavior increases the amount of damage from lesser date moth.

As the season progresses, second-and third-generation larvae cause damage to the developing fruit in the Kimri stage. Early instar larvae use silk to tie fruit to the strands before severing the attachment between the fruit and the stalk. The larvae then enter the fruit from the calyx end. At this time, larvae feed on fruit and seed evacuating all contents, leaving only the outer skin which dries and turns reddish-brown in color. After about 4 weeks, the infested fruits become dark in color, dry out, and fall from the tree, although some damaged fruits remain attached to the bunch stalks by the silken threads. In severe infestations, bunches appear to have empty strands (Fig. 6.7c) as most of the damaged fruit drops to the ground (Fig. 6.7d).

The larvae continue to feed on decaying tissues, and rot-causing microorganisms that enter the infested fruits seem to help the caterpillar digest the fruit tissues that otherwise would be unpalatable because of the high tannin content (Talhouk 1983).

Fruit losses can range between 50 and 75 % (Michael 1970; Dowson 1982; Elwan 2000) and infestations may cause fruit decay and fermentation, leading to an increase in densities of sap beetle (Nitidulidae) (Blumberg 2008). In addition to damage caused by lesser date moth in the field, it has been reported as a pest of stored dates (Wiltshire 1957; Dowson 1982; Shayesteh et al. 2010).

6.3.5 Management

Field sanitation and mechanical fruit protection can reduce damage. Removal of fallen fruits and pruning of bunches that stay on the palms after harvest also reduce future populations of the moth. Bunch covering with light cloths, mass trapping, and mating disruption techniques can be used as components of an IPM program.

6.3.5.1 Cultural Control

Harhash et al. (2003) evaluated the efficiency of some agricultural practices and date palm varieties on the incidence of *B. amydraula* in El-Beheira Governorate, Egypt. They evaluated covering flowers with paper, covering date bunches with porous plastic cloth, thinning fruit from bunches, and using special metal rings to separate fruit in bunches. All cultural practices tested reduced infestation by the moth and improved date yield; however covering bunches with porous cloth was the best. The lesser date moth lays eggs on the bunch; therefore, preventing oviposition by covering bunches early reduces pest damage. Bagging bunches just after pollination with a light cloth minimizes the number of infested fruit and thus the damage to those fruit. In addition, bagging bunches reduced damage from wasps and birds (Latifian 2012).

Another practice is sanitation of the date garden by removing fruit that fall to the floor after harvest, thereby reducing future pest harborages (Harhash et al. 2003). Also, since overwintering larvae remain on bunches left on the palms, bunch removal reduces the number of overwintering sites, and the number of overwintering larvae, that would become the first-generation adults the following year (Latifian 2012).

6.3.5.2 Trapping and Monitoring

Until recently, densities of the lesser date moth were monitored by inspection of fruits for larval damage. In addition, searches were made for pupae in the axils of fronds or in hiding places on other parts of the palm trunk. Currently, adult densities are estimated using light traps fitted with a 150 Watt mercury bulb (Kakar et al. 2010).

In addition, the identification of two sex pheromone components of B. amydraula provides the first opportunity to monitor and possibly control this pest, with semiochemicals (Levi-Zada et al. 2011). Studies are underway to use this new technology.

Field experiments with pheromone lures and insecticides have shown promising results for the management of the lesser date moth (Kaakeh 2006). The female sex pheromones of the lesser date moth have been identified from laboratory females, with the component being (Z4, Z7)-4,7-decadien-1-yl acetate and Z5-decen-1-yl acetate. The pheromone blend was attractive to male moths and can be used as bait in sticky traps (Levi-Zada et al. 2011). The investigators reported that this was the first pheromone to be identified from a Batrachedridae species. It also was determined that the primary component was a novel molecular structure among sex pheromones in moths.

Revaluation of the discovered sex pheromone indicated that the optimal attractive blend is a three-component mixture of (Z4, Z7)-4,7-decadien-1-yl acetate, Z5-decen-1-yl acetate, and Z5-decen-1-ol in a ratio of 1:2:2. This blend gave a fivefold higher trap catch of B. amydraula males compared to the previously published binary blend of (Z4, Z7)-4,7-decadien-1-yl acetate and Z5-decen-1-yl acetate in a ratio of 1:2 (Levi-Zada et al. 2011). The alcohol Z5-decen-1-ol is an essential synergistic component of the sex pheromone of *B. amydraula*. All other identified compounds are inert, being neither synergists nor inhibitors of the pheromone. For practical fieldwork, large delta sticky traps baited with 1 mg of the ternary blend of (Z4, Z7)-4,7-decadien-1-yl acetate, Z5-decen-1-yl acetate, and Z5-decen-1-ol in a ratio of 1:2:2, hung at 2 m-height, are recommended for monitoring and control of this pest, either by mating disruption or mass trapping (Levi-Zada et al. 2013).

6.3.5.3 Host Plant Resistance

A considerable amount of research has been conducted on the susceptibility of different date varieties to lesser date moth. Percentage of infested fruits and infestation severity were recorded on nine date palm cultivars in Behbahan, Iran, based on non-infested and infested date fruits. The cultivars were classified as susceptible (Khasi, Kabkab, and Mezafati) and non-susceptible (Shahani, Zahidi, Piarom, Khazravi, Gantar, and Haj-Mohammadi) (Khajehzadeh and Latifian 2013). In another study, Harhash et al. (2003) evaluated a number of cultural practices (see Sect. 6.3.5.1) on the date palm varieties, Sammany, Hayany, and Halawy. They determined that the variety Sammany was more resistant to the infestation than the other two varieties. This resistance was attributed to differences in the wax content and the histology of the green fruit between the varieties. The variety Sammany had higher wax content than Hayany or Halawy. In addition, the epidermal layers in Sammany fruit acted as natural barriers against penetration of the larvae, resulting in slower infestations. In another study, the date variety Ghasb was found to be more tolerant to infestation by the lesser date moth than Zahidi and Deyri varieties (Shayesteh et al. 2010). Thus, resistance of date palm cultivars to infestation by the lesser date moth should be considered when designing an IPM program for the management of this pest.

6.3.5.4 Biological Control

Parasitoids and Predators

Currently, there are no parasitoids or predators being used for management of the lesser date moth. Compared with other date palm pests, there are only a few natural enemies that have been reported in the literature. The parasitic wasp *Parasierola swirskiana* Argaman attacks lesser date moth larvae and is found commonly in date palm groves in the Arava Valley in Israel, Jordan, and Afghanistan (Argaman 1991–1992). Augmentative release of *P. swirskiana* on first-generation *B. amydraula* larvae was evaluated as an alternative to chemical treatments (Eitam 2001). Several biological traits of this larval ectoparasitoid, including adult female longevity of more than 70 days, short-generation developmental time (mean of 14 days), laying of several eggs per host (mean of 5), and brood guarding, make it a potential candidate for biological control of the lesser date moth (Eitam 2001).

Another egg parasitoid, *Trichogramma cacoeciae* Marchal, was evaluated but was not effective on its own. However, it did show promise when combined with the fungus *Beauveria bassiana* (Balsamo-Crivelli) Vuillemin (Navon et al. 1999). The larval ectoparasitoid *Bracon* sp. (Hammad et al. 1983; Blumberg et al. 2001) and the ovo-larval endoparasitoid *Phanerotoma flavitestacea* Fischer (Hammad et al. 1983) also attack the lesser date moth. In addition, there are a number of unidentified ants and spiders that were reported to feed upon diapausing larvae of the *B. amydraula* (Talhouk 1991).

Bacillus thuringiensis

Bt is used in organic date gardens to control *B. amydraula* (Navon et al. 1999). These products are used in spray or dust formulations, and the latter is capable of penetrating date bunches to improve larval mortality (Blumberg et al. 2001). This fungus is more effective early in the season, when larva feed on multiple small fruits to complete development. As larvae feed on a single, larger fruit later in the season, they are exposed to less Bt (Blumberg et al. 2001). Several trials using different biopesticides were conducted in Saudi Arabia during two campaigns (2004 and 2005) in order to assess their efficacy against first-generation *B. amydraula*. The following products were applied: *B. thuringiensis kurstaki* (Condor), spinosad (Tracer 480), Carpovirusine (CYD-X), Sunspray 98.8 % (summer oil), and matrine3 (herbal source). Each application was made at the rate of 3–7 l per tree, depending on the number of bunches and tree size at a time corresponding to the oviposition and hatching periods (a few days after fruit set). All tested biopesticides effectively controlled lesser date moth, and the efficacy was improved when sanitation measures were applied in the orchard to remove over wintering larvae in fruit stalks and dry fruit (Habib and Essaadi 2007).

6.3.5.5 Chemical Control

Presently, lesser date moth populations are managed primarily with synthetic insecticides. These are used prophylactically due to the lack of reliable monitoring methods and other control strategies. Chemical control is not effective unless the insecticide is applied early in the season at the time of pollination. This is when first instar larvae, which are susceptible to pesticides, are exposed to contact insecticides applied to the date clusters (Al-Samarraie et al. 1989).

As far as specific chemicals, there has been a considerable amount of research done on this topic. Several insecticides have been evaluated in the past and one of the most effective materials was endosulfan (Thionex) (Blumberg et al. 1977). However, the moth became resistant to endosulfan in the early 1980s and it was replaced with the organophosphate, chlorpyrifos (Venezian and Blumberg 1982). Malathion 50 % EC, diazinon 60 EC, carbaryl 85 % WP, and Actellic® 50 % EC were applied with high-pressure ground application equipment and this significantly increased yields (El-Bashir and EL-Makaleh 1983).

Other trials evaluated a formulation of equal parts of wheat flour and pollen grains, mixed with chlorpyrifos, fenitrothion, or pirimiphos-methyl. These were dusted onto young date clusters at pollination. All insecticides were effective, but fenitrothion and chlorpyrifos were superior (Al-Samarraie et al. 1989). More recently, Trebon® 20 % EC (etofenprox) at the rate of 80 ml/100 L of water and Evisect S® 50 % WP (thiocyclam hydrogen oxalate) at the rate of 150 ml/100 L of water gave more than 92 % and 77 % reduction in fruit infestation, respectively, after 14 days of application (Al-Khatri et al. 2009).

6.4 Raisin Moth (*Cadra = Ephestia figulilella*)

The raisin moth *Cadra* (= *Ephestia*) *figulilella* (Gregson) (family: Pyralidae) is a serious primary insect pest of ripening and ripened dates in the old and new world (Donahaye and Calderon 1964; Carpenter and Elmer 1978). *Cadra figulilella* is a major pest of dates, but it also is found occasionally in dried fruits and nuts, fallen and rotten figs on the ground, and injured or neglected grapes clusters attached to the vines (Khajepour et al. 2011). It inflicts significantly more damage to dates during maturation, which may occur from the very beginning of ripening through harvest (Kehat and Greenberg 1969).

A prolific insect, *C. figulilella* infests fruit not only while it is on the tree but also after it has been picked (Lindgren et al. 1948; Carpenter and Elmer 1978; Warner et al. 1990b). However, there is no evidence that it is able to maintain or increase its numbers in storage conditions (Cox 1974). If date fruits are not fumigated before storage, the larvae continue to feed and develop into adults (Kehat and Greenberg 1969) causing damage to the dates during storage (Kehat et al. 1969).

6.4.1 Distribution and Host Range

The raisin moth first was identified in Fresno County, California (USA), in 1928 on Muscat raisins. It reached high densities in 1930, and since then it has spread to other parts of the world, causing severe quantitative and qualitative losses in dates. *Cadra figulilella* originally spread to tropical areas and then was spread into cooler regions by trade (Donohoe et al. 1949). The movement into cooler areas may be attributed to its photoperiod-induced diapause (Cox 1975a).

The raisin moth is found infesting dates in the field in Australia, Iraq, Iran, Israel, North Africa, and the United States (Cox 1975b). It also is known as a storage pest in Egypt (Carpenter and Elmer 1978), North Africa, and the Middle East (Moore 2001). Furthermore, the pest also is well known in other countries where date palms are grown (Hussain 1963; Strümpel 1969; Michael 1970; Michael and Habib 1971). Khajepour et al. (2011) reported that this pest has a cosmopolitan distribution and causes severe quantitative and qualitative losses throughout the world.

Cadra figulilella is a polyphagous insect, and the larvae feed on a variety of dried and drying fruits, fallen figs, ripe carob pods, grapes in the field, and raisins on drying trays or in storage facilities. Cottonseed cake, cacao beans, prunes, peaches, apricots, pears, figs, and cashew kernels are among the host foods (Donohoe and Barnes 1934). Fallen mulberries are important because they are available in early spring when other hosts are scarce.

6.4.2 Biology, Ecology, and Natural History

The eggs of *C. figulilella* are 0.40×0.33 mm long and ellipsoid to ovoid in shape (Sedlacek et al. 1996) with a star-shaped sculpturing. They are slightly yellow-orange in color, and while they generally are not glued to host material, they are sticky (Burks and Johnson 2012). Eggs are laid singly in small batches on or near their food source, and they hatch in 3–6 days (Anonymous 1999).

The larvae are small, creamy-white in color and they turn greenish-gray as they mature. The body of *C. figulilella* is streaked with six rows of lavender-colored dots running horizontally along the back. The head is reddish-brown, and the peritreme of spiracles, setae, pinacula, and prothoracic and anal plates are reddish-or yellowish-brown (Fig. 6.8). The larval period lasts for 32 days under optimum conditions, and they reach 16 mm at maturity. Larvae molt between four and eight times, but most commonly they molt six times (Anonymous 1999). The larval developmental period is dependent upon temperature; thus it is longer during the winter season (Simmons and Nelson 1975). *C. figulilella* overwinters in the larval stage and they pupate in the spring into a pre pupal stage which lasts for 1 day.

The pupae are yellowish, or reddish-brown with slightly raised spiracles (Fig. 6.8). They have a rounded cremaster with eight hooks. Pupae are enclosed in a silken cocoon, which is spun in hidden places on the palm or in the topsoil. Pupation takes about 10 days. In April, adults begin to emerge and they are most numerous in May (Anonymous 1999).

Fig. 6.8 Developmental stages of *Cadra figulilella*. Larva (left), pupae (center), and adult (right) (Simmons and Nelson 1975; with permission from Judy A. Johnson, USDA-ARS, California, USA)

The adult has a body length of about 1 cm, with a wingspan of about 2 cm. The moth is brownish-gray in color with a brownish-gray forewing. The postmedian and antemedian fasciae are darker and margined with whitish color (Fig. 6.8). The antemedian line is sometimes broad and more distinct with two small spots present on some specimens. The hind wing is shining white, with veins and a terminal line that is pale brown. The antennae are simple filiform (Kehat and Greenberg 1969). The female pheromone glands are located on the ventral aspect of the membrane between the 8th and 9th abdominal segment (Smithwick and Brady 1977a, b). Females call for males only at night and, while calling, they beat their antennae (175 beats$\cdot$ min^{-1}) (Krasnoff et al. 1983). The male moths live for an average of 11 days and females live an average of 16 days. The adults stay in protected areas in the day, becoming active in the early evening (Carpenter and Elmer 1978).

Cadra figulilella females lay an average of 160 eggs on the surface of the date fruit, on the second and third nights after eclosion from the pupal stage. Most eggs are laid in the first few hours of scotophase. The eggs hatch within 4 days of oviposition, and neonate larvae bore into the fruit. This provides entry for secondary organisms to infect the fruit. The combination of direct damage by raisin moths and infection by secondary organisms can result in fruit damage up to 50 % (Blumberg 2008). Before pupation, larvae leave the host and seek suitable harborage in which to pupate. In the date garden, this typically is under the bark near the base of a tree or vine, or in adjacent soil near the surface (Donohoe et al. 1949). This species overwinters in the last larval instar in topsoil or under the loose bark of the host plant (Donohoe et al. 1949), and it is presumed that they are in diapause (Cox 1975a).

Raisin moth larvae can be found year-round in decaying date fruit; thus alternate hosts are not needed. However, ripening dates in bunches are susceptible to pest attack (Kehat et al. 1969). The moth requires 54–65 days for full development at 30 °C and 75 % RH (Kehat and Greenberg 1969), while the temperature threshold of eggs is 15–36 °C with 70 % RH (Cox et al. 1981). This pest has four to five generations per year. While all developmental stages can overwinter in stored fruit, only the larvae overwinter in the date garden (Assari and Khajepour 2013).

Cadra figulilella females release pheromones which attract males for courtship and mating. They have an elaborate courtship and mating behavior (Krasnoff and Vick 1984; Phelan and Baker 1990). The male approaches the female, and as he comes close to the female, he curls his abdomen over its head. When males approach from the front of the female, they establish the head-to-head posture. Just prior to this, the males expose their abdominal hair pencils (Phelan and Baker 1990). Male approaching from the rear causes the female to turn and face him. At this time, the female abdomen is elevated. Males hold their antennae outside the female antennae at a 30–40′ angle from the ground. Males then engage in a head bump, which is followed by a copulatory thrust. Most of these thrusts (75 %) are successful on the first attempt (Phelan and Baker 1990). After successful copulations, females performed drag-walks, which is a postcopulatory behavior in which the female drags the male behind her. Krasnoff and Vick (1984) speculated that attraction to species-specific male pheromone represented a form of sexual selection. Females courted by interspecific males avoided normal courtship behavior (Phelan and Baker 1990).

Courtship can serve as mechanism for sexual selection, and it provides a behavior tool that can be useful in phylogenetic determinations. Raisin moth has attracted attention in this area because of its elaborate courtship sequence. This is important, because many of the other Phycitinae moths share components of their female sex pheromone. This results in cross-attraction between species (Kuwahara et al. 1971; Brady and Daley 1972; Struble and Richards 1983; Phelan and Baker 1986). Thus courtship behavior can be helpful in distinguishing among species.

Raisin moth adults are susceptible to low temperature, and only limited flight activity can be observed at less than 13 °C (Donohoe et al. 1949). Field studies in central California in the 1930s found adult activity from April to November, with peak abundance in September and October (Donohoe et al. 1949). These authors observed three complete flights and a partial fourth, and they attributed this relatively long period of abundance to varied and plentiful host material.

After ripened dates are picked and until new ripening fruits are available, raisin moth infests dropped, decaying fruits in the date garden (Kehat et al. 1969; Kehat and Greenberg 1969). Kehat et al. (1992) used pheromone traps to identify peak male activity from May to June. They speculated this was the generation that came from overwintering larvae and progeny from this generation infested dates on the trees. The second peak in male activity (September to October) occurs during the fruit ripening period (Kehat et al. 1992). Ahmad and Ali (1995) used pheromone traps to calculate the number of degree days (DD) for each generation. Starting on

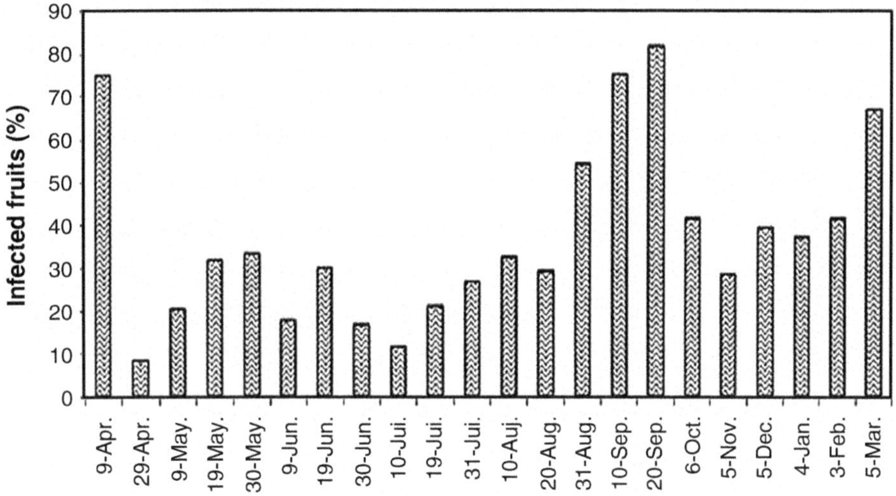

Fig. 6.9 Percentage of *Cadra figulilella* infected date fruits in Shahdad and Chahar Farsakh (Iran) during 2008 (Assari and Khajepour 2013; with permission from Taylor and Francis)

January 1, they determined 348 DD were required for the first generation, while 236 DD were required starting on March 1. Assari and Khajepour (2013) used pheromone traps to estimate the seasonal densities of raisin moth in three Qasab variety date gardens in the Kerman Province of Iran. The average infestation was 40.2 % for the first year and 41.2 % in the second year (Figs. 6.9 and 6.10). Peak densities were present from August through April, which is interesting since this includes the winter months when there are no fruit on the palms. This may reflect the time when there is more moisture in the date garden relative to the summer months. Perring et al. (2005) observed higher numbers of raisin moth in date gardens in California, during wet periods.

6.4.3 Damage and Losses

Fruit damage by *C. figulilella* is caused by the larval stages. Larvae feed internally on the fruit, which becomes filled with frass. This leaves considerable contamination in the form of webbing and frass, and even if the insect is killed by fumigation in post harvest processing, the contamination is still present. Date fruit can be infested as soon as it starts ripening on bunches; fruit in the field are susceptible for 6–8 weeks, until they are harvested (www.agri.huji.ac.il/mepests/ CadraFigulilella.html). Kehat et al. (1992) found only 1.3 % of the fruits were infested by *C. figulilella* early in the fruit ripening process, and it peaked at 11.7 % in mid-September (11.7 %). Assari and Khajepour (2013) reported that up to 90 % of the dates in Shahdad (Kerman Province of Iran) were contaminated by the pest.

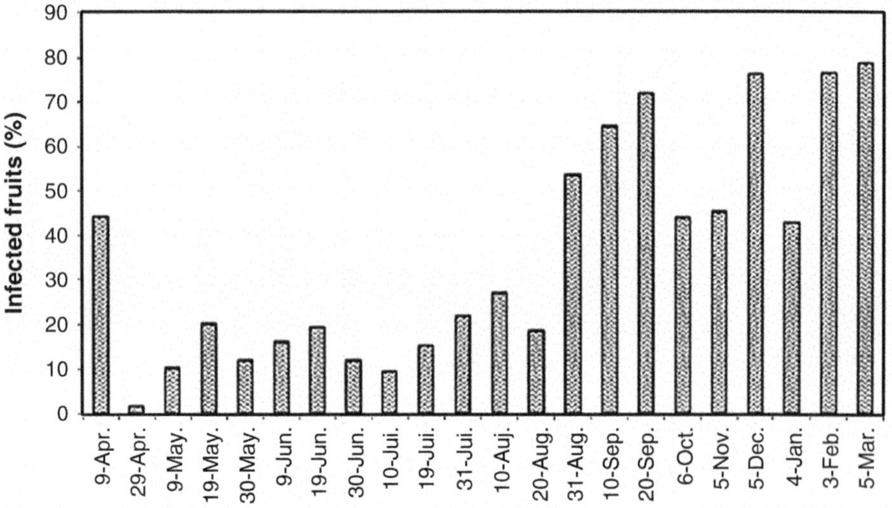

Fig. 6.10 Percentage of *C adra figulilella* infected date fruits in Shahdad and Chahar Farsakh (Iran) during 2009 (Assari and Khajepour 2013; with permission from Taylor and Francis)

6.4.4 Management

6.4.4.1 Cultural Control

Raisin moth is considered a "garbage insect" because it inhabits mummified or spoiling fruit in and around the date garden. A principal strategy for reducing these pest populations is to clean up and destroy infected fruit and other alternate host plants from surrounding areas. This measure should be adopted early in the fall before raisin moth larvae leave the date fruit for their overwintering sites under bark or in the soil (Coviello 2000). Additionally, removing the waste dates from the ground will serve to reduce overwintering raisin moth densities that will mature and infest the crop the following season. In some cases, this goal could be achieved by grazing animals, chiefly donkeys and horses, as they are good grazers of these types of food (Blumberg 2004).

The existence of larvae in rotting dates year-round explains how the species survives from one crop season to another. This highlights the importance of keeping the floor of the date garden clean (Kehat and Swirski 1964). Early harvesting of date fruit also may reduce damage caused by the raisin moth. In Israel, early harvesting of Medjool, Deglet Noor, Hayany, and Barhee varieties was effective in reducing infestation by the raisin moth. This practice effectively shortened the period of time in which ripening fruit was available for moth attack. In this study, early harvesting nearly replaced the need for chemicals and other control practices (Kehat et al. 1969).

6.4.4.2 Physical Control

In date gardens where early fruit harvest is not usual or late maturing varieties are cultivated, covering bunches with plastic nets at the beginning of the ripening season may provide good protection from infestation against raisin moth. The first use of this technology was to protect the ripening dates from vertebrate pests (birds, lizards, etc). Later, they were modified to protect fruit from date moths and sap beetles (Kehat et al. 1969). The plastic netting or wire with 2.0×2.0 mm mesh size was effective at reducing damage by raisin moth (Blumberg 2004; Kehat et al. 1969). The date bunches frequently are covered in mid-August when the fruit starts to change color. Moreover, mesh coverings provide a method for preserving dates that normally would drop to the garden floor prior to harvest (Kehat et al. 1969; Perring and Nay 2015).

6.4.4.3 Trapping and Monitoring

Semiochemical monitoring is an important sampling tool, and it has been used widely to determine the presence of the *C. figulilella*. Additional work needs to be done, since density estimates can be used to develop an economic threshold (Bjostad et al. 1993) in support of IPM programs (Ridgway et al. 1990; Hedin 1991).

For raisin moth management on dates, adult trap catches have been useful (Ridgway et al. 1990) and "traps and lures" methods are commercially available for this purpose (www.pestproducts.com/phemoth.htm). Due to the fact that raisin moth damage occurs only during the 6–8 weeks of fruit ripening, pesticide application early in the infestation is critical, so that larvae can be killed before they move inside the fruit.

Kehat et al. (1992) studied different doses of (Z,E)-9,12-tetradecadienyl acetate (Z9,E12-14:Ac) for raisin moth monitoring. They found no significant differences between male and female captures in traps impregnated with 1 or 10 mg pheromone/dispenser. However, significantly more male captures were recorded in traps baited with 1 mg, compared to 0.1 mg of pheromone per dispenser. No differences were observed between male captures in traps hung near bunches (at a height of 8 m) or on traps hung at a height of 2–3 m on palm trunks, although the 2–3 m locations are recommended (www.agri.huji.ac.il/mepests/ CadraFigulilella.html).

6.4.4.4 Biological Control

Biological control of *C. figulilella* is problematic because of the paucity of local natural enemies and the frequent application of chemical insecticides. There are no parasitoids specific to the raisin moth that have been reported. However, a number of hymenopteran parasitoids were found in a culled fig warehouse, including *Venturia canescens* Gravenhorst, *Habrobracon hebetor* Say, and a *Goniozus* Förster sp. (Johnson et al. 2000). Bentley et al. (2013) reported that several parasites and some diseases attack raisin moth. Among them, the parasitoid *Bracon hebetor* Say actively stings and parasitizes larvae under bark on warm days in winter. It also efficiently parasitizes raisin moth larvae in storage (Coviello 2000).

6.4.4.5 Reproductive Control

Mating disruption, using a synthetic pheromone mimic, is one of the most promising alternatives to the use of chemical pesticides in stored product systems. These mimics overwhelm the release of natural pheromone by virgin females, ultimately interfering with the male's ability to locate females. It results in fewer matings and a reduction in viable eggs laid (Jones 1998). The pheromone (9Z, 12E)-tetradecadienyl acetate is a male attractant pheromone of many pyralid moths, known as TDA or ZETA (Levinson and Buchelos 1981). Mafra-Neto and Baker (1996) and Shani and Clearwater (2001) evaluated this pheromone against raisin moth and found it reduced the production of offspring at all experimental sites.

It is reported in unpublished notes from the Dried Fruit Insect Laboratory (Fresno, California, USA) that blue neon light proved most attractive to the raisin moth in field experiments. This type of light attraction should be evaluated further.

6.4.4.6 Chemical Control

The use of chemical insecticides for the control of raisin moth is a very successful approach (Kehat and Greenberg 1969), although they should be used only when needed. Blumberg (2008) noted that chemical control can be expensive, and difficult due to tree height and may cause spotting of the fruit. For successful control of raisin moth larvae, it is important to apply chemicals before larvae are inside the dates. In the absence of pheromone monitoring, the first spray should be applied as a preventative treatment when the fruit color changes. This should be followed by an application 3–4 weeks later (Kehat and Swirski 1964; Kehat et al. 1966; Blumberg 2008). Besides the control of raisin moth, these applications also will prevent infestations by the Parlatoria date scale, *Parlatoria blanchardi* (Targioni-Tozzetti) (Kehat and Swirski 1964) and *Carpophilus* spp. (Kehat et al. 1966).

The first experiments on chemical control for raisin moth on dates were conducted in date gardens in the southern Arava Valley in Israel in the late 1960s (Kehat et al. 1969). The first applications with azinphos-methyl, organophosphates, and diazinon were made at the stage when the fruit changed color, while the second application was applied 3 weeks later. These applications reduced fruit damage. However, the continuous use of these chemicals caused residues of organophosphates on the harvested fruit, and the insects developed resistance in the 1980s (Blumberg 2008). Kehat and Greenberg (1969) conducted two experiments at Eilat and Yotvata, Israel, to determine the percentage contamination of date fruit after treatment with various compounds including diazinon, azinphos-methyl, fenthion, malathion, aminocarb, dimethoate, trichlorfon, and cryolite. Diazinon and azinphos-methyl exhibited excellent control of the pest. Malathion was more effective in one of the experiments, suggesting further examination. Fenthion and aminocarb also were effective, but less so than diazinon or azinphos-methyl. Cryolite was moderately effective, and dimethoate was the least effective. Malathion also has been shown to be effective in other studies (Vincent and Lindgren 1971).

Kehat and colleagues conducted experiments in 1980s indicating that using a single treatment of azinphos-methyl, combined with fruit washing after harvest, can considerably reduce residues (Kehat et al. 1985a, b; Blumberg et al. 1987). About 20 years later, after the consistent use of these chemicals, resistance to azinphos-methyl, diazinon, and malathion was clearly demonstrated in the raisin moth and other date moths (Blumberg 2008). Regardless, azinphos-methyl and other members of the same group were banned by the European Trade Corporation (EUREP-GAP) in 2002 (Anonymous 2001). These materials have been replaced by pyrethroids such as cyhalothrin and cypermethrin in the management of date moths (Ucko and Bitton 2000). Perring et al. (2005) evaluated pyriproxyfen, methoxyfenozide, novaluron, spinosad, malathion dust (5 %), and GF-968® (10 %) in a commercial date plantation in the Coachella Valley (Riverside County, California, USA) that had high numbers of raisin moth. Only malathion (0 % infestation) was significantly different from the non-treated control (14.3 % infestation).

6.5 Future Research

It is clear from this review of the three lepidopteran pests of dates presented in this chapter that the development of a sustainable IPM program is in its infancy. Pest densities are mitigated largely through the use of prophylactic use of synthetic chemicals, with the occasional natural product being used. These insecticides will continue to be only a part of the solution and continued research into products with newer, more specific chemistries, is needed. At the same time, while there has been substantial effort in the biological control of carob moth, there has been relatively little biological control research for lesser date moth and raisin moth. More research is needed for all three of these pests.

Another promising avenue of control is the development of tools based on insect communication. The registration of a pheromone mimic for the carob moth that has been shown to be effective and economically viable gives great promise for reducing the grower's sole reliance on insecticides. Adapting this same strategy for lesser date moth and raisin moth would do the same for those insects, particularly when coupled with in season and overwintering sanitation of the date gardens and pest exclusion techniques like bagging the fruit. There is also potential in the further development of sterile insect strategies for all three pests.

Fundamental to the implementation of an integrated control program is a way to sample insect densities and relate these densities to damage. The development of these two avenues of research, i.e., sampling strategies and economic thresholds, is the area that could prove to be the most fruitful. Investment in this aspect of IPM is difficult for a commodity like dates for several reasons. First, date fruits grow in bunches at the top of tall palm trees. Since the damage is to the fruit, it is difficult to sample the insects. This is where pheromone-based traps and sampling strategies based on dates that can be collected from the garden floor can be most useful. Second, there is a time lag between pest infestation of the crop and the harvest, sometimes as long as 4 months. Thus elucidating the relationship between insect numbers at the beginning of the season and the final infestation at the end of the season is challenging. Yet, it is critical that this work is accomplished if practitioners wish to follow a truly sustainable pest management program.

6.6 Summary

The carob moth, lesser date moth, and raisin moth are the three most damaging lepidopterous pests of dates. With seasonal life cycles that result in high insect densities at the time fruits are developing on date palms, yield losses due to these pests can be extensive. Date growers have relied most heavily on pesticidal control, yet research has enabled an understanding of the biology of these pests, and through this understanding, there are additional management options that have been developed. These strategies, which are described in this chapter, include using natural enemies, cultural and physical control tactics, and technologies that rely on the mating biology of the insects. With continued research on these and other areas, particularly sampling techniques, management of lepidopterous pests in dates will become more economically and environmentally sustainable.

Acknowledgments The authors wish to thank the California Date Commission, USA, the USDA IR-4 program, and ISCA Tech, Inc., Riverside, California, USA, for their support of carob moth research. In particular, we thank Mr. Albert Keck, Dr. Agenor Mafra-Neto, and Dr. Justin E. Nay for their valuable contributions. In addition we wish to thank Ms. Bernice Ridgeway and Maria Mendoza, University of California, Riverside, California, USA, library staff, for their assistance in procuring literature. We also give special thanks to Eng. Hassan Al-Fardan, researcher at the Biological Control Laboratory, Date Palm Research Center in Qatif, Ministry of Agriculture, Saudi Arabia, for providing the pictures of damage by the lesser date moth. In addition, we thank Dr. Judy A. Johnson, USDA-ARS, Parlier, California, for granting permission to reproduce images of the raisin moth.

References

Abdul-Jabar, A., Ali, A., & El-Haidari, H. S. (1982). Control of the date moth (*Batrachedra amydraula* Meyrick) by pesticides in Bahrain. *Date Palm Journal, 1*, 34–36.

Abo-El-Saad, M., & El-Shafie, H. (2013). Insect pests of stored dates and their management. In M. Siddiq, M. A. Salah, & A. A. Kader (Eds.), *Dates: Postharvest science, processing technology and health benefits* (319 pp). Chichester: Wiley.

Adham, Z. (1965). Appearance of *Ectomyelois ceratoniae* in Saudi Arabia. *FAO Plant Protection Bulletin, 13*, 113–114.

Agenjo, R. (1959). La polilla de las garrofas, plaga actual de las naranjas (Lep. Phycit.). *Graellsia Madrid, 17*, 7–17.

Ahmad, T. R., & Ali, M. A. (1995). Forecasting emergence and flight of some Ephestia spp. (Lep. Pyralidae) based on pheromone trapping and degree-day accumulations. *Journal of Applied Entomology, 119*, 611–614.

Aitken, A. D. (1963). A key to the larvae of some species of Phycitinae (Lepidoptera, Pyralidae) associated with stored products, and of some related species. *Bulletin of Entomological Research, 54*, 175–188.

Aldini, M. Z., Goldansaz, S. H., Ashoori, A., & Ghasempour, A. R. (2010). A comparison of the calling behavior and some biological characters of three different geographic populations of Ectomyelois ceratoniae under laboratory conditions. *Iranian Journal of Plant Protection Science, 41*, 81–93.

Alfken, J. D. (1928). On the occurrence of the Pyralid, M. ceratoniae, in Bremen. *Mitteilungen der Gesellschaft fur Vorratsschutz, 4*, 55–56.

Al-Izzi, M. A. J., & Al-Maliky, S. K. (1996). Action of tannic acid on consumption and utilisation of food by *Ectomyelois ceratoniae* Zeller (Lepidoptera: Pyralidae). *Journal of Stored Products Research, 32*, 195–199.

Al-Izzi, M. A. J., Al-Maliky, S. K., Younis, M. A., & Jabbo, N. F. (1985). Bionomics of *Ectomyelois ceratoniae* (Lepidoptera: Pyralidae) on pomegranates in Iraq. *Environmental Entomology, 14*, 149–153.

Al-Izzi, M. A. J., Al-Maliky, S. K., & Jabbo, N. F. (1987). Culturing the carob moth, *Ectomyelois ceratoniae* (Zeller) (Lepidoptera, Pyralidae), on artificial diet. *Journal of Economic Entomology, 80*, 277–280.

Al-Izzi, M. A. J., Al-Maliky, S. K., Younis, M. A., & Jabbo, N. F. (1988). Effect of supplemental dietary L-lysine and soybean flour on growth and fertility of *Ectomyelois ceratoniae* Zeller (Lepidoptera, Pyralidae). *Journal of Economic Entomology, 81*, 970–972.

Al-Izzi, M. A. J., Al-Maliky, S. K., & Jabbo, N. F. (1990). Effect of gamma rays on males of *Ectomyelois ceratoniae* Zeller (Lepidoptera, Pyralidae) irradiated as pupae or adults. *Annales de la Societe Entomologique de France, 26*, 65–69.

Al-Izzi, M. A. M., Al-Maliky, S. K., & Khalaf, M. Z. (1992). Low-temperature for predicting induction and termination of diapause in *Apanteles angaleti* (Muesebeck) (Hymenoptera, Braconidae). *Annales de la Societe Entomologique de France, 28*, 385–389.

Al-Izzi, M. A. M., Al-Maliky, S. K., & Khalaf, M. Z. (1993). Effects of gamma-irradiation on inherited sterility of pomegranate fruit moth, *Ectomyelois ceratoniae* Zeller. *Insect Science and its Application, 14*, 657–679.

Al-Jamali, N. A. (2006). Factors affecting the efficiency of pheromone traps in attracting the male adults of pomegranate fruit worm *Ectomyelois ceratoniae* (Zeller) (Lepidoptera: Pyralidae). *Arab Journal of Plant Protection, 24*, 32–36.

Al-Khatri, S. A. Al-Abri, N. M., Al-Busaidi, A. Y., & Al-Rabani, M. A. (2009). Evaluation of some pesticides against lesser date moth *Batrachedra amydraula* Meyrick. *Agriculture and livestock research annual report 2009. Ministry of Agriculture, Sultanate of Oman* (pp. 179–181).

Al-Maliky, S. K., & Al-Izzi, M. A. J. (1986). Parasites of *Ectomyelois ceratoniae* with biological studies on Apanteles sp. group ultor in Iraq. *Entomophaga, 31*, 313–319.

Al-Maliky, S. K., & Al-Izzi, M. A. J. (1990). *Apanteles angaleti* (Muesebeck), Hymenoptera: Braconidae. A new record species in Iraq. *Bulletin of the Iraq Natural History Museum, 8*, 193–194.

Al-Maliky, S. K., Al-Izzi, M. A. J., & Jabbo, N. F. J. (1988). Effects of temperature and photoperiod on the development and oviposition of *Apanteles* sp. group ultor (Hym.: Braconidae), a larval parasite of the carob moth *Ectomyelois ceratoniae* (Lep.: Pyralidae). *Entomophaga, 33*, 193–200.

Alrubeai, H. F. (1987). Growth and development of *Ectomyelois ceratoniae* (Lepidoptera, Pyralidae) under laboratory mass-rearing conditions. *Journal of Stored Products Research, 23*, 133–135.
Alrubeai, H. F. (1988). Susceptibility of *Ectomyelois ceratoniae* to *Bacillus thuringiensis* isolates under laboratory and field conditions. *Journal of Agriculture and Water Resources Research, Plant Production, 7*, 125–136.
Al-Samarraie, A., Al-Hafdh, I., Abdul-Majed, K., & Basumy, M. A. (1989). The chemical control of the lesser date moth, *Batrachedra amydraula* Meyr, and residue levels of organophosphate insecticides in dates. *Pesticide Science, 25*, 227–230.
Anonymous. (1926). Departmental activities: Entomology. *Journal of the Department of Agriculture, Union of South Africa, 12*(3), 195–201.
Anonymous. (1938) Pests and diseases of the carob tree. *Cyprus Agricultural Journal, 33*, 27–28.
Anonymous. (1999). The IPM partner – Guidelines for stored product insect monitoring (5 pp). STORGARD®, TRÉCÉ® Incorporated, 1143 Madison Lane, Salinas, California, USA.
Anonymous. (2001). EUREPGAP protocol for fresh fruit and vegetables (15 pp). Version September 2001, Revision 02. EUREPGAP c/o FoodPLUS GmbH, Cologne, Spichernstr, Köln, Germany.
Argaman, Q. (1991–1992). *Parasierola swirskiana* n. sp. (Hymenoptera: Bethylidae), from the lesser date moth, *Batrachedra amydraula* Meyrick (Lepidoptera: Batrachedridae). *Israel Journal of Entomology, 25–26, 195–198.*
Arnal, E., Clavijo, J., Soto, E., & Ramos, F. (1998). *Batrachedra nuciferae* Hodges 1966 (Lepidoptera: Momphidae). Nueva plaga del Cocotero en Venezuela. *Boletín. de Entomología Venezolana, 13*, 69–71.
Arutyunyan, R. G. (1988). Morphological differences in Pyralids (Lepidoptera: Phycitidae). *Biologicheskii Zhurnal Armenii, 41*, 681–686.
Ashman, F. (1968). The control of infestation in carobs (*Ceratonia siliqua L.*) with special reference to Cyprus. *European and Mediterranean Plant Protection Organization Publication Series A*, 64E, 117–120.
Ashraf, Z., & Hamidi-Esfahani, Z. (2011). Date and date processing: A review. *Food Reviews International, 27*, 101–133.
Assari, M. J., & Khajepour, S. (2013). Population fluctuation of *Ephestia figulilella* in Kerman province of Iran. *Archives of Phytopathology and Plant Protection, 46*, 862–867.
Aubert, J. F. (1964). Ichneumonides de France et du Bassin méditerranéen appurtenant à un genre nouveau et neur espèces nouvelles. *Bulletin de la Société Entomologique de Mulhouse*, 1964, 35–40.
Aubert, J. F. (1966). In Service d'Identification des Entomophages. Liste d'identification No. 7. *Entomophaga, 11*(1), 135–151.
Aubert, J. F., Halperin, J., & Gerling, D. (1984). The Ichneumonidae of Israel. *Entomophaga, 29*, 211–235.
Avidov, Z., & Gothilf, S. (1960). Observations on the honeydew moth (*Cryptoblabes gnidiella* Milliere) in Israel. I. Biology, phenology and economic importance. *Ktavim, 10*, 109–124.
Aytas, M., Yumruktepe, R., & Aktura, T. (1996). Studies on the chemical control of carop moth, *Ectomyelois ceratoniae* Zeller [Apomyelois ceratoniae] on citrus in the Eastern Mediterranean region. *Zirai Mucadele Arastirma Yilligi, 28–29*, 81–82.
Bachrouch, O., Mediouni-Ben, J. J., Wissem, A. W., Talou, T., Marzouk, B., & Abderrabe, M. (2010). Composition and insecticidal activity of essential oil from *Pistacia lentiscus* L. against *Ectomyelois ceratoniae* Zeller and *Ephestia kuehniella* Zeller (Lepidoptera: Pyralidae). *Journal of Stored Products Research, 46*, 242–247.
Baez, M. (1998). *Mariposas de Canarias* (216 pp). Alcorcón: Editorial Rueda.
Baker, T. C., Francke, W., Lofstedt, C., Hansson, B. S., Du, J. W., Phelan, P. L., Vetter, R. S., & Youngman, R. (1989). Isolation, identification, and synthesis of sex pheromone components of the carob moth, *Ectomyelois ceratoniae*. *Tetrahedron Letters, 30*, 2901–2902.

Baker, T. C., Francke, W., Millar, J. G., Lofstedt, C., Hansson, B., Du, J.-W., Phelan, P. L., Vetter, R. S., Youngman, R., & Todd, J. L. (1991). Identification and bioassay of sex pheromone components of carob moth, *Ectomyelois ceratoniae* (Zeller). *Journal of Chemical Ecology, 17*, 1973–1988.

Barrevald, W. H. (1993). Date palm products (FAO Agricultural Services Bulletin No. 101). Rome: Food and Agriculture Organization of the United Nations. Available at: http://www.fao.org/docrep/t0681E/t0681e00.htm. Accessed on 6 Mar 2015.

Basirat, M., & Mehrnejad, M. R. (2005). The study of lower threshold temperature and thermal constant for two insect pests of stored nuts *Plodia interpunctella* (Lep.: Pyralidae) and *Apomyelois ceratoniae* (Lep.: Pyralidae). *Journal of the Entomological Society of Iran, 24*, 19–33.

Beccaloni, G. W., Scoble, M. J., Robinson, G. S., & Pitkin, B. (2011). LepIndex: The global lepidoptera names index. In: F. A. Bisby, Y. R. Roskov, T. M. Orrell, D. Nicolson, L. E. Paglinawan, N. Bailly, P. M. Kirk, T. Bourgoin, G. Baillargeon, & D. Ouvrard (Eds.), Species 2000 and ITIS catalogue of life: 2011 annual checklist in species 2000. Reading Available at: http://www.catalogueoflife.org/annual-checklist/2011/details/database/id/18. Accessed on 6 Mar 2015.

Ben Jemaa, J. M., Haouel, S., Bouaziz, M., & Khouja, M. L. (2012). Seasonal variations in chemical composition and fumigant activity of five *Eucalyptus* essential oils against three moth pests of stored dates in Tunisia. *Journal of Stored Products Research, 48*, 61–67.

Ben Jemaa, J. M., Haouel, S., & Khouja, M. L. (2013). Efficacy of *Eucalyptus* essential oils fumigant control against *Ectomyelois ceratoniae* (Lepidoptera: Pyralidae) under various space occupation conditions. *Journal of Stored Products Research, 53*, 67–71.

Bentley, W. J., Coviello, R. L., Vasquez, S. J., & Haviland, D. R. (2013). Raisin moth. In L. J. Bettiga (Ed.), *Grape pest management* (University of California, Agriculture and Natural Resources Communication Services Publ. N. 3343 3rd ed., pp. 380–380). Richmond: University of California, Agriculture and Natural Resources Communication Services.

Bessi, H., Ferchichi, C., Bellagha, S., Grissa, K., Bikoba, V., & Mitcham, E. (2013). Effects of fumigation with ethyl formate on the quality of date fruit. In H. Abdullah & M. N. Latifah (Eds.), Proceedings of the VII international postharvest symposium, 25–29 Jun 2012, Kuala Lumpur, Malaysia. *ISHS Acta Horticulturae* 1012, 693–698.

Biliotti, E., & Daumal, J. (1970). Bionomics of *P. flavitestacea*. Details of a method of permanent rearing for work on the biological control of *E. ceratoniae*. *Annales de Zoologie Ecologie Animale, 1*, 379–394.

Bitaw, A. A., & Saad, A. A. (1990). Survey of natural enemies of date palm pests in Libya. *Arab Journal of Plant Protection, 8*, 12–15.

Bitaw, A. A., Bin Saad, A. A., & El-Mokhtar, O. (1988). Natural enemies of date palm pests in Jamahiriya. *Arab and Near East Plant Protection Newsletter, 7*, 26.

Bjostad, L. B., Hibbard, B. E., & Cranshaw, W. S. (1993). Application of semiochemicals in integrated pest management programs. In S. O. Duke, J. J. Menn, & J. R. Plimmer (Eds.), Pest control with enhanced environmental safety, 203rd National meeting of the American Chemical Society Symposium Series No. 524, Apr 5–10 1992, San Francisco, California, USA, 199–218.

Blumberg, D. (1975). Preliminary notes on the phenology and biology of *Batrachedra amydraula* Meyrick (Lepidoptera: Cosmopterigidae), a new pest of date palms in Israel. *Phytoparasitica, 3*, 55–57.

Blumberg, D. (2004). Date palm pests and their control measurement. In Z. Bernstein (Ed.), *Date growing in Israel* (398–433 pp). Tel-Aviv: The Plant Production and Marketing Board.

Blumberg, D. (2008). Date palm arthropod pests and their management in Israel. *Phytoparasitica, 36*, 411–448.

Blumberg, D., Swirski, E., & Greenberg, S. (1977). Field tests for the control of the lesser date moth. *International Pest Control, 19*, 18–20.

Blumberg, D., Kehat, M., Greenberg, S., Gantz, S., Biton, S., & Adato, I. (1987). Effectiveness of azinphosmethyl for controlling sap beetles (*Carpophilus* spp.) (Coleoptera: Nitidulidae) in date palms: Further tests. *Alon Hanotea, 41*, 1152–1157.

Blumberg, D., Navon, A., Kehat, M., Eliahu, M., Levski, S., & Nakash, J. (2001). Date palm pests in Israel at the beginning of the third millennium. *Alon Hanotea, 55*, 42–48.

Bodenheimer, F. S. (1930). The insect pest fauna of Palestine with particular regard to the chief pests of the Mediterranean region. *Monographien zur angewandten Entomologie, Verlagsbuchhandlung Paul Parey, Berlin, Germany, 16*, 438.

Bouka, H., Chemseddine, M., Abbassi, M., & Brun, J. (2001). The date moth in the area of Tafilalet in the southeast of Morocco. *Fruits, 56*, 189–196.

Brady, U. E., & Daley, R. C. (1972). Identification of a sex pheromone from the female raisin moth, *Cadra figulilella*. *Annals of the Entomological Society of America, 65*, 1356–1358.

Bridwell, J. C. (1919a). Some notes on Hawaiian and other Bethylidae with descriptions of new species. *Proceedings of the Hawaiian Entomological Society, 4*, 21–38.

Bridwell, J. C. (1919b). Miscellaneous notes on Hymenoptera with descriptions of new genera and species. *Proceedings of the Hawaiian Entomological Society, 4*, 109–165.

Burks, C. S., & Johnson, J. A. (2012). Biology, behavior and ecology of stored fruit and nut insects. In D. W. Hagstrum, T. W. Phillips, & G. Cuperus (Eds.), *Stored product protection* (Kansas State Research and Extension Service Publication No. S156, pp. 21–32). Manhattan: Kansas State Research and Extension Service.

Calcat, A. (1959). Disease and pests of date palms in the Sahara and North Africa. *Bulletin of Entomological Research, 11*, 287–303.

Calderon, M., Navarro, S., & Donahaye, E. (1969). *Ectomyelois ceratoniae* (Zell.) (Lep., Phycitidae), a major pest of stored almonds in Israel. *Journal of Stored Products Research, 5*, 427–428.

Carpenter, J. B., & Elmer, H. S. (1978). *Pests and diseases of the date palm* (United States Department of Agriculture, Agricultural Research Service Handbook No. 527, 42 pp). Washington, DC: United States Department of Agriculture, Agricultural Research Service.

Catling, H. D. (1970). Notes on new minor pests of citrus in Southern Africa. *South African Citrus Journal, 444*, 13–14.

CDFA (California Department of Food and Agriculture). (1983). Carob moth in California. In R. Gill & S. Kaiser (Eds.), *California plant pest and disease report no 2* (pp. 11–14). Sacramento: California Department of Food and Agriculture.

Cock, M. J. W. (1987). Pyralidae: *Ectomyelois ceratoniae* (Zeller), a possible predator of limacodid pupae. In M. J. W. Cock, H. C. J. Godfray, & J. D. Holloway (Eds.), *Slug and nettle caterpillars: The biology, taxonomy, and control of limacodidae of economic importance on palms in South-east Asia* (199 pp). Wallingford: CABI Publishing, CAB International.

Corbet, A. S., & Tams, W. H. T. (1943). Keys for the identification of the Lepidoptera infesting stored food products. *Proceedings of the Zoological Society of London, Series B, 113*, 55–145.

Cosse, A. A., Endris, J. J., Millar, J. G., & Baker, T. C. (1994). Identification of volatile compounds from fungus-infected date fruit that stimulate upwind flight of female *Ectomyelois ceratoniae*. *Entomologia Experimentalis et Applicata, 72*, 233–238.

Coviello, R. L. (2000). Lepidopterous insects. In L. P. Christensen (Ed.), *Raisin production manual* (University of California, Agriculture and Natural Resources Communication Services Publication No. 3393, pp. 182–186). Oakland: University of California, Agriculture and Natural Resources Communication Services.

Cox, P. D. (1974). The influence of temperature and humidity on the life-cycles of *Ephestia figulilella* Gregson and *Ephestia calidella* (Guenée) (Lepidoptera: Phycitidae). *Journal of Stored Products Research, 10*, 43–55.

Cox, P. D. (1975a). The influence of photoperiod on the life-cycles of *Ephestia calidella* (Guenée) and *Ephestia figulilella* Gregson (Lepidoptera, Phycitidae). *Journal of Stored Products Research, 11*, 75–85.

Cox, P. D. (1975b). The suitability of dried fruits, almonds, and carobs for the development of *Ephestia figulilella* Gregson, *E. calidella* (Guenée) and *E. cautella* (Walker) (Lepidoptera: Phycitidae). *Journal of Stored Products Research, 11*, 229–233.

Cox, P. D. (1976). The influence of temperature and humidity on the life-cycle of *Ectomyelois ceratoniae* (Zeller) (Lepidoptera: Phycitidae). *Journal of Stored Products Research, 12*, 111–117.

Cox, P. D. (1979). Influence of photoperiod on the life cycle of *Ectomyelois ceratoniae* (Zeller) (Lepidoptera: Pyralidae). *Journal of Stored Products Research, 15*, 111–115.

Cox, P. D., Crawford, L. A., Gjestrud, G., Bell, C. H., & Bowley, C. R. (1981). The influence of temperature and humidity on the life cycle of *Corcyra cephalonica* (Stainton) (Lepidoptera: Pyralidae). *Bulletin of Entomological Research, 71*, 171–181.

Daumal, J., Jourdheuil, P., & Marro, J. P. (1973). Acclimatization of *Phanerotoma flavitestacea* (Hymenoptera: Braconidae) parasite of *Ectomyelois ceratoniae* (Lepidoptera: Pyralidae) on the French Mediterranean Coast. *Annales de Zoologie Ecologie Animale, 5*, 593–608.

de Jong, Y. S. D. M. (2013). Fauna europaea version 2.6. Available at: www.faunaeur.org. Accessed on 6 Mar 2015.

De Stefani, T. (1919). Some insects infesting carob pods. *Bollettino Stazione Sperimentale di Frutticoltura e di Agrumicoltura, Acireale, Italy, 37*, 5–6.

De Villiers, E. A., & Wolmarans, M. M. (1980). Chemical control of false codling moth and other nut borers on macadamia. *Citrus and Subtropical Fruit Research Institute Information Bulletin, 97*, 9–11.

Delvare, G., Talaee, L., & Goldansaz, S. H. (2011). New Chalcididae (Hymenoptera: Chalcidoidea) of economic importance from Iran. *Annales Zooligici, 61*, 789–801.

Deseo, K. V., Briolini, G., & Bruno, P. (1981). Observations on pests of walnut. *Informatore Fitopatologico, 31*, 3–6.

Dhouibi, M. H. (1992). The effect of bactospeine XLV on the larvae of the carob moth *Ectomyelois ceratoniae* Zeller (Lepidoptera: Pyralidae). *Mededelingen van de Faculteit Landbouwwetenschappen, Universiteit Gent, 57*, 505–514.

Dhouibi, M. H., & Jemmazi, A. (1993). Biological control of the carob moth *Ectomyelois ceratoniae* (Lepidoptera: Pyralidae) by *Habrobracon hebetor* (Hymenoptera: Braconidae) in pomegranate orchards. *Mededelingen Faculteit Landbouwkundige en Toegepaste Biologische Wetenschappen Universiteit Gent, 58*, 427–436.

Dobie, P. (1978). The effects of a previous field infestation upon *Ephestia cautella* (Walker) and *Lasioderma serricorne* (F.) infesting carobs. *Journal of Stored Products Research, 14*, 35–39.

Donahaye, E., & Calderon, M. (1964). Survey of insects infesting dates in storage in Israel. *Israel Journal of Agricultural Research, 14*, 97–100.

Donohoe, H. C., & Barnes, D. F. (1934). Notes on host materials of *Ephestia figulilella*. *Journal of Economic Entomology, 27*, 1075–1077.

Donohoe, H. C., Simmons, P., Barnes, D. F., Kaloostian, G. H., Fisher, C. K., & Heinrich, C. (1949). Biology of the raisin moth (United States Department of Agriculture, Agricultural Research Service, Technical Bulletin No. 994, Washington DC, USA, 23 pp).

Doumandji, S. (1981). Biologie et Ecologie de la Pyrale des Caroubes dans le Nord de L'Algerie, *Ectomyelois ceratoniae* (Zeller), (Lepidoptera: Pyralidae) (146 pp). PhD thesis, L'University Pierre et Marie Curie, Paris, France.

Doumandji-Mitiche, B., & Doumandji, S. (1982). Les ennemis naturels *d'Ectomyelois ceratoniae* Zeller dans les palmeries de Briskra. *Bulletin de Zoologie Agricola, 4*, 1–10.

Doumandji-Mitiche, B., & Idder, A. (1986). Studies of releases of *Trichogramma embryophagum* Hartig (Hymenoptera, Trichogrammatidae) against date pyralids *Ectomyelois ceratoniae* Zeller (Lepidoptera, Pyralidae) in a plantation at Ouargla. *Annales de l'Institut National Agronomique El-Harrach, 10*, 167–180.

Dowson, V. H. W. (1982). Date production and protection (Food and Agriculture Organization of the United Nations Plant production and Protection Paper No. 35, Rome, Italy, 294 p).

Durrant, J. H. (1915). *Myelois phoenicis*, sp. n. bred from dates in Algeria and England (Lepid, Pyral). *Entomologist's Monthly Magazine, 51*, 304–305.
Dyar, H. G. (1911). Two species of Phycitinae new to our fauna. *Proceedings of the Entomological Society of Washington DC, USA, 13*, 30.
Ehteshami, F., Aleosfoor, M., Allahyari, H., & Alichi, M. (2013). First record of Goniozus legneri (Hym.: Bethylidae), the larval ectoparasitoid of carob moth, in Iran. *Journal of the Entomological Society of Iran, 33*, 89.
Eitam, A. (2001). Oviposition behavior and development of immature stages of *Parasierola swirskiana*, a parasitoid of the lesser date moth *Batrachedra amydraula*. *Phytoparasitica, 29*, 405–412.
El-Bashir, S., & El-Makaleh, S. (1983). Control of the lesser date moth *Batrachedra amydraula* Meyrick in the Tihama region of the Yemen Arab Republic. In *Proceedings of the first symposium on date palms*, 23–25 March 1982, Saudi Arabia (418–422 pp).
Elsayed, G., & Bazaid, S. A. (2011). Field investigation of pomegranate fruit worms in Taif and laboratory evaluation of *Bacillus thuringiensis* against *Ectomyelois ceratoniae*. *Archives of Phytopathology and Plant Protection, 44*, 28–36.
El-Shafie, H. A. F. (2012). Review: list of arthropod pests and their natural enemies identified worldwide on date palm, *Phoenix dactylifera* L. *Agriculture and Biology Journal of North America, 3*, 516–524.
Elwan, A. A. (2000). Survey of the insect and mite pests associated with date palm trees in Al-Dakhliya region, Sultanate of Oman. *Egypt Agricultural Research, 78*, 653–664.
Entomological Society of America. (2014). ESA common names database. Available from: http://www.entsoc.org/common-names. Accessed on 6 Mar 2015.
Farahani, H. K., Goldansaz, S. H., & Sabahi, G. (2009). Report of two fly parasitoids of carob moth, *Ectomyelois ceratoniae* (Lepidoptera: Pyralidae). *Journal of Entomological Society of Iran, 29*, 57–58.
Farahani, H. K., Goldansaz, S. H., Sabahi, G., & Baur, H. (2010a). First report of *Pachycrepoideus vindemmiae* (Hym.: Pteromalidae) from Iran. *Journal of Entomological Society of Iran, 29*, 117–118.
Farahani, H. K., Goldansaz, S. H., Sabahi, G., & Hortsmann, K. (2010b). First report of two Ichneumonid wasp species from Iran. *Journal of Entomological Society of Iran, 30*, 59–60.
Farahani, H. K., Goldansaz, S. H., Sabahi, G., & Shakeri, M. (2010c). Larval parasitoids of the carob moth, *Ectomyelois ceratoniae* Zeller (Lep.: Pyralidae) in three regions of Iran: Varamin, Qom, and Saveh. *Iranian Journal of Plant Protection Science, 41*, 337–344.
Farahani, H. K., Bell, H., & Goldansaz, S. H. (2012a). Biology of *Apanteles myeloenta* (Hymenoptera: Braconidae), a larval parasitoid of carob moth *Ectomyelois ceratoniae* (Lepidoptera: Pyralidae). *Journal of Asia-Pacific Entomology, 15*, 607–610.
Farahani, H. K., Goldansaz, S. H., & Allahyari, H. (2012b). Biology of *Venturia canescens*, a larval parasitoid of carob moth, *Ectomyelois ceratoniae* under laboratory conditions. *Iranian Journal of Plant Protection Science, 43*, 111–119.
Farahani, H. K., Goldansaz, S. H., & Sabahi, G. (2012c). A survey on the overwintering larval parasitoids of *Ectomyelois ceratoniae* in three regions in Iran. *Crop Protection, 36*, 52–57.
Farahani, H. K., Bell, H., & Goldansaz, S. H. (2013). Is host age an important factor in the bionomics of *Apanteles myeloenta* (Hymenoptera: Braconidae)? *European Journal of Entomology, 110*, 277–283.
Farzaneh, A. (1984). *Spectrobates ceratoniae* Z. (Lep.: Pyralidae) in Iran. *Proceedings of the International Congress of Entomology, 17*, 577.
Fischer von Röslerstamm, J. E. (1839). Abbildungen zur Berichtigung und Ergänzung der Schmetterlingskunde, besonders der Microlepidopterologie, Leipzig, Germany (304 pp).
Gençer, N. S., Coşkuncu, K. S., & Kumral, N. A. (2005). Determination of harmful and beneficial fauna in fig orchards in Bursa Province. *Ondokuz Mayis Üniversitesi, Ziraat Fakültesi Dergisi, 20*, 24–30.
Georgevits, R. P. (1981). Seed insects of *Robinia pseudoacacia*. *Dasikon Ereunon, 2*, 223–255.

Gharib, A. (1968). *Batrachedra amydraula* Meyr. (Super-family; Gelechoidea) – (Momphidae (Cosmopterygidae). *Entomlogie et Phytopatologie Appliqees, 27*, 63–67.

Ghavami, S. (2006). Effects of three artificial diets on biological characteristics of the carob moth, *Ectomyelois ceratoniae* (Lep.: Pyralidae). *Journal of Entomological Society of Iran, 25*, 63–76.

Gilligan, T. M., & Passoa, S. C. (2014). LepIntercept – An identification resource for intercepted lepidoptera larvae. Available at: http://idtools.org/id/leps/lepintercept/ceratoniae.html. Accessed on 9 Apr 2015.

Goater, B. (1986). *British pyralid moths. A guide to their identification* (175 pp). Colchester: Harley Books.

Goldansaz, S. H., Talaei, L., Poorjavad, N., & Dehghani, Y. H. (2012). Inhibition of carob moth damage using *Ferula assafoetida* essential oil in pomegranate orchards of Iran. In P. Melgarejo & D. Valero (Eds.), Options Mediterraneennes. Serie A, Seminaires Méditerranéens (II International Symposium on the Pomegranate, October 19–21, 2011, Madrid, Spain, 103, 129–131.

Gonzalez, R. H. (2002). Development of first and second generation of apple codling moth *Cydia pomonella* (L.) and its influence on management programs. *Revista Fruticola, 23*, 5–16.

Gonzalez, R. H. (2003). Las polillas de la fruta en Chile (Lepidoptera: Tortricidae, Pyralidae). *Universidad de Chile, Serie Ciencias Agronomicas, 9*, 188.

Gonzalez, R. H., & Cepeda, D. (1999). The carob-bean moth, *Spectrobates ceratoniae* (Zeller) (Pyralidae), a barrier to the implementation of mating disruption for control of codling moth on walnuts. *Revista Fruticola, 20*, 57–67.

Gothilf, S. (1968). The biology of the carob moth, *Ectomyelois ceratoniae* (Zell) in Israel. Part 1. Mass culture on artificial diet. *Israel Journal of Entomology, 3*, 109–119.

Gothilf, S. (1969a). Natural enemies of the carob moth *Ectomyelois ceratoniae* (Zeller). *Entomophaga, 14*, 195–202.

Gothilf, S. (1969b). A contribution to the biology of *Phanerogam flavitestacea* FI. A parasite of *Ectomyelois ceratoniae* (Zell.). *Israel Journal of Entomology, 4*, 55–71.

Gothilf, S. (1969c). The biology of the carob moth *Ectomyelois ceratoniae* in Israel. Part 2. Effect of food, temperature, and humidity on development. *Israel Journal of Entomology, 4*, 107–116.

Gothilf, S. (1970). The biology of the carob moth *Ectomyelois ceratoniae* in Israel. Part 3. Phenology of various hosts. *Israel Journal of Entomology, 5*, 161–175.

Gothilf, S. (1978). Establishment of imported parasite *Pentalitomastix plethoricus* (Hymenoptera, Encyrtidae) on *Ectomyelois ceratoniae* in Israel. *Entomophaga, 23*, 299–302.

Gothilf, S. (1984). Biology of *Spectrobates ceratoniae* on almonds in Israel. *Phytoparasitica, 12*, 77–87.

Gothilf, S., & Mazor, M. (1987). Release and recovery of imported parasites of the carob moth *Spectrobates ceratoniae* (Lepidoptera: Pyralidae) in Israel. *Israel Journal of Entomology, 21*, 19–24.

Gothilf, S., Levy, E. C., Cooper, R., & Lavie, D. (1975). Oviposition stimulants of the moth *Ectomyelois ceratoniae*, the effect of short chain alcohols. *Journal of Chemical Ecology, 1*, 457–464.

Habib, D. M., & Essaadi, S. H. (2007). Biocontrol of the lesser date moth *Batrachedra amydraula* Meyrick (Cosmopteridae = Batrachedridae) on date palm trees. *Acta Horticulturae, 736*, 391–397.

Halperin, J. (1986a). Braconidae (Hymenoptera) associated with forest and ornamental trees and shrubs in Israel. *Phytoparasitica, 14*, 119–135.

Halperin, J. (1986b). Occurrence of the carob moth in pistachio. *Alon Hanotea, 40*, 923–926.

Hameed, A. A., Al-Taweel, A. A., Al-Jboory, I. J., & Al-Zaidy, S. M. (2011). Using the parasitoid, *Bracon hebetor* Say and the pheromone traps to control the moth insects, *Ephestia* spp. in date stores in Iraq. *Egyptian Journal of Biological Pest Control, 21*, 377–383.

Hammad, S. M., Kadous, A. A., & Ramadan, M. M. (1983). Predators and parasites of date palm insects in Al-Hassa and Al-Qatif regions (Eastern province, Saudi Arabia). In Proceedings of the first symposium on date palms, 23–25 Mar 1982, Saudi Arabia (pp. 322–341).

Hampson, G. F. (1896). *Fauna of British India, including Ceylon and Burma* (Vol. 4, 91 pp, Vol. 4). London: Taylor and Francis Group.

Hampson, G. F. (1903). The moths of India. *Journal of the Bombay Natural History Society, 15*, 19–37.

Harhash, M., Mourad, A. K., & Hammad, S. M. (2003). Integrated crop management of the lesser date moth *Batrachedra amydraula* Meyr. (Lepidoptera: Cosmopteridae) infesting some date-palm varieties in Egypt. *Communications in Agriculture and Applied Biological Sciences, 68*, 209–221.

Harpaz, I., & Wysok, M. (1984). Susceptibility of the carob moth, *Ectomyelois ceratoniae*, to *Bacillus thuringiensis*. *Phytoparasitica, 12*, 189–191.

Harvey, A. (2013). Carob moth in Afghanistan. *Outlooks on Pest Management, 24*, 213–216.

Hedin, P. A. (Ed.). (1991). *Naturally occurring pest bioregulators* (456 pp). Washington, DC: American Chemical Society.

Heinrich, C. (1956). American moths in the subfamily Phycitinae. *United States Natural Museum Bulletin, 207*, 1–581.

Hodges, R. W. (1998). Gelechioidea. In N. P. Kistensen (Ed.), *Handbook of zoology, Volume IV, Arthropoda: Insecta, Part 35. Lepidoptera, moths and butterflies, Volume 1, evolution, systematics, and biogeography* (pp. 131–158). Berlin: Walter de Gruyter GmbH.

Huisman, K. (1974). Interesting captures of Lepidoptera. *Entomologische Berichten, 34*, 153–154.

Hung, C. C., Chiang, B. Y., & Wang, W. L. (2003). Development and fecundity of the carob moth, *Apomyelois ceratoniae*, reared on different foods and its eclosion, mating, and ovipositing behavior. *Plant Protection Bulletin (Taichung), 45*, 185–197.

Hussain, A. A. (1963). Notes on borers of date palm in Iraq. *Bulletin of Entomological Research, 54*, 345–348.

Hussain, A. A. (1974). *Date palms and dates with their pests in Iraq*. Bagdad: University of Bagdad.

Idder, M. A., Idder-Ighili, H., Saggou, H., & Pintureau, B. (2009). Infestation rate and morphology of the carob moth, *Ectomyelois ceratoniae* (Zeller), on different varieties of the palm date, *Phoenix dactylifera* (L.). *Cahiers Agricultures, 18*, 63–71.

Idder, M. A., Doumandji-Mitiche, B., & Pintureau, B. (2013). Biological control in Algerian palm groves. In N. Bouguedoura, M. Bennaceur, & J. C. Pintaud (Eds.), Proceedings of the first international symposium on date palm, Algiers, Algeria. ISHA Acta Horticulturae, 994, 347–354.

Jacobs, S. N. A. (1933). *Aphomia gularis* (Zell.) and other rare warehouse moths. *The Entomologist, 66*, 195.

Jemni, M., Oton, M., Ramirez, J. G., Artes-Hernandez, F., Chaira, N., Ferchichi, A., & Artes, F. (2014). Conventional and emergent sanitizers decreased *Ectomyelois ceratoniae* infestation and maintained quality of date palm after shelf-life. *Postharvest Biology and Technology, 87*, 33–41.

Johnson, J. A., Valero, K. A., Hannel, M. M., & Gill, R. F. (2000). Seasonal occurrence of postharvest dried fruit insects and their parasitoids in a culled fig warehouse. *Journal of Economic Entomology, 93*, 1380–1390.

Johnstone, D. R., Walker, P. T., & Huntington, K. A. (1972). Ultra-low volume, hand-operated motorised sprayers for insecticide application in citrus. *International Pest Control, 14*(8–15), 19.

Jones, O. T. (1998). The commercial exploitation of pheromones and other semiochemicals. *Pest Science, 54*, 293–296.

Kaakeh, W. (2006). Management of the lesser date moth, *Batrachedra amydraula*, in the United Arab Emirates date palm plantations. In 3rd international date palm conference, 10 Dec 2006, Abu Dhabi, United Arab Emirates (55 pp).

Kakar, M. K., Nizamani, S. M., Rustamani, M. A., & Khuhro, R. D. (2010). Periodical lesser date moth infestation on intact and dropped fruits. *Sarhad Journal of Agriculture, 26*, 393–396.

Karami, E., Mirabzadeh, A., Karahroudi, Z. R., & Ioni, S. (2011). Comparison of the effect of cutting off stamens of pomegranates and releasing *Trichogramma embryophagum* with *Ectomyelois ceratoniae* (Lep: Pyralidae) in Saveh region, Iran. *Munis Entomology and Zoology, 6*, 306–311.

Kashkuli, A., & Eghtedar, E. (1976). Biologie und Okologie von *Spectrobates ceratoniae* (Lep. Pyralidae) in der Provinz Fars. *Entomologie et Phytopathologie Appliquées, 41*, 21–32.

Kehat, M., & Greenberg, S. (1969). The biology and phenology of *Arenipses sabella* Hmps. and *Cadra figulilella* (Gregson) (Lepidoptera, Pyralidae) on dates in Israel. *Bulletin of Entomological Research, 58*, 411–419.

Kehat, M., & Swirski, E. (1964). Chemical control of the date palm scale, *Parlatoria blanchardi*, and the effect of some insecticides on the lady beetle *Pharoscymnus aff. numidicus* Pic. *Israel Journal of Agricultural Research, 14*, 101–110.

Kehat, M., Stoller, S., Meisner, J., & Greenberg, S. (1966). Development of population and control of *Carpophilus dimidiatus* F., *C. hemipterus* L. and *Coccotrypes dactyliperda* F. in dates. *Israel Journal of Agricultural Research, 16*, 173–176.

Kehat, M., Blumberg, D., & Greenberg, S. (1969). Experiments on the control of the raisin moth, *Cadra figulilella* Gregs. (Phyticidae, Pyralidae), on dates in Israel. *Israel Journal of Agricultural Research, 19*, 121–128.

Kehat, M., Blumberg, D., & Greenburg, S. (1976). Fruit drop and damage in dates. The role of *Coccotrypes dactyliperda* F. and nitidulid beetles and prevention by mechanical measures. *Phytoparasitica, 4*, 93–99.

Kehat, M., Blumberg, D., Greenberg, S., & Gordon, D. (1985a). Laboratory and field evaluation of insecticides for the control of sap beetles (Coleoptera: Nitidulidae) in dates. *Alon Hanotea, 39*, 797–801.

Kehat, M., Greenberg, S., Biton, S., Gantz, S., & Adato, I. (1985b). Field evaluation of insecticides for the control of sap beetles (Coleoptera: Nitidulidae) in dates. *Alon Hanotea, 40*, 787–790.

Kehat, M., Eitam, A., Blumberg, D., Dunkelblum, E., Anshelevich, L., & Gordon, D. (1992). Sex pheromone traps for detecting and monitoring the raisin moth, *Cadra figulilella*, in date palm plantations. *Phytoparasitica, 20*, 99–106.

Khajehzadeh, Y., & Latifian, M. (2013). Response of different date palm cultivars to lesser date moth *(Batrachedra amydraula* Meyr.) under natural infestation in Behbahan. *Seed and Plant Improvement Journal, 29*, 311–330.

Khajepour, S., Izadi, H., & Asari, M. J. (2011). Evaluation of two formulated chitin synthesis inhibitors, hexaflumuron and lufenuron against the raisin moth, *Ephestia figulilella*. *Journal of Insect Science, 12*, 1–7.

Khoualdia, O., R'Houma, A., Marro, J. P., & Brun, J. (1996). Release of *Phanerotoma ocuralis* Kohl to combat the carob moth *Ectomyelois ceratoniae* Zeller on dates in an experimental plot at Tozeur in Tunisia. *Fruits (Paris), 51*, 129–132.

Khoualdia, O., Takrouni, M. L., Mahmoud, O. B., Rhouma, A., Alimi, E., Ridha, B. H., Abid, M., & Brun, J. (2002). Control of date moth in Southern Tunisia: Trials of two natural products, spinosad and azadirachtin. *Phytoma, 551*, 15–17.

Klimesch, J. (1968). Die Lepidoptera fauna Mazedonien 4. Microlepidoptera. *Posebna Izdanja Prirod Muz Skopje, 5*, 1–203.

Krasil'Nikova, G. A. (1964). A new pest of the pomegranate (*Punica granatum*) *Ectomyelois ceratoniae* (Lepidoptera, Pyralidae). *Turkmenian Academy of Science, Biological Series, 1*, 40–46.

Krasil'Nikova, G. A. (1965). Food specialization of *Ectomyelois ceratoniae* and *Euzophera punicaella* caterpillars (Lepidoptera, Pyralididae). *Izvestiya Akademii Nauk Turkmenskoi SSR Seriya Biologicheskikh Nauk, 1*, 55–60.

Krasil'Nikova, G. A. (1966). Conditions promoting adaptation of some pyralids to synanthsopism. *Entomologicheskoe Obozrenie, 45*, 764–773.

Krasnoff, S. B., & Vick, K. W. (1984). Male wing-gland pheromone of *Ephestia elutella*. *Journal of Chemical Ecology, 10*, 667–679.

Krasnoff, S. B., Vick, K. W., & Mankin, R. W. (1983). Female calling behavior in *Ephestia elutella* and *E. figulilella* (Lepidoptera: Pyralidae). *Florida Entomologist, 66*, 249–254.

Ksentini, I., Jardak, T., & Zeghal, N. (2010a). *Bacillus thuringiensis*, deltamethrin and spinosad side-effects on three *Trichogramma* species. *Bulletin of Insectology, 63*, 31–37.

Ksentini, I., Monje, J. C., Jardak, T., & Zeghal, N. (2010b). Naturally occurring egg parasitoids of the genus *Trichogramma* (Hymenoptera: Trichogrammatidae) in a pomegranate orchard in Tunisia. *Entomological Science, 13*(1), 99–106.

Ksentini, I., Jardak, T., & Zeghal, N. (2013). A laboratory comparison of four *Trichogramma* species (Hymenoptera: Trichogrammatidae) as potential biocontrol agents for the carob moth, *Ectomyelois ceratoniae* (Lepidoptera: Pyralidae). *Biocontrol Science and Technology, 23*, 465–469.

Kugler, J., & Nitzan, Y. (1977). Biology of Clausicella suturata (Diptera, Tachinidae), a parasite of *Ectomyelois ceratoniae* (Lepidoptera: Phycitidae). *Entomophaga, 22*, 93–106.

Kuwahara, Y., Kitamura, C., Takahashi, S., Hara, H., Ishii, S., & Fukami, H. (1971). Sex pheromone of the almond moth and the Indian meal moth: Cis-9, trans-12 tetradecadienyl acetate. *Science, 171*, 801–802.

Lange, C. E. (1991). A Nosema-type microsporidian in *Ectomyelois ceratoniae* (Lepidoptera: Pyralidae). *Journal of Invertebrate Pathology, 58*, 348–352.

Latifian, M. (2012). The effects of cultural management on the lesser date moth (*Batrachedra amydraula* Myer) infestation. *Emirates Journal of Food and Agriculture, 24*, 224–229.

Leonard, M. D. (1933). Notes on insect conditions in Puerto Rico for the fiscal year, July 1931 thru June 1932. *Journal of the Department of Agriculture, Puerto Rico, 17*, 97–137.

Lepigre, A. (1963). Experimental control of date pyralid (*Myelois ceratoniae* Zeller: Pyralidae) on trees. *Annales des Epiphyties, 14*, 85–101.

Leraut, P. (2002). Contribution a l'etude des Phycitinae (Lepidoptera, Pyralidae). *Nouvelle Revue d'Entomologie, 19*, 141–177.

Levinson, H. Z., & Buchelos, C. T. (1981). Surveillance of storage moth species (Pyralidae, Gelechiidae) in a flour mill by adhesive traps with notes on the pheromone-mediated flight behaviour of male moths. *Zeitschrift für Angewandte Entomologie, 92*, 233–251.

Levinson, H. Z., & Gothilf, S. A. (1966). A semisynthetic diet for axenic growth of the carob moth *Ectomyelois ceratoniae* (Zell.). *Rivista di Parassitologia, 26*, 19–26.

Levi-Zada, A., Fefer, D., Anshelevitch, L., Litovsky, A., Bengtsson, M., Gindin, G., & Soroker, V. (2011). Identification of the sex pheromone of the lesser date moth, *Batrachedra amydraula*, using sequential SPME auto-sampling. *Tetrahedron Letters, 52*, 4550–4553.

Levi-Zada, A., Sandusky, A., Dobrinin, S., David, M., Ticuchinski, T., Fefer, D., Greenberg, A., & Blumberg, D. (2013). Reevaluation of the sex pheromone of the lesser date moth, *Batrachedra amydraula*, using autosampling SPME-GC/MS and field bioassays. *Chemoecology, 23*, 13–20.

Lindgren, D. L., Bliss, D. E., & Barnes, D. F. (1948). Insect infestations and fungus spoilage their relation and control. *Date Growers Institute Report, 25*, 12–17.

London Natural History Museum Global Lepidoptera Names Index. (2014). Available at: http://www.nhm.ac.uk/research-curation/research/projects/lepindex/search/detail.dsml?TaxonNo=9040.0. Accessed on 6 Mar 2015.

Lounsbury, C. P. (1919). Worms in walnuts. *South African Fruit Grower and Smallholder, 6*, 307–309.

Lucas, D. (1950). Contribution a l'etude des Lepidopteres nord-africains. *Bulletin de la Société Entomologique de France, 55*, 141–144.

Madkouri, M. (1978). Study of the host-parasite relations in the field and laboratory. *Awamia, 56*, 37–70.

Mafra-Neto, A., & Baker, T. C. (1996). Timed, metered sprays of pheromone disrupt mating of *Cadra cautella* (Lepidoptera: Pyralidae). *Journal of Agricultural Entomology, 13*, 149–168.

Mafra-Neto, A., De Lame, F. M., Fettig, C. J., Munson, A. S., Perring, T. M., Stelinski, L. L., Stoltman, L. L., Mafra, L. E. J., Borges, R., & Vargas, R. I. (2013). Manipulation of insect behavior with Specialized Pheromone and Lure Application Technology (SPLAT®). In J. J. Beck, J. R. Coats, S. O. Duke, & M. E. Koivunen (Eds.), Pest management with natural products (American Chemical Society Symposium Series, Washington DC, USA, 1141, pp. 31–58).

Mamay, M., Ikinci, A., Unlu, L., & Dogan, E. (2014). Infestation rate and host preference of Carob moth [*Apomyelois (= Ectomyelois) ceratoniae* Zeller (Lepidoptera: Pyralidae)] on different pomegranate cultivars. *Turkish Journal of Entomology, 38*, 101–110.

Mansour, S. A. (1984). Preliminary studies on insect attractants in pomegranate orchards. *Bulletin de la Societe Entomologique d'Egypte, 65*, 49–58.

Mart, C., & Kilincer, N. (1993a). Population fluctuations and number of generations of carob moth, *Ectomyelois ceratoniae* Zeller (Lepidoptera, Pyralidae) on pomegranate in Southeast Anatolia region. *Turkiye Entomoloji Dergisi, 17*, 209–216.

Mart, C., & Kilincer, N. (1993b). Studies on the development of the carob moth, *Ectomyelois ceratoniae* Zeller (Lepidoptera, Pyralidae) at the different temperatures. *Turkiye Entomoloji Dergisi, 17*, 77–86.

Martin, H. E. (1958). Pests and diseases of date palm in Libya. *FAO Plant Protection Bulletin, 6*, 120–123.

Martinez-Sanudo, I., Mazzon, L., Vecchia, P. D., Bagnoli, B., Lucchi, A., Marchesini, E., & Mori, N. (2013). Pyralidae Phycitinae in Italian vineyards: Behavioural and molecular genetic investigations. In A. Calonnec, C. Duso, C. Gessler, M. Maixner, D. Thiery, & T. Zahavi (Eds.), Proceedings of the IOBC/WPRS Working Group "Integrated Protection and Production in Viticulture", 2–5 October 2011, Lacanau, France. IOBC/WPRS Bulletin, 85, 211–215.

McFarlane, J. (1962). A note on the field infestation of pods of the tamarind tree, Tamarindus indica L., in Jamaica. *Entomologist's Monthly Magazine, 97*, 198–199.

Mediouni, J., & Dhouibi, M. H. (2007). Mass-rearing and field performance of irradiated carob moth *Ectomyelois ceratoniae* in Tunisia. In M. J. B. Vreysen, A. S. Robinson, & J. Hendrichs (Eds.), *Area-wide control of insect pests* (pp. 265–273). Dordrecht: Springer Science+Business Media.

Mehaoua, M. S., Hadjeb, A., Lagha, M., Bensalah, M. K., & Ouakid, M. L. (2013). Study of the toxicity of azadirachtin on larval mortality and fertility of carob moth's female *Ectomyelois ceratoniae* (Lepidoptera, Pyralidae) under controlled conditions. *American-Eurasian Journal of Sustainable Agriculture, 7*, 1–9.

Mehrnejad, M. R. (1993). Biology of the carob moth (*Apomyelois ceratoniae*) a new pest of pistachio in Rafsanjan. *Applied Entomology and Phytopathology, 60*, 1–2.

Mehrnejad, M. R. (1995). Carob moth, a pest of pistachio nut in Iran. In N. Kaska, A. B. Kuden, L. Ferguson, & T. Michailides (Eds.), First international symposium on pistachio nut, 20–24 September 1994. Adana, Turkey. ISHS Acta Horticulturae, 419, 365–372.

Mehrnejad, M. R. (2001). The current status of pistachio pests in Iran. *Cahiers Options Mediterraneennes, 56*, 315–322.

Mehrnejad, M. R., & Panahi, B. (2006). The influence of hull cracking on aflatoxin contamination and insect infestation in pistachio nuts. *Applied Entomology and Phytopathology, 73*, 39–42.

Mesbah, H. A., El Kady, M. B., & El Sayed, N. A. A. (1998). Survey and population studies on pests of date palm trees in Siwa oasis (Western Desert, Egypt). *Journal of the Egyptian German Society of Zoology, 25*, 99–117.

Meyer, M. (1996). The Lepidoptera of the Macaronesian region 7: the pyralid fauna of Madeira and the Azores (Microlepidoptera, Pyralidae). Annotated checklist and the state of recent revision. *Verhandlungen Westdeutscher Entomologentag*, 1996, 45–59.

Meyer, M., Nuss, M., & Spiedel, W. (1997). Commented checklist of the Pyraloidea of the Azores, with description of three new species (Lepidoptera: Pyraloidea). *Beitraege Zur Entomologie, 47*, 13–34.

Meyrick, E. (1935). *Exotic microlepidoptera* (2nd ed., Vol. 4, 553 pp). Hampton: E.W. Classey Limited.

Michael, P. (1968). Insect pests: The carob moth. *Journal of Agriculture, Western Australia, Ser, 4*(9), 81–82.

Michael, I. F. (1970). Economic importance and control of *Batrachedra amydraula* Meyr. (the lesser date moth) in the U.A.R. *Date Grower's Institute Report, 47*, 9–10.

Michael, I. F., & Habib, A. A. (1971). Biology of *Batrachedra amydraula* Meyr., the lesser date moth. *Date Growers Institute Report, 48*, 6–8.

Millar, J. G. (1990). Synthesis of 9Z, 11E, 13-tetradecatrienal, the major component of the sex-pheromone of the carob moth, *Ectomyelois ceratoniae* (Lepidoptera, Pyralidae). *Agricultural and Biological Chemistry, 54*, 2473–2476.
Mills, K. A., Wontner-Smith, T. J., Cardwell, S. C., & Bell, C. H. (2003a). The use of phosphine as an alternative to methyl bromide for the disinfestation of palm dates. In P. F. Credland, D. M. Armitage, C. H. Bell, P. M. Cogan, & E. Highley (Eds.), Advances in stored product protection, 8th International Working Conference on Stored Product Protection (IWCSPP), York, England (717–724 pp).
Mills, K. A., Wontner-Smith, T. J., Cardwell, S. C., & Bell, C. H. (2003b). The use of carbon dioxide as an alternative to methyl bromide for the disinfestation of palm dates. In P. F. Credland, D. M. Armitage, C. H. Bell, P. M. Cogan, & E. Highley (Eds.), Advances in stored product protection, 8th International Working Conference on Stored Product Protection (IWCSPP), York, England (729–735 pp).
Mineo, G. (1967). La *Myelois ceratoniae* Zeller infesta anche le olive. *Università di Palermo Istituto di Entomologia Agraria, 7*, 5–7.
Mirkarimi, A. (2000). Biological control of carob moth with mass release of *Trichogramma embryophagum* Hartig for Pomegranate worm control, the *Ectomyelois* (*Spectrobates*) *ceratoniae* Zell. *Iranian Journal of Agricultural Sciences, 31*, 103–110.
Mirkarimi, A. (2002). The effect of stuffing Pomegranate neck (Ceelyx) on reduction of Pomegranate neck worm *Spectrobates ceratoniae* Zell. (Lep. Pyralidae: Phycitinae) damage. *Iranian Journal of Agricultural Sciences, 33*, 375–383.
Mnif, I., Elleuch, M., Chaabouni, S. E., & Ghribi, D. (2013). *Bacillus subtilis* SPB1 biosurfactant: Production optimization and insecticidal activity against the carob moth *Ectomyelois ceratoniae*. *Crop Protection, 50*, 66–72.
Moawad, G. M. (1979). Ecological studies on the pomegranate fruit moth, *Ectomyelois ceratoniae* Zeller (Lepidoptera, Pyralidae) in Saudi Arabia. *Indian Journal of Agricultural Sciences, 49*, 739–741.
Moezipour, M., & Shojaei, S. S. (2008). Developmental biology and fertility life table parameters of *Trichogramma brassicae* (Bezdenko), on eggs of *Sitoteroga* [*Sitotroga*] *cerealella* at different temperatures, in laboratory conditions. *Pakistan Entomologist, 30*, 5–10.
Moezipour, M., Kafil, M., & Allahyari, H. (2008). Functional response of *Trichogramma brassicae* at different temperatures and relative humidities. *Bulletin of Insectology, 61*, 245–250.
Moezzipour, M., Goldansaz, S. H., Radjabi, G., Attaran, M., & Seyyedoleslami, H. (2009). Fertility life table parameters of *Trichogramma brassicae* Bezd., collected from pomegranate orchards of Saveh, at three different constant temperatures and relative humidity levels. *Journal of Agricultural Sciences, 1*, 51–61.
Moore, D. (2001). Insects of palm flowers and fruits. In F. W. Howard, D. Moore, R. M. Giblin-Davis, & R. G. Abad (Eds.), *Insects on palms* (pp. 233–266). Wallingford: CABI Publishing, CAB International.
Mozaffarian, F., Sarafrazi, A., & Ganbalani, G. N. (2007a). Host plant-associated population variation in the carob moth *Ectomyelois ceratoniae* in Iran: A geometric morphometric analysis suggests a nutritional basis. *Journal of Insect Science, 7*, 1–11.
Mozaffarian, F., Sarafrazi, A., & Ganbalani, G. N. (2007b). Sexual dimorphism in the wing shape and size of the carob moth, *Ectomyelois ceratoniae* (Lepidoptera: Pyralidae). *Journal of Entomological Society of Iran, 26*, 61–73.
Mozaffarian, F., Sarafrazi, A., Ganbalani, G. N., & Ariana, A. (2007c). Morphological variation among Iranian populations of the Carob Moth, *Ectomyelois ceratoniae* (Zeller, 1839) (Lepidoptera: Pyralidae). *Zoology in the Middle East, 41*, 81–91.
Mozaffarian, F., Mardi, M., Sarafrazi, A., & Ganbalani, G. N. (2008). Assessment of geographic and host-associated population variations of the carob moth, *Ectomyelois ceratoniae*, on pomegranate, fig, pistachio and walnut, using AFLP markers. *Journal of Insect Science, 8*, 1–9.
Munroe, E. G. (1983). Pyralidae. In R. W. Hodges, T. Dominick, D. R. Davis, D. C. Ferguson, J. G. Franclemont, E. G. Munroe, & J. A. Powell (Eds.), *Check list of the Lepidoptera of America*

north of Mexico including Greenland (pp. 67–85). London: E.W. Classey Ltd., Wedge Entomological Research Foundation.

Navarro, S., Donahaye, E., & Calderon, M. (1986). Development of the carob moth, *Spectrobates ceratoniae*, on stored almonds. *Phytoparasitica, 14*, 177–186.

Navon, A., Nakashe, Y., Blumberg, D., Levski, S., Gan-Mor, S., Bechar, A., Ronen, B., & Ucko, O. (1999). The effectiveness of *Bacillus thuringiensis* in granular feeding baits for use against the lesser date moth, *Batrachedra amydraula* in date palms, in comparison with chemical and biological control agents (1995–1999). *Alon Hanotea, 53*, 484–489.

Nay, J. E. (2006). Biology, ecology, and management of the carob moth, *Ectomyelois ceratoniae* (Zeller) (Lepidoptera: Pyralidae), a pest of dates, *Phoenix dactylifera* L., in Southern California (278 pp). PhD Dissertation, University of California, Riverside, California. USA.

Nay, J. E., & Perring, T. M. (2005). Impact of ant predation and heat on carob moth (Lepidoptera: Pyralidae) mortality in California date gardens. *Journal of Economic Entomology, 98*, 725–731.

Nay, J. E., & Perring, T. M. (2006). Effect of fruit moisture content on mortality, development, and fitness of the carob moth (Lepidoptera: Pyralidae). *Environmental Entomology, 35*, 237–244.

Nay, J. E., & Perring, T. M. (2008a). Influences of host plant stages on carob moth (Lepidoptera: Pyralidae) development and fitness. *Environmental Entomology, 37*, 568–574.

Nay, J. E., & Perring, T. M. (2008b). Blue Palo Verde, *Parkinsonia florida* (Benth. ex Grey) S. Wats. (Fabaceae), a new host of the introduced carob moth, *Ectomyelois ceratoniae* (Zeller) (Lepidoptera: Pyralidae), in Southern California. *Pan-Pacific Entomologist, 84*, 143–145.

Nay, J. E., & Perring, T. M. (2009). Effect of center cut strand thinning on fruit abscission and *Ectomyelois ceratoniae* (Lepidoptera: Pyralidae) infestation in California date gardens. *Journal of Economic Entomology, 102*, 948–953.

Nay, J. E., Boyd, E. A., & Perring, T. M. (2006). Reduction of carob moth in 'Deglet Noor' dates using a bunch cleaning tool. *Crop Protection, 25*, 758–765.

Nay, J. E., Park, Y. L., & Perring, T. M. (2007). Effect of bunch sanitation on spatial distribution of abscised fruit and phycitine moths (Lepidoptera: Pyralidae) in California date gardens. *Journal of Economic Entomology, 100*, 1773–1780.

Neunzig, H. H. (1979). Systematics of immature phycitines (Lepidoptera: Pyralidae) associated with Leguminous plant in the southern United States (United States Department of Agriculture, Agriculture Research Service, Technical Bulletin No. 1589, Washington DC, USA, 119 pp).

Neunzig, H. H. (1990). *The moths of America north of Mexico including Greenland. Fascicle 15.3 Pyraloidea, Pyralidae (Part), Phycitinae (Part)* (165 pp). Washington, DC: Wedge Entomological Research Foundation.

Nixon, G. E. J. (1976). A revision of the North-western European species of the merula, laeteus, vipio, ultor, ater, butalidis, popularis, carbonarius, and validus groups of *Apanteles* Förster (Hymenoptera: Braconidae). *Bulletin of Entomological Research, 65*, 687–732.

Nixon, R. W., & Carpenter, J. B. (1978). Growing dates in the United States.(United States Department of Agriculture Information Bulletin No. 207, 63 pp). Washington, DC: United States Government Printing Office.

Norouzi, A., Talebi, A. A., & Fathipour, Y. (2008). Development and demographic parameters of the carob moth *Apomyelois ceratoniae* on four diet regimes. *Bulletin of Insectology, 61*, 291–297.

Norouzi, A., Talebi, A. A., Fathipour, Y., & Lozan, A. I. (2009). *Apanteles laspeyresiellus* (Hymenoptera: Braconidae), a new record for Iran insect fauna. *Journal of Entomological Society of Iran, 28*, 79–80.

Ould, E. M. D., Abdi, M., & Doumandji, S. (2007). Impact of treatment with an acridicide on the entomofauna associated with date palms in a palm grove in the Ouargla basin (Northeastern Northern Algerian Sahara). *Rivista Italiana EPPOS, 43*, 25–36.

Ozturk, N., Olculu, M., & Ulusoy, M. R. (2011). The adult population dynamics of the Carob moth (*Ectomyelois ceratoniae* Zell. (Lepidoptera: Pyralidae) in citrus orchards in Adana and Mersin provinces. *Bitki Koruma Bulteni, 51*, 347–35.

Ozturk, N., Ulusoy, M. R., & Bayhan, E. (2005). Pest and natural enemy species determined in pomegranate orchards in the Eastern Mediterranean Region, Turkey. *Turkiye Entomoloji Dergisi, 29*, 225–235.

Palm, E. (1986). *North European pyralids – with particular reference to the Danish fauna (Lepidoptera: Pyralidae). Danmarks Dyreliv 3* (287 pp). Copenhagen: Fauna Books.

Palmony, Y. (1969). The Pyralidae of the lake Tiberias region: A faunistic pheno-ecological survey. *Israel Journal of Entomology, 4*, 293–321.

Park, J. J., & Perring, T. M. (2010). Development of a binomial sampling plan for the carob moth (Lepidoptera: Pyralidae), a pest of California dates. *Journal of Economic Entomology, 103*, 1474–1482.

Perring, T. M., & Nay, J. E. (2015). Evaluation of bunch protectors for preventing insect infestation and preserving yield and fruit quality of dates, *Phoenix dactylifera* L. *Journal of Economic Entomology, 108*, 654–661.

Perring, T. M., Farrar, C. A., & Nay, J. E. (2005). *Carob moth and raisin moth control on dates. Western Region IR4 efficacy program final report* (4 pp). Riverside: Department of Entomology, University of California.

Perring, T., Braverman, M., & Mafra-Neto, A. (2014). Pheromone-based mating disruption reduces pest attack on dates and results in new registration. *United States Department of Agriculture, IR-4 Project Newsletter, 45*(8–9), 13.

Peyrovi, M., Goldansaz, S. H., & Jahromi, K. T. (2011). Using *Ferula assafoetida* essential oil as adult carob moth repellent in Qom pomegranate orchards (Iran). *African Journal of Biotechnology, 10*, 380–385.

Phelan, P. L., & Baker, T. C. (1986). Cross-attraction of five species of stored product Phycitinae (Lepidoptera: Pyralidae) in a wind tunnel. *Environmental Entomology, 15*, 369–372.

Phelan, P. L., & Baker, T. C. (1990). Comparative study of courtship in twelve phycitine moths (Lepidoptera: Pyralidae). *Journal of Insect Behavior, 3*, 303–326.

Pintureau, B., & Daumal, J. (1979). Les *Ectomyelois* de l'Ancien Monde (Lep. Pyralidae). *Bulletin de la Societe Entomologique de France, 84*, 84–88.

Poorjavad, N., Goldansaz, S. H., Hosseininaveh, V., Nozari, J., Dehghaniy, H., & Enkegaard, A. (2011). Fertility life table parameters of different strains of *Trichogramma* spp. collected from eggs of the carob moth *Ectomyelois ceratoniae*. *Entomological Science, 14*, 245–253.

Priesner, E. (1989). Two component sexual attractant for male *Batrachedra pinicolella* (Zeller) (Lepidoptera: Batrachedridae). *Zeitschrift für Naturforschung, 44*, 1061–1062.

Rahmani, S., Marouf, A., & Amir-Maafi, M. (2008). Demography of *Batrachedra amydraula* (Lep; Batrachedridae) under laboratory conditions. *Applied Entomology and Phytopathology, 76*, 79–90.

Real, P. (1948). Les Myelois parasites des dates (Lepidoptera: Phycitinae). *Revue de Pathologie Vegetale et d'Entomologie Agricole de France Paris, 27*, 59–64.

Reuveni, O. (1986). Dates. In S. P. Monselise (Ed.), *CRC handbook of fruit set and development* (pp. 119–144). Boca Raton: CRC Press.

Ridgway, R. L., Silverstein, R. M., & Inscoe, M. N. (1990). *Behavior-modifying chemicals for insect management: Applications of pheromones and other attractants* (761 pp). Madison Avenue: Marcel Dekker.

Roesler, R. U. (1968). Das neue systematische Verzeichnis der deutschen Phycitinae (Lepidoptera, Pyralidae). *Nachrichtenblatt der Bayerischen Entomologen, 17*, 1–10.

Roesler, R. U., & Küppers, R. V. (1981). Beitrage zur Kenntnis der Insektenfauna Sumatras. Teil 9. Die Phycitinae (Lepidoptera: Pyralidae) von Sumatra; Taxonomie Teil 13, Okologie und Geobiologie. *Beitraege zur Naturkundlichen Forschung in Suedwestdeutschland Beihefte, 4*, 1–282.

Saadat, D., Seraj, A. A., Goldansaz, S. H., & Karimzadeh, J. (2014). Environmental and maternal effects on host selection and parasitism success of Bracon hebetor. *BioControl, 59*, 297–306.

Sarjami, M. S., Ghanbalani, G. N., Goldansaz, H., & Zakaria, R. A. (2009). Calling behavior of the carob moth, *Ectomyelois ceratoniae* (Zeller) (Lepidoptera: Pyralidae), laboratory and field experiments. *Munis Entomology and Zoology, 4*, 472–485.

Schiliro, E., & Bellini, G. (1978). Damage to orange fruits caused by *Myelois ceratoniae*. *Tecnica Agricola, 30*, 425–430.

Sedlacek, J. D., Weston, P. A., & Barney, R. J. (1996). Lepidoptera and Psocoptera. In B. Subramanyam & D. W. Hagstrum (Eds.), *Integrated management of insects in stored products* (pp. 41–70). New York: Marcel Dekker.

Serghiou, C. S. (1983). The citrus mealybug, *Planococcus citri* Risso – carob moth, *Ectomyelois ceratoniae* Zeller, pest complex on grapefruit and its chemical control. *Technical Bulletin, Agricultural Research Institute, Nicosia, Cyprus, 56*, 1–17.

Shaffer, J. C. (1995). Phycitinae. In J. B. Heppner (Ed.), *Atlas of neotropical Lepidoptera. Checklist: Part 2. Hyblaeoidea – Pyraloidea – Tortricoidea* (Vol. 3, pp. 93–105). Gainesville: Association for Tropical Lepidoptera and Scientific Publishers.

Shafik, M. (1938). Constant temperature hot air sterilizer for the control of *Ephestia* and *Myelois* on Dates. *Bulletin de la Société Fouad Ier d'Entomologie, 22*, 233–264.

Shakeri, M. (1993). First report of attack of *Spectrobates ceratoniae* Zell. to figs in Iran. *Applied Entomology and Phytopathology, 60*, 29.

Shani, A., & Clearwater, J. (2001). Evasion of mating disruption in *Ephestia cautella* (Walker) by increased pheromone production relative to that of undisrupted populations. *Journal of Stored Products Research, 37*, 237–252.

Shayesteh, N., Marouf, A., & Amir-Maafi, M. (2010). Some biological characteristics of the *Batrachedra amydraula* Meyrick (Lepidoptera: Batrachedridae) on main varieties of dry and semi-dry date palm of Iran. In: 10th international working conference on stored product protection. Julius-Kühn-Archiv, 425, 151–154.

Sheppard, R. A. (1926). Insect pests imported on miscellaneous plant products. In 56th annual report of the Entomological Society of Ontario, 1925, Canada (pp. 50–54).

Simmons, P., & Nelson, H. D. (1975). Insects on dried fruits (United States Department of Agriculture, Agricultural Research Service, Handbook No. 464, 25 pp). Washington, DC: Agricultural Research Service, U.S. Department of Agriculture.

Sinev, S. Y. (1986). A guide to the insects of the European part of the USSR. Lepidoptera. Family Phycitidae. *Opredeliteli po Faune SSSR, 4*, 251–340.

Singh, M. P. (1991). Observations on the pest complex of *Balanites aegyptiaca* (L.) Delile in the arid zones of India. *Entomon, 16*, 249.

Smithwick, E. B., & Brady, U. E. (1977a). Site and production of sex pheromone in developing female Indian meal moth, *Plodia interpunctella*. *Journal of Georgia Entomological Society, 12*, 1–13.

Smithwick, E. G., & Brady, U. E. (1977b). Histology of the sex pheromone gland in developing female Indian meal moths, *Plodia interpunctella*. *Journal of Georgia Entomological Society, 12*, 13–29.

Sohn, J. C., Park, K.-T., Lee, S. K., & Cho, S. (2009). A taxonomic review of the genus *Batrachedra* (Lepidoptera: Gelechioidea: Batrachedridae) in Korea. *Journal of Asia-Pacific Entomology, 12*, 101–105.

Sonda, M. (1963). An interception of the locust bean moth, *Ectomyelois ceratoniae* (Zeller), in Japan, on English walnuts from Italy (Lepidoptera, Pyralididae). *Kontyu, 31*, 292–293.

Soofbat, M., Nouri, G., Goldansaz, S. H., & Asghari-zakaria, R. (2007). Effects of age and temperature on calling behavior of carob moth, *Ectomyelois ceratoniae*, Zell. (Lepidoptera: Pyralidae) under laboratory conditions. *Pakistan Journal of Biological Sciences, 10*, 2976–2979.

Sorhagen, L. (1881). Aus meinem entomologischen Tagebuche. *Berliner Entomologische Zeitschrift, 25*, 17–40.

Sternlicht, M. (1971). Mites and moths attacking citrus trees. *Mites and Moths Attacking Citrus Trees, 51*, 1018–1020.

Struble, D. L., & Richards, K. W. (1983). Identification of sex pheromone components of the female dried fruit moth, *Vitula edmandsae serratilineella*, and a blend for attraction of male moths. *Journal of Chemical Ecology, 9*, 785–801.

Strümpel, H. (1969). Die tierischen Schädlinge der Dattelpalme Phoenix dactylifera L. und ihrer Früchte in Nordafrika. *Zeitschrift fuer Angewandte Entomologie, 64*, 233–240.
Sumpich, J., & Skyva, J. (2008). Faunistic records from the Czech Republic – 255. Lepidoptera: Tineidae, Gracillariidae, Gelechiidae, Tortricidae, Alucitidae, Pyralidae. *Klapalekiana, 44*, 83–85.
Swezey, O. H. (1923). The Erythrina Twig-Borer (*Terastia meticulosalis*) in Hawaii (Pyralidae, Lepidoptera). *Proceedings of the Hawaiian Entomological Society, 5*, 297–298.
Taki, M., Dabbagh, G. R., Torabi, R., & Kavianpoor, M. R. (2014). Some mechanical methods to behavior manipulation of the carob moth, *Ectomyelois ceratoniae* in pomegranate orchards. *International Journal of Biosciences, 4*, 67–73.
Talhouk, A. S. (1983). The present status of date palm pests in Saudi Arabia. In Proceedings of the first symposium on date palms, 23–25 Mar 1982, Saudi Arabia, 432–438 pp).
Talhouk, A. S. (1991). On the management of the date palm and its arthropod enemies in the Arabian Peninsula. *Journal of Applied Entomology, 111*, 514–520.
The Species 2000 and ITIS Catalogue of Life. (2011). Available at: http://www.catalogueoflife.org/annual-checklist/2011. Accessed on 6 Mar 2015.
Thompson, W. R. (1946). A catalogue of the parasites and predators of insect pests. *Commonwealth Bureau of Biological Control, 1*, 259–385.
Todd, J. L., Millar, J. G., Vetter, R. S., & Baker, T. C. (1992). Behavioral and electrophysiological activity of (Z, E)-7,9,11-dodecatrienyl formate, a mimic of the major sex-pheromone component of carob moth, *Ectomyelois ceratoniae. Journal of Chemical Ecology, 18*, 2331–2352.
Tokmakogiu, C., Soylu, Z., Devecioughu, H., & Devecioglu, H. (1967). Studies on the biology and methods of controlling carob worm, *Myelois ceratoniae. Bitki Koruma Bülteni, 7*, 91–106.
Trabut. (1923). Diseases and enemies of the fig tree. *Agricultural Bulletin of Algeria, Tunisia, Morocco, 29*, 117–124.
Ucko, O., & Bitton, S. (2000). *Recommendations for the control of pests and diseases of the date palm.* Israel: Israel Ministry of Agriculture, Extension Service and Diklaim Organization (in Hebrew).
USDA - United States Department of Agriculture. (2014). National Agricultural Statistics Service, Noncitrus fruits and nuts, date, 2013. Available at: http://www.nass.usda.gov/Statistics_by_Subject/result.php?50B6F2CA-1B72-321C-84FD-E739C4D2DB04&sector=CROPS&group=FRUIT%20%26%20TREE%20NUTS&comm=DATES. Accessed on 6 Mar 2015.
Vaughan, H. (1870). Descriptions of three species of Phycidae (from Britain) new to science. *Entomologist's Monthly Magazine, 7*, 130–132.
Venezian, A., & Blumberg, D. (1982). Phenology, damage and control of the lesser date moth, *Batrachedra amydraula* Meyrick in date palms in Israel. *Alon Hanotea, 36*, 785–788.
Vetter, R. S., Tatevossian, S., & Baker, T. C. (1997). Reproductive behavior of the female carob moth, (Lepidoptera: Pyralidae). *Pan-Pacific Entomologist, 73*, 28–35.
Vetter, R. S., Millar, J. G., Vickers, N. J., & Baker, T. C. (2006). Mating disruption of carob moth, *Ectomyelois ceratoniae*, with a sex pheromone analog. *Southwestern Entomologist, 31*, 33–47.
Vincent, L. E., & Lindgren, D. L. (1954). Malathion dust for control of nitidulid beetles and pyralid moths infesting Deglet Noor and Khadrawi dates in the Coachella Valley. *Date Growers Institute Report, 31*, 28.
Vincent, L. E., & Lindgren, D. L. (1971). Fumigation of dried fruit insects with hydrogen phosphide and ethyl formate. *Date Growers Institute Report, 48*, 4–5.
Warner, R. L. (1988). Contributions to the biology and the management of the carob moth, *Ectomyelois ceratoniae* (Zeller) in 'Deglet Noor' date gardens in the Coachella Valley of California (260 pp). PhD thesis, University of California, Riverside, California, USA.
Warner, R. L., Barnes, M. M., & Laird, E. F. (1990a). Reduction of insect infestation and fungal infection by cultural practice in date gardens. *Environmental Entomology, 19*, 1618–1623.
Warner, R. L., Barnes, M. M., & Laird, E. F. (1990b). Chemical control of a carob moth, *Ectomyelois ceratoniae* (Lepidoptera: Pyralidae), and various nitidulid beetles (Coleoptera: Nitidulidae) on 'Deglet Noor' dates in California. *Journal of Economic Entomology, 83*, 2357–2361.

Warren, W., & Rothschild, N. C. (1905). Lepidoptera from the Sudan. *Novitates Zoologicae, 12*, 21–23.
Wertheimer, M. (1958). Un des principaux parasites du palmier-dattier algerien le Myelois decolor. *Fruits, 13*, 309–323.
Widiez, M. (1932). The fig cultivation in North Africa. *Agriculture Pratique des Pays Chauds, 24*, 415–445.
Wilkinson, D. S. (1925). Entomological notes. *Cyprus Agricultural Journal, 20*, 9–10.
Wilkinson, D. S. (1937). A new species of Apanteles (Hym.: Brac.) bred from *Myelois ceratoniae* attacking carobs in Cyprus. *Bulletin of Entomological Research, London, 28*, 463–166.
Wiltshire, E. P. (1957). *The Lepidoptera of Iraq* (Revised edition, 162 pp). London: Nicholas Kaye Limited.
Wysoki, M. (1977). Insect pests of macadamia in Israel. *Phytoparasitica, 5*, 187–188.
Wysoki, M. (1986). New records of Lepidopterous pests of macadamia in Israel. *Phytoparasitica, 14*, 147.
Yoshiyasu, Y., & Kitatsuji, R. (2008). Discovery of *Ectomyelois ceratoniae* (Lepidoptera, Pyralidae) in a peach orchard in Japan. *Scientific Reports of Kyoto Prefectural University, Life and Environmental Sciences, 60*, 31–37.
Yousefi, M., Sendi, J. J., & Salehi, L. (2004). Biology of the Carob moth, *Spectrobates ceratoniae* Zeller (Lep.: Pyralidae) in different temperature regiments under laboratory conditions. *Journal of Agriculture Sciences, 1*, 29–38.
Zaid, A., & De Wet, P. F. (1999). Botanical and systematic description of the date palm. In A. Zaid (Ed.), Date palm cultivation (FAO Plant Production and Protection Paper No. 156, pp. 29–44). Rome: Food and Agriculture Organization of the United Nations.
Zaid, A., De Wet, P. F., Djerbi, M., & Oihabi, A. (1999). Diseases and pests of date palm. In: Zaid, A. (Ed.), Date palm cultivation (FAO Plant Production and Protection Paper No. 156, pp. 223–278). Rome: Food and Agriculture Organization of the United Nations.
Zand, A. J., Fesharaki, S. S. H., Karimzadeh, J., Besharatnejad, M. H., & Gavanji, S. (2013). Study of damage of *Ectomyelois ceratoniae* in three pomegranate cultivars. *Technical Journal of Engineering and Applied Sciences, 3*, 344–348.
Zare, D., Sendi, J. J., Nodoushan, A. J., & Khosravi, R. (2013). Life table parameters and biological characteristics of *Apomyelois ceratoniae* Zeller (Lepidoptera: Pyralidae) on three cultivars of pomegranate. *Archives of Phytopathology and Plant Protection, 46*, 766–773.
Zaviezo, T., Romero, A., Castro, D., & Wagner, A. (2007). First record of *Goniozus legneri* (Hymenoptera: Bethylidae) in Chile. *Ciencia E Investigation Agraria, 34*, 57–61.
Zeller, P. E. (1839). Versuch einer naturgemässen Eintheilung der Schaben. *Isis von Oken, 32*, 167–220.
Zimmerman, E. C. (1958). Insects of Hawaii (volume 8), Lepidoptera: Pyraloidea. *Ectomyelois ceratoniae*, 360–361. University of Hawaii Press, Honolulu, Hawaii.
Zirari, A., & Laaziza Ichir, L. (2010). Effect of different kinds of bunch coverings on date palm fruit (*Phoenix dactylifera* L.) moth's infestation rate. In A. Zaid, & G. A. Alhadrami (Eds.), Proceedings of the 4th International Date Conference. IAHS Acta Horticultura, 1009–1014.
Zolfagharieh, H., Babaii, M., Omidpour, A., Mostafavi, H., & Tabatabaii, Z. (2011). Evaluation of potential of sterile insect technique in controlling *Ectomyelois ceratoniae* Zeller. In M. K. Sheikh, A. N. Mokashi,, & A. K. Rokhade (Eds), II international symposium on pomegranate and minor – including mediterranean fruits). ISHS Acta Horticulturae, 890, 561–564.

Chapter 7
Major Hemipteran Pests

Hamadttu A.F. El-Shafie, Jorge E. Peña, and Mohammed Zaidan Khalaf

Abstract This chapter highlights the geographical distribution, biology, ecology and economic importance of six major Hemipteran pests of date palm. In addition, 14 species are listed as minor pests, however, some of which are potentially serious and may become major ones. Sucking of date palm sap by hemipterous pests usually results in chlorosis or discoloration of leaves, fruit malformation, yellowing, pre-mature withering and dropping of leaves. Moreover, the copious amount of honeydew secretion by these pests attracts and encourages sooty growth on surfaces of fronds that eventually impairs photosynthesis. The severity of infestation by these pests differs according to the locality, date palm cultivars, environmental conditions and their management practices. In many date palm growing countries, control of hemipteran pests depends largely on the application of synthetic insecticides, which aggravated the situation by killing their natural enemies. Future management strategy for these pests on date palm should rely on finding alternative means of control to replace the existing practice of excessive use of non-selective insecticides. Such alternatives should include pest population monitoring, estimation of economic threshold, use of biological control agents, selected cultural practices and use of resistant cultivars. These components of pest control should be used in a more compatible pattern as an integrated management program.

H.A.F. El-Shafie (✉)
Date Palm Research Center of Excellence, King Faisal University,
Hofuf, Al-Ahsa, Saudi Arabia
e-mail: elshafie62@yahoo.com

J.E. Peña
Department of Entomology and Nematology, Tropical Research and Education Center,
University of Florida, Homestead, FL, USA
e-mail: jepena@ufl.edu

M.Z. Khalaf
Integrated Pest Control Research Center, Directorate of Agricultural Research,
Ministry of Science and Technology, Karrada, Jadyria, Baghdad, Iraq
e-mail: mkhalaf34@yahoo.co.uk

W. Wakil et al. (eds.), *Sustainable Pest Management in Date Palm: Current Status and Emerging Challenges*, Sustainability in Plant and Crop Protection,
DOI 10.1007/978-3-319-24397-9_7

7.1 Introduction

The frond (leaf) of date palm consists of a blade or lamina, a petiole (midrib or rachis) and a base. The petioles of palm fronds are thick, tough and are normally rigid, but, show flexibility during strong wind. The leaflets or pinnae are in symmetrical ranks from either side of the rib with each blade folded lengthwise in a uniform pattern. The length of date palm frond can reach 5 cm and thrives for up to 6 years. Fronds can be grouped according to their location on the tree and therefore their age i.e. old, middle age and young. Due to these morphological features, sampling of insects on these parts can be precisely defined, standardized and applied (Al-Shayeb and Seaward 2000).

A date palm has an average of 112 green leaves (Nixon and Wedding 1956). Most species of insects that feed on palm fronds show preference for the abaxial (lower) surfaces because of protection from abiotic and biotic factors. In addition, the thicker wax layer of the adaxial (upper) surfaces of fronds is more difficult for insects to penetrate (Howard et al. 2001).

A basic understanding of the date palm morphology is very essential for proper monitoring, sampling of date palm insect pests and studying their phenology. Howard et al. (2001) recognized Hemiptera as one of six orders of insects that include notable palm pests worldwide. In general, pests in the order Hemiptera that feed on date palms attack their fronds, trunks and fruits, fall into six families; Tropiduchidae and Issidae (date bugs) (Shah et al. 2012), Asterolecanidae (pit scales), Phoenicoccidae, Diaspididae (date scales) and Pseudococcidae (mealybugs). According to Gullan and Cook (2007), the order Hemiptera is divided into two suborders:

- Sternorrhyncha which include the scale insects (Coccoidea), aphids, whiteflies and psyllids
- Auchenorrhyncha including cicadas, hoppers and fulgorids

The members of the superfamily Coccoidea are important pests of agricultural, horticultural and forest crops (Gullan and Cook 2007). The taxonomy of the Coccoidea is mainly based on the microscopic cuticular features of the adult female, which is paedomorphic, maturing to a wingless juvenile form with functional mouthparts, whereas the adult male (when present) goes through "prepupal" and "pupal" stages and turns into an alate form with non-functional mouthparts (Kondo et al. 2008). Adult females are small insects (1.0–2.0 mm); covered by an infuse scale, legs absent, antennae reduced to stumps; abdominal segments five to six fused into a pygidium, which is important in the diagnosis of species. The pygidium bears a complex of specialized structures, tubular ducts, marginal lobes, plates, gland spines and the anus lies on the dorsal surface (http://wbd.etibioinformatics.nl/bis/diaspididae.php). Moghaddam (2013) listed 13 families of scale insects in Iran represented by 275 species.

Scale insects are cosmopolitan and highly polyphagous (McClure 1990). The adults and nymphs, suck the host sap causing direct tissue damage through toxicity of their saliva (McClure 1990). These insects cause necrosis and chlorosis of leaves, severe infestations may lead to death of the host plant. Miller and Davidson (2005) listed 13 species of Diaspidids attacking palms in the genus *Phoenix*, including *Parlatoria blanchardi* Targioni-Tozzetti (Hemiptera: Diaspididae). For instance, Blumberg (2008) stated that in 1950s, the white scale *P. blanchardi* and the green scale *Palmaspis phoenicis* caused severe damage on date palm in Israel. Bénassy (1990) also reported that one of the pests that can cause yield reduction in cultivated date palms is the parlatoria date scale, *P. blanchardi*, which originated in the oases of Mesopotamia; the *Parlatoria* scale was introduced from the Middle East into many date-growing countries through the propagation and replanting of infested date palm offshoots. Another group of important date palm insect pests is the pit scales, which produce gall-like pits in the bark of their host.

The red date scale *Phoenicococcus marlatti* Cockerell (Phoenococcidae) ravage date palm in the southwestern states of USA. This scale is usually found at the base of leaf petioles or under fibrous covering of the trunk. The superfamily Fulgoroidea is represented by the Issid date bug, *Asarcopus palmarum* that feeds on plant juices and like many Hemiptera, produce honeydew. According to Gnezdilov (2009) much knowledge is needed on the morphology and molecular analysis of these planthoppers.

This chapter gives an overview of the host range and distribution of important Hemipteran pests of date palm, the description of damage they inflict, their biology and bionomics and possible control measures adopted against them.

7.2 Dubas Date Bug (*Ommatissus lybicus*)

7.2.1 Distribution and Host Range

Ommatissus lybicus (De Bergevin) Asche and Wilson (Family: Tropiduchidae) is known as "dubas" in Arab countries, and "dubas bug" or "Old World date bug" in the literature (Kranz et al. 1978). The bug is commonly called planthopper in some countries of the Middle East (Howard 2001). The name of dubas bug came from the Arabic word "dibis" which means honeydew (Hussain 1963). Sheragoo is the local name to describe the secretion of honeydew by dubas in Pakistan (Shah et al. 2012). More details on the taxonomy of the genus *Ommatissus* were given by Asche and Wilson (1989). Dubas bug is a key pest of date palms in many date palm-growing countries in the world. The bug is recorded from the Near East, North Africa and Southeast Russia, Iran, and Spain. The pest is now reported from Egypt, Israel, Saudi Arabia, Oman, Iraq, Yemen, United Arab Emirates (UAE), Qatar and Pakistan (Klein and Venezian 1985; Al-Azawi 1986; El-Haidri and Al-Hafidh 1986; Kinawy 2005). It is likely that it originated from the Tigris-Euphrates River Valley (Dowson 1936). The bug moved to other areas probably through transportation of date palm offshoots infested with its egg stage (Howard et al. 2001). Dubas bug has shown high specificity to date palm (*Phoenix dactylifera* L.) (Howard et al. 2001; Gassouma 2004; Blumberg 2008). However, it can also be found on *Chamaerops humilis* L.

(Lepesme 1947). In addition, Shah et al. (2012) collected dubas from *Nannorrhops ritchieana* (Family: Palmaceae) about 15–18 km apart from date palm groves in Panjgur, Pakistan.

7.2.2 Biology and Seasonal Incidence

The adult is yellowish brown, to greenish (Fig. 7.1), and a strong chitinous saw-like tool surrounding the ovipositor used to burrow a lane tunnel in date palm tissues is a distinguishing feature of females. Females are usually longer than the males, reaching 5–6 mm compared to the 3–3.5 mm of males. There are two black spots on the front side of the head in both sexes. The eggs are elongate in shape, 0.5–0.8 mm long and light green in color when laid, but turn yellowish white, then bright yellow before hatching. There is a distinct suture at the front end of each egg. There are five nymphal instars, each of which is yellow to greenish yellow, distinguished by a bundle of waxy caudal filaments and a number of dorsal gray lines along their bodies (Al-Abbasi 1988) (Fig. 7.2). The nymphal duration is about 6 weeks.

The female makes small holes in frond tissues especially in the midrib of the leaflets to lay eggs. Each tunnel contains only one egg and the top of the laid egg usually protrudes from the tunnel. In case of severe infestations, eggs are not only laid on both lower and upper surface of the leaflets but also on the fruit stalks, and a single female can lay between 100 and 180 eggs (Hussain 1963; Jassim 2007). Most of the eggs are laid in straight compact rows parallel with the long axis of the leaflets in a pattern resembling straight sewing machine stitches (Dowson 1936; Klein and Venezian 1985).

Dubas bugs have two generations per year, a spring and an autumn generation. It hibernates in the winter and aestivates in the summer as eggs in the axial veins and

Fig. 7.1 An adult of dubas bug, *Ommatissus lybicus* (Photo: Mohammed Z. Khalaf)

Fig. 7.2 *Ommatissus lybicus* nymph showing caudal filaments and dorsal lines (Photo: Mohammed Z. Khalaf)

Fig. 7.3 Scars of egg laying by *Ommatissus lybicus* on the midrib of a date palm frond (Photo: Mohammed Z. Khalaf)

midrib of the date palm fronds (Fig. 7.3). The adults can fly, and their habitat is the same as the nymphs (monoecious). On an average, dubas completes total life cycle of the first and second generations in 217.25 and 136.35 days, respectively (Shah et al. 2012).

In Oman, the emergence of nymphs during spring and summer generations occurs in February and August, respectively (Thacker et al. 2003) (Fig. 7.4). However, Al-Mahmooli et al. (2005) reported the emergence of the spring generation in January.

The adults and nymphs hide at the base of newly formed leaves to avoid direct sunlight (Klein and Venezian 1985). Abd-Allah et al. (1998) reported a total

Fig. 7.4 First instar nymphs of *Ommatissus lybicus* on date palm frond (heavy infestation) (Photo: Mohammed Z. Khalaf)

nymphal development of the spring generation from 45 to 52 days with an average of 48.4 days. The life span of the adult stage was 82 days for the male and 72 for the female, during which a female laid up to 143 eggs. Incubation period of the eggs averaged 120 days for the spring generation and 95 days for the autumn generation. Under laboratory conditions, 572.5, 648.2 and 1184.4° days were required for the development of egg, nymph and adult stages, respectively. The lower thermal threshold for the three stages were 12.9, 12.9 and 13.2 °C, respectively. Accordingly, 25–27.5 °C is considered to be optimal for the development of *O. lybicus* (Mokhtar and Nabhani 2010). The highest number/percentage of eggs were laid on second and fourth frond row in spring by both the summer generations, respectively (Hussain 1963; Jassim and Al-Zubaidy 2010). Several factors like density of adult females per unit area, female fecundity, host plant, and environmental factors such as sun light, humidity and wind velocity, significantly affect the egg laying behavior of the dubas bug (Shah et al. 2013). Furthermore, the varying spatial distribution of eggs, nymphs and honeydew quantity was observed among the fronds of palm trees in Iraq with the highest eggs density (26,934 eggs/frond) in the fifth row, nymphal population (7035/frond) in the fourth row, and the highest honeydew quantity (546.4 g/frond) in seventh row (Khalaf and Khudhair 2015).

7.2.3 Damage, Economic Threshold and Losses

The dubas bug is a phloem-sap feeder (Howard 2001) and both adults and nymphs feed on the sap of the leaflets and midribs of the date palm frond. At high levels of infestation, the dubas bug can also attack the fruit stalks and fruits, depleting a large amount of sap from the date palms. The infested fronds become chlorotic and

Fig. 7.5 Heavily infested date palm frond showing copious amount of honeydew secreted by *Ommatissus lybicus* (Photo: Mohammed Z. Khalaf)

Fig. 7.6 Droplets of *Ommatissus lybicus* honeydew on apple leaves (intercropped with date palms) (Photo: Mohammed Z. Khalaf)

necrotic due to feeding and egg laying activities (Fig. 7.4). In heavy infestation, dubas produce copious amount of honeydew that covers the surface of the leaves (Fig. 7.5) and may encourage the growth of black mold that collect dust and reduces the leaf photosynthesis by blocking stomata and hindering gaseous leaf exchange (Elwan and Al-Tamiemi 1999; Mokhtar and Al-Mjeni 1999; Gassouma 2004). Klein and Venezian (1985), however, reported that no sooty mold on honeydew was produced by dubas in Israel and they attributed the lack of sooty mold to the low relative humidity in the region. The droplets of honeydew may cause an indirect damage by falling on citrus trees and other crops, commonly intercropped with date palm as in most Arab countries (Fig. 7.6) or may even fall on pipes of irrigation water (Fig. 7.7).

Fig. 7.7 Copious amount of *Ommatissus lybicus* honeydew on irrigation pipe (heavy infestation) (Photo: Mohammed Z. Khalaf)

Dubas bug population on date palm can be estimated either by direct count or indirectly by estimating the honeydew they produced in a specified period of time. Direct counting of dubas bug nymphs and adults seems to be a difficult task due to the height of the date palm tree and the small size of the bug (Mokhtar et al. 2003). Therefore, the amount of honeydew produced was used as a way to estimate the economic threshold for dubas in date palm, The water sensitive paper (WSP) saturated with bromocresol blue was used in this method, the color of the paper is yellow but changes to blue upon contacting water. The proposed threshold was the number of insects that produce honeydew covering 10 % of the leaf surface area in a week or it can be 11 (first nymphal instar)/leaf, 4 (second instar)/leaf or 1 (third, fourth and fifth instar)/leaf of date palm. This threshold was found to be statistically effective than using direct insect count methods. Yellow sticky traps were used to monitor the dubas population at weekly intervals. The rapid rise in feeding activity as indicated by honeydew on WSP could be used to predict the most appropriate time to apply a chemical control against dubas (Thacker et al. 2003). In Iraq, dubas infestation level is determined (Blow 2006) by:

1. Counting nymphs and eggs on leaflets from four fronds taken from five trees in a hectare
2. Infestation levels were deemed:
   (a) Low – less than ten eggs or nymphs/frond
   (b) Moderate – between 10 and 15 eggs or nymphs/frond
   (c) High – >15 eggs or nymphs/frond

There is a difference among different date palm varieties with respect to infestation by dubas bug. Among the three varieties of date palm (Zahdi, Khustawi, Daery),

Zahidi was the favorite variety for eggs laying in the field while Daery was the less favorite one (Jassim and Al-Zubaidy 2010). Based on the honeydew secretion, the infestation of dubas on Medjhool cultivar was high compared to Deglet Noor cultivar (Klein and Venezian 1985). In Panjgur, Pakistan, the commercial cultivars Kehraba and Sabzoo were more susceptible and infested by dubas bug compared to Mozavati cultivar (Shah et al. 2013).

No quantitative estimates of dubas damage on date palm are available, but Dowson (1936) reported extensive damage in the Basra area of Iraq in 1934 and very severe damage in an area of about 800 ha. Large populations of dubas may ruin the date crop and severe infestation may lead to 25–60 % loss in yield (Kranz et al. 1978; Shah et al. 2013). Dubas bugs are not known to be a vector of date palm diseases (Gassouma 2004).

7.2.4 Management

7.2.4.1 Cultural Control

Farmers must maintain enough space between date palm trees and do the silvicultural services such as pruning of old leaves, removal of excess offshoots and suckers. Dubas bug reaches high epidemic numbers only in densely spaced groves because of the ideal microclimate resulting in lower temperature coupled with high relative humidity. Sun scorching and desiccation in well spaced groves (Talhouk 1983) usually kills the delicate newly hatched nymphs. Shah et al. (2013) recommended the removal of lower frond rows (two to three) before egg hatching as a part of IPM approach for the dubas management under the agro-climatic conditions of Panjgur in Pakistan. The removal of the lower fronds will reduce egg, hence reduce crop losses due to attack of this pest.

7.2.4.2 Biological Control

There is no organized biological control against dubas bug in most Arab countries, where the pest represents real thread for date palm. However, there is potential to use predators, parasitoids, and pathogens to manage this pest particularly in organic farming ecosystems and urban landscapes where the use of non-selective insecticides is undesirable.

Hussain (1963) reported a small chalcidoid egg parasitoid on the eggs of dubas bug. He also reported *Chrysoperla carnea* (Stephens), *Coccinella septempunctata* (L.), *C. undecimpunctata* (L.) and *Chilocorus bipustulatus* (L.) feeding on dubas infesting date palm. These coccinellids are general predators and are not expected to provide adequate control of dubas bug due to its seasonality and well-protected prey eggs inside the frond tissue (Blumberg 2008). *Chrysoperla carnea* and the coccinellid *Exochomus nigripennis* Erichson were also found (both larvae and adults) attacking dubas bug (Talhouk 1983). The coccinellids *Cheilomenes propinqua* Chevrolat, the praying mantid *Calidomantis savignyi* (Saussure), and the ants *Crematogaster* spp. were reported to prey on dubas nymphs and adults in Yemen (Hubaishan et al. 2007; Ba-Angood et al. 2009). The specialized egg parasitoid *Pseudoligosita babylonica* Viggiani was reported from Iraq and Yemen (Al-Jboory 2007; Hubaishan and Bagwaigo 2010). The parasitoid spends winter and summer inside the eggs of dubas, and the female is capable of parasitizing the eggs drilling with its ovipositor through the lower surface of the pinnae indicating high specificity and specialization of the parasitoid (Al-Jboory 2007). In addition to the predators and parasitoid from the class Insecta, an efficient predacious mite (whirling mite), *Anystis agilis* (Banks) was reported to feed voraciously on dubas nymphs in Iraq with an average consumption rate of 21.13 dubas nymphs/adult mite/24 h (Al-Jboory 2007).

7.2.4.3 Chemical Control

Ground application of insecticides is ineffective against dubas and impractical in dense date palm groves. Aerial application is commonly used when the height of the date palm trees makes it difficult to apply insecticides from the ground. In Oman, the Ministry of Agriculture is responsible for the chemical control of dubas and the organophosphate pesticide fenitrothion was applied by either using ground-based mist blowers or helicopters fitted with Micronair spraying heads. The aerial applications were more effective than ground spraying (Thacker et al. 2003). About 400 tons of insecticides have been used as aerial spray to control dubas bug in Oman during 1999–2006, with an estimated cost of 9 million Omani Rials (Anonymous 2010). Several pyrethroid and organophosphate insecticides were sprayed. Dichlorvos at the rate of 3.75 L/ha, malathion 96 % ULV at 2 L/ha and deltamethrin at 100 ml/100 L were the most common used for ground spraying. Deltamethrin and diazinon insecticides were more toxic to eggs, larvae and dubas adults under laboratory conditions (Hamad and Al-Rawy 2009). In addition, endotherapy (injection) with thiamethoxam (1 g a.i. per tree) proved to be effective against this pest in Iraq (Al-Jboory et al. 2001). In 1960s, dichlorvos (DDVP) was applied successfully as ultra-low volume (ULV) aerial spray in Iraq against this pest (El-Haidari et al. 1968). Aerial spraying against dubas stopped in 2004 and 2005 due to war in Iraq in 2003. Consequently, the bug populations increased and date production declined. In late 2005, the Ministry of Agriculture (MoA) with the Multi-National Force-Iraq (MNF-I) began planning to conduct aerial spraying to control the bug. They used AN-2 aircraft fitted with Micronair AU 5000 ULV sprayer and MI-2 helicopters where 77,000 ha of date palms were sprayed until 2006. A variety of organophosphorus and pyrethroid insecticides were used at a rate of 2 L/ha (Blow 2006). Blow (2006) concluded that knowledge of the phenology of dubas and prevailing climatic conditions is essential for a successful application of insecticides against this pest.

7.3 Green Pit Scale (*Palmaspis phoenicis*)

7.3.1 Distribution and Host Range

The origin of the genus *Asterolecanium* (*Palmaspis*) was reported to be central Asia particularly Iran (Ezz 1973). The green pit scale *Palmaspis phoenicis* (Ramachandra Rao) (Family: Asterolecaniidae) is reported from Iran, Iraq, Saudi Arabia, Qatar, and Sudan (Carpenter and Elmer 1978; Howard et al. 2001). In Israel, it occurs in all the date-growing areas (Kehat et al. 1964). It is also found in Egypt (Ezz 1973). Gharib (1974) employed the name *Palmaspis phoenicis* (Ramachandra Rao) for this insect. The green scale is found in the UAE and because of the similarity of the terrain, growing conditions and exchange of planting materials throughout the year, it is likely that the pest exists in all Gulf or Middle East counties (El-Bouhssini et al. 2012). The insect is major pest in Sudan, and it was first time reported by Ali (1989) in El Golid area in the far north of the country as a result of an accidental introduction of infested planting materials in 1970s from the Kingdom of Saudi Arabia.

7.3.2 Biology and Seasonal Incidence

A detailed description of the insect morphology is given by Kehat and Amitai (1967). In Israel, the green scale has three generations per year mainly in autumn, winter, and summer (Kehat and Amitai 1967). In Sudan, the population of the green pit scale insect is found all the year round with three peaks mainly in March, June, and October. The north wind, infested offshoots and male pollination strands are also important factors in the epidemiology of the scale (Ali 1989; Ahmed 2007). Gassouma (2003) stated that more than six generations, mostly overlapping, might be produced annually under Sudan conditions. Nevertheless, few data exist concerning the biology and seasonal occurrence of the green pit scale and hence more research is needed to uncover this important aspect, which will definitively improve the management of such an important insect pest.

7.3.3 Damage, Economic Threshold and Losses

The green pit scale infests almost all green parts in the crown of the date palm particularly the leaflets, petioles, fruit bunches, and fruits. The color of infested leaflets becomes pale and yellow (Fig. 7.8), and heavy infestation may lead to complete death of the tree (Kehat et al. 1964). Other symptoms of damage include chlorosis and fruit malformation (Fig. 7.9), degeneration of leaflets, malformation and low quality of fruits (Ahmed 2007) (Fig. 7.10).

Fig. 7.8 Damage by *Palmaspis phoenicis* (yellowing and drying of fronds) on date palm in Northern Sudan (Photo: Mahadi A.R. Ahmed, with permission)

Fig. 7.9 Chlorosis and yellowing symptoms on date palm pinnae caused by *Palmaspis phoenicis* (Northern Sudan) (Photo: Mahadi A.R. Ahmed, with permission)

Four fronds representing the four main directions should be chosen from a randomly selected date palm. Two leaflets collected from each frond should be inspected and examined under a binocular microscope at biweekly intervals. The leaflet was divided into three sections; bottom, middle, and upper and 3 cm^2, each from these sections were taken for sampling. The average number of living females and nymphs per cm^2 was used to express the intensity of infestation (Ahmed et al. 2013).

In 2008, the green pit scale infested about 12.5 % of the total number of date palms. The chlorosis and yellowing of the leaves and fruit malformation before

Fig. 7.10 Fruit malformation caused by *Palmaspis phoenicis* infestation; uninfested dates (*upper*), infested dates (*lower*) (Photo: Mahadi A.R. Ahmed, with permission)

maturity resulted in about 90 % production loss (Ali and El-Nasr 1992). Fageer and Ahmed (2010) evaluated the economic impact of the green sale infestation on cost of date production. He concluded that heavy infestation by green pit scale has resulted in an increase of cost of production and a decrease in farmers return.

7.3.4 Management

The key for a successful management of any pest is the correct identification. It took 15 years to correctly identify the green pit scale in Sudan. The green scale was mistaken for the white scale which is indigenous to the area. Sometimes the two species are found on the same pinna coexisting together. The indigenous *Parlatoria* is flat while the exotic green scale is a hump-shaped scale (Gassouma 2003). In Sudan, Farmers Field Schools (FFS) and participatory research approach are essential for adapting any integrated management program against the green pit scale (Fig. 7.11).

7.3.4.1 Cultural Control

Pruning and removal of old infested fronds, adequate and regular irrigation, optimal fertilization, and strict effective internal quarantine measures play an important role in the management of green pit scale in date palms. Some farmers are resorted to burning severely infested palms, however, this practice has resulted in killing date palm trees and it has caused wide fire hazards that destroyed several thousands palms. Over pruning in an attempt to control the green pit scale has resulted in reduction of yield.

Fig. 7.11 Farmers Field School (FFS) in Northern Sudan on how to manage *Palmaspis phoenicis* and other key pests of date palm (Photo: Mahadi A.R. Ahmed, with permission)

Ahmed et al. (2013) studied a mechanical method of control and some cultural practices to manage the green pit scale in Sudan. They recommended the following:

1. Pruning, removal, and burning of old infested fronds as sanitary measure
2. Construction of a basin around each palm to improve irrigation
3. Pre-watering of trees 24 h before the application of insecticides

7.3.5 Host Plant Resistance

There seem to be differences among date palm cultivars with regard to infestation by the green pit scale. Barakawi is the dominant variety in the area on which the pest was noticed for the first time, and it is more susceptible than other varieties particularly Meshrig Wad Laggi that has been found relatively more resistant to the green scale (Gassouma 2003; Ahmed et al. 2013). Thus, date palm varieties resistant to green pit scale attack may be included in the integrated date palm production system.

7.3.5.1 Legislative Control

A very good and simple internal quarantine procedure has been used to contain the green pit scale inside Sudan. This quarantine consists in 13 checkpoints that inspect vehicles carrying any date palm planting materials outside the infested areas.

Fig. 7.12 Adult of the nitidulid beetle, *Cypocephalus dudichi* feeding on *Palmaspis phoenicis* in Sudan (Photo: Daffallah E. Yousof; with permission)

Moreover, around the main infested area of Al Ghaba there are two natural barriers, which prevent any crawlers from crossing to uninfested areas. These barriers are two stretches of bare sand each about 4 km wide to the north and to the south of Sudan (Gassouma 2003).

7.3.5.2 Biological Control

A number of general predators were reported from Sudan to attack the green scale. These include the coccinellid, *C. bipustulatus*, which was initially introduced in 1986 from France but failed to establish (Ahmed 2007). Yousof et al. (2013) reported three predators and one parasitoid on *P. phoenicis* in Sudan; these were *Cypocephalus dudichi* (L.) (Fig. 7.12), *Pharoscymnus numidicus* (Pic) (Fig. 7.13) the lacewing, *Chrysoperla* sp. and the parasitoid *Metaphycus* sp. (Fig. 7.14). These predators were reported to feed on the crawlers as well as on young females.

Parasitoids may have a role in suppression of green scale population in Israel as indicated by exit holes in the cadavers of parasitized insects (Blumberg 2008). The parasitoid, *Metaphycus* sp. caused 16 % parasitism which is relatively high from October to November, which coincides, with the optimal period for development and survival of predators in Sudan when temperatures are favorable (Yousof et al. 2013).

7.3.5.3 Chemical Control

Organophosphate insecticides used in combination with mineral oils gave good results against the scale (Kehat et al. 1964). Application of contact insecticides can only be effective when applied against the first instar nymphs before the crawlers settle on the leaflets of fruits (Carpenter and Elmer 1978). In Sudan, aerial application of diazinon 60 EC, dimethoate 32 %, and folimat 80 % reduced the green scale

Fig. 7.13 Adult of the ladybird beetle, *Pharoscymnus numidicus*, a predator of *Palmaspis phoenicis* (Photo: Daffallah E. Yousof; with permission)

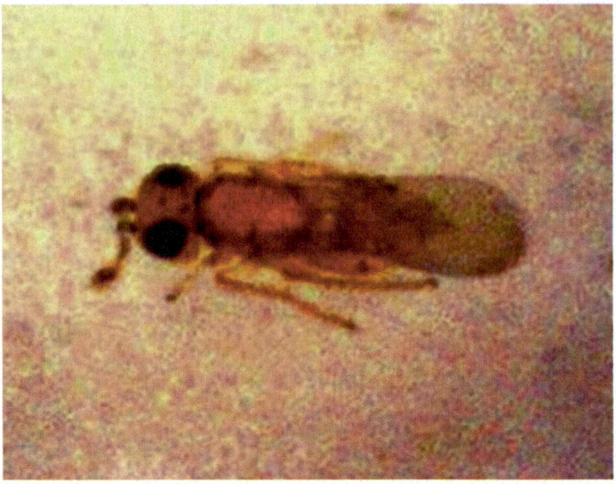

Fig. 7.14 The hymenopteran *Metaphycus* sp., parasitoid of *Palmaspis phoenicis* (Photo: Daffallah E. Yousof, with permission)

population to the minimum, but resulted in a population increase 1 year after the treatment (Ahmed et al. 2013).

Ahmed (2007) recommended the following insecticides with two different application methods to manage *P. phoenicis* in Sudan:

- *Soil Application*
  - (a) Actara 25 WG (thiamethoxam) 18 g product/palm (4.5 g a.i.)
  - (b) Rinfidor 20 % SL (imidacloprid) 35 ml product/palm (7 g a.i.)
- *Trunk Injection*
  - (a) Actara 25 WG (thiamethoxam) 10 g product/palm (2.5 g a.i.)
  - (b) Confidor 200 SL (imidacloprid) 20 ml product/palm (4 g a.i.)

Date palms treated with systemic insecticides by trunk injection continued to develop normally without having any phytotoxicity effects. In addition, no insecticide residues were detected from either method in dates, soil or grasses. The injection and soil application methods further caused no harmful effects on natural enemies and non-target organisms in the date palm agro-ecosystem.

Thiamethoxam (Actara 25 WG) injected in date palm at the rate of 1 and 2 g a.i./tree translocated rapidly into date palm trunk and reached the leaves in a short time (Al-Sammariae et al. 2006). Trunk injection is recommended as an alternative method to soil application particularly in situation where date palm is not irrigated frequently. As an alternative to chemical insecticides, Eldoush et al. (2011) applied the powder of the argel plant *Solenostemma argel* and *Calotropis procera* as soil treatments at a rate of 100 g powder/date palm tree. The products reduced the population of the green pit scale and enhanced the yield of date palm. These two plants are available to date palm farmers free of charge or at a minimum cost. Some date palm farmers grow the argel plant while *C. procera* is a common weed found in marginal and fallow lands.

7.4 Issid Date Bug (*Asarcopus palmarum*)

7.4.1 Distribution

According to Blumberg (2008) the issid date bug, *Asarcopus palmarum* Horváth (Family: Issidae) is recorded from Israel, Egypt, North Africa, and USA feeding primarily on date palms (Howard et al. 2001). In early 1990s, large numbers of the bug were observed in Israel feeding at protected sites on date palm, such as bases of fruit bunches and fronds (Blumberg 2008).

7.4.2 Biology

The females are up to 3 mm long by 1.5 mm wide (males are smaller), reddish-brown in color, have no wings and all stages have red eyes (Blumberg 2008; Downer 2009). Insects are active throughout the year (Downer 2009).

7.4.3 Damage

The issid bug feeds on the inflorescences stalks and on soft protected tissues of date palm at the frond bases (Blumberg 2008). However, it can also feed further along the rachis of unexpanded fronds. Adults are difficult to detect because they live between unexpanded fronds inside the crown of the palm. When populations are dense they may produce copious honeydew which attracts ants. Damage is first seen as speckling or small yellow blotches on newly emerging fronds (Howard et al. 2001; Downer 2009). In time, palms become stunted and frond loss increases giving the palm a tattered or sickly appearance.

7.4.4 Management

Although issid bug causes some damage to date palm in USA, the damage is insignificant and does not reach economic threshold so it is not necessary to take control measures against this pest (Blumberg 2008).

7.5 White Date Scale (*Parlatoria blanchardi*)

7.5.1 Distribution and Host Range

Parlatoria blanchardi (Targioni-Tozzetti) (Family: Diaspididae) is cosmopolitan and found wherever date palm is grown. It is one of the oldest pests of date palm and is thought to have originated from the oases of Mesopotamia or Iraq (Calcat 1959; Carpenter and Elmer 1978; Bénassy 1990). *Parlatoria* scale was accidentally introduced in the USA, particularly in Arizona, California, and Texas and it was eradicated in 1914 from Arizona, 1919 from Texas and in the 1930s from California (Bénassy 1990; Gill 1997). Due to the international movement and trade in date palm offshoots, the pest is found now in Europe, Central Asia, Middle East, Africa, Western Hemisphere and Oceania. The white scale has been reported on date palm in Iraq, Saudi Arabia, Syria, Palestine, Turkey, Soviet Turkmenistan, India, Pakistan, North Africa, Egypt, Australia, and Iran (Gharib 1973).

It is also found on Canary Island palm, *Phoenix canariensis* Chabaud, *Washingtonia* palm, *Washingtonia filifera* (Lindl.) H. Wendl., and doum palm, *Hyphaene thebaica* (L.) Mart. (Stickney 1934). In addition, the scale has been reported on plants belonging to the families Apocynaceae, Oleaceae, and Rhamnaceae (Anonymous 2014). Other hosts of *Parlatoria* scale included species of the genera *Jasminum*, *Latania*, *Pritchardia* and *Vinca* (Borchsenius 1966).

7.5.2 Biology and Seasonal Incidence

Stickney (1934) gave a detailed description of the external anatomy of *Parlatoria* scale. Reproduction is sexual and oviparous throughout the year (Bénassy 1990). Each female lives 5–25 days and lays 4–13 eggs in a cluster underneath her scale cover (Abivardi 2001). The developmental period differs depending upon climatic conditions of the area. In Iran, female development ranged from 85 to 100 days in the spring and 120–150 days for the winter generation (Gharib 1973). Bénassy (1990), reported the life cycle duration as about 75 days in summer to 150–180 days in winter. Pairing is followed by a pre-oviposition period of 10–15 days in spring and 5–7 days in summer.

Upon completion of egg laying, the female dies leaving the eggs protected under her cover. The crawlers leave the scale after 10–15 days in spring and 3–5 days in summer, settling on the lower surfaces of the leaflets in shady places in the lower part of the canopy. Although, *Parlatoria* scale is a xerophilous (adapted to very dry climate/habitat) and thermophilous (warmth loving) insect, it avoids the scorching sun by settling inside folds at the base of the pinnae (Talhouk 1983). The first instar crawler is the dispersal stage and can crawl for a short distance before selection of the feeding site (Stansly 1984). High mortality of crawlers was reported as a result of the combined effect of high temperature, low relative humidity and wind (Bénassy 1990).

The life cycle of the female lasts 85–100 days in the spring and summer generations and 120–150 days in the winter generation; the life cycle of the male, which includes a pupal stage, averages 30–45 days. Female adults live 5–25 and males 2–4 days (Gharib 1973; Bénassy 1990). Two forms of male exist in this species; the alate form (winged) which are usually found in summer while the apterous form appears in spring (Abd-El-Kareim 1998). *Parlatoria* scale has three to five overlapping generations annually (Gharib 1973; Kehat et al. 1974b; Bénassy 1990; Khoualdia et al. 1993). In Iran, it has three to four generations per year (Abivardi 2001). The sex ratio favors females, with males generally constituting less than 25 % of the total population (Bénassy 1990). The population of the white scale seems to be negatively affected by extreme temperature in Israel as three overlapping generations are observed annually (Kehat 1967).

7.5.3 Damage, Economic Threshold and Losses

The white scale attacks the entire date palm but is pre-ferentially found on the abaxial leaf surfaces. During severe infestations, it is found on fruit stalks and fruit bunches (Boyden 1941; Khoualdia et al. 1993; Blumberg 2008). Fruit losses due to direct feeding of the white scale reached 70–80 % (Smirnoff 1957). Because of feeding and injection of toxic saliva, a discolored area of injured tissue develops in areas where crawlers settle and feed. Scale feeding lead to impairment of transpiration, depletion of nutrients, destruction of chlorophyll, and hindering the photosynthesis which eventually lead to marked reduction in date palm productivity (Bénassy 1990).

The infestation often results in chlorosis or discoloration of leaves, yellowing and pre-mature withering and dropping of leaves. Infested fruits become small, stunted, shriveled, distorted with low marketed value and sometimes unfit for human consumption. In heavy infestation, encrustations of scale on the leaf surface look dirty white or gray in color (Bénassy 1990; Abivardi 2001).

Four leaves representing the main four directions should be chosen from a randomly selected palm. Three leaflets from each leaf (base, middle and tip) should be detached at random and examined under binocular microscope for presence of living scale females and crawlers. The scales in an area of 1 cm^2 from each of the three parts of leaflet (base, middle and tip) should be counted (Eldoush et al. 2011). The infestation intensity is expressed as an average number of scales/cm^2 (Bénassy 1990). A ranking scale can also be followed, for example a rank of one (8 scales/cm^2), rank of two (15 scales/cm^2), and rank of three (35 scales/cm^2) to indicate low, moderate and high infestation rate.

The pest has devastated palms in Algeria during 1920s killing around 100,000 date trees (Rosen 1990). It causes economic damage in India, Pakistan, Israel, Saudi Arabia, Sudan, Mauritania, Egypt, Libya, Tunisia, Algeria and Morocco (Bénassy 1990). This species coexists with another related species, *Palmaspis phoenicis* which make the situation more serious, both species may be found on the same leaflet of an infested frond in Sudan and Iran (Gharib 1974). In Saudi Arabia, the varieties Sokkari and Mactomi were the most susceptible while Magvesi was the most resistant (Dabbour 1981). In the region of Biskra (Algeria), the scale insects thrived well on the cultivars Ghars and Deglet Noor than the cultivar Degla Baida (Matallah and Biche 2013). The severity of damage by *Parlatoria* scale seems to be variable depending on the locality, date palm cultivar, environmental conditions, and management practices. The state of whether the scale is indigenous or recently introduced into an area plays role as an important factor in the population control program of this pest.

7.5.4 Management

A management strategy involving an educational system for field personnel, establishing a survey system for the pest, restructuring the farm crop management, reduction of pesticides, and enhancing the beneficial agents has been used to manage *P. blanchardi* in Israel (Kehat et al. 1974a).

7.5.4.1 Cultural Control

Pruning and disposing of infested leaves as phytosanitary measures proved to be effective in controlling *Parlatoria* scale particularly under the conditions of minor infestations (Siddig 1975; El-Bouhssini et al. 2012).

7.5.4.2 Trapping and Monitoring

Parlatoria scale may be found concealed in frond axils and underneath fibrous sheath at the leaf base making detection of initial infestation very difficult. Careful examination of leaf pinnae and bases for discoloration and white scale cover using a field hand lens may be useful tool in identification of infestation. Sex pheromone and sticky traps can also be used for monitoring of male (Abd-El-Kareim 1998).

7.5.4.3 Host Plant Resistance

The tolerance of some date palm varieties to infestations by this insect pest can be an important component of IPM program (Swaminathan and Verma 1991).

7.5.4.4 Legislative Control

Inspection of offshoots before transporting into new areas also restricts outbreaks. Application of an effective and strict internal quarantine measures will help to restrict the spread the insect by human activities.

7.5.4.5 Biological Control

Since *Parlatoria* scale is the oldest and wide spread pest of date palm (El-Bouhssini et al. 2012). It has also a numerous and a wide range of predators and parasitoids (Blumberg 2008). The parasitic Hymenoptera (2 species), predatory Coccinellidae (25 species), predatory Cybocephalidae (5 species), *C. carnea* were reported in Israel to attack the pest (Blumberg 2008). As reported by FAO (1995), the natural enemies of *P. blanchardi* in the date palm agro-ecosystems were mostly coccinellid beetles. Biological control against *P. blanchardi* can be used to reduce the pest population to acceptable levels in those areas where chemical control cannot be used (Smirnoff 1957; Carpenter and Elmer 1978). The use of classical biological control to manage the white scale in Morocco and Sudan in 1970s and 1980s was not successful (Zaid et al. 2002). However, the release of the predators *C. bipustulatus* and *C. bipustulatus* var. *iranensis* into date palm grooves in Mauritania, Niger, and Tunisia, respectively, provided an acceptable level of control for *P. blanchardi* (Stansly 1984; Khoualdia et al. 1993, 1997). The predator, *Pharoscymnus horni* (Weise) was effective against the white scale in India (Swaminathan and Verma 1991). *Rhyzobius lophanthae* (Blaisdell) is an exceptional biological control agent because of its high fecundity, lack of parasitoids, absence of diapause, and resistance to low temperatures especially in the immature stages (Stathas 2000). Predators have been recorded to provide up to 45 % mortality of the white scale populations (Gharib 1973). The effectiveness of biological control agents against *P. blanchardi* appears to be dependent upon the predators' ability to adapt a particular region, its fecundity and consumption rate. Talhouk (1983) reported that it was possible to bring down the infestation of the white scale in Saudi Arabia to sub-economical level using the predator *Cybocephalus* sp. The predacious mite *Hemisarcoptes coccophagus* Meyer proved to be an effective biocontrol agent against white scale in Niger (Kaufmann 1977).

7.5.4.6 Chemical Control

Chemical insecticides are less than the ideal solution to scale insect problem, because they have more effect on the natural enemies while the scales on the body of white scale insects serve as barrier and protect them from the applied chemicals (El-Bouhssini et al. 2012). The existence of the infestation inside the frond axils may also serve to reduce the effectiveness of chemical control against the white scale (Kehat 1967; Hodges et al. 2003). Foliar applications of a variety of oil-based compounds have proven effective against the immature stages of the pest (Al-Hafidh et al. 1981). Earlier chemical control of white scale relied mainly on organophosphates. In 1960s, dimethoate and 2 % mineral oil gave satisfactory control of this pest (Kehat et al. 1964). Sprays of oil emulsion with dimethoate, malathion or methyl-parathion, and of methyl-parathion alone, also gave significant mortality of the pest. Oil emulsion with dimethoate increased the yield/tree up to 74 % (Siddig 1975). Oils, either ultra-fine horticultural oils or a product containing fish oils are effective against scale insects in general (Hodges et al. 2003). The oils cover the insects and suffocate them besides making the surface of the plant difficult for crawlers to settle (Howard and Weissling 1999). Imidacloprid used as a soil drench can be very effective; however, it should be mixed at a very high concentration (Howard and Weissling 1999).

7.6 Red Date Scale (*Phoenicococcus marlatti*)

7.6.1 Distribution and Host Range

Phoenicococcus marlatti Cockerell (Family: Phoenicococcidae) is endemic to the Middle East. It is also found in Europe, Asia, North America and the Caribbean Region (McDaniel 1974; El-Haidari 1981; Moustafa 2012; Howard et al. 2001), Central America, and South America. However, Howard et al. (2001) reported that while *P. blanchardi* persists in California and Florida, *P. marlatti* has been eradicated from California (Espinosa et al. 2012).

It appears that its hosts are mostly palms (Arecaceae), particularly those of the genus *Phoenix*, i.e. *P. canariensis*, *P. dactylifera*, *P. reclinata*, *P. roebelinii*. The other palm species include, rattan palms, *Calamus* sp., and *Daemonorops* spp., the California fan palm, *W. filifera*, Additionally, it has been reported from *Pandanus* spp. (Pandanaceae) and from *Eucalyptus* (Myrtaceae) (Espinosa et al. 2012; Moustafa 2012).

7.6.2 Biology and Seasonal Incidence

Espinosa et al. (2012) cite Avidov and Harpaz (1969) and stated that four overlapping generations occur in the U.S. annually. The life cycle is completed within 60–158 days depending on prevailing temperature. Reproduction is either ovoviviparous or viviparous and male nymphs usually pass through five instars while female nymphs have only three instars (Stickney et al. 1950). Some eggs hatch within the female, resulting in viviparous reproduction. The remaining eggs are laid external to the body (Avidov and Harpaz 1969). In Egypt, egg and adult densities reached its maximum in May 2010 and to its minimum in February 2010 (Moustafa 2012).

7.6.3 Damage

Phoenicococcus marlatti is found in great numbers on the white tissues at the bases of fronds and fruit bunches (Sickney et al. 1950). Heavy infestation weak the young palms, resulting in complete dryness and eventually the death of tree (Moustafa 2012).

7.6.4 Management

7.6.4.1 Biological Control

The predators *Pharoscymnus anchorago* F., and *R. lophanthae* were reported from North Africa and Spain and proved to be efficient against this pest (Stickney et al. 1950; Vives 2002). Moustafa (2012) reported *Pharoscymnus varius* (Kirsch) and *Scymnus punetillum* Weise preying on *P. marlatti* in Egypt.

Blumberg (2008) reported as an active predator, *P. anchorago* in North Africa while *R. lophanthae* has been reported by Vives (2002). Momen et al. (2009) concluded that females mites of *Typhlodromus negevi* Swirski and Amitai developed on 6.3 days when fed on a *P. marlatti* eggs diet compared to 7.7 days development when fed on eggs of *Bemisia tabaci* (Gennadius). However, the reproductive net capacity of this predator was higher when feeding on pollen and lower when feeding entirely on eggs as prey.

7.6.4.2 Chemical Control

Due to the cryptic nature of the scale, chemical control is not effective and rather difficult to apply (Blumberg 2008). Hazir and Buyukozturk (2013) suggested the application of organophosphorus insecticides against active crawlers. The same authors cite Zaid et al. (2002) who suggested the use of malathion and parathion in infested nurseries at an application rate of 80–100 g and 26 g per 100 l water, respectively. The red date scale can be successfully managed by frequent application of insecticidal soaps and horticultural oils showering the infested palms (Anonymous 2011; Hazir and Buyukozturk 2013).

7.7 Pink Pineapple Mealybug (*Dysmicoccus brevipes*)

7.7.1 Distribution and Host Range

The pineapple mealybug, *Dysmicoccus brevipes* (Cockerell) (Family: Pseudococcidae) is a cosmopolitan pest occurring in Australia and Asia especially in tropical and subtropical areas. It is also found in Nearctic, Oriental, Palearctic and Neotropical regions of the world (Blumberg 2008). Ben-Dov (1994) reported that in the Middle East this mealybug can be found in Egypt and Israel.

Worldwide, pineapple industries regard this mealybug as a major threat to the crop because of its association with the devastating disease, Pineapple Mealybug Wilt (PMW). Within the plant order Arecales and family Arecaceae, Howard et al. (2001) collected this species from *Areca catechu* (L.), *Carpentaria acuminata* (H. Wendl. & Drude) Becc., *Cocos nucifera* (L.), *Elaeis guineensis* Jacq., *P. dactylifera*, *Rhapis* sp., *Roystonea* spp. and *Sabal bermudana* L.H. Bailey.

Burke et al. (1994) reported *D. brevipes* infesting *Prosopis glandulosa* Torr. in Texas (USA). This species originated in tropical America, where the pineapple itself originated. The mealybug was spread on planting material to plantations around the world (Petty et al. 2002). Its more than 50 host species include sugarcane, perennial grasses, e.g. *Panicum barbinode* and *Tricholaena rosea*, and sisal *Agave sisalana* (Petty et al. 2002). Dhileepan (1991) emphasized that in Kerala (India) *D. brevipes* was found on fruit bunches in oil palm plantations with an incidence of 3.2–100 %. The mobile immature crawler stage, with flattened bodies and long hairs may be wind dispersed for hundreds of meters within plantations (Rohrbach et al. 1988). The big-headed ant, *Pheidole megacephala* (F.), also plays an important role in its dispersal (Petty et al. 2002). In pineapple, roots and lower leaf axils are pre-dominantly infested (Petty et al. 2002).

7.7.2 Biology

In pineapple, reproduction is parthenogenetic and ovoviviparous. However, a bisexual race occurs in Ivory Coast, Madagascar, Dominican Republic and Martinique; apart from this race, pink mealybugs only produce females (Petty et al. 2002). According to Petty et al. (2002), until about 1959, *D. brevipes* was confused with *D. neobrevipes* Beardsley, the gray pineapple mealybug. Petty et al. (2002) cite that Ito (1938) noted biological and behavioral differences, and Beardsley (1959) explained the morphological differences. For instance, pink mealybug has a parthenogenic reproduction, prefers to attack roots and lower parts of pineapple, it can survive in graminaceous plants, while the gray mealybug, cannot.

At 23.5 °C, developmental stages of the pink mealybug have the following durations (Ito 1938): three larval (crawler) instars –34.03 days; adult pre-larva position –26.58 days; adult larva position –24.84 days; adult post-larva position

–4.70 days; total adult lifespan –56.23 days; total lifespan –90 days (Petty et al. 2002). On average, 234 yellowish, less than 1 mm long crawlers are produced per reproductive mealybug, with a 692 maximum, recorded in pineapple (Petty et al. 2002).

The 3–4 mm long adult female is elongate oval shaped, distinctly segmented and coated with a white waxy secretion. Around their bodies lower periphery are 34 waxy filaments – lateral filaments are 0.25 by body width and caudal filaments 0.5 by body width (Petty et al. 2002). Colen et al. (2000) studied the development of *D. brevipes* at temperatures fluctuating from 20 to 35 °C and demonstrated that nymphal development was not observed at 35 °C and higher male longevity was obtained at 20 and 25 °C. The thermal threshold for first and second instar nymphs, cocoons and nymphal period was 12.1, 13.5, 12.8 and 12.8 °C, respectively.

7.7.3 Symbiotic Interaction: Mealybug and Ant

All information on the importance of the mealybug-ant relationship is based on information from pineapple plantations (Petty et al. 2002). Symbiotically related ant species in pineapple systems include *P. megacephala*; fire ant, *Solenopsis geminata* (F.); Argentine ant, *Iridomyrmex humilis* (Mayr); *Technomyrmex albipes* (Fr. Smith); *Camponotus friedae* Forel; *Anoplolepis gracilipes* (Smith) [formerly known as *Anoplolepis longipes* (Jerdon)]; and *Araucomyrmex* spp. Caretaking by these species includes the building and maintenance of shelters, protecting mealybugs from natural enemies and inclement weather (Rai and Sinha 1980, cited by Petty et al. 2002).

Ants remove mealybug excreted honeydew, preventing an unhygienic build-up of *Capnodium* sp. sooty mould (Beardsley et al. 1982; Duodu and Thompson 1992). Mealybug leaf infestation of Queen cultivar pineapples decreased by 90 % within 20 weeks of *P. megacephala* control with hydramethylnon bait-toxin, and root infestation by almost 100 % within 12 weeks (Petty and Tustin 1993). Under these conditions, a high positive linear correlation prevailed between ant infestation and mealybug leaf infestation ($r=0.978$, $P<0.01$), and root infestation ($r=0.769$, $P<0.01$). The relationship between ant distribution and mealybug infestation was similar. When uncontrolled, ant numbers did not correlate, i.e. were no longer a limiting factor, and pineapple growth in 52 weeks was reduced by 17 %.

7.7.4 Damage

Blumberg (2008) reported bug survival all the year round in the arid Arava Valley of Israel, at the base of date palm without harming the tree. During summer, the mealybugs migrate to the crown of the tree to infest ripening dates (Blumberg 2008).

7.7.5 *Management*

Prior to the environmental conditions getting favorable for the development of the mealybugs, the infestation of the roots and basal ten leaves should be assessed well in time (Petty and Webster 1981, cited by Petty et al. 2002). In South Africa, infestations increase during the drier autumn and winter months.

By assessing symbiotic ant infestations, mealybugs may be indirectly monitored. A groundnut butter-soybean oil mixture is highly attractive to big-headed ants, which may be counted on wooden laths with the mixture, placed at suitable intervals around pineapple plantation blocks. Even a small number of these ants is sufficient to warrant further investigation and possibly control measures (Petty et al. 2002). Once established, wilt is difficult to control because of vegetative propagation and latent symptoms in planting material (Rohrbach and Apt 1986). Management of the PMW problem requires the integration of species monitoring and the application of chemicals safer to the natural and biological control agents. Cultural control measures should also be applied if possible (Petty et al. 2002).

7.7.5.1 Biological Control

A number of potentially effective coccinellid beetles and encyrtid parasitoids have been identified. *Anagyrus ananatis* Gahan effectively controls mealybugs in the absence of ants (Gonzalez-Hernandez 1995). It originated from Brazil, is very host specific, and has a 20 day generation period. *Exochomus concavus* Fursch, under laboratory conditions, consumed 61 and 94 (155 total) mealybugs per larva and adult, respectively (Petty 1985, cited by Petty et al. 2002). Noyes (2007) cited by Blumberg (2008) reported many encyrtid and a few signiphorid parasitoids attacking *D. brevipes*. In addition, coccinellid predators were reported to attack this pest. Estrada et al. (2009) recorded an average of 1.68 insects per pineapple plant treated with *Beauveria bassiana* strain D0106 compared to 3.26 per plant receiving no treatment. The mealybug population on pineapple in Hawaii is managed by the following natural enemies: the cecidomyiid *Dicrodiplosis pseudococci* (Felt), the encyrtids *Anagyrus coccidivorus* Dozier and *Hambletonia pseudococcina* Compere (Blumberg 2008).

7.7.5.2 Chemical Control

Petty et al. (2002) mentioned several chemical products to control ants in pineapple, hydramethylnon bait toxin/Amdro® (American Cyanamid Company, Wayne, New Jersey, USA). Amdro® 1500 ppm formulation at the rate of 2.24 kg/ha in 1980 controlled the big-headed ant more efficiently compared to the standard treatment of that time, Mirex. No data on ant control on date palms was found in this review. However, in other crops, i.e. vineyards, use of liquid ant bait stations are known to provide a more sustainable ant control strategy (Nelson and Daane 2007).

7.7.5.3 Other Tactics

Miranda and Blanco (2013) also reported results of studies on the effect of biological (i.e. *B. bassiana*, *Metarhizium anisopliae*), organic insecticides (soaps), and botanical extracts against this mealybug on pineapple in Costa Rica. The best results were obtained with the botanical extract (chili pepper, garlic, onion, mustard and gavilana) and with liquid soaps.

7.8 Minor Insect Pests

Minor pests, which were reported to cause damage on date palm as well as potential ones that are expected to emerge as major pests are listed in Table 7.1.

7.9 Future Research

Detailed knowledge on bionomics and phenology of hemipterans on date palm is required for establishment of proper programs for their management. The cooperation between farmers and researchers as participatory approach is needed in conduction of field research, which is more efficient and appropriate than any other approaches in increasing the framers awareness about date palm insect pests. Future research priorities should focus on methods of sampling and determination of economic threshold adapted for the farmers as important components of integrated pest management program. Research is also needed on the role of natural enemies in the management of these pests in the date palm agroecosystem. The symbiotic relationship between the hemipteran pests and ants that ward off natural enemies should also be studied to improve management of these pests by using biological control. The isolation and use of entomopathogenic agents from local environment to replace the current use of synthetic insecticides is a big challenge. It is known that some date palm cultivars with folded frond pinnae provide shelter and hiding places for dubas and scale insects. Thus, breeding for cultivar with open pinnae can help solving the problem.

7.10 Summary

More than 20 hemipteran insect pests were reported on date palm; however, only six species can be considered key pests capable of inflicting economic damage. Green pit scale insect is a major devastating pest threatening date palm cultivation in Sudan, while it is a minor pest in other part of the world. Likely, dubas bug is a key pest of date palm in Iraq and Oman where aerial spray over vast areas and with large

Table 7.1 Minor hemipteran pests reported on date palm

Common name	Scientific name and family	Remarks	Reference
Long scale	*Fiorinia phoenicis* Balachowsky, Diaspididae	Egypt, UAE, Oman, potential pest	El-Bouhssini et al. (2012) and Radwan (2012)
Oriental yellow scale	*Aonidiella orientalis* (Newstead), Diaspididae	Pantropical	Talhouk (1983) and Hill (2008)
Purple scale	*Chrysomphalus aonidum* (L.), Diaspididae	Pantropical	Hill (2008)
Black thread scale	*Ischnaspis longirostris* (Sign.), Diaspididae	Pantropical	Hill (2008)
Citrus mealybug	*Planococcus citri* (Risso), Pseudococcidae	Libya, Qatar	Bitaw and Ben Saad (1990)
Hibiscus mealybug	*Maconellicoccus hirsutus* (Green), Pseudococcidae	Saudi Arabia, infests fruits	Talhouk (1993) and Elwan (2000)
Spherical mealybug	*Nipaecoccus viridis* (Newstead), Pseudococcidae	Qatar	Mokhtar (2009)
Cotton mealybug	*Phenacoccus solenopsis* Tinsley, Pseudococcidae	Pakistan, India	Kumar and Kontodimas (2012)
Giant mealybug	*Pseudospidoproctus hypheniacus*, Margarodidae	UAE, infests bases of fronds	Elwan (2000)
Coconut scale	*Aspidiotus destructor* Signoret, Diaspididae	India, infests foliage and fruits	Hill (2008)
Apache cicada	*Diceroprocta apache* (Davis), Tibicinidae	USA, infests date strands	Carpenter and Elmer (1978)
Date palm offshoot hopper	*Perindus binudatus* Emeljanov, Diaspididae	Qatar, Oman	Mokhtar (2009)
Arabian cicada	*Platypleura arabica* Myers, Cicadidae	Sultanate of Oman	Elwan (2000)
Leafhopper	*Cicadulina bipunctata* (Melichar), Cicadellidae	Saudi Arabia, vector of Al-Wajam disease	Alhudaib et al. (2007)

amount of chemical insecticides is carried out against it, while it occurs sporadically as minor pest in Saudi Arabia and other date palm growing countries. Thus, the severity of infestation by these pests seems to be largely affected by climatic conditions. Hemipteran pests of date palm are delicate insects of small size; however, they possess highly evolved adaptations enabling them to escape sun scorching, desiccation, and natural enemies by either production of protective scales, wooly cushion or seeking refuge in well protected sites on date palms. Hemipterans cause damage to date palm by directly sucking large amount of sap or secretion of copious amount of honeydew, which indirectly contaminate fruits, attract sooty mold fungi that block stomata and hamper respiration and photosynthesis by fronds. The symbiotic relationship between ants and hemipteran pests complicates management as the former ward off natural enemies and cause nuisance for workers who perform the different cultural operations on date palms. The international movement and commerce in date palm offshoots is the reason behind the wide geographical distribution of these pests. Thus, strict quarantine measures should be adopted during importation and exportation of date palm planting materials. During pruning and pollination processes, the worker can accidentally move the scale insect from infested orchards to uninfested ones. Therefore, field sanitation and internal quarantine protocols should be made to avoid new infestation. Densely cultivated palms usually suffer from dubas infestation due to the shade and high relative humidity, which favor the build-up of the insect. Adjacent date palm also favor the spread of scale insects. Management of hemipteran pests in date palm depends mainly on the use of chemical insecticides without effectiveness due to the difficulty in reaching the hiding and feeding sites of these pests. Chemical treatments, when necessary, should be timed with the most vulnerable stages of the pests. For example, scale insects should be treated while most of its population is in the crawler or mobile stage before settlement and formation of protective scales. Dubas bug should be treated while in the nymph stages before adult emergence. Future research on hemipteran pests of date palm should focus on understanding their biology and phenology and finding alternatives to chemical insecticides for sustainable integrated management programs. Currently, more research is directed towards red palm weevil, neglecting other serious pests including hemipterans. The indiscriminate use of non-selective insecticides against red palm weevil and other pests in most Gulf countries, where large number of date palms exist, will upset the natural balance in the ecosystem and will favor the emergence of hemipterans as major pests after the extermination of their natural enemies.

Acknowledgments The authors wish to thank Dr. Mahadi Abdel Rahman Ahmed, Dongola Agricultural Research Corporation, Khartoum, Sudan for providing information and pictures on damage of the green pit scale insect. The provision of pictures pertaining to the natural enemies of the green pit scale provided by Dr. Daffallah Elrayah Yousof, Dongola Agricultural Research Corporation, Northern State, Khartoum, Sudan is gratefully acknowledged.

References

Abd-Allah, F. F., Al-Zidiali, T. S., & Al-Khatri, S. A. (1998, February 23–25). Biology of *Ommatissus lybicus* Bergevin under field and laboratory conditions during spring 1995. In *Proceedings international conference on integrated pest management* (pp. 75–79), Sultan Qaboos University, Muscat, Sultanate of Oman.

Abd-El-Kareim, A. I. (1998). Swarming activity of the adult males of *Parlatoria* date scale in response to sex pheromone extracts and sticky color traps. *Archives of Phytopathology and Plant Protection, 31*, 301–307.

Abivardi, C. (2001). *Iranian entomology: An introduction* (Applied entomology, Vol. 2, pp. 445–1033). Berlin/Heidelberg: Springer.

Ahmed, M. A. (2007). The Efficacy of four systemic insecticides using two methods of application against the green date palm pit scale insect *Asterolecanium phoenicis* Rao. (*Palmaspis phoenicis*) (Homoptera: Asterolecaniidae) in Northern Sudan. *ISHS Acta Horticulturae, 736*, 369–389.

Ahmed, M. A., Abdelbagi, A. O., Elshafie, A. F. H., Fageer, E. A., & Abass, I. A. (2013). Efficacy of imidacloprid (Confidor 200 SL) and improved cultural practices in the control of the green date palm pit scale insect (*Asterolecanium phoenicis* Rao) (*Palmaspis phoenicis*) (Homoptera: Asterolecaniidae) in Northern Sudan. *Scientific Research and Essays, 8*, 1752–1758.

Al-Abbasi, S. H. (1988). Biology of *Ommatissus binotatus* De Berg. (Homoptera: Tropiduchidae) under laboratory conditions. *Date Palm Journal, 2*, 412–423.

Al-Azawi, A. F. (1986). A survey of insect pests of date palms in Qatar. *Date Palm Journal, 4*, 247–266.

Al-Hafidh, E. M. T., Swair, I. A., & Abdul-Sattar, A. (1981). Ecological studies and chemical control of *Parlatoria* date scale insect (*Parlatoria blanchardi* Targ.) (Homoptera, Diaspididae) on date palms in Iraq. *Date Palm Journal, 1*, 117–127.

Alhudaib, K., Arocha, M., Wilson, M., & Jones, P. (2007). Al-Wijam, a new Phytoplasma disease of date palm in Saudi Arabia. *Bulletin of Insectology, 60*, 285–286.

Ali, A. A. (1989). Studies on *Asterolecanium phoenicis* (Rao) A date palm scale insect in El-Golid area. MSc Thesis, Faculty of Agriculture University of Khartoum, Sudan.

Ali, A. A., & El-Nasr, Y. E. S. (1992). New record of green pit scale insect, *Asterolecanium phoenicis* Rao, on date palm in the Sudan. *FAO Plant Protection Bulletin, 40*, 115.

Al-Jboory, I. J. (2007). Survey and identification of the biotic factors in the date palm environment and its application for designing IPM-program of date palm pests in Iraq. *University of Aden Journal of Natural and Applied Sciences, 11*, 423–457.

Al-Jboory, I. J., Al-Sammariae, A. I., Whaib, J. F., & Ahmed, W. A. (2001). Evaluation of thiamethoxam in different application techniques to control Dubas bugs (*Ommatissus binotatus lybicus* DeBerg.). *Arabian Journal of Plant Protection, 19*, 107–112.

Al-Mahmooli, H. I. S., Deadman, M. L., & Al-Wahabi, A. K. (2005, October 31 to November 02). Evaluation of azadirachtin formulation for the control of dubas bug in Oman. In *Proceedings of the BCPC international congress of crop science and technology* (pp. 551–556), volume 1, Glasgow, Scotland, UK.

Al-Sammariae A. I., Ausama, I. A., Ali, E. A., Nawal, A. H., & Emad, T. Y. (2006, February 19–21). The translocation, distribution and concentrations of Thiamethoxam (Actara) in date palm. In *Proceedings of the third international date palm conference*, Abu Dhabi, UAE.

Al-Shayeb, S. M., & Seaward, M. R. D. (2000). Sampling standardization of date palm, *Phoenix dactylifera* L. leaflets as a biomonitor of metal pollutants in arid environment. *Asian Journal of Chemistry, 12*, 977–989.

Anonymous. (2010). Developing integrated management of dubas bug *Ommatissus lybicus* de Bergevin in the Sultanate of Oman. Call for Proposal, IPM of Dubas Research Program, The Research Council, Sultanate of Oman, 17 pp.

Anonymous. (2011). The red date scale, *Phoenicococcus marlatti* Cockerell (Insecta: Hemiptera: Phoenicoccidae). Available at: http://entnemdept.ufl.edu/creatures/orn/palms/red_date_scale.htm. Accessed on 02 Apr 2015.

Anonymous. (2014). *Hosts of a scale query results. Scalenet: A database of the world.* Available at: http://www.sel.barc.usda.gov/scalecgi/hostsof.exe?Family=Diaspididae&genus=PARLATORIA&species=BLANCHARDI&subspecies=. Accessed on 04 Apr 2015.

Asche, M., & Wilson, M. R. (1989). The palm-feeding plant hopper genus *Ommatissus* (Homoptera: Fulgoroidea: Tropiduchidae). *Systematic Entomology, 14*, 127–147.

Avidov, Z., & Harpaz, I. (1969). *Plant pests of Israel*. Jerusalem: Israel Universities Press. 600 pp.

Ba-Angood, S. A., Al Ghurabi, A. S., & Hubaishan, M. A. (2009). Biological and chemical control of the old world bug (dubas bug) *Ommatissus lybicus* DeBerg on date palm trees in the coastal areas of Hadramout Governorate, Republic of Yemen. *Arab Journal of Plant Protection, 27*, 1–9.

Beardsley, J. W. (1959). On the taxonomy of pineapple mealybugs in Hawaii, with a description of a previously unnamed species (Homoptera: Pseudococcidae). *Proceedings of the Hawaiian Entomological Society, 17*, 29–37.

Beardsley, J. W., Su, T. H., McEwwen, F. L., & Gerling, D. (1982). Field investigations on the interrelationships of the big-headed ant, the gray pineapple mealybug wilt disease in Hawaii. *Proceedings of the Hawaiian Entomological Society, 24*, 51–67.

Bénassy, C. (1990). Date palm. In D. Rosen (Ed.), *Armored scale insects, their biology, natural enemies and control* (World crop pests, Vol. 4B, pp. 585–591). Amsterdam: Elsevier-Academic Press.

Ben-Dov, Y. (1994). *A systematic catalogue of the mealybugs of the world (Insecta: Homoptera: Coccoidea: Pseudococcidae and Putoidea) with data on geographical distribution, host plants, biology and economic importance*. Andover: Intercept Limited. 686 pp.

Bitaw, A. A., & Ben Saad, A. A. (1990). Survey of date palm trees insect pests in Libya. *Arab Journal of Plant Protection, 8*, 72–76.

Blow, J. A. (2006). *Agricultural aerial spraying in a combat zone: Control of the dubas bug (Ommatissus lybicus) in date palm in Iraq*. Available at: http://www.docstoc.com/docs/48537745. Accessed on 20 Apr 2014.

Blumberg, D. (2008). Review: Date palm arthropod pests and their management in Israel. *Phytoparasitica, 36*, 411–448.

Borchsenius, N. S. (1966). *A catalogue of the armored scale insects (Diaspidoidea) of the world*. Russian, Nauka, Moscow, Leningrad, Russia, 449 pp.

Boyden, B. L. (1941). *Eradication of the Parlatoria date scale in the United States* (Miscellaneous Publication No. 433). Washington, DC: United States Department of Agriculture. 62 pp.

Burke, H. R., Jackman, J. A., & Rose, M. (1994). *Insects associated with woody ornamentals*. Department of Entomology, Texas A&M University, Extension Publication No EEE-00019, USA, 166 pp.

Calcat, A. (1959). Diseases and pests of date palm in the Sahara and North Africa. *Bulletin of Entomological Research, 11*, 287–303.

Carpenter, J. B., & Elmer, H. S. (1978). *Pests and diseases of the date palm* (42 pp). United States Department of Agriculture Handbook No. 527, Washington, DC, USA.

Colen, K., Santa-Cecilia, L. C., Moraes, J. C., & Reis, P. R. (2000). Effects of different temperatures upon the biology of the mealybug *Dysmicoccus brevipes* (Cockerell, 1893) (Hemiptera: Pseudococcidae. *Revista Brasileira de Fruticultura, 22*, 248–252.

Dabbour, A. I. (1981). Distribution of the scale insect *Parlatoria blanchardi* Targ. on date-palm trees. *Mededelingen van de Faculteit Landbouwwetenschappen. Rijksuniversiteit Gent, 46*, 553–558.

Dhileepan, K. (1991). Insects associated with oil palm in India. *FAO Plant Protection Bulletin, 39*, 94–99.

Downer, J. (2009). *What is wrong with pygmy date palms?* Landscape Notes. University of California, Cooperative Extension, Venture County, CA, USA, 22, 2.

Dowson, V. H. W. (1936). A serious pest of date palm, *Ommatissus binotatus* Fieb. (Homoptera: Tropiduchidae). *Tropical Agriculture (Trinidad), 13*, 180–181.

Duodu, Y. A., & Thompson, W. (1992). Management of ant-mealybug complex in pineapple fields in Guyana. *FAO Plant Protection Bulletin, 40*, 82–88.

El Haidari, H. S. (1981). New records of mites and insects infesting date palms in the Near East and North Africa. *Date Palm Journal, 1*, 133–134.

El-Bouhssini, M., Brownbridge, M., Gassouma, S. (2012). Pests of the date palm (*Phoenix dactylifera*). Regional Workshop on date palm development in the Arabian Peninsula, 29–31 May 2004, Abou-Dhabi, UAE, 21 pp.

Eldoush, K. O. A. M., Taha, A. K., Idris, T. E. I. M., Sidahmed, O. A. A., Musa, F. E., & Mardi, H. G. (2011). Application of plant based extract for the control of the green pit scale insect (*Asterolecanium phoenicis* Rao) with yield enhancement on date palm. *Emirates Journal of Food and Agriculture, 23*, 404–412.

El-Haidari, H. S., Mohammed, I. I., & Daoud, A. A. K. (1968). Evaluation of DDVP against the dubas bug, *Ommatissus binotatus libycus* Berg., on date palms in Iraq (Hemiptera: Homoptera: Tropiduchidae). *Bulletin of the Entomological Society of Egypt Economic Services, 2*, 91–94.

El-Haidri, H. S., & Al-Hafidh, E. M. T. (1986). Palm and date arthropods pests in the Near East and North Africa (in Arabic). Project for palm and date research center in the near east and North Africa. *FAO Bulletin, Baghdad, Iraq*, 17 pp.

Elwan, A. A. (2000). Survey of the insect and mite pests associated with date palm trees in Al-Dakhliya region, Sultanate of Oman. *Egyptian Journal of Agricultural Research, 78*, 653–664.

Elwan, A. S. A., & Al-Tamiemi, S. S. (1999). Life cycle of dubas bug, *Ommatissus binotatus lybicus* De Berg. (Homoptera: Tropiduchidae) in Sultanate of Oman. *Egypt Journal of Agricultural Research, 77*, 1547–1553.

Espinosa, A., Hodges, A., Hodges, G., & Mannion, C. (2012). *Red date scale, Phoenicococcus marlatti (Cockerell) (Insecta: Hemiptera: Phoenicoccidae)*. Gainesville: Department of Entomology and Nematology, Florida Cooperative Extension Service Publication No. EENY 454, Institute of Food and Agricultural Sciences, University of Florida. 4 pp.

Estrada, H. A., Alvarado, E., Tabora, P., & Castillo, H. (2009). Evaluation of natural insecticides for the control of the pineapple (*Ananas comosus* L.) mealybug, *Dysmicoccus brevipes* Cockerell (Hemiptera: Pseudococcidae). *Universidad EARTH, Guacimo, Costa Rica, Tierra Tropical, Sostenibilidad, Ambiente y Sociedad* 5, pp. 159–168.

Ezz, I. A. (1973). *Asterolecanium phoenicis* (Homoptera-Coccoidea), a date palm pest recorded for the first time in Egypt. *Agricultural Research Review, 51*, 47.

Fageer, E. A., & Ahmed, M. A. (2010). Economic impacts of date palm infestation by the green pit scale insect (*Palmaspis phoenicis* Ramachandra Rao) in Northern Sudan. *ISHS Acta Horticulturae, 882*, 457–464.

FAO. (1995, April 22–26). *Food and Agriculture Organization report of the Expert Consultation on date palm – Pest problems and their control in the Near East*, Al-Ain, United Arab Emirates.

Gassouma, M. S. (2003). *Palmaspis phoenicis*, Asterolecaniidae, Coccoidea, Homoptera, Insecta. Available at: http://ecoport.org/ep?SearchType=slideshowView&slideshowId=137. Accessed on 24 Apr 2014.

Gassouma, M. S. (2004, May 29–31). Pests of date palm (*Phoenix dactylifera* L.). In *Proceedings of regional workshop on date palm development in the Arabian Peninsula*, Dubai, UAE, 32 pp.

Gharib, A. (1973). *Parlatoria blanchardi* Targ. (Homoptera – Diaspididae). *Entomologie et Phytopathologie Appliquées, 34*, 10–17.

Gharib, A. (1974). *Palmaspis phoenicis* (Ramachandra Rao) = *Asterolecanium phoenicis* (Homoptera – Asterolecanidae). *Entomologie et Phytopathologie Appliquées, 36*, 1–11.

Gill, R. J. (1997). *The scale insects of California, Part 3. The armored scales (Homoptera: Coccoidea: Coccidae)* (Technical Series in Agricultural Biosystematics and Plant Pathology No. 3). Sacramento: California Department of Food and Agriculture. 307 pp.

Gnezdilov, V. M. (2009). Revisionary notes on some tropical Issidae and Nogodinidae (Hemiptera: Fulgoroidea). *Acta Entomologica Musei Nationalis Pragae, 49*, 75–92.

Gonzalez-Hernandez, H. (1995). The status of the biological control of pineapple mealybugs in Hawaii. PhD Thesis, The University of Hawaii, USA, 128 pp.

Gullan, P. J., & Cook, L. G. (2007). Phylogeny and higher classification of the scale insects (Hemiptera: Sternorrhyncha: Coccoidea). *Zootaxa, 1668*, 413–425.

Hamad, B. S., & Al-Rawy, M. A. (2009). Toxicity of deltamethrin and Basudin to *Chrysoperla mutata* MacL. and dubas nymphs *Ommatissus lybicus* DeBerg. *Arab Journal of Plant Protection, 27*, 210–213.

Hazir, A., & Buyukozturk, H. D. (2013). *Phoenix* spp. and other ornamental palms in Turkey: The threat from red palm weevil and red palm scale insects. *Emirates Journal of Food and Agriculture, 25*, 843–853.

Hill, D. S. (2008). *Pests of crops in warmer climates and their control*. Delhi: Springer. 708 pp.

Hodges, G. S., Howard, F. W., Buss, E. A. (2003). Update on management methods for cycad aulacaspis scale. Department of Entomology and Nematology, Florida Cooperative Extension Service Publication No. ENY 680, Institute of Food and Agricultural Sciences, University of Florida, Gainesville, FL, USA.

Howard, F. W. (2001). Sap feeders on palms. In F. W. Howard, D. Moore, R. M. Giblin-Davis, & R. G. Abad (Eds.), *Insects on palms* (pp. 199–232). Wallingford: CABI Publishing, CAB International.

Howard, F. W., & Weissling, T. J. (1999). Questions and answers about the cycad aulacaspis scale insect. *Proceedings of the Florida State Horticultural Society, 112*, 243–245.

Howard, F. W., Moore, D., Giblin-Davis, R. M., & Abad, R. G. (2001). *Insects on palms*. Wallingford: CABI Publishing, CAB International. 400 pp.

Hubaishan, M. A., & Bagwaigo, A. A. (2010). The effect of the egg parasitoid *Pseudoligosita babylonica* on dubas infestation in Doaan, Shichawi and Asid (Hadramout, Yemen). *ISHS Acta Horticulturae, 882*, 451–455.

Hubaishan, M. A., Gamal, S. B., Ashour, M. A., Amjad, B. (2007). Distribution and control of dubas date bug in Republic of Yemen. In: Proceedings of the 4th symposium on date palm in Saudi Arabia, 5–8 May 2007, King Faisal University, Al-Hassa, Saudi Arabia, pp. 1388–1405.

Hussain, A. A. (1963). Biology and control of dubas bug, *Ommatissus binotatus* var. lybicus de berg. (Homoptera: Tropiduchidae) infesting date palm in Iraq. *Bulletin of Entomological Research, 53*, 737–745.

Ito, K. (1938). Studies on the life history of the pineapple mealybug, *Pseudococcus brevipes* (Ckll.). *Journal of Economic Entomology, 2*, 291–298.

Jassim, H. K. (2007). Studies on the biology of date palm dubas bug, *Ommatissus lybicus* (De Bergevin) Asche and Wilson (Homoptera: Tropiduchidae) and its biocontrol by some isolates of entomopathogenic fungi, *Beauveria bassiana* (Balsamo) Vuil and *Lecanicillium = Verticillium) lecanii* (Zimm.) Zare and Oami. PhD Dissertation, College of Agriculture, Baghdad University, Iraq, 160 pp.

Jassim, H. K., & Al-Zubaidy, H. K. (2010). Egg distribution of old world bug (Dubas bug) *Ommatissus lybicus* (DeBerg.) Asche and Wilson (Homoptera: Tropiduchidae) on fronds rows and effect of dust storm on three varieties of date palm trees. *ISHS Acta Horticulturae, 882*, 465–470.

Kaufmann, T. (1977). *Hemisarcoptes* sp. and biological control of the date palm scale, *Parlatoria blanchardi* Targioni, in the Sahel region of Niger. *Environmental Entomology, 6*, 882–884.

Kehat, M. (1967). Some notes on the life cycle of the date palm scale, *Parlatoria blanchardi* Targ. in Israel. *Israel Journal of Agricultural Research, 17*, 175–179.

Kehat, M., & Amitai, S. (1967). The morphology and phenology of the scale *Asterolecanium phoenicis* Rao. on date palms in the Bet Shean Valley. *Israel Agricultural Research, 17*, 85–93.

Kehat, M., Swirski, E., & Bankier, E. (1964). Trials in the control of the date palm scale *Asterolecanium phoenicis* Rao. on date palms. *Israel Journal of Agricultural Research, 14*, 19–26.

Kehat, M., Swirski, E., Blumberg, D., & Greenberg, S. (1974a). Integrated control of date palm pests in Israel. Special Publication of Agricultural Research Organization Series No. 251-E, The Volcani Center, Bet Dagan, Israel 65, pp. 197–114.

Kehat, M., Swirski, E., Blumberg, D., & Greenberg, S. (1974b). Integrated control of date palm pests in Israel. *Phytoparasitica, 2*, 141–149.

Khalaf, M. Z., & Khudhair, M. W. (2015). Spatial distribution of dubas bug, *Ommatissus lybicus* (Homoptera: Tropiduchidae) in date palm frond rows. *International Journal of Entomological Research, 3*, 9–13.

Khoualdia, O., Rhouma, A., & Hmidi, M. S. (1993). Contribution to the bio-ecological study of the white scale *Parlatoria blanchardi* Targ. (Homoptera, Diaspididae) of date palm in Djerid (Southern Tunisia). *Annales de l'Institut National de la Recherche Agronomique de Tunisie, 66*, 89–108.

Khoualdia, O., Rhouma, A., Brun, J., & Marro, J. P. (1997). Biological control of white scale. Introduction of an exotic predator in the palm grove of Segdoud. *Phytoma, 49*, 41–42.

Kinawy, M. M. (2005). Date palm and date pests in Oman. Royal Court Affairs, Royal Gardens and Far, Sultanate of Oman, 431 pp.

Klein, M., & Venezian, A. (1985). The dubas date Tropiduchid, *Ommatissus binotatus lybicus*, a threat to date palms in Israel. *Phytoparasitica, 13*, 96–101.

Kondo, T., Gullan, P. J., & Williams, D. J. (2008). Coccidology. The study of scale insects (Hemiptera: Sternorrhyncha: Coccoidea). *Revista Corpoica – Ciencia y Tecnologia Agropecuaria, 9*, 55–61.

Kranz, J., Schmutterer, H., & Koch, W. (1978). *Diseases, pests and weeds in tropical crops* (pp. 304–305). Chichester: Wiley.

Kumar, S., & Kontodimas, D. C. H. (2012). Temperature dependent development of *Phenacoccus solenopsis* under laboratory conditions. *Entomologia Hellenica, 21*, 25–38.

Lepesme, P. (1947). *Les Insectes des Palmiers*. Paris: Editions P. Lechevalier. 904 pp.

Matallah, S., & Biche, M. (2013). Biological behavior of *Parlatoria blanchardi* Targioni 1892 (Homoptera: Diaspididae) towards three cultivars of date palm tree in the region of Biskra, Algeria. *ISHS Acta Horticulturae, 994*, 389–394.

McClure, M. (1990). Patterns of host specificity. In D. Rosen (Ed.), *Armored scale insects, their biology, natural enemies and control* (pp. 301–303). Amsterdam: Elsevier-Academic Press.

McDaniel, B. (1974). The armored scale insects of Texas (Homoptera: Coccoidea: Diaspididae). *The Southwestern Naturalist, 18*, 417–442.

Miller, D. R., & Davidson, J. A. (2005). *Armored scale insect pests of trees and shrubs*. Ithaca: Cornell University Press. 442 pp.

Miranda, A., & Blanco, H. (2013). Control of *Dysmicoccus brevipes* (Hemiptera: Pseudococcidae) on the pineapple fruit, San Carlos, Costa Rica. *Agronomia Costarricense, 37*, 103–111.

Moghaddam, M. (2013). An annotated checklist of the scale insects of Iran (Hemiptera, Sternorrhyncha, Coccoidea) with new records and distribution data. *ZooKeys, 334*, 1–92.

Mokhtar, A. M. (2009). Insect pests of date palm, current challenges and future perspectives. Available at: www.qu.edu.qa/qulss/09/talks/Mokhtar-QULSS-DatePalm2009.pdf. Accessed on 23 May 2014.

Mokhtar, A. M., & Al-Mjeni, A. M. (1999). A novel approach to determine the efficacy of control measures against dubas bug, *Ommatissus lybicus* de Berg. on date palm. *Agricultural Science, 4*, 1–4.

Mokhtar, A. M., & Al-Nabhani, S. S. (2010). Temperature-dependent development of dubas bug. *Ommatissus lybicus* (Hemiptera: Tropiduchidae), an endemic pest of date palm, *Phoenix dactylifera*. *European Journal of Entomology, 107*, 681–685.

Mokhtar, A. M., ALMjeni, A. M., & Salem, H. E. M. (2003). An attempt to estimate honeydew production of different stages of *Ommatissus lybicus* de Berg. (Homoptera: Tropiduchidae) in vitro. *Mansoura University Journal of Agricultural Sciences, 28*, 5669–5677.

Momen, F., Abdel-Khalek, A., & El-Sawi, S. (2009). Life tables of the predatory mite *Typhlodromus negevi* feeding on prey insect species and pollen diet (Acari: Phytoseiidae). *Acta Phytopatologica et Entomologica Hungarica, 44*, 353–361.

Moustafa, M. (2012). Host plant, distribution and natural enemies of the red date scale insect *Phoenicococcus marlatti* (Hemiptera: Phoenicococcidae) and its infestation status in Egypt. *Journal of Basic and Applied Zoology, 65*, 4–8.

Nelson, E. H., & Daane, K. M. (2007). Improving liquid bait programs for Argentine ant control: Bait station density. *Environmental Entomology, 36*, 1475–1484.

Nixon, R. W., & Wedding, R. T. (1956). Age of date leaves in relation to efficiency of photosynthesis. *Proceedings of the American Society for Horticultural Science, 67*, 265–269.

Noyes, J. S. (2007). *Universal Chalcidoidea Database*. Available at: http://www.nhm.ac.uk/research-curation/research/projects/chalcidoids/introduction.html. Accessed on 11 Apr 2015.

Petty, G. J. (1985) The pineapple mealybug. *Farming in South Africa*, H.15, Department of Agricultural Development, Pretoria, Republic of South Africa, 8 pp.

Petty, G. J., & Tustin, H. (1993). Ant (*Pheidole megacephala*), in pineapple plantations with the proprietary bait Amdro. *Fruits, 50*, 343–346.

Petty, G. J., & Webster, G. S. (1981). Importance of spray volume for control of mealybug, *Dysmicoccus brevipes* (Ckll.) and its distribution on the pineapple plant. *Subtropica, 2*, 6–8.

Petty, G. J., Stirling, G., & Bartholomew, D. (2002). Pests of pineapple. In J. E. Peña, J. L. Sharp, & M. Wysoki (Eds.), *Tropical fruit pests and pollinators: Biology, economic importance, natural enemies and control* (pp. 157–195). Wallingford: CABI Publishing, CAB International.

Radwan, S. G. (2012). Seasonal fluctuation of fiorinia date scale, *Fiorinia phoenicis* Balachowsky (Hemiptera: Diaspididae) population on date palm trees at Qalyubia Governorate, Egypt. *The Journal of Basic and Applied Zoology, 65*, 47–54.

Rai, B. K., & Sinha, A. K. (1980). Pineapple: Chemical control of mealybug and associated ants in Guyana. *Journal of Economic Entomology, 73*, 41–45.

Rohrbach, K. G., & Apt, W. J. (1986). Nematode and disease problems of pineapple. *Plant Disease, 70*, 81–87.

Rohrbach, K. G., Beardsley, J., German, T., Reimer, N., & Sanford, W. G. (1988). Mealybug wilt, mealybugs and ants on pineapple. *Plant Disease, 72*, 558–565.

Rosen, D. (1990). Biological control: Selected case histories. In D. Rosen (Ed.), *Armored scale insects, their biology, natural enemies and control* (World Crop Pests, Vol. 4B, pp. 497–505). Amsterdam: Elsevier.

Shah, A., Ul-Mohsin, A., Naeem, M., Nasir, M. F., Ul-Haq, M. I., & Hafeez, Z. (2012). Biology of dubas bug *Ommatissus lybicus* (Homoptera: Tropiduchidae), a pest on date palm during spring and summer seasons in Panjgur, Pakistan. *Pakistan Journal of Zoology, 44*, 1603–1611.

Shah, A., Ul-Mohsin, A., & Hafeez, Z. (2013). Egg distribution behavior of dubas bug (Ommatissus lybicus: Homoptera: Tropiduchidae) in relation to seasons and some physico-morphic characters of date palm leaves. *Journal of Insect Behavior, 26*, 371–386.

Siddig, S. A. (1975). Field control of the scale insect Parlatoria blanchardi Targ. (Diaspididae) infesting date palm in the Sudan. *Journal of Horticultural Science, 50*, 13–19.

Smirnoff, W. A. (1957). La cochenille du palmier-dattier (*Parlatoria blanchardi* Targ.) en Afrique du Nord. Comportement, importance économique, prédateurs et lutte biologique. *Entomophaga, 2*, 1–99.

Stansly, P. A. (1984). Introduction and evaluation of *Chilocorus bipustulatus* (Col.: Coccinellidae) for control of *Parlatoria blanchardi* (Hom., Diaspididae) in date palm groves of Niger. *Entomophaga, 29*, 29–39.

Stathas, G. J. (2000). The effect of temperature on the development of predator *Rhyzobius lophanthae* Blaisdell (Coleoptera: Coccinellidae) and its phenology in Greece. *BioControl, 45*, 439–451.

Stickney, F. S. (1934). The external anatomy of *Parlatoria* date scale, *Parlatoria blanchardi* Targioni-Tozzetti with studies of the head skeleton and associated parts. Technical Bulletin No. 42, USA, 66 pp.

Stickney, E. S., Barnes, D. W., & Simons, P. (1950). *Date palm insects in the United States*. United States Department of Agriculture Circular No. 846, Washington, DC, USA, 57 pp.

Swaminathan, R., & Verma, S. K. (1991). Studies on the incidence of date palm scale, *Parlatoria blanchardi* (Targ.) in Western Rajasthan. *Journal of Entomology, 16*, 217–221.

Talhouk, A. S. (1983). The present status of date palm pests in Saudi Arabia. In *Proceedings of the first symposium on the date palm*, 23–25 Mar 1982, Saudi Arabia, pp. 432–438.

Talhouk, A. S. (1993). Some insect species injurious to crops in Middle East countries. *Journal of Applied Entomology, 116*, 105–111.

Thacker, J. R. M., Al-Mahmooli, I. H., & Deadman, M. L. (2003). Population dynamics and control of the dubas bug *Ommatissus lybicus* in the Sultanate of Oman. In *Proceedings of the BCPC international congress – Crop science and technology*, Glasgow, UK, pp. 987–992.

Vives, S. G. (2002). Cría masiva de *Rhyzobius lophanthae* Blaisdell (Coleoptera: Coccinellidae) depredador de la cochinilla roja de las palmeras (*Phoenicococcus marlatti* Cockerell). *Boletín de Sanidad Vegetal Plagas, 28*, 167–176.

Yousof, D. E., Mahmoud, M. E. E., & Mohamed, A. H. (2013). Prospects of biological control of date palm green pit scale insect Asterolecanium phoenicis Rao. (Homoptera: Asterolecaniidae) in Sudan. *Persian Gulf Crop Protection, 2*, 42–48.

Zaid, A., de Wet, P. F., Djerbi, M., & Oihabi, A. (2002). Diseases and pests of date palm. In A. Zaid & E. Arias-Jiménez (Eds.), *Date palm cultivation*. FAO Plant Production and Protection Paper No. 156, Revision 1, Rome (pp. 347–352). Rome: FAO.

Chapter 8
Sap Beetles

Mevlut Emekci and Dave Moore

Abstract Nitidulid beetles have a worldwide occurrence, on a variety of plants with fruits containing a significant level of carbohydrates. They are mostly found in rotting and fermenting fruits and thus are regarded as secondary pests of date palms. However, in Australia, they are the primary pests of stone fruits. Depending on the climate, sap beetles overwinter as pupae and/or adults in mass accumulations. Adults are strong fliers capable of long distance flights. Eggs are laid in single small clusters in or on the fruits. Larvae feed on the flesh of the fruits for about 1½ weeks. Mature larvae fall onto the ground and pupate inside the soil. Five to eight generations are produced per year under optimal conditions. Hygiene and other cultural practices such as removing fallen fruits and early harvest are important measures to reduce the beetles number. Chemical pesticides applied against other primary pests in date palms also incidentally control sap beetles. In stone fruits, mass trapping by means of aggregation pheromone combined with fermenting food baits has been shown to be very effective. Covering the fruit bunches with plastic netting is an important physical control option. Attempts for biological control by means of parasitoids, predators, nematodes and other microbial agents so far have been neither successful nor economical. Detailed knowledge of the ecology of natural enemy complexes is needed. Developments in production techniques of beneficials would improve their value in terms of both economy and effectiveness, within sustainable IPM systems.

M. Emekci (✉)
Department of Plant Protection, Faculty of Agriculture, Ankara University, Diskapi, Ankara 06110, Turkey
e-mail: mevlut.emekci@agri.ankara.edu.tr

D. Moore
CABI, Bakeham Lane, Egham TW20 9TY, Surrey, UK
e-mail: d.moore@cabi.org

W. Wakil et al. (eds.), *Sustainable Pest Management in Date Palm: Current Status and Emerging Challenges*, Sustainability in Plant and Crop Protection,
DOI 10.1007/978-3-319-24397-9_8

8.1 Introduction

Nitidulids (a Latin-derived word, meaning shiny), or sap beetles (Nitidulidae: Coleoptera), are polyphagous insects reported from more than 200 species of host plants, including cotton, Chinese date, grape, apple, pear, peach and many other crop plants. They are small to minute coleopterans, with the adult body size ranging from less than 1 to ~15 mm. The most abundant and speciose taxa such as *Carpophilus* Stephens and *Epuraea* Erichson range between 1.5 and 6.0 mm in length (Cline 2005).

Sap beetles have different body forms adapted to different ecologies and life history strategies (Cline 2005), and can be easily recognized. They possess three-segmented club like antennae, five segmented tarsi, and, usually, truncated elytra that leaves the apex of the pygidium or sometimes even the preceding 1 or 2 tergites exposed (Larson 2013). Nitidulidae represent the second most diverse family of cucujoid beetles following Coccinellidae (containing >6000 species) with more than 4000 described species (Lawrence 1982 cited in Cline 2005). This number is expected to increase several folds with the addition of taxonomic descriptions from less explored areas such as the Neotropics, Afrotropics, or Southeast Asia (Cline 2005).

Most species of sap beetles are attracted to the sap emerging from wounds occurring in trees and, as their common name implies, feed on this sap (Larson 2013). However, the habitats of the Nitidulidae are quite variable (Hayashi 1978; Parsons 1943). Besides sap feeding, other important feeding behaviors and life history strategies include fungivory, frugivory, saprophagy, anthophily, phytophagy, predation, necrophagy and inquilinism with social Hymenoptera (Aksit et al. 2003; Tezcan et al. 2003; Cline 2005; Larson 2013).

Several species of nitidulid beetles attack ripened fruits and typically feed in their pulp. They have previously been considered as minor pests (Dowd 1991; Moore 2001) but in recent years they achieved pest status in certain situations, and now are of major significance in USA and Israel. Consequently, sap beetles range from mere nuisance to major pests in date palm. At one point it seemed that their control was rarely needed or that control measures for other pests also controlled sap beetles incidentally. Recently, *Carpophilus* spp. are believed to have become a serious problem in Australian orchards, due to the decreasing use of broad-spectrum insecticides for the control of other stone fruit pests (James and Vogele 2000).

Sap beetles, on the other hand, are the most common insect visitors to flowers of annonaceous crops which are highly dependant on these beetles for fruit set (Gazit et al. 1982). Thus in Israel, growers try to enhance *Carpophilus* spp. as beneficial insects, using fermenting fruits and aggregation pheromones in *Annona* plantations, in order to improve fruit pollination (D. Blumberg, pers. comm., 2014).

8.2 Distribution and Host Range

Sap beetles characteristically have short-lived larval (James and Vogele 2000), and long-lived adult stages and thus are able to survive on a wide variety of microhabitats, mainly composed of ripe and rotting fruits and decaying plant tissues (Peng and Williams 1990; Myers 2001). According to Lawrence (1982), there are more than 4000 species of nitidulids described (Cline 2005) and more than half of the genera show a cosmopolitan distribution (Rondon et al. 2004).

Despite the large number of species determined, only a small number of species are considered of agricultural importance (Peng and Williams 1990; Myers 2001). These pests present problems worldwide in the field or in storage of a variety of important agricultural crops, including figs (Aksit et al. 2003; Turanli 2003) stone fruits (James et al. 1997; Tezcan et al. 2003), pome fruits, oranges (Leschen and Marris 2005), grapes (Blumberg et al. 2005), pomegranates (Oztürk et al. 2005), strawberries (Loughner et al. 2007) and maize (Nout and Bartelt 1998).

Similarly, a few species of nitidulids are associated with date palm orchards in date growing areas of the world (Table 8.1). The species found in date palm groves show some variation according to the region or to the specific date cultivars.

The pineapple beetle, *Urophorus* (= *Carpophilus*) *humeralis* (F.) (Fig. 8.1), has a shiny black appearance. It is the largest nitidulid infesting date palms, with a body length of 3.3–4 mm, with a faint brown area at the base of each forewings (Carpenter and Elmer 1978). In California (USA) in addition to dates, this beetle has been found in waste grapefruits. In Turkey, it is the most common nitidulid beetle infesting organic cherry orchards (Tezcan et al. 2003). This species is also found in sugarcane, rotten breadfruit and decaying cucumbers in Guam (USA) (Swezey 1942, cited in Stickney et al. 1950).

The adult of the driedfruit beetle, *Carpophilus hemipterus* (L.) (Fig. 8.2), which is a black beetle of about 0.3 cm in body length, can be easily recognized by two distinctive amber-brown spots on each forewings, one near the tip, and a smaller one at the base outer margin. The driedfruit beetle is widely distributed in warm parts of the world (Stickney et al. 1950; Williams et al. 1983). In Turkey, this beetle has been found in fig orchards (Aksit et al. 2003; Turanli 2003) and organic cherry orchards (Tezcan et al. 2003). Other produce attacked are listed in Table 8.1.

The adults of the corn sap beetle, *Carpophilus dimidiatus* (F.) (Fig. 8.3), have a body length of 0.16–0.32 cm, have no elytral spots and their color ranges from brownish-yellow through various shades of brown to black tinged with red. This species shows a cosmopolitan distribution (Stickney et al. 1950; Williams et al. 1983). The host range of *C. dimidiatus* recorded to date is given in Table 8.1.

The yellowish nitidulid, *Epuraea luteola* Erichson (Fig. 8.4) is shorter and broader than the corn sap beetle *C. dimidiatus*, with a blunt appearance and rectangular forewings that reach to the apex of the abdomen. Its body length is around 2.4 mm (Stickney et al. 1950). It is common in figs, various waste fruits, and grape marc in California. According to Swezey (1942), cited in Stickney et al. (1950), this species is found in rotten materials such as pandanus (screwpine) fruit, breadfruit and sugarcane in Guam. Many sap beetles may be viewed as secondary, nuisance pests in rotting produce of very limited economic value. However, they are also known as primary pests in a restricted number of other crops.

The dried sap beetle, *Carpophilus mutilatus* Erichson (Fig. 8.5), has a close resemblance to *C. nepos* (Ewing and Cline 2004). Adults measure 1.5–1.8 mm in body length, and show a color pattern of uniformly light tan to brown (Leschen and Marris 2005). The head and pronotum are light reddish brown, with the pronotum

Table 8.1 Sap beetles reported in date palm orchards worldwide and other recorded hosts/produce

Species	Common name	Locality	References	Other hosts
Carpophilus hemipterus	Driedfruit beetle; fig beetle; two-spotted beetle	Iraq USA (California) Libya Israel Egypt Somalia Palestine Oman New Zealand	El-Haidari et al. (1981), Lindgren and Vincent (1949), Najla et al. (2005), Kehat et al. (1976), Abdel-Rahman (2007), Omar (1971), Abdul-Wahab and Eissa (1986), Ali et al. (2002), Hussien (2007), Al-Azawi (1986), Al-Azawi (1986), Al-Zadjali et al. (2006), and Archibald and Chalmers (1983)	Fig, cherry, apply (dried/rotten), rockmelon, flour, pumpkin flowers, papaya, lemon peel, persimmon, pineapple, bagged sugar, iris bulbs, oranges, water chestnut, sweet orange, peach/nectarine, grape/raisins, garlic, rotten garlic bulbs, rice, dried citrus peel, molasses, tomato, shallots, orchids, lote fruit, Indian jujube, Chinese date, tea, onion, bhindi, dried apricots, taro, chocolates
Carpophilus mutilatus	Confused sap beetle	Iraq Libya Israel USA (California) Australia	El-Haidari et al. (1981), Kehat et al. (1976), Najla et al. (2005), Kehat et al. (1976), Lindgren and Vincent (1949), and Leschen and Marris (2005)	Peach, grapes, figs, oranges, nuts, cherry, peanuts, sugar, decaying fruits and grains
Epuraea luteola	The yellowish nitidulid	Iraq USA (California) Israel Libya India	El-Haidari et al. (1981), Lindgren and Vincent (1949), Kehat et al. (1976), Najla et al. (2005), and Batra and Sohi (1972)	Figs, various waste fruits, grape mare, rotten materials such as pandanus (screwpine) fruit, breadfruit, sugarcane
Carpophilus dimidiatus	The corn sap beetle	Libya Egypt New Zealand	Najla et al. (2005), Omar (1971), Abdul-Wahab and Eissa (1986), and Archibald and Chalmers (1983)	Almonds, figs, bagged sugar, stored copra, stored pepper, stored maize, cocoa beans, bean seeds, corn, chocolate, candle nuts, yams, hyacinth bean, coconut palm, hazel nuts, lychee, dried bananas, Brazil nuts, lote fruit, Indian jujube, Chinese date, garlic seeds, dried apricots, spices, jalapeno chillies, rice, blackberry, peanut

Carpophilus obsoletus		Iraq Libya Qatar Morocco	El-Haidari et al. (1981), Najla et al. (2005), Al-Turaihi (2007), and Sedra (2003)	Figs, pineapple, cocoa, bananas, peanut, papaya, coconut, coconut palm, dasheen, Chinese date plum, taro, dried kaki, lychee, bananas, sweet almond, dried grape berry, dried persimmon, sweet almond, various waste fruits, grape marc, rotten pandanus (screwpine) fruit, breadfruit, sugarcane, yam, coffee beans, tea leaves, garlic, rice, rotten wheat
Carpophilus nepos	Former freeman sap beetle, *C. freemani*	Libya	Najla et al. (2005)	Oranges, grapefruits, tangerines, apples, strawberries, peaches, apricots, kiwi, grapes, jackfruit, figs, vegetables, maize nuts, bones
Urophorus humeralis	The pineapple beetles	Libya Iraq USA (California) Israel Iran Indonesia and Tropical Africa	Najla et al. (2005), El-Haidari et al. (1981), Lindgren and Vincent (1949), Blumberg et al. (1993), and Williams et al. (1983)	Cherry sugarcane, rotten breadfruit, decaying cucumbers, waste grapefruits

Fig. 8.1 *Urophorus humeralis* dorsal view (Brown 2010 (*left*); Birgit Rhode, Landcare Research, NZ © (*right*))

Fig. 8.2 *Carpophilus hemipterus* dorsal view [Simon Hinkley and Ken Walker, Museum of Victoria (*left*); Birgit Rhode, Landcare Research, NZ © (*middle*); Dave M. Larson, Canada (*right*)]

darker in the center (Ewing and Cline 2005). *Carpophilus mutilatus* was synonymized with *C. dimidiatus* from 1913 to 1954 (Ewing and Cline 2004). It has a cosmopolitan distribution (Williams et al. 1983; Leschen and Marris 2005), and is commonly found in decaying fruits and grains (Ewing and Cline 2005). Peach, grapes, figs, sugarcane, oranges, nuts, cherry and peanuts are among the other important hosts (Archibald and Chalmers 1983; Aksit et al. 2003; Tezcan et al. 2003; Turanli 2003; Leschen and Marris 2005).

Carpophilus obsoletus Erichson (Fig. 8.6), which damages inflorescences of date palms (Sedra 2003), is a 1.8–2.2 mm long beetle. Body color varies between chocolate to dark brown, with a lighter elytra. According to Parsons (1943), Williams et al. (1983), and Connell (1991), *C. obsoletus* is distributed in Africa, China,

Fig. 8.3 *Carpophilus dimidiatus* dorsal view [Birgit Rhode, copyright Landcare Research, NZ (*left*); Joe MacGown, Mississippi Entomological Museum (*middle*); Dave M. Larson, Canada (*right*)]

Fig. 8.4 *Epuraea luteola*, dorsal view (C. Harding, MAF Plant Health and Environment Laboratory, New Zealand)

Fig. 8.5 *Carpophilus mutilatus*, dorsal view [Simon Hinkley and Ken Walker, Museum Victoria (*left*); Birgit Rhode, © Landcare Research, New Zealand (*right*)]

Fig. 8.6 *Carpophilus obsoletus*, dorsal view [Simon Hinkley and Ken Walker, Museum of Victoria (*left*); Birgit Rhode, © Landcare Research, New Zealand (*right*)]

Fig. 8.7 *Carpophilus nepos*, dorsal view [Birgit Rhode, © Landcare Research, New Zealand (*left*); Michele B. Price, University of Minnesota, USA (*right*)]

Central America, Europe, Iran, Japan, Madagascar, and USA. It is reported as a pest of figs in Turkey (Aksit et al. 2003; Turanli 2003). Archibald and Chalmers (1983) and other authors have listed their hosts (Table 8.1).

Carpophilus nepos Murray (formerly known as Freeman sap beetle, *Carpophilus freemani* Dobson), is characterized by a red-brown dorsum and a dull yellow elytra (Ewing and Cline 2005) (Fig. 8.7). Adults are 1.5–2.6 mm in length (Price and Cira 2007). They are cosmopolitan except for polar and colder temperate regions (Connell 1991) and are widely distributed in the Americas, Africa (Williams et al. 1983) and South Asia. They damage all kind of fruits containing a high level of carbohydrates including dates, oranges, grapefruits, tangerines, apples, strawberries, peaches, apricots, kiwi, grapes, jackfruit, figs (Paolo Audisio, pers. comm., 2014), vegetables, maize, nuts and bones (Tate and Ogawa 1975; Smilanick 1979; Rodriguez-Del-Bosque et al. 1998; Hill 2002; Leschen and Marris 2005).

In California, 53 species of nitidulids belonging to 21 genera have been recognized (Evans and Hogue 2006). Seven species of nitidulids, namely *C. mutilatus* (mistakenly identified as *C. dimidiatus*) (Bartelt et al. 1994b), *C. hemipterus*, *U. humeralis*, *E. luteola* (Lindgren and Vincent 1953), *C. nepos*, and *C. obsoletus* (Bartelt et al. 1994b) can be found in California date groves. Their status has changed over time: *C. hemipterus* was the most important species before the 1950s and *C. mutilatus* became the major species after 1950s (Lindgren and Vincent 1953). Fruit dissections made in the date-maturing seasons from July to November between 1947 and 1951 revealed a 75–90 % infestation rate of *C. mutilatus* (as *C. dimidiatus*) in the date variety Deglet Noor (Lindgren and Vincent 1953).

Although most date production is from the Old World, nitidulid records from that region are very limited, and this is attributed to their so-called "lesser importance" compared with the major pests of date palms. In the Middle East region, there has been a continuous accumulation of research dealing with sap beetles in Israel. In other countries, however, there are few recent studies, concentrating mainly on the occurrence of sap beetles.

In Iraq, *C. hemipterus*, *C. mutilatus*, *C. obsoletus*, *U. humeralis* and *E. luteola* were recorded for the first time attacking dates on the palm in 1979. Of these, *U. humeralis* and *E. luteola* were also detected in fallen citrus fruits and on rotten cucumbers (El-Haidari et al. 1981).

In Israel, nitidulid beetles are represented by 44 species belong to four subfamilies. Four species, which belong to the Carpophilinae and Epuraeinae subfamilies, are considered as date plantation pests of economic importance. The Carpophilinae subfamily contains ten species, of which *C. hemipterus*, *C. mutilatus* (easily confused with *C. dimidiatus*) and *U. humeralis* are of economic importance. The Epuraeinae subfamily contains five species, of which *E. luteola* is the most common (Blumberg et al. 1993; Blumberg 2008). Besides date palms, the two subfamilies also infest figs, pomegranates, nectarines, citrus, apples and vines in Israel (Blumberg et al. 2005).

In Libya, the most abundant nitidulids in date palm groves are recorded as *C. hemipterus*, *C. dimidiatus*, *U. humeralis* and *E. luteola* (Najla et al. 2005). Four more nitidulids, namely *C. obsoletus*, *C. nepos*, *C. mutilatus* and *E. luteola* are also recorded in date palm orchards, although with a lower frequency (Najla et al. 2005).

Other reports are more limited in numbers of species detected. In the United Arab Emirates (UAE), *C. mutilatus* was detected for the first time feeding on dates, both on the palm and on the ground in the Al-Ain, Sha'am, Shamal and Al Dhaid areas, in late September, 1980 (El-Haidari 1981). In the Yemen Arab Republic, a *Carpophilus* sp. was found for the first time in fallen dates in the Qadaba area of Tihamna region, at the end of April, 1980 (El-Haidari 1981).

8.3 Damage, Economic Threshold and Losses

Several nitidulid species are of cosmopolitan distribution and infest a variety of pre- and post-harvest agricultural products (Bartelt and Hossain 2010). Among the affected crops, date palms, figs, stone fruits, strawberries and maize are the most significant (Peng and Williams 1990; Myers 2001; Bartelt and Hossain 2010). Both adults and larvae cause damage by direct feeding on the fruit pulp (Hinton 1945; Blumberg 2008). Beside the direct damage associated with feeding, the beetles also vector, and additionally allow access, to microorganisms that speed up the natural fermentation (Lussenhop and Wicklow 1990; Dowd 1991, 1995) and thus cause fruit degradation (Qasem and Christensen 1960; Kable 1969). Some of these microorganisms result in the production of mycotoxins, which may carry a significant health threat to consumers and may also lead to the rejection of crops.

Mycotoxins, produced by fungi especially from the genera *Aspergillus*, *Penicillium* and *Alternaria*, can cause, as toxic fungal metabolites, chronic and acute health problems (Ozer et al. 2012). Amongst them, aflatoxins are probably the best known secondary metabolites produced by aflatoxigenic *Aspergillus* spp. in/on foods and feeds (Dowd et al. 1998). Dried figs, dates and citrus fruits are among the fruits readily contaminated by aflatoxins, which affect their international trade. Aflatoxigenic *Aspergillus* spp. were detected in 40 % of 25 date varieties at first stage of maturation (Kimri, stage in which the fruit is quite hard, the color is apple green and it is not suitable for eating) (Shenasi et al. 2002). Mycotoxin contamination of dates is not generally associated with dried fruits (Ozer et al. 2012).

Some nitidulids transmit plant diseases such as oak wilt (Dorsey and Leach 1956; Cease and Juzwik 2001; Juzwik et al. 2004), and brown rot of stone fruits (Anonymous 2003).

Though sap beetles are generally considered to be secondary infestation insects (Omar 1971; Abdul-Wahab and Eissa 1986; Abdel-Rahman 2007), they can be primary pests for crops such as date palms (Kehat et al. 1983; Blumberg 2008), figs (Aksit et al. 2003; Turanli 2003), stone fruits (James et al. 1993) and strawberries (Cline 2005), since they infest mature fruits. *Carpophilus* spp. alone can cause economic losses of up to 30 % in stone fruits at harvest by chewing and penetrating the fruit near the pedicel (Hossain et al. 2000). In 1945, nitidulids and associated fungi caused date fruit loss between 50 % and 75 %, depending on the variety (Barnes and Lindgren 1946).

Sap beetles do not infest green, unripe date fruits on the bunches until ripening starts (Kehat et al. 1976). However, as fruit ripening progresses, they cause significant damage to fruits either on the bunches or on those dropped to the ground (Nixon 1951; Kehat et al. 1976), thus promoting the fruit rotting process (Kehat et al. 1976). Sap beetles usually enter the fruit at the calyx end and feed on the pulp, causing damage (Blumberg 2008). In Israel, nitidulid infestations start with the ripening process, in late August. Sap beetles readily infest the dropped and rotting fruits throughout the summer (Kehat et al. 1976). In Oman, windfall dates are readily attacked by *C. hemipterus* (El-Haidari 1993). Sap beetles are, therefore, considered as pests of ripe and ripening fruit only, and not of unripe fruit (Kehat et al. 1976). As ripening progresses, nitidulid presence gradually increases on fruits in bunches that are covered by a plastic mesh. Finally, infestation levels between fruits that fall off from late September till harvest time in early October and those that remain on the bunch show a similar pattern (Kehat et al. 1976). According to Hussein et al. (1971), 89 % of untreated dates were infested by *C. hemipterus* in Aswan, Egypt

(Hussein et al. 1971 cited in Carpenter and Elmer 1978). Damage to ripening fruits by nitidulids in Israel is apparently caused mainly by *E. luteola* and *U. humeralis* (Kehat et al. 1983).

Climatic conditions affect the extent of damage. Damp weather (Nixon 1951) or high humidity conditions (Kehat et al. 1976) result in greater damage, therefore early picking is carried out in some areas to avoid this problem. In laboratory choice tests with different humidity levels, it was shown that *C. hemipterus* and *C. dimidiatus* prefer moist zones (Amos and Waterhouse 1967). In Qatar, *Carpophilus* species infest mature dates with high moisture contents, both in the field and in storage (Al-Azawi 1986).

The unripe green dates that drop in early-mid summer (caused mainly by infestation of the date stone beetle, *Coccotrypes dactyliperda* F.) serve as hosts for nitidulid beetles, thus promoting both an increase in population and the subsequent crop damage (Blumberg et al. 1993). The larvae developing inside the fruits contaminate them with their excretions. Pest penetration into the fruit inevitably leads to microbial spoilage, ending with rotting or fermenting that downgrades the product. Damage caused by sap beetles is thus represented by both reduced yield and lower fruit quality. Control of these pests in date palm plantations is hence of great economic importance (Kehat et al. 1976; Blumberg et al. 1993).

Economic thresholds have not been developed for sap beetles in date palms, since control measures aimed at other pests have also controlled sap beetles adequately. There are few guidelines dealing with the timing of the treatment in other crops for sap beetles. In sweet corn for example, the presence of adult sap beetles in the silk is proposed as the action threshold (Bessin et al. 1994).

8.4 Biology and Seasonal Incidence

In the Nitidulidae family, life cycles vary between species and depend on environmental conditions such as temperature and moisture. In rainy seasons, sap beetles are more likely to build up high populations (Carpenter and Elmer 1978).

Nitidulids encountered in date palms areas can tolerate extreme summer temperatures and complete their life cycle in 2 weeks at 32 °C (Blumberg 2008), reaching high populations very quickly (Lindgren and Vincent 1949). The adults are strong fliers and search for ripe or fermenting fruits, migrating into ripening crops from harvested plots, from nearby untreated areas and other crops that act as infestation sources (Hossain et al. 2009). Females lay their eggs, up to 2000 or more (Avidov and Harpaz 1969), on ripening or rotting fruits lying either individually or in small clusters on shaded ground under citrus and date palm trees (George 1978; Blumberg 2008) (Fig. 8.8).

The eggs of *Carpophilus lugubris* (Murray) take about 2–5 days to hatch at 24 °C (Myers 2001). The eggs of *U. humeralis*, *C. mutilatus* and *C. hemipterus* are able to complete development at constant temperatures ranging from 20 to 37.5 °C, while eggs of *C. hemipterus* can also complete development at higher temperatures

Fig. 8.8 Nitidulid eggs (*left*), larvae (*middle*) and pupae (*right*) (source: Karakoyun 2011)

of 40–42.5 °C (James and Vogele 2000). Increases in temperature reduce the egg incubation period. Egg hatching times of *C. hemipterus*, *C. mutilatus*, *E. luteola* and *U. humeralis*, at around 3.0–3.9 days at 21.1 °C, are shortened to 1.1–1.6 days at 32.2 °C (Lindgren and Vincent 1953). Among the three nitidulids, *Carpophilus hemipterus* has the fastest egg development followed by *U. humeralis* and *C. mutilatus* (James and Vogele 2000). Lindgren and Vincent (1953), and George (1978) observed faster egg development in *E. luteola* than in *C. hemipterus*.

After hatching, larvae feed on fruit pulp and develop rapidly at higher temperatures (Lindgren and Vincent 1953; Blumberg 2008). Typically, sap beetles have three or four larval instars (Kelts 2005). The size of larvae when hatched is around 1.6 mm long. The fully grown larvae measure about 6.4 mm in length. They are white or yellowish. The head and the end of abdomen are amber brown in color, with two prominent spine-like projections, with two smaller ones in front of them (Fig. 8.8). They are sparsely hairy (Stickney et al. 1950).

Larval development of *C. hemipterus*, *C. mutilatus*, *E. luteola* and *U. humeralis* is completed in less than a week at temperatures higher than 26 °C, and in 3.2–5.7 days at 32.2 °C (Lindgren and Vincent 1953). George (1978), however, reported the larval development period ranging between 5 and 16 days at 27 °C for *E. luteola*, *C. nepos*, *U. humeralis*, *C. mutilatus* and *C. hemipterus*. Pupal development of the same species is also completed in comparatively short time periods, in the range of 2.9–5.3 days at 32.2 °C (Lindgren and Vincent 1953). Larval and pupal development of *E. luteola* at 32.2 °C is given as 12.6 and 2.4 days, respectively (Stickney et al. 1950).

Life history data of some important sap beetles of dates are given in Table 8.2. Larval and pupal development in *C. hemipterus* is completed at temperatures ranging from 20 to 40 °C. For other species, the pre-adult development are completed at temperatures of up to 37.5 °C for *C. mutilatus* and up to 32.5 °C for *U. humeralis* (James and Vogele 2000). The duration of larval and pupal development of *U. humeralis* and *C. mutilatus* are similar at most of the temperatures between 20 and 40 °C. Mature larvae leave the fruits and drop to the soil to pupate in pupal cells formed a few centimeters below the soil surface, near their food sources. The pupae, which are around 0.3–0.4 cm long, are white or pale yellow, and somewhat spiny

Table 8.2 Life history parameters of some nitidulids infesting date palms

Species	Temp. (°C)	RH (%)	Diet	Egg (h)	Larvae (d)	Pupae (d)	Development time from egg to adult	Female longevity (d)	Egg production	References
Carpophilus hemipterus	27	60	Fig	1.9	~7–17	~11–13	19–27	92	79.4	George (1978)
	26.7	Moist sand	Date	1.6	4.9	6.1	15.8			Lindgren and Vincent (1953)
	27.5	>60	Synthetic media	~2.5	~17.9	~6.6	~25.1			James and Vogele (2000)
	32.2		Moist date	1.0	12.4	5.8	19.2			Stickney et al. (1950)
	32.2			1	12	5.8	19.2	103	>1000	Simmons and Nelson (1975)
Carpophilus mutilatus	26.7	Moist sand	Date	2.2	7.1	7.3	20.7			Lindgren and Vincent, (1953)
	27	60	Fig	2.3	~7–13	~12	22–26	91	98.2	George (1978)
	27.5	>60	Synthetic media	~1.9	~11.8	~6.4	~16.8			James and Vogele (2000)
Carpophilus dimidiatus	26.7		Moist date	2.0	17.8	7.5	27.3			Stickney et al. (1950)
	32.2						15		175–225	Simmons and Nelson (1975)
Urophorus humeralis	26.7	Moist sand	Date	2.0	6.3	6.3	20.6			Lindgren and Vincent (1953)

(continued)

Table 8.2 (continued)

Species	Temp. (°C)	RH (%)	Diet	Egg (h)	Larvae (d)	Pupae (d)	Development time from egg to adult	Female longevity (d)	Egg production	References
	26.7		Moist date	2.2	13.8	8.0	24.0			Stickney et al. (1950)
	27	60	Fig	2.09	~7–12	~10–12	20–24	99	107.9	George (1978)
	27.5	>60	Synthetic media	~2.1	~16.4	~6.3	~25.4			James and Vogele (2000)
	32.2						16.5	89	880	Simmons and Nelson (1975)
Carpophilus nepos	27	60	Fig	2.1	~7–10	~9	16–20	98	127.0	George (1978)
Epuraea luteola	27	40	Artificial						1000.6	Dowd and Webder (1991)
	21.1						25.4			Simmons and Nelson (1975)
	26.7	Moist sand	Date	1.4	5.1	2.6	11.9			Lindgren and Vincent (1953)
	26.7		Moist date	1.3	12.2	3.3	16.8			Stickney et al. (1950)
	27	60	Fig	1.7	~5–13	~6–12	13–20	49	151.6	George (1978)
	32.2						9.6			Simmons and Nelson (1975)

(Stickney et al. 1950; Blumberg 2008). They can also overwinter as pupae in the soil (Blumberg 2008) or as adults. Reproduction can occur during the winter months (Bartelt et al. 1994b).

The strawberry sap beetle *Stelidota geminata* (Say) can also feed, complete development, overwinter and breed in high numbers in a wide variety of other plant hosts such as apple, blueberry, sweet corn, cherry, raspberry, strawberry and ecological sites such as woods (Loughner et al. 2007). Thus, to provide an environmentally and ecologically sound pest management program against sap beetles, habitat type surrounding date palm groves must also be taken into account.

The average development time from egg to adult in *U. humeralis* at 32.2 °C is about 20.4 days (Stickney et al. 1950). On average, the females live 89 days, with a mean fecundity of 882 eggs produced during an oviposition period lasting around 76 days (Schmidt 1935; cited in Stickney et al. 1950).

The mated females of *C. hemipterus* have a mean longevity of 103 days and lay an average of 1071 eggs in 79 days. Egg survival rate is about 89 %. The average development from egg to adult in *C. hemipterus* at 32.2 °C is about 19.2 days (Simmons et al. 1931, cited in Stickney et al. 1950). The average development from egg to adult in *E. luteola* is about 16.6 days, at 32.2 °C (Stickney et al. 1950).

Carpophilus dimidiatus grown on cracked rough rice on a layer of damp sand lived an average of 63 days in summer with an egg production of 175–225 eggs per female (Balzer 1942, cited in Stickney et al. 1950). The average development from egg to adult in *C. dimidiatus* at 32.2 °C is about 21.7 days (Stickney et al. 1950). *Carpophilus dimidiatus* fed on rolled oats and yeast completes its development successfully between 20 and 32.5 °C. The longest development time was recorded at 20 °C (68.9 days), with no development at 17.5 °C. At 30 °C, development is fastest at relative humidities higher than 70 %, in 23.6–24.7 days. Optimum temperatures for egg production is 22.5 and 25 °C, with fecundity higher at 70 % than at 50 % RH. Fecundity is reduced at higher (30 °C) and lower (20 °C) temperatures, or lower RH (50 %) (Porter 1986). The population potential of nitidulids is very high and a single female has the capacity to produce 79–151 offsprings successfully (George 1978).

The Development Zero[1] (DZ) values for *C. mutilatus* from egg to adult is 15.3 °C and is similar to that of *U. humeralis*, which is 15.4 °C (James and Vogele 2000). The value for *C. hemipterus*, however, is approximately 1 °C lower at 14.6 °C. The DZ values for individual stages of the same species range from 14.0 to 16.0 °C, and slightly decrease as the development progress. Egg to adult development of the three species requires a degree-day (DD) sum around 260.4–320. In this sum, *C. hemipterus* and *U. humeralis* require the least and the most DD for full development, respectively. Among the pre-adult stages of three species, the larvae take most time for development (James and Vogele 2000).

Egg to adult survivorship in *C. hemipterus* (29.3 %) at constant temperatures ranging from 20 to 42.5 °C is higher than that of *C. mutilatus* (14.2 %) and *U. humeralis* (13.8 %). Greatest survival in *C. hemipterus* occurs at 20 °C, while optimal survival of *U. humeralis* and *C. mutilatus* occurs at 25 and 27.5 °C, respectively.

[1] Temperature at which development is arrested.

Temperature mainly affects larval, rather than egg or pupal survival rates of *C. hemipterus*, *C. mutilatus* and *U. humeralis*, with greatest survival occurring at 25–30 °C (James and Vogele 2000).

8.5 Seasonal Incidence

There are significant differences in seasonal incidence of nitidulid beetles in date palm groves in different countries or even in different regions of the same country. In Egypt, the driedfruit beetle, *C. hemipterus* is considered to be a secondary pest (Abdel-Rahman 2007) infesting maturity stage Khalal (= Balah stage, in which the fruit is physiologically mature, hard ripe and the color changes completely from green to greenish yellow, yellow, pink, red or scarlet depending on the cultivar) of Maghal and Saidi cultivars (Hussien 2007). Larvae, pupae and adults of *C. hemipterus* are found in Bahria, Maghal and Zaghlool cultivars between July to October (Hussien 2007). Infestation rates of *C. hemipterus* in fallen and bunch dates increase gradually. However, the number of insects per fruit is approximately double in fallen fruits compared with fruit on the tree (Abdel-Rahman 2007).

Phenologies vary amongst different species in arid zones in Israel: *Carpophilus* species are found during autumn, winter, spring and early summer. Their populations decline during summer. *Urophorus humeralis* is the most common species during summer and sometimes autumn. *Epuraealuteola* is abundant, mainly during late summer and autumn. Therefore, *Carpophilus* species dominate in the spring followed by *U. humeralis* in the summer and *E. luteola* in late summer and autumn. Since date fruits ripen during the autumn, between September and November, damage to dates by sap beetles is apparently caused mainly by *E. luteola* and *U. humeralis* and to a much lesser extent by *Carpophilus* spp. which, consequently are not the major pests of date fruit in Israel (Kehat et al. 1983; Blumberg et al. 2005). Nitidulid beetle infestations of fruits on bunches enclosed within the commercial nettings start with the beginning of ripening by the end of August and heavy infestations occur in dropped fruits, which remain in the nets, throughout the summer. As ripening progresses, beetle infestation level in bunching fruits increases and eventually at picking time, no significant differences in infestation levels occur between fruits which fall off from late September to picking, and those which remained on the bunches till picking (Kehat et al. 1976). Population density of nitidulids in other fruits including figs, pomegranates, nectarines, citrus, apple and vines showed that *C. mutilatus* dominates while *E. luteola* had the lowest populations (Blumberg et al. 2005).

In Libya, sap beetles are present all year round in date palm groves found in coastal regions: *C. hemipterus* is the most common species followed by *C. dimidiatus*, *U. humeralis* and *E. luteola*. Three more nitidulids, namely *C. obsoletus*, *C. nepos* and *C. mutilatus* are also found in Libya, where *C. hemipterus* has 2–3 generations, while *U. humeralis* shows 2 generations and *E. luteolus* has 3–4 generations, in different localities (Najla et al. 2005).

In Oman, *Carpophilus* spp. (mainly *C. hemipterus*) are considered as pest of stored dates (Al-Zadjali et al. 2006). *Carpophilus hemipterus* infests windfall dates (El-Haidari 1993), while *U. humeralis* is reported to feed on fruits of Nighal date palm cultivar at the beginning of the Rutab stage. This is also known as the soft ripe stage, in which the tip at the apex starts ripening, the color changes to brown or black and the fruit become soft and can be consumed (Saaidi 1992).

In Qatar, adults of *Carpophilus* spp. attack mature dates with high moisture content, left on trees and in storehouses. *Carpophilus hemipterus* is distributed in Iraq, Palestine, Egypt, Somalia and Libya (Al-Azawi 1986).

In California, *C. mutilatus* is generally the dominant species over other nitidulids in date palm groves, representing more than 96 % of the adult catches. This is followed by the other species, *C. hemipterus*, *C. nepos*, *C. obsoletus*, *U. humeralis* and *E. luteola*, each of which has a trap catch below 2 %. A similar dominance of *C. mutilatus* is true for fallen dates, with dominating adult abundance (65 % of the adult nitidulids) and adult infestation rate (57 % of the sampled dates) in date fruits (Bartelt et al. 1994b). *Carpophilus nepos*, *C. obsoletus*, *U. humeralis* and *E. luteola* fly more readily in summer than *C. mutilatus* or *C. hemipterus* (Bartelt et al. 1994b). Pheromone traps alone do not give accurate population estimates, because they reflect flight activity only and many beetles remain in date palm groves in large numbers, all year round, without flying (Bartelt et al. 1994b).

8.6 Management

Control of sap beetles in most of the date palm growing countries is ensured mainly by insecticide applications, which generally aimed at other concurrent key pests. Sap beetles are difficult to control in ripened and dried dates because of the risk of insecticide residues. Besides toxic chemicals, agricultural practices and agro-technical measures, such as mass trapping, covering bunches with plastic nets, phytosanitary measures, early harvesting of certain cultivars, biological control, organic farming and IPM, are currently used to reduce pest populations and their damage (Blumberg 2008).

Correct agronomic practice is of critical importance. Implicit in this is to ensure that the intrinsic capacity of plant cultivars to resist pest attack is exploited fully. Despite the obvious advantages of using larger range of cultivars (Latifian et al. 2012), the great diversity of date palms cultivars (and perhaps wild varieties) does not seem to have been exploited extensively for pest management. Correct management, in terms of irrigation, types and levels of fertilizer use or organic manures and pruning does not appear to have been extensively studied for the effect on sap beetles control in date groves.

8.6.1 Physical Control

Technologies, such as the use of pre-harvest perforated polyethylene or cloth bunch covers have shown to be effective against fruit infesting pests including sap beetles, and thus are adapted as common practices in date palm growing areas (Pegna et al.

2012). Mesh size is important and dense plastic netting having a mesh opening of 0.8 × 0.5 mm is very effective in reducing the infestation rates below 2.2 %, and fruit drop as well (Kehat et al. 1976).

8.6.2 Trapping and Monitoring

Fermenting food baits of various types are effective in mass trapping of adult nitidulids. Among the food baits, fermenting whole-wheat bread dough (Bartelt et al. 1994a; Bartelt and Weisleder 1996), fermenting apple juice, liquid synthetic food attractant (Hossain et al. 2012), fermenting apple juice absorbed in polyacrylamide granules (James et al. 1998), fermenting fig (Warner 1961) and synthetic fermenting fruit (Smilanick et al. 1978) have been advocated. In Libya, locally made food bait traps alone caught beetles in thousands when placed in date palm groves (Najla et al. 2005). An inexpensive and easy-to-produce synthetic co-attractant, which is highly effective and reliable in field conditions, has been developed (Bartelt and Hossain 2006). Instead of using large amounts of ripe fruits as co-attractant, which is laborious and messy, synthetic co-attractants can be used in attract-and-kill stations (Hossain et al. 2012).

Male-produced aggregation pheromones to which both males and females respond have been identified for a number of *Carpophilus* species. Aggregation pheromones composed of mixtures of unsaturated hydrocarbons with alkyl branching and three or four conjugated double bonds attract large numbers of *Carpophilus* species in the field. Since *Carpophilus* species often use the same pheromone components, cross-attraction among species is fairly common (Bartelt 1997). Thus, cross attraction of *C. mutilatus*, *C. obsoletus* and *U. humeralis* to the *C. hemipterus* pheromone has long been known in the date palm groves (Bartelt et al. 1992). Combining the use of the aggregation pheromones of *C. davidsoni*, *C. hemipterus* and *C. mutilatus* along with the synthetic co-attractant for mass trapping provided better control than pesticide sprays in orchards (Hossain et al. 2007).

Though active under field conditions, aggregation pheromones are more effective when combined with food volatiles (Bartelt et al. 1990, 1992, 1993). Previous work by Bartelt et al. (1994b) and Bartelt and Weisleder (1996) report a "dramatic" synergism in which the use of pheromone and host odors together were over 100 times more attractive in the field than pheromone or host odor alone. Such scent combinations was shown to attract ca. 114,000 *C. mutilatus* per trap per 3 day period in a heavily infested California date garden (Bartelt et al. 1994a). Pheromones for *C. hemipterus*, *C. mutilatus* and *C. nepos* can be combined into a single lure without loss of activity, which may facilitate their practical application in date palm groves (Bartelt et al. 1995).

Different factors such as pheromone dose, trap height and age of pheromone formulation play important roles in trap efficacy and their effectiveness can vary according to species (Bartelt et al. 1994a). Species generally tend to respond best to their synthetic pheromones and at their highest doses up to 15,000 μg per septum (Bartelt et al. 1994a). Longevity of pheromone rubber septa is negatively affected by time over a week, mainly due to declining release rates that fall anywhere from 18 to 90 % of its original value. Trap catches for major nitidulids, though increased with elevation, are not affected greatly by variation in trap heights of 0.3–3.0 m (Bartelt et al. 1994a). In Israel, however, suspension height of the traps used on date palm groves significantly affects the capture rates, with *E. luteolus* being the most abundant species at a ground level of 0.2 m and *C. hemipterus* most abundant at bunch level of 16 m (Blumberg et al. 2005). Positioning the traps at the highest locations is recommended for better captures of certain species (Bartelt et al. 1994a).

Due to the fact that well-fed beetles do not fly readily (Bartelt et al. 1990) and the aroma from ripening fruit is more attractive than the synthetic pheromones in the traps (Hossain et al. 2013), nitidulid pheromone traps are not always effective in situations where abundant and high-quality food is present (Bartelt et al. 1990, 1994a). Thus, use of pheromone-baited traps might be compromised by presence of food sources that could prove distractive (Bartelt et al. 1994a).

James et al. (2001) proposed that attract-and-kill stations be positioned 15–20 m outside cherry and stone fruit orchards since most of the *Carpophilus* species, instead of entering the traps for unknown reasons, tend to land on the nearby trees and cause damage over >40 % to fruits around the traps. The pheromone traps can, consequently, attract insects into the area to be protected. Altering the style of a trap can drastically change the efficiency of catch (Blumberg 2008).

8.6.3 Biological Control

8.6.3.1 Parasites

Microsporidia are little studied, but Yaman (2007) reported *Nosema meligethi* Issi and Raditscheva in the nitidulid *Meligethes aeneus* (F.) in Turkey. It was uncommon, but the results suggested that its occurrence was greater in areas where insecticides were not used.

8.6.3.2 Parasitoids

There are a few hymenopteran wasps attacking nitidulid species (Williams et al. 1984). Although these parasitoids were easily reared in the laboratory (Williams et al. 1992), their applications in date palms groves failed to result in field establishment (Blumberg et al. 1984). *Cerchysiella utilis* (Noyes), a solitary internal larval parasitoid of *Carpophilus* spp. was successfully reared in laboratory on *C. hemipterus* and *C. mutilatus*, but not *U. humeralis* or *E. luteolus*. It is considered rather an inefficient parasitoid due to the low infestation rates in date palm groves in Israel (Blumberg 2008).

Similarly, *Microctonus nitidulidis* (Loan), which parasitizes the adult stage, and *Brachyserphus* (= *Cryptoserphus*) *abruptus* (Say), which attacks the larvae, were imported from USA and released in 1980s. However, also these species failed to become established in date palm groves in Israel (Blumberg et al. 2001).

For other crops, biological control studies against sap beetles have been promising. In strawberry fields, *B. abruptus* is found effective in reducing the number of sap beetle larvae with a level of 18 % parasitism (Williams et al. 1995). It also parasitizes sap beetle larvae in sweet corn (Alston et al. 2014). Perennial crops tend to have good landscape diversity which often supports higher levels of attack by parasitoids (Scheid et al. 2011).

8.6.3.3 Predators

There is some information concerning the biology and efficacy of the predators of sap beetles, but not from date palm. Adults and immature stages of the staphylinid predator *Atheta coriaria* (Kraatz) have been reported to consume eggs and early instars of several nitidulids, including *C. hemipterus*, *C. lugubris*, *C. freeman*, *U. humeralis* and *E. luteola*. The predator produced a generation in 13 days at 26.7 °C while consuming 20 *C. hemipterus* eggs per day (Miller and Williams 1983). In north of Israel, large amounts of *Atheta* sp. were found along with *C. mutilatus* and *U. humeralis* in an *Annona* plantation (D. Blumberg, pers. comm., 2014).

Peregrinator biannulipes (Montrouzier and Signoret), a reduviid predator (Hemiptera), that feeds on a range of pests including *C. mutilatus*, was found to be a promising candidate to suppress copra pests in Philippines, including *Carpophilus* spp. (Zipagan and Pacumbaba 1994)

Spiders are often neglected in terms of their biological control capacity, despite being recognized as important in limiting arthropod pest densities in agro-ecosystems (Riechert and Lockley 1984). Perennial crops are likely to harbor strong assemblages as shown by Mushtaq et al. (2000) who found 8 families, 25 genera and 73 species of foliage spider from date palms in Pakistan. Although generalist predators, and hence liable to consume beneficial as well as pest organisms, the overall impact of spiders in pest control is positive.

The storage mite, *Cosmoglyphus hughesi* Samšinák (Acaridae: Acari) (= *Caloglyphus hughesi*) has been reported in laboratory experiments as feeding on the nitidulid, *Aethina tumida* Murray (Strauss et al. 2010). Similar predation by mites might be expected in date palm groves, but there are no observations available.

An unappreciated control is facilitated by frugivores. Fermenting fruits are attractive to many animals and eating of these infested fruits will remove many larvae from the date groves. The grazing of ruminants, and a diversity of other vertebrates in date groves, removes an unquantified proportion of nitudilids. This is equivalent to components of garden hygiene, removing fallen fruit from the area.

8.6.3.4 Entomopathogenic Fungi (EPF)

All plants studied to date contain endophytes (Sun and Guo 2012) and long-lived plants usually contain a significant number and diversity of endophytes, as occurs in forest trees. These can include species used as mycoinsecticides such as *Beauveria bassiana* (Balsamo) Vuillemin (Thomas 2014). Palms are no different from other plant groups, not only do they contain rich assemblages of endophytes (Taylor et al. 1999; Pinruan et al. 2010), but EPF species such as *B. bassiana* and *Lecanicillium* spp. can be incorporated into palms (Gómez-Vidal et al. 2006). Endophytes can moderate insect attacks by a variety of means (Azevado et al. 2000) leading to the possibility of their practical use in IPM strategies in date palm groves.

Although many EPF isolates are likely to kill sap beetles and *B. bassiana* conidia are reported to cause adult mortality at a 90 % rate in *C. lugubris* in 3 days of exposure (Dowd and Vega 2003), they are difficult targets for conventional mycoinsecticides. Developing sap beetle larvae are protected by the fruits they invade, and adults on the aerial parts of date palms are difficult to access with sprays.

Another potential use for microbial organisms is through auto-dissemination, in which adults contaminate themselves at bait stations, then transferring the microbes to oviposition sites. The auto-dissemination approach makes the use of microbial agents economical, which is otherwise costly when applied in conventional ways through field spraying (Dowd and Vega 2003). Thus, laboratory assays with auto-inoculative release have indicated that auto-dissemination of *B. bassiana* has potential for sap beetle control (Vega et al. 1995).

A commercial strain of *Bacillus subtilis* (Ehrenberg) Cohn inhibits the growth and aflatoxin production by *Aspergillusflavus* in developing maize kernels (Cuero et al. 1991). Field studies yielded positive results on the auto-inoculative dispersal of *B. subtilis* by natural populations of dusky sap beetle, *C. lugubris* in maize (Dowd et al. 1998) and this method can also be adopted to reduce aflatoxigenic *Aspergillus* and aflatoxins in date fruits.

8.6.3.5 Entomopathogenic Nematodes (EPN)

Several species of Steinernematidae and Heterorhabditidae are in use commercially against several agricultural pests in many countries. Steinernematid and Heterorhabditid nematodes were found effective against sap beetles both in laboratory and in the field (Vega et al. 1994; Glazer et al. 1999). However, field tests with four strains of *Heterorhabditis* applied at the rate of 100 infective juveniles (IJs)/cm^2 in a date palm orchard in Israel failed to suppress adult emergence of *U. humeralis*, while a similar test applied at 75 and 150 Ijs/cm^2 in a fig orchard yielded a reduction in adult emergence at both concentrations by 50–70 % (Glazer et al. 2007). A mermithid nematode parasite, *Hexamermis* spp., which can induce infection rates of up to 89 %, was found promising in controlling adult dusky sap beetles,

C. lugubris in strawberry fields. However, these nematodes are difficult to mass produce (Dowd et al. 1995). Conservation methods are probably more appropriate for mermithids than augmentation.

8.6.4 Chemical Control

Sap beetles are usually kept under control by insecticide applications used to control other key-pests. In most date palm growing countries, no insecticidal treatments are made specifically against sap beetles. Thus, two insecticide applications to control other fruit infesting insects, the first in mid-summer and the second 3–4 weeks before harvest, also effectively eliminate sap beetles (Blumberg 2008).

Control of nitidulids with residual contact insecticides such as gamma-isomer of benzene hexachloride, chlordan and parathion in date palms goes back to 1940s (Lindgren and Vincent 1949). In Israel, malathion was the dominant insecticide used during the 1960s onward, being replaced by azinphos-methyl (25 % WP) during the early 1980s, with a single application 2 months before harvest at the dose of 0.2 % (Blumberg 2008; Kehat et al. 1976, 1986). Based on various published studies, Blumberg (2008) listed number of insecticides found to be effective against sap beetles in Israel, including cypermethrin, cyhalothrin, lambda-cyhalothrin, bifenthrin (synthetic pyrethroids), diflubenzuron, hexaflumuron, teflubenzuron (IGRs), imidacloprid (neonicotinoid) and thiacloprid (chloronicotinoid). The IGRs, which cause egg sterilization and larval mortality, were recommended for commercial use (Blumberg et al.1985; Ascher et al. 1986; Blumberg 2008). Kehat et al. (1966) reported a significant reduction in fruit infestation by *Carpophilus*larvae by malathion applications.

8.6.5 Other Tactics

Orchard sanitation by removal and destruction of fallen fruits at weekly intervals is very important in reducing the populations of sap beetles attacking ripening fruits in date palm groves (Warner et al. 1990; Blumberg 2008). Moreover, a decrease in the availability of food sources increases the effectiveness of the population monitoring traps. Thus a double benefit is gained by orchard sanitation (Bartelt et al. 1994a).

Early harvesting is another method used to prevent infestation and damage by sap beetles in various date cultivars, such as Deglet Noor, Medjool, Barhee and Hayany (Blumberg 2008). Similarly, in India, date palm fruits are harvested prematurely before the onset of monsoon to prevent spoilage in June at the Khalal stage and thus are free from sap beetle infestations (Muhammad M. Khan, pers. comm., 2014).

8.7 Future Research

Overall, more detailed ecological understanding of date groves and the date palms is required. Several natural enemies of date palm pests, including general predators and beneficial microbes, such as entomopathogenic agents and endophytes, are largely unrecognized. Equally, the vegetative diversity at the landscape scale will be very influential in the overall ecological interactions between beneficials and potential pest species. Landscape scale ecological research is needed to understand the intricate pest control required in date palm groves. Some specific research areas are listed below:

1. Further and continuous research is needed to determine the exact species composition, life history, host range and seasonal fluctuations of nitidulids in date palm groves in the Arab peninsula and North Africa, in particular.
2. More research on the trapping of nitidulids is needed. In addition to other aspects of improving trap efficiency, traps should also be improved to achieve attract-and-kill strategies of nitidulids and auto-inoculative dispersal of microbial agents.
3. Commercial sources for obtaining biological control agents against sap beetles are needed.
4. Day-degree models should be developed to assist timing of control actions.
5. Sterile insect technique may be a solution to physically isolated date palm groves.
6. Work on molecular biology and genetics is needed to develop new strategies such as gene silencing or biopesticides that are not only effective and targeted at nitidulids, but also environmentally benign and safe for food products.
7. The economic feasibility of biorational pest management methods should be determined so that date palm farmers can select the most cost-effective management methods

8.8 Summary

Sap beetles are polyphagous insects with a worldwide distribution, attacking more than 200 species of host plants. With over 4000 described species, they are the second most diverse family of cucujoid beetles, after Coccinellidae.

They are small to minute beetles in size ranging from less than 1 to ~15 mm. However, the most abundant and rich taxa such as *Carpophilus* and *Epuraea* range in the size of 1.5–6.0 mm in length. Sap beetles are easily recognized by their three-segmented club like antennae, five segmented tarsi of which segments 1–3 are usually dilated and segment 4 is small, and, usually, truncated elytra that expose the apex of the pygidium or sometimes even the preceding 1 or 2 tergites.

As their common name implies, sap beetles are attracted to the sap oozed from wounds occurring in trees. However, with their short development time (14–27 days), comparatively long lived adults (2–3 month) and different body forms, sap beetles adapt to different ecologies and life history strategies. These include fungivory, frugivory, saprophagy, anthophily, phytophagy, predation, necrophagy, and inquilinism with social Hymenoptera. They also are known to transmit a variety of microorganisms that cause plant diseases.

In agricultural production, sap beetles are often considered a minor pest; however, practically all kind of fruits containing a good level of carbohydrates including date palms, figs, oranges, grapefruits, mandarin oranges (tangerines), apples, strawberries, peaches, apricots, kiwi, grapes and several others are readily attacked by these beetles. The internal contamination of fruits by adults and larvae cause serious problems in international trading after the ban for methyl bromide fumigation by the international community. Moreover, they facilitate the spread of mycotoxin producing fungi to fruits, an important issue necessitating their control.

Carpophilus hemipterus (the driedfruit beetle, fig beetle or two-spotted beetle), *C. mutilatus* (confused sap beetle), *C. dimidiatus* (the corn sap beetle), *C. obsoletus*, *U. humeralis* (the pineapple beetle), *C. nepos*, *E. luteolus* (the yellowish nitidulid) are known to attack date palm trees in date producing countries worldwide. Seasonal occurrence varies in different countries or even at different regions in the same country.

Adults are strong fliers and capable to fly up to 4 km. Adult sap beetles attack fruits throughout the growing season. After mating, eggs are laid in single or in small clusters in or on the fruits. Following hatching of eggs, larvae burrow inside the fruits, feeding on the flesh for approximately 1½ weeks. Mature larvae fall onto the ground, burrow inside the soil, and pupate. In California fig groves, they produce 5–8 generations per year with a potential of up to 26 generations.

The control activities against sap beetles are mainly based on chemical pesticides against key pests in date palms groves. However, multiple applications, which are often applied near harvest, pose a high risk of contamination of fruits by insecticides. In stone fruits, mass trapping by means of aggregation pheromone combined with fermenting food baits is a very effective way of control. Hygiene and other cultural practices such as removing fallen fruits and early harvest are very important in reducing the beetle populations. Physical control measures involve covering date bunch with plastic mesh covers. Attempts at biological control against sap beetles have been very limited, and so far no success has been achieved to control the beetles by means of parasitoids, predators, nematodes or other microbial agents. Developments in the production techniques of microbial agents and entomopathogenic nematodes would make them promising agents for sap beetle control in terms of both economy and effectiveness.

Acknowledgments Thanks are due to Lesley Ragab and Nazife Eroglu Yalcin for providing some old, but useful, papers for this review. The authors would also like to thank to Richard A. B. Leschen, Birgit Rhode (Landcare Research, New Zealand); Simon Hinkley, Ken Walker (Museum of Victoria, Australia); Joe MacGown, (Mississippi Entomological Museum, USA); C. Harding (MAF Plant Health and Environment Laboratory, New Zealand); Dave Larson (Canada); Samuel D. J. Brown (Lincoln University, New Zealand), Michele B. Price (University of Minnesota, USA) and N. Siray Karakoyun (Ministry of Food, Agriculture and Livestock, Turkey) for providing the nitidulid pictures used in this review. The authors are also grateful to Daniel Blumberg (Israel) for his comments on an earlier version of the manuscript.

References

Abdel-Rahman, G. A. (2007). Insect pest control in organic date palm production in Egypt. In *Proceedings of the fourth symposium on date palm in Saudi Arabia, May 5–8, 2007* (pp. 1356–1360). Hofuf: King Faisal University.

Abdul-Wahab, A. E., & Eissa, I. S. (1986). Biological studies on some dry-dates pests in Egypt. In *Proceedings of the second symposium on date palm, March 3–6, 1986* (pp. 284–290). Kingdom of Saudi Arabia: King Faisal University.

Aksit, T., Özsemerci, F., & Cakmak, I. (2003). Studies on determination of harmful fauna in the fig orchards in Aydin Province (Turkey). *Turkish Journal of Entomology, 27*(3), 181–189.

Al-Azawi, A. F. (1986). A survey of insect pests of date palms in Qatar. *The Date Palm Journal, 4*, 247–266.

Ali, M. A., Metwally, M. M., & Hussain, A. E. (2002). Pest suppression of date palm insect populations and effects on yield as a component of sustainable development of El-Bahria Oases, Egypt. *Bulletin of the Entomological Society of Egypt, 79*, 89–102.

Alston, D., Bunn, B., & Nischwitz, C. (2014). Sweet corn production. In B. Bunn, C. Creasey, D. Goldsberry, & M. Murray (Eds.), *Utah vegetable production and pest management guide* (pp. 87–111). Logan: Utah State University IPM Program, Utah State University Extension. Available at: http://utahpests.usu.edu/IPM/files/uploads/Publications/UT-veg-guide-2014.pdf. Accessed on 1 Aug 2014.

Al-Turaihi, E. H. (2007). The impacts of good agricultural practices (GAPs) on pests and diseases of date palm. In *Proceedings of the fourth symposium on date palm in Saudi Arabia, May 5–8, 2007* (pp. 1692–1702). Hofuf: King Faisal University.

Al-Zadjali, T., Abd-Allah, F. F., & El-Haidari, H. S. (2006). Insect pests attacking date palms and dates in Sultanate of Oman. *The Egyptian Journal of Agricultural Research, 84*, 51–58.

Amos, T. G., & Waterhouse, F. L. (1967). Phasic behaviour shown by two *Carpophilus* species (Coleoptera, Nitidulidae) in various humidity gradients and its ecological significance. *Oikos, 18*, 345–350.

Anonymous. (2003). Common pests of summer fruit in Western Australia. Western Australian Department of Agriculture, Bulletin No. 4583, pp. 23.

Archibald, R. D., & Chalmers, I. (1983). Stored product Coleoptera in New Zealand. *New Zealand Entomologist, 7*, 371–397.

Ascher, K. R. S., Nemny, N. E., Blumberg, D., & Goldenberg, S. (1986). Egg-sterilizing effect of benzoylphenylureas via the adult stage of the nitidulid beetle *Carpophilus hemipterus*. *Phytoparasitica, 14*, 187–192.

Avidov, Z., & Harpaz, I. (1969). *Plant pests of Israel* (600 pp). Israel: Israel Universities Press.

Azevado, J. L., Maccheroni, W., Jr., Pereira, J. O., & de Araújo, W. L. (2000). Endophytic microorganisms: A review on insect control and recent advances on tropical plants. *Electronic Journal of Biotechnology, 3*, 40–65.

Balzer, A. I. (1942). Life-history of the corn sap beetle in rice. *Journal Economic Entomology, 35*, 606–607.

Barnes, D. F., & Lindgren, D. L. (1946). The beetle infestation in dates. *Date Growers' Institute's Annual Report, 23*, 34–35.

Bartelt, R. J. (1997). Calibration of commercial solid-phase microextraction device for measuring headspace concentrations of organic volatiles. *Analytical Chemistry, 68*, 364–372.

Bartelt, R. J., & Hossain, M. S. (2006). Development of synthetic food-related attractant for *Carpophilus davidsoni* and its effectiveness in the stone fruit orchards in southern Australia. *Journal of Chemical Ecology, 32*, 2145–2162.

Bartelt, R. J., & Hossain, M. S. (2010). Chemical ecology of *Carpophilus* sap beetles (Coleoptera: Nitidulidae) and development of an environmentally friendly method of crop protection. *Terrestrial Arthropodan Reviews, 3*, 29–61.

Bartelt, R. J., & Weisleder, D. (1996). Polyketide origin of pheromones of *Carpophilus davidsoni* and *C. mutilatus* (Coleoptera: Nitidulidae). *Bioorganic and Medicinal Chemistry, 4*, 429–438.

Bartelt, R. J., Dowd, P. F., Plattner, R. D., & Weisleder, D. (1990). Aggregation pheromone of driedfruit beetle, *Carpophilus hemipterus*: wind tunnel bioassay and identification of two novel tetraene hydrocarbons. *Journal of Chemical Ecology, 16*, 1015–1039.

Bartelt, R. J., Dowd, P. F., Vetter, R. S., Shorey, H. H., & Baker, T. C. (1992). Responses of *Carpophilus hemipterus* (Coleoptera: Nitidulidae) and other sap beetles to the pheromone of *C. hemipterus* and host-related coattractants in California field tests. *Environmental Entomology, 21*, 1143–1153.

Bartelt, R. J., Carlson, D. G., Vetter, R. S., & Baker, T. C. (1993). Male-produced aggregation pheromone of *Carpophilus mutilatus* (Coleoptera: Nitidulidae). *Journal of Chemical Ecology, 19*, 107–118.

Bartelt, R. J., Vetter, R. S., Carlson, D. G., & Baker, T. C. (1994a). Influence of pheromone dose, trap height, and septum age on effectiveness of pheromones for *Carpophilus mutilatus* and *C. hemipterus* (Coleoptera: Nitidulidae) in a California date garden. *Journal of Economic Entomology, 87*, 667–675.

Bartelt, R. J., Vetter, R. S., Carlson, D. G., & Baker, T. C. (1994b). Responses to aggregation pheromones for five *Carpophilus* species (Coleoptera: Nitidulidae) in a California date garden. *Environmental Entomology, 23*, 1534–1543.

Bartelt, R. J., Vetter, R. S., Carlson, D. G., Petroski, R. J., & Baker, T. C. (1995). Pheromone combination lures for *Carpophilus* (Coleoptera: Nitidulidae) species. *Journal of Economic Entomology, 88*, 864–869.

Batra, R. C., & Sohi, B. S. (1972). The yellowish nitidulid beetle on date palm. *FAO Plant Protection Bulletin, 20*, 91–92.

Bessin, R., Johnson, D., Strang, J, Dunwell, W., & Jones, T. (1994). Kentucky sweet corn insect, integrated pest management scout manual No. IPM-10 (pp. 26). USA. Available at: http://www.uky.edu/Ag/IPM/manuals/ipm10swt.pdf. Accessed on 1 June 2014.

Blumberg, D. (2008). Date palm arthropod pests and their management in Israel. *Phytoparasitica, 36*, 411–448.

Blumberg, D., Demeter, M., Kehat, M., & Williams, R. N. (1984). Biological studies of *Zeteticontus utilis* (Hymenoptera: Encyrtidae), a parasite of *Carpophilus* spp. (Coleoptera: Nitidulidae). *Annals of the Entomological Society of America, 77*, 130–133.

Blumberg, D., Doron, S., & Bitton, S. (1985). Effect of triflumuron on two species of nitidulid beetles, *Carpophilus hemipterus* and *Urophorus humeralis. Phytoparasitica, 13*, 9–19.

Blumberg, D., Kehat, M., Goldenberg, S., Bartelt, R. J., & Williams, R. N. (1993). Responses to synthetic aggregation pheromones, host-related volatiles, and their combinations by *Carpophilus* spp. (Coleoptera:Nitidulidae) in laboratory and field tests. *Environmental Entomology, 22*, 837–842.

Blumberg, D., Navon, A., Kehat, M., Eliahu, M., Levski, S., & Nakash, J. (2001). Date palm pests in Israel at the beginning of the third millennium. *Alon Hanotea, 55*, 42–48. (in Hebrew, with English Abstract).

Blumberg, D., Mendel, Z., Navon, A., Eliahu, M., Yablovich, Z., & Nakash, J. (2005). Study of the frequent occurrence of sap beetles (Coleoptera: Nitidulidae) in dates and other fruit orchards in Israel using traps baited with fruits and aggregation pheromones. *Alon Hanotea, 59*, 31–39. (in Hebrew, with English Abstract).

Brown, S. D. J. (2010). New records of adventive Nitidulidae in New Zealand. *The Weta, 39*, 12–16.

Carpenter, J. B., & Elmer, H. S. (1978). *Pests and diseases of the date palm* (United States Department of Agriculture, Agricultural Research Service Handbook No. 527, 42 pp). Washington DC: United States Department of Agriculture, Agricultural Research Service.

Cease, K. R., & Juzwik, J. (2001). Predominant nitidulid species (Coleoptera: Nitidulidae) associated with spring oak wilt mats in Minnesota. *Canadian Journal of Forest Research, 31*, 635–643.

Cline, A. R. (2005). *Revision of Pocadius Erichson (Coleoptera: Nitidulidae)* (p. 376). PhD Dissertation, Louisiana State University and Mechanical College, Baton Rouge.

Connell, W. A. (1991). Sap beetles (Nitidulidae, Coleoptera). In J. Richard (Ed.), *Insect and mite pests in food: An illustrated key* (pp. 151–174). U.S. Government Printing Office, Washington, DC.

Cuero, R. G., Duffus, E., Osuji, G., & Pettit, R. (1991). Aflatoxin control in preharvest maize: Vectors of chitosan and two microbial control agents. *Journal of Agricultural Science, 117*, 165–169.

Dorsey, C. K., & Leach, J. G. (1956). The bionomics of certain insects associated with oak wilt with particular reference to the Nitidulidae. *Journal of Economic Entomology, 49*, 219–230.

Dowd, P. F. (1991). Nitidulids as vectors of mycotoxin-producing fungi. In O. Shotwell & C. R. Hurburgh Jr. (Eds.), *Aflatoxin in corn: New perspectives* (Research Bulletin 599, 335–342 pp). Iowa State University, Ames, IA.

Dowd, P. F. (1995). Sap beetles and mycotoxins in maize. *Food Additives and Contaminants, 12*, 497–508.

Dowd, P. F., & Vega, F. E. (2003). Autodissemination of *Beauveria bassiana* by sap beetles (Coleoptera: Nitidulidae) to overwintering sites. *Biocontrol Science and Technology, 13*, 65–75.

Dowd, P. F., & Weber, C. M. (1991). A labor-saving method for rearing a corn sap beetle, *Carpophilus freemani* Dobson (Coleoptera: Nitidulidae), on pinto bean-based diet. *Journal of Agricultural Entomology, 8*(3), 149–153.

Dowd, P. F., et al. (1995). Occurrence of a Mermithid nematode parasite of *Carpophilus lugubris* (Coleoptera: Nitidulidae) in Central Illinois. *Environmental Entomology, 24*, 1245–1251.

Dowd, P. F., Vega, F., Nelsen, T. C., & Richard, J. L. (1998). Dusky sap beetle mediated dispersal of *Bacillus subtilis* to inhibit *Aspergillus flavus* and aflatoxin production in maize *Zea mays* L. *Biocontrol Science and Technology, 8*, 221–235.

El-Haidari, H. S. (1981). New records of mites and insects infesting date palms in the Near East and North Africa. *The Date Palm Journal, 1*(1), 133–134.

El-Haidari, H. S. (1993). *Report on insect pests of date palms in Sultanate of Oman*. Oman: Ministry of Agriculture and Fisheries in Oman and Islamic Development Bank.

El-Haidari, H. S., El-Bana, M. M., & Khudhair, S. A. (1981). New records of insects attacking date palm treated with growth regulators in Iraq. *The Date Palm Journal, 1*, 134–136.

Evans, A. V., & Hogue, J. N. (2006). *Field guide to beetles of California* (441 pp). Berkeley: University of California Press.

Ewing, C. P., & Cline, A. R. (2004). New records and taxonomic updates for adventive sap beetles (Coleoptera: Nitidulidae) in Hawai'i. *Bishop Museum Occasional Papers, 79*, 42–47.

Ewing, C. P., & Cline, A. R. (2005). Key to adventive sap beetles (Coleoptera: Nitidulidae) in Hawaii, with notes on records and habits. *The Coleopterists Bulletin, 59*, 167–183.

Gazit, S., Galon, I., & Podoler, H. (1982). The role of nitidulid beetles in natural pollination of annona in Israel. *Journal of the American Society for Horticultural Science, 107*, 849–852.

George, L. D. (1978). *Comparative nitidulid life histories of Haptoncus luteolus, Urophorus humeralis, Carpophilus freemani, C. mutilatus and C. hemipterus on figs* (pp. 58). M.A. Dissertation, California State University, Fresno, USA.

Glazer, I., Salame, L., Goldenberg, S., & Blumberg, D. (1999). Susceptibility of sap beetles (Coleoptera: Nitidulidae) to entomopathogenic nematodes. *Biocontrol Science and Technology, 9*, 259–266.

Glazer, I., Eliyau, M., Salame, L., Nakash, Y., & Blumberg, D. (2007). Evaluation of the efficacy of the entomopathogenic nematodes *Heterorhabditis* sp. against sap beetles (Coleoptera: Nitidulidae). *BioControl, 52*, 259–270.

Gómez-Vidal, S., Lopez-Llorca, L. V., Jansson, H.-B., & Salinas, J. (2006). Endophytic colonisation of date palm (*Phoenix dactylifera* L.) leaves by entomopathogenic fungi. *Micron, 37*, 624–632.

Hayashi, N. (1978). A contribution to the knowledge of the larvae of Nitidulidae occurring in Japan (Coleoptera: Cucujoidea). *Insecta Matsumurana (New Series), 14*, 1–97.

Hill, D. S. (2002). *Pests of stored foodstuffs and their control* (476 p). Boston: Kluwer Academic Publishers.

Hinton, H. E. (1945). *A monograph of the beetles associated with stored products* (443 pp). London: Jarrold and Sons.

Hossain, M. S., Mansfield, C., & Brown, C. (2000). Control of *Carpophilus* spp. by using an aggregation pheromone. *Australian Fresh Stone Fruit Quarterly, 2*, 13–14.

Hossain, M. S., Williams, D. G., Hossain, A. B. M. M., & Norng, S. (2007). Comparison of trap designs for use with aggregation pheromone and synthetic co-attractant in a user-friendly attract and kill system to control *Carpophilus* spp. (Coleoptera: Nitidulidae). *Australian Journal of Entomology, 46*, 244–250.

Hossain, M. S., Hossain, A. B. M. M., Williams, D. G., Ekanayake, S., Rettke, M., & Lo, P. (2009). *Investigation of an area wide approach to control Carpophilus beetle in stone fruit (SF05006/ SF05002)* (124 pp). Sydney: Horticulture Australia Ltd.

Hossain, M. S., Hossain, A. B. M. M., Norng, S., & Williams, D. G. (2012). Quantity of food attractant and size of the container orifice affect catch in *Carpophilus* spp. traps. *Australian Journal of Entomology, 51*, 60–66.

Hossain, M. S., Hossain, A. B. M. M., Williams, D. G., & Chandra, S. (2013). Management of *Carpophilus* spp. beetles (Nitidulidae) in stone fruit orchards by reducing the number of attract-and-kill traps in neighbouring areas. *International Journal of Pest Management, 59*, 135–140.

Hussien, A. E. (2007). Insect pests attacking date palms and dates in some Egyptian regions. In *The fourth symposium on date palm in Saudi Arabia, May 5–8, 2007* (pp. 1320–1329). Alahsa: King Faisal University.

Hussein, F., Goubar, K. A., & Abdelmessieh, N. K. (1971). Control of insects infesting dry-dates grown at Asswan. *Beitrage zur Tropischen und Subtropischen Landwirtschaft und Tropenveterinarmedizin, 9*, 241–246.

James, D. G., & Vogele, B. (2000). Development and survivorship of *Carpophilus hemipterus* (L.), *Carpophilus mutilatus* Erichson and *Carpophilus humeralis* (F.) (Coleoptera: Nitidulidae) over a range of constant temperatures. *Australian Journal of Entomology, 39*, 180–184.

James, D. G., Bartelt, R. J., Faulder, R. J., & Taylor, A. (1993). Attraction of Australian *Carpophilus* spp. (Coleoptera: Nitidulidae) to synthetic pheromones and fermenting bread dough. *Journal of the Australian Entomological Society, 32*, 339–345.

James, D. G., Faulder, R. J., Vogele, B., Bartelt, R. J., & Moore, C. J. (1997). Phenology of *Carpophilus* spp. (Coleoptera: Nitidulidae) in stone fruit orchards as determined by pheromone-trapping: implications for prediction of crop damage. *Australian Journal of Entomology, 36*, 165–173.

James, D. G., Moore, C. J., Faulder, R. J., & Vogele, B. (1998). An improved coattractant for pheromone-trapping of *Carpophilus* spp. (Coleoptera: Nitidulidae). *Australian Journal of Entomology, 37*, 357–361.

James, D. G., Vogele, B., Faulder, R. J., Bartelt, R. J., & Moore, C. J. (2001). Pheromone-mediated mass trapping and population diversion as strategies for suppressing *Carpophilus* spp. (Coleoptera: Nitidulidae) in Australian stone fruit orchards. *Agricultural and Forest Entomology, 3*, 41–47.

Juzwik, J., Skalbeck, T. C., & Neuman, M. F. (2004). Sap beetle species (Coleoptera: Nitidulidae) visiting fresh wounds on healthy oaks during spring in Minnesota. *Forest Science, 50*, 757–764.

Kable, P. F. (1969). Brown rot of stone fruits on the Murrumbidgee irrigation areas: 1. Aetiology of the disease in canning peaches. *Australian Journal of Agricultural Research, 20*, 301–316.

Karakoyun, S. (2011). *Investigation on the effects of sulfuryl fluoride against Carpophilus hemipterus (L.) (Coleoptera: Nitidulidae)* (pp. 73). PhD Dissertation, Ankara University Graduate School of Natural and Applied Sciences, Department of Plant Protection, Ankara, Turkey.

Kehat, M., Stoller, S., Meisner, J., & Greenberg, S. (1966). Development of population and control of *Carpophilus dimidiatus* F., *C. hemipterus* L. and *Coccotrypes dactyliperda* F. in dates. *Israel Journal of Agricultural Research, 16*, 173–176.

Kehat, M., Blumberg, D., & Greenberg, S. (1976). Fruit drop and damage in dates: The role of *Coccotrypes dactyliperda* F. and nitidulid beetles, and prevention by mechanical measures. *Phytoparasitica, 4*, 93–99.

Kehat, M., Blumberg, D., & Williams, R. N. (1983). Seasonal abundance of sap beetles (Coleoptera: Nitidulidae) in date plantations in Israel. *Phytoparasitica, 11*, 109–112.

Kehat, M., Greenberg, S., Biton, S., Ganz, S., & Adato, I. (1986). Field evaluation of insecticides for the control of sap beetles Coleoptera: Nitidulidae in dates. *Alon Hanotea, 40*, 787–790.

Kelts, C. A. (2005). *Evaluation of attractants and monitoring for sap beetle control in strawberries* (77 p). MSc Dissertation, Graduate School of the University of Florida, USA.

Larson, D. (2013). *Key to Saskatchewan species of Nitidulidae (sap beetles) and Kateretidae (short-winged flower beetles)*. Available at: http://www.entsocsask.ca/documents/insect_lists/Key%20to%20species%20of%20Nitidulidae.pdf. Accessed on 1 June 2014.

Latifian, M., Rahnama, A. A., & Sharifnezhad, H. (2012). Effects of planting pattern on major date palm pests and diseases injury severity. *International Journal of Agriculture and Crop Sciences, 4*, 1443–1451.

Lawrence, J. F. (1982). Coleoptera. In S. P. Parker (Ed.), *Synopsis and classification of living organisms* (Vol. 2, 482–553 pp). New York: McGraw-Hill.

Leschen, R. A. B., & Marris, J. W. M. (2005). *Carpophilus (Nitidulidae) of New Zealand and notes on Australian species* (MAF Contract Report (FMA121), Australia. 41 p).

Lindgren, D. L., & Vincent, L. E. (1949). Investigations of the life history and control of date insects and the date mite. *Date Growers' Institute's Annual Report, 26*, 8.

Lindgren, D. L., & Vincent, L. E. (1953). Nitidulid beetles infesting California dates. *Hilgardia, 22*, 97–118.

Loughner, R. L., Loeb, G. M., Demchak, K., & Schloemann, S. (2007). Evaluation of strawberry sap beetle (Coleoptera: Nitidulidae): Use of habitats surrounding strawberry plantings as food resources and overwintering sites. *Environmental Entomology, 36*, 1059–1065.

Lussenhop, J., & Wicklow, D. T. (1990). Nitidulid beetles (Nitidulidae: Coleoptera) as vectors of *Aspergillus flavus* in preharvest maize. *Transactions of the Mycological Society of Japan, 31*, 63–74.

Miller, K. V., & Williams, R. N. (1983). Biology and host preference of *Atheta coriaria* (Coleoptera: Staphylinidae), an egg predator of nitidulidae and muscidae. *Annals of the Entomological Society of America, 76*, 158–161.

Moore, D. (2001). Fungal control of pests. In D. Pimentel (Ed.), *Encyclopedia of pest management* (320–324 pp). New York: Marcel Dekker Inc.

Mushtaq, S., Hadait, S., & Ali, L. (2000). Diversity and relative abundance of spiders of date palm groves. *International Journal of Agriculture and Biology, 2*, 52–54.

Myers, L. (2001). *Sap beetles of Florida-Nitidulidae (Featured Creatures.)*. Department of Entomology and Nematology, Institute of Food and Agricultural Sciences Publication No. EENY-256, University of Florida, Gainesville, USA. Available at: http://entnemdept.ufl.edu/creatures/field/corn/sap_beetles.htm. Accessed on 1 June 2014.

Najla, M. Z., Kerra, H. M., Edongali, E. A., & Abogila, O. S. (2005). Population dynamics of the sap beetles (Nitidulidae) associated with date palm trees in coastal regions of Libya. *Egyptian Journal of Agricultural Research, 83*, 1343–1353.

Nixon, R. W. (1951). The date palm: "Tree of Life" in the subtropical deserts. *Economic Botany, 5*, 274–301.

Nout, M. J. R., & Bartelt, R. J. (1998). Attraction of a flying nitidulid (*Carpophilus humeralis*) to volatiles produced by yeasts grown on sweet corn and a corn-based medium. *Journal of Chemical Ecology, 24*, 1217–1239.

Omar, M. T. A. (1971). *Insect pests attacking dry-dates with special reference to Cadra calidella (Guen.)* (pp. 65). MSc Dissertation, Faculty of Agriculture, Al- Azhar University, Egypt.

Ozer, H., Imge, H., Basegmez, O., & Ozay, G. (2012). Mycotoxin risks and toxigenic fungi in date, prune and dried apricot among Mediterranean crops. *Phytopathologia Mediterranea, 51*, 148–157.

Oztürk, N., Ulusoy, M. R., & Bayhan, E. (2005). Pest and natural enemy species determined in pomegranate orchards in the Eastern Mediterranean Region, Turkey. *Turkish Journal of Entomology, 29*(3), 225–235. (in Turkish).

Parsons, C. T. (1943). A revision of nearctic Nitidulidae (Coleoptera). *Bulletin of the Museum of Comparative Zoology, 92*, 121–248.

Pegna, F. G., Battaglia, M., & Bergesio, C. (2012). *Italian machinery and equipment for date palm field operations* (69 pp). Italy: University of Florence.

Peng, C., & Williams, R. N. (1990). Pre-oviposition period, egg production and mortality of six species of hibernating sap beetles (Coleoptera: Nitidulidae). *Journal of Economic Entomology, 23*, 453–457.

Pinruan, U., Rungjindamai, N., Choeyklin, R., Lumyong, S., Hyde, K. D., & Gareth Jones, E. B. (2010). Occurrence and diversity of basidiomycetous endophytes from the oil palm, *Elaeis guineensis*, in Thailand. *Fungal Diversity, 41*, 71–88.

Porter, J. (1986). Some studies on the life history and oviposition of *Carpophilus dimidiatus* (F.) (Coleoptera: Nitidulidae) at various temperatures and humidities. *The Journal of Stored Products Research, 22*, 135–139.

Price, M. B., & Cira, T. (2007). *Kateretidae and Nitidulidae of Wisconsin*. Available at: http://entomology.wisc.edu/irc/wibeetles/Price/K&N%20Intro.htm. Accessed on 1 June 2014.

Qasem, S. A., & Christensen, C. M. (1960). Influence of moisture content, temperature, and time on the deterioration of stored corn by fungi. *Phytopathology, 48*, 544–549.

Riechert, S. L., & Lockley, T. (1984). Spiders as biological control agents. *Annual Review of Entomology, 29*, 299–320.

Rodriguez-Del-Bosque, L. A., Leos-Martinez, J., & Dowd, P. F. (1998). Effect of ear wounding and cultural practices on abundance of *Carpophilus freemani* (Coleoptera: Nitidulidae) and other microlepidopterans in maize in northeastern Mexico. *Journal of Economic Entomology, 91*, 796–801.

Rondon, S. I., Price, J. F., & Cantliffe, D. J. (2004). *Sap Beetle (Coleoptera:Nitidulidae) management in strawberries*. Department of Entomology and Nematology, Institute of Food and Agricultural Sciences Publication No. HS993, University of Florida, Gainesville, USA. Available at: http://edis.ifas.ufl.edu/pdffiles/HS/HS23400.pdf. Accessed on 1 June 2014.

Saaidi, M. (1992). *FAO report on date palm protection and training activities* (121 pp). Oman: Ministry of Agriculture and Fisheries.

Scheid, B. E., Theis, C., & Tscharntke, T. (2011). Enhancing rape pollen beetle parasitism within sown flower fields along a landscape complexity gradient. *Agriculture and Forest Entomology, 13*, 173–179.

Schmidt, C. T. (1935). Biological studies on the Nitidulid beetles found in pineapple fields (Nitidulidae Coleoptera). *Annals of the Entomological Society of America, 28*, 475–511.

Sedra, M. H. (2003). *Le palmier dattier base de la mise en valeur des oasis au Maroc Techniques phoénicicoles et Création d'oasis* (265 pp). Morocco: INRA, Rabat-Instituts.

Shenasi, M., Aidoo, K. E., & Candlish, A. A. G. (2002). Microflora of date fruits and production of aflatoxins at various stages of maturation. *International Journal of Food Microbiology, 79*, 113–119.

Simmons, P., & Nelson, H. D. (1975). *Insects on dried fruits* (United States Department of Agriculture, Agricultural Research Service Handbook No. 464, 25 pp). Washington, DC: United States Department of Agriculture, Agricultural Research Service.

Simmons, P., Reed, W. D., & McGregor, E. A. (1931). Circular No. 157: Fig Insects in California. United States Department of Agriculture, Bureau of Entomology, Washington, D.C., USA.

Smilanick, J. M. (1979). Colonization of ripening figs by *Carpophilus* spp. *Journal of Economic Entomology, 72*, 557–559.

Smilanick, J. M., Ehler, L. E., & Birch, M. C. (1978). Attraction of *Carpophilus* spp. (Coleoptera:Nitidulidae) to volatile compounds present in figs. *Journal of Chemical Ecology, 4*, 701–707.

Stickney, F. S., Barnes, D. W., & Simmons, P. (1950). *Date palm insects in the United States, 57 pp* (U.S. Department of Agriculture Circular No. 846). Washington, DC: U.S. Department of Agriculture.

Strauss, U., Human, H., Crewe, R. M., & Pirk, C. W. W. (2010). The first report of storage mites, *Caloglyphus hughesi* (Acaridae) on laboratory-reared *Aethina tumida* Murray (Coleoptera: Nitidulidae) in South Africa. *African Entomology, 18*, 379–382.
Sun, X., & Guo, L. D. (2012). Endophytic fungal diversity: Review of traditional and molecular techniques. *Mycology, 3*, 65–76.
Swezey, H. O. (1942). Miscellaneous families of Guam Coleoptera. In Insects of Guam – I. *Bernice P. Bishop Museum Bulletin, 172*, 150–171.
Tate, K. G., & Ogawa, J. M. (1975). Nitidulid beetle as vectors of *Monilinia fructicola* in California stone fruits. *Phytopathology, 65*, 977–983.
Taylor, J. E., Hyde, K. D., & Jones, E. B. G. (1999). Endophytic fungi associated with the temperate palm, *Trachycarpus fortunei*, within and outside its natural geographical range. *New Phytologist, 142*, 335–346.
Tezcan, S., Düzbastilar, M., & Lechanteur, F. (2003). Evaluation of the Nitidulidae (Coleoptera) fauna of ecologically managed cherry orchards in Izmir and Manisa Provinces of Turkey. *Alatarim, 2*, 16–20. (in Turkish).
Thomas, S. E. (2014). *Fungi and their potential as biological control agents of beech bark disease* (194 p). PhD Dissertation, Royal Holloway, University of London, UK.
Turanli, F. (2003). Studies on infestation levels of pests on dried fig in Aydin and Izmir provinces. *Turkish Journal of Entomology, 27*, 171–180. (in Turkish).
Vega, F. E., Dowd, P. F., & Nelsen, T. C. (1994). Susceptibility of driedfruit beetle (*Carpophilus hemipterus* L.; Coleoptera: Nitidulidae) to different *Steinernema* species (Nematoda: Rhabditae: Steinernematidae). *Journal of Invertebrate Pathology, 64*, 276–277.
Vega, F. E., Dowd, P. F., & Bartelt, R. J. (1995). Dissemination of microbial agents using an autoinoculating device and several insect species as vectors. *Biological Control, 5*, 545–552.
Warner, R. M. (1961). Area baiting program 1960 results. *Proceedings of California Fig Institute Research, 15*, 36–40.
Warner, R. L., Barnes, M. M., & Laird, E. F. A. D. (1990). Chemical control of a carob moth, *Ectomyelois ceratoniae* (Lepidoptera: Pyralidae), and various nitidulid beetles (Coleoptera) on 'Deglet Noor' dates in California. *Journal of Economic Entomology, 83*, 2357–2361.
Williams, R. N., Fickle, D. S., Kehat, M., Blumberg, D., & Klein, M. G. (1983). *Bibliography of the genus Carpophilus Stephens (Coleoptera: Nitidulidae)*. Ohio State University Agricultural Research and Development Center, Research circular, 278, (95 p). Ohio State University, Agricultural Research and Development Center.
Williams, R. N., Weiss, M. J., Kehat, M., & Blumberg, D. (1984). The hymenopterous parasites of Nitidulidae. *Phytoparasitica, 12*, 53–64.
Williams, R. N., Blackmer, J. L., Richmond, D. S., & Ellis, M. S. (1992). Nitidulidae (Coleoptera) diversity in three natural preserves in Portage County, Ohio, USA. *Ohio Journal of Science, 92*, 82–87.
Williams, R., Fickle, D., & Ellis, S. (1995). Strawberry sap beetle control by parasite. *Strawberry IPM Update, 1*, 9–10.
Yaman, M. (2007). Distribution of *Nosema meligethi* (Microsporidia) in populations of *Meligethes aeneus* (Coleoptera: Nitidulidae) in Turkey. *Entomological Research, 37*, 298–301.
Zipagan, M. B., & Pacumbaba, E. P. (1994). *Peregrinator biannulipes* Montrouzier and Signoret, a reduviid predator of insect pests of copra. In 25th pest management council of the Philippines, Inc. Anniversary and annual scientific meeting, May 3–6, 1994. Cagayan de Oro City, Philippines.

Chapter 9
Pests of Stored Dates

Charles S. Burks, Muhammad Yasin, Hamadttu A.F. El-Shafie, and Waqas Wakil

Abstract Dates are a major food crop across a large band of Africa and Eurasia, and to a lesser extent elsewhere. In most of its growing range, dates are threatened with infestation in the field by a complex of pests including nitidulid beetles and pyralid moths of the Subfamily Phycitinae. They are further threatened with loss in processing and storage by stored product moths along with some beetles. This review examines the biology and life history of these key pests, and best practices for sustainable management of these pests. Future research directions are also briefly considered.

9.1 Introduction

The total world production of dates was 7.78 million tons in 2013 (FAO 2015). The top ten producers, representing 88 % of this production, are in North Africa, the Arabian Peninsula, and Southwest Asia. Outside this region the largest producers are China, with 150,000 tons, and the USA, with 22,000 tons (Wright 2007). Before reaching the consumer, dates are stored at the farm level in centrally cooled storage facilities, or in regular storage in packinghouses. Storage of dates, like any other agricultural commodity, is an important factor in preservation of value and profitability.

C.S. Burks (✉)
USDA-ARS, Commodity Protection and Quality, South Riverbend Avenue, Parlier, CA, USA
e-mail: charles.burks@ars.usda.gov

M. Yasin • W. Wakil
Department of Entomology, University of Agriculture, Faisalabad, Punjab, Pakistan
e-mail: yasin_1876@yahoo.com; waqaswakeel@hotmail.com; arid1972@yahoo.com

H.A.F. El-Shafie
Date Palm Research Center of Excellence, King Faisal University,
Hofuf, Al-Ahsa, Saudi Arabia
e-mail: elshafie62@yahoo.com

W. Wakil et al. (eds.), *Sustainable Pest Management in Date Palm: Current Status and Emerging Challenges*, Sustainability in Plant and Crop Protection,
DOI 10.1007/978-3-319-24397-9_9

During storage, insect infestation of dates is one of the primary causes of post-harvest losses in quality and quantity. The extent of damage inflicted by stored-product pests on dates is influenced by many factors including date variety, whether dates are stored loose or pressed together, and the storage conditions, particularly temperature and relative humidity. Pressed dates are more likely to withstand stored-product insect pests than loose dates, due perhaps to the absence of free space in the former, which appears to hinder the free movement of insects, their behavior and finally their population build-up.

Direct damage of dates is caused by feeding. Reductions in market value of dates can also result from the presence of insect body parts (antennae, legs, wings, cast skins etc.), frass, webbing, and other excretions. In the U.S., for example, most quality standards for dried fruits do not allow for the presence of any live insects (USDA-AMS, 2013).

For dried fruits and nuts there is the potential for insect pest infestation of the fruit in the orchard to remain with the fruit until it reaches the end consumer. In addition, such products are at additional risk from traditional stored product insect pests during processing and storage through the marketing channels. We consider insect pests of dates in two groups, based on whether they first infest prior to processing or in processing and storage. Among insects commonly infesting dates prior to harvest or processing, we include nitidulid beetles and several pyralid moths from the subfamilies Galleriinae and Phycitinae (Table 9.1).

Among the post-harvest pests we include beetles of the families Silvanidae, Tenebrionidae, and Laemophloeidae (Table 9.2). In a previous review of insect pests of dried fruits and nuts (Burks and Johnson 2012), three of the phycitine moths from Table 9.1 (almond moth, Indian meal moth, Mediterranean flour moth) were include as species that infest post-harvest.

Table 9.1 Insect pests of post-harvest concern infesting dates

Order	Family	Scientific name	Common name
Coleoptera	Nitidulidae	*Carpophilus hemipterus* (L.)	Driedfruit beetle
		Carpophilus mutilatus Erichson	Confused sap beetle
		Urophorus humeralis (F.)	Pineapple beetle
Lepidoptera	Pyralidae, subfamily Galleriinae	*Aphomia sabella* Hampson	Greater date moth
	Pyralidae, subfamily Phycitinae	*Ectomyelois ceratoniae* (Zeller)	Carob moth
		Cadra figulilella (Gregson)	Raisin moth
		Cadra cautella (Walker)	Almond moth
		Plodia interpunctella (Hübner)	Indian meal moth
		Ephestia kuehniella (Zeller)	Mediterranean flour moth

Table 9.2 Main post-harvest insect species infesting dates

Order	Family	Scientific name	Common name
Coleoptera	Silvanidae	*Oryzaephilus surinamensis* (L.)	Sawtoothed grain beetle
	Tenebrionidae	*Tribolium castaneum* (Herbst)	Red flour beetle
		Tribolium confusum Jacquelin du Val	Confused flour beetle
	Laemophloeidae	*Cryptolestes ferrugineus* (Stephens)	Rusty grain beetle

Here we include these species with those infesting prior to harvest, since there are persistent reports that this occurs for dates in their various production areas (Carpenter and Elmer 1978; Bachrouch et al. 2010; Jemâa et al. 2012; Aldawood et al. 2013; Marouf et al. 2013). The species in Tables 9.1 and 9.2 are principal insect pests of dates in storage, reported from large portions of the areas where dates are produced. A previous review (El-Shafie 2012) offered a more comprehensive listing of insect attacking dates during harvest and storage.

Life history information is given separately for the species in Tables 9.1 and 9.2, and then control measures and their integration are considered categories as a whole for each of these two groups of pests. Preference is given to common names and taxonomic designations used in the Common Names of Insects Database of the Entomological Society of America (ESA 2015) for species which occur in that database.

9.2 Insect Pests of Post-harvest Concern

In this group, there is a marked contrast between the life histories of the nitidulid beetles and the moths of the family Pyralidae. A couple of key differences are: (1) the nitidulid beetles feed as adults and potentially live a long time, whereas the pyralid moths do not feed as adults and have short adults lives; and (2) the nitidulid beetles have an aggregation pheromone that is attractive to both sexes, whereas the pyralid moths generally have sex pheromones produced by females and attractive to males only.

These differences in life history fit broad patterns commented on previously by stored product entomologists (Trematerra 2012), and could have implications for monitoring and pest management for these species. The greater date moth is a possible exception to the latter generalization, as described below.

9.2.1 *Nitidulid Beetles*

Several beetles of the family Nitidulidae have been reported as serious pests of dates (Carpenter and Elmer 1978; Williams et al. 1983; Blumberg 2008; Perring and Nay 2015). These include the driedfruit beetle, *Carpophilus hemipterus* (L.); the confused sap beetle, *Carpophilus mutilatus* Erichson; and the pineapple beetle, *Urophorus* (= *Carpophilus*) *humeralis* (F.). In addition to the three described here, other widely distributed nitidulid species noted in association with dates include the yellow nitidulid, *Epuraea luteolus* (Erichson), and the corn sap beetle, *Carpophilus dimidiatus* (F.) (Carpenter and Elmer 1978; Blumberg 2008). Before the introduction of the carob moth in the 1980s, nitidulid beetles were considered the most economically important pest of dates in the U.S. (Nay et al. 2007).

These species are all cosmopolitan and polyphagous. They are negatively phototaxic, attracted by fermentation, and are found in or on the ground under a wide variety of fruits and other media in intermediate states of decay. In Central California (USA) they are generally considered of no concern in stone fruit and table grapes because they do not infest intact (i.e., marketable) fruit. In dried fruits, in contrast, nitidulid beetles are pests of widespread concern (Simmons and Nelson 1975; Carpenter and Elmer 1978). The relative abundance of species present varies with year and season (Blumberg 2008). In the field, the activity and crop damage caused by the sap beetles noticeably increased during rainy years (Lindgren and Vincent 1953; Warner et al. 1990a, b). Sap beetles can cause severe damage to ripened dates in the field or during storage, where several generations can develop when dates are stored for an extended period of time (Blumberg 2008).

9.2.1.1 Driedfruit Beetle (*Carpophilus hemipterus*)

Biology and Seasonal Incidence

Adult driedfruit beetles (3–4 mm) are oval and black in color with two large conspicuous yellow-brown spots on the wing covers, which can distinguish it from other species of sap beetles. The wing covers are short and leave part of the abdomen exposed. The legs and antennae are often reddish or amber colored. The 11-segmented antennae are slender except for the last few segments, which are distinctly enlarged into a club (Blumberg 2008).

The biology of the driedfruit beetle is described in detail in Simmons et al. (1931). Eggs are white, cylindrical, 2–3 mm long laid singly, on ripe or fermenting fruit, in the field or in stored-products. Lifetime fecundity of driedfruit beetle females is up to 2,134 eggs at 28.1 °C, with an average of 1,070 eggs in her lifetime. Eggs of the driedfruit beetle and other nitidulids are white, sausage-shaped, and 0.8–1.2 mm long by 0.2–0.3 mm wide (Gautam et al. 2014). The eggs have a single aeropyle (respiratory opening) at the anterior end (Gautam et al. 2014). The pre-oviposition, oviposition, and post-oviposition periods are 3, 61, and 9 days, respectively (Mason 2010). The egg incubation period is 1–4 days. Larvae are small (5–7 mm in length) whitish or yellowish, with a light brown head and two hardened spine-like projections at the end of their abdomens that are species specific. The larval development is completed within 4–14 days. Larvae are very active and hide themselves in response to disturbance (Simmons and Nelson 1975; Blumberg 2008). Like other nitidulids described here, *C. hemipterus* pupae are exarate (i.e. the appendages are not fused to the body). Pupation usually takes place in heavy soil in field populations, and within the infested commodity in case of stored-products. Pupation occurs after a pre-pupal period of 3–8 days, and adults emerge 4–16 days later. Thus, the entire development cycle is about 12 days at warm temperatures (32.2 °C) and up to 42 days at cooler temperatures (18.3 °C). Though adults can live for more than a year, driedfruit beetle males generally live 146 days while females average 103 days. The driedfruit beetle and other nitidulid beetles leave host material to pupate in the soil, and they generally overwinter in the soil in this stage before completing pupation and adult eclosion in the spring (Simmons et al. 1931).

Adults are active during the day and fly only when the temperature is above 18°C (Simmons and Nelson 1975). Adult sap beetles are attracted to volatiles from fermenting fruit and grains can migrate up to 2 miles in 4 days (Mason 2010). Males and females of *C. hemipterus* locate each other and food sources through a synergistic attraction to food volatiles (alcohol, esters and acids) and aggregation pheromones (a blend of 13–15 carbon molecules) produced by males and attract both males and females. The pheromone blends are cross-attractive between the driedfruit beetle and other nitidulid species discussed in this section (Bartelt et al. 1992).

The driedfruit beetle has greater tolerance to heat, as well as faster development, compared to *C. mutilatus* and *C. humeralis* (James and Vogele 2000), possibly explaining the frequent prevalence of the driedfruit beetle in date growing areas. In warm areas or within heated buildings, breeding is continuous and there are several generations per year. When temperatures are too low for breeding, the beetles hibernate as mature larvae, pupae or adults in the soil, stored commodity, or fruit left on the ground (Mason 2010).

Humidity is very important to the survival of driedfruit beetles and low humidity affects negatively both larval development and oviposition. Larval development times increase with decreasing humidity so that the mean period from egg to adult increases from 19 days at 92 % RH to 24 days at 66 % RH.

9.2.1.2 Confused Sap Beetle (*Carpophilus mutilatus*)

Biology and Seasonal Incidence

The adult of *C. mutilatus* is a small (1.5–1.8 mm) beetle. Both the head and pronotum are light reddish brown in color, the elytra (forewing) is with background lighter than pronotum and has light and dark setae (Walker 2007b). During the average adult period of 6–12 months, a single female can lay 500–1,000 or sometimes up to 2,000 eggs (Avidov and Harpaz 1969) on the trees or in rotting fruit lying on shaded ground. The eggs hatch within 1–2 days and emerging larvae are yellow or white in color having brown head, the body length of last larval instar measures 5–7 mm which completes its development in about 33 days at 20 °C (James and Vogele 2000). The pupae are white or yellow in color and 3–4 mm in length (Parsons 1943; Blumberg 2008). The developmental time from egg to adult emergence varies from 47 to 65 days at 20 °C, to as little as 14–18 days at 32.5 °C (James and Vogele 2000).

In Israel, *C. mutilatus* can be found throughout the autumn, winter and spring, while in the summer season their populations decline greatly (Kehat et al. 1983). James et al. (1994) found that, in New South Wales (Australia), *C. mutilatus* was first trapped in stone fruit orchards in the end of spring (September) when average daily temperature was 17.5 °C. Trap catches reached to the highest during October and declined to very few from November to December that again rose from February to March most probably due to the increased availability of host fruits during this time of the year.

9.2.1.3 Pineapple Beetle (*Urophorus = Carpophilus humeralis*)

Biology and Seasonal Incidence

The pineapple beetle is a shiny, polished and brown to black beetle with an oval to broadly oblong body. The elytra are reduced exposing 2–3 abdominal segments and generally have a small pale spot within the humerus region. The body length of 3.0–4.5 mm with width of 1.6–1.9 mm makes this beetle amongst the largest beetles infesting dried fruits (Simmons and Nelson 1975; Ewing and Cline 2005; Walker 2007b). An adult can survive up to 89 days (Simmons and Nelson 1975; Carpenter and Elmer 1978; Ewing and Cline 2005; Walker 2007b). The ovipositing beetles are site-specific and lay eggs only in the decomposing material (Schmidt 1935). A single female can lay an average of 880–1,000 eggs (Simmons and Nelson 1975). Eggs take 1–2 days for hatching, the emerging neonate larvae feed on date pulp for 15 days or more and the last larval instar move into the soil for pupation. The pupae average 4.4 mm in length and 2.0 mm in width. Initially, they are white in color and turned to creamy and later on tan colored near the adult emergence (Parsons 1943). The pupation period lasts averages 5.3 days (Schmidt 1935). The developmental period from egg to adult emergence takes 47–65 days at 20 °C and 14–18 days at 32.5 °C (James and Vogele 2000). After adult emergence the beetle start infesting dates either on palms or on the ground, both larvae and adults feed and develop under dark, moist and moldy environment.

The population of *U. humeralis* in date plantation of Israel was pre-dominant during summer when fruits started to ripen (Kehat et al. 1983). In California date gardens, the nitidulid beetles were more abundant when mean daily temperatures were about 32 °C (June–September), Additionally, the high humidity during this period appeared to be extremely important in supporting population buildup of sap beetles (Bartelt et al. 1994).

9.2.2 *Pyralid Moths*

Several species of Pyralidae are widespread and serious pests of dates. The greater date moth, *Aphomia sabella* Hampson, of the subfamily Galleriinae, is a monophagous herbivore specialized for development on date palm. The remaining five species, of the subfamily Phycitinae, are generalist feeders that often exploit dates as a resource. The greater date moth, carob moth, and raisin moth are principally field pests that can be carried over into harvested commodities, whereas the almond moth, Indian meal moth, and Mediterranean flour moth are principally indoor pests that can also occur outdoors in suitable climates. Two other species of Phycitinae have also occasionally been noted in association with dates: the tobacco moth, *Ephestia elutella* (Hübner) (Ahmad and Ali 1995), and the driedfruit moth, *Ephestia calidella* (Guenée) (Boshra and Mikhaiel 2006). The raisin moth, almond moth, Indian meal moth, and Mediterranean flour moth have cross-attracting sex pheromone blends (Phelan and Baker 1986), and therefore may appear together in the same monitoring traps. Except for the greater date moth, these species can be separated by using taxonomic keys that can be downloaded from the Internet (Ferguson 1987; Weisman 1987; Solis 2006).

9.2.2.1 Greater Date Moth (*Aphomia* = *Arenipses sabella*)

Distribution and Host Range

The greater date moth is found in almost all the date growing areas of North Africa, Northern India, the Middle East, Iraq, Saudi Arabia, Oman, Egypt, Algeria and Iran (Kehat and Greenberg 1969; Carpenter and Elmer 1978; Al-Azawi 1986; Talhouk 1991; Aldryhim 2008). It most likely originated from Mesopotamia (Buxton 1920). Earlier records also show its presence at the head of Persian Gulf, and in Algeria, Persia, and Arabia (Buxton 1920). It reached England through trade where its larvae collected from date gardens of London were reared under laboratory conditions and adult specimen was put in the British Museum (Buxton 1920). Date palms and date fruit are the only known host for this species (Blumberg 2008), and it has been considered one of the more serious pests of dates in Saudi Arabia (Talhouk 1991) and in Sinai, Egypt (El-Sherif et al. 1996).

Biology and Seasonal Incidence

The adult moths are grey or yellowish brown in color with a body length of 18–22 mm, and a wing span of 33–35 mm in males and 40–42 mm in females. The front wing are brownish to yellow bearing black scales and rare wings are light brown. Both head and thorax are light brown, while abdomen is silvery white (Aldryhim 2008). A single female can lay 200–400 creamy white and spherical eggs, singly or in clusters and mostly at night. Eggs hatch out in 4–5 days at 30 °C (Aldryhim 2008; Blumberg 2008). The caterpillar is dark pink with hairs on the body and 28–35 mm in length (Aldryhim 2008). The pupae are light brown, elongate and about 18 mm long (Aldryhim 2008). The developmental period of larvae and pupae lasts for 34–40 days at 27 °C, while adult survive for 7–8 days (Imam 2012).

The close tie of the greater date moth with its host have made studies of its biology difficult. Laboratory experiments found that copulation of the greater date moth occurred only in the presence of date fruit and cut date branches, and the researchers concluded that sexual communication and copulation in the field most likely occurs in the crowns of date palms (Levi-Zada et al. 2014). Studies using gas chromatography and mass spectrometry found materials released by males in a circadian rhythm, and detected by female antennae. No such compounds were found emitted by females. No flight was observed to live moths or putative semiochemicals. These data suggest that this species lacks a female-emitted sex pheromone. This deviation from the more usual situation of males responding to a female-emitted sex pheromone has been observed in a few other moth species (Levi-Zada et al. 2014).

Continuous rearing in the laboratory suggests the capacity for three to four generations between April and September (Kehat and Greenberg 1969). In the field, though, there are only two periods of abundance per year; one in early spring and one in late fall (Hussain 1963), suggesting that biotic or abiotic conditions are unfavorable during the summer (Kehat and Greenberg 1969). The greater date moth overwinters as a larva in the leaves near the crown of palm trees, and emerges in spring to feed and complete development (Hussain 1963; Kehat and Greenberg 1969; Carpenter and Elmer 1978; Imam 2012). The spring brood is mostly found during April in the form of massive larval population, the unripe fallen fruits on the ground provide food to the larvae of summer brood, while, autumn brood shift to the ripening dates on the palms (Blumberg 2008). Pupae also occur in stored dates (Carpenter and Elmer 1978), but there is little evidence that this species reproduces in storage.

9.2.2.2 Carob Moth (*Ectomyelois ceratoniae*)

Distribution and Host Range

The carob moth is widely distributed in the Mediterranean region, and in other parts of the world with a similar climate (Calderon et al. 1969; Gothilf 1970; Cox 1979). Specific reports of crops attacked include: citrus, acacia, carob, and almonds in Israel (Calderon et al. 1969; Gothilf 1970); walnut, fig, pear, apple, and pistachios in Iraq (Al-Izzi et al. 1985; Mehrnejad 1994); walnut, pistachio, and fig in Iran (Mozaffarian et al. 2008); apple, and loquat in southern South America (Legner and Silveira-Guido 1983; Zaviezo et al. 2007); and almond in Australia (Michael 1968). In both Iraq (Al-Izzi et al. 1985) and Iran (Mohseni 2009), pomegranate is an important crop and the carob moth is a particularly important pest of pomegranates. In the U.S. the carob moth was associated with tamarind in southern Florida (USA) since at least the 1940s (Neunzig 1979), but it was first detected in California in 1982

(Warner et al. 1990b). The carob moth quickly became a pest and, by the 2000s, it was the most economically important pest in the USA date industry in the Coachella Valley (USA) (Nay et al. 2007). At the current writing, however, the carob moth has not been identified as an agricultural pest in the Central Valley of California, which is located 300 km to the northwest and in which almond, pistachio and walnut are important crops, and pomegranates and figs are also grown commercially. While mature or senescing fruit is usually attacked, imperceptibly diseased almonds are also attacked prior to hull suture (Navarro et al. 1986). While the carob moth infests fruit in the field, population growth in almonds can continue in storage (Navarro et al. 1986).

Biology and Seasonal Incidence

The carob moth is slightly larger than many of the stored product moths of the subfamily Phycitinae, with a wingspan of around 2 cm. Like other members of the Pyralidae, the moth parts form a tusk-like “snout”. The wing pattern consists of alternating light and dark zig-zag lines.

The basic life cycle of the carob moth is similar to the other moths of the subfamily Phycitinae discussed here (Gothilf 1970; Al-Izzi et al. 1985; Navarro et al. 1986). They are generalist feeders that develop on a wide variety of hosts. Adults are short-lived (2 weeks or less) and do not feed, although they may take liquids. Females emit a sex pheromone, comprised of (Z,E)-9,11,13-tetradecatrienal as a major component and (Z,E)-9,11-tetradecadienal plus (Z)-9- tetradecadienal as minor components (Baker et al. 1991), to attract males for copulation. Females call in about the middle of the night (Vetter et al. 1997). Oviposition peaks the second night after adult emergence, and declines rapidly and is finished by 10 days of adult life (Navarro et al. 1986). Females have an average lifetime fecundity of 110 eggs (Navarro et al. 1986; Alrubeai 1987). Oviposition is generally in some type of crack, fissure, or opening in the host material, and chemical cues have been identified both to guide females to the host and to stimulate egg laying (Gothilf 1970). Larvae are unable to penetrate intact hosts, but are able to exploit small openings in the host (Gothilf 1970). Larval development on carob occurs over 17 days at 30 °C and 70 % RH, with an additional 7 days in the pupal stage for a developmental period of 23 days from egg to larva (Cox 1976). Developmental time is, however, affected markedly by host, and a developmental period of 32–33 days was estimated for California dates under summer positions (Nay and Perring 2008). A diapause at the end of the last larval instar has been documented (Cox 1979), although carob moth in the field overwinters in various larval stages (Gothilf 1970, 1984; Nay and Perring 2008) in fruit in the tree or on the ground (Nay and Perring 2008). Photoperiod can apparently also affect the larval development period in larvae that do not diapause (Al-Izzi et al. 1985). Unlike phycitine moths more traditionally associated with stored products (the Mediterranean flour moth, the almond moth, and the Indian meal moth), there is no post-feeding wandering phase at the end of the last larval instar. Larvae chew an exit hole, then pupate inside the larval host and later emerge as adults. Multiple larvae can develop in an individual host (Gothilf 1970).

The carob moth differs from phycitine moths more traditionally associated with stored products (the Mediterranean flour moth, the almond moth, and the Indian meal moth) in that it is adapted to find and exploit injured or senescing fruit, usually while it is still on the tree. In Israel, carob moth females are attracted to injuries to seed pods in host legumes (carob, acacia) made by midges and brucchid beetles prior to pod split (i.e., prior to when the pods would otherwise be susceptible) (Gothilf 1970). They are also attracted by and oviposit on carobs infested with the fungus *Phomopsis* sp., although they apparently do not successfully develop on such carobs (Gothilf 1970). In contrast, they are also attracted to intact (pre-hull split) almonds infected with *Colletotrichum gloeosporioides* (Penz.) Penz. & Sacc., the causal agent of almond anthracnose. Larvae are able to enter and use such almonds for development, and almond orchards with anthracnose support higher carob moth abundance (Gothilf 1970). In dates, the carob moth is attracted also by fungal infection as well as by maturing fruit (Cossé et al. 1994).

9.2.2.3 Raisin Moth (*Cadra* = *Ephestia figulilella*)

Distribution and Host Range

The raisin moth is slightly smaller compared to related stored product moths (wing span 1.5–2.0 cm). Adults can be distinguished from other stored product moths by light pink scales. It is found in the Mediterranean, and similar climates in the Americas and Australia (Donohoe et al. 1949; Cox 1974; Horak 1994), and it is a pest of dates in both the Eastern and Western hemispheres (Kehat et al. 1992; Nay et al. 2007; Assari and Khajepour 2013). In Australia, the raisin moth is associated primarily with raisin production (Horak 1994). In Central California the raisin moth has been noted on ripe grapes, raisins, figs on the ground, and fallen apricots, plums, peaches, and nectarines (Donohoe et al. 1949). A similar host variety has been noted in the Middle East (Khajepour et al. 2012). This species is primarily a field pest; reproduction in storage is apparently limited (Simmons and Nelson 1975; Soderstrom et al. 1987), and compared to other stored product moths, this species is less common in warehouses and marketing channels.

Biology and Seasonal Incidence

In the field, dropped decaying dates are infested by the raisin moth throughout the year, but infestation of dates in the tree does not occur until they begin to ripen (Kehat and Greenberg 1969). Eggs are small, round and sticky, but generally not glued to host material. Larval development requires higher moisture compared to related stored product pests (Donohoe et al. 1949; Cox 1974; Simmons and Nelson 1975), and development from egg to adult on dates requires 40–60 days at 30 °C (Kehat and Greenberg 1969; Cox 1975b). Diapause at the end of the last larval instar is induced by shortened photoperiod (Cox 1975a), and in central California larvae overwinter in diapause in the soil or under the bark at the base of grape vines or other host plants (Donohoe et al. 1949). In Central California and the Middle East adult flight occurs from April through November, with adult abundance generally greater later in the year (Donohoe et al. 1949; Kehat et al. 1992; Ahmad and Ali 1995). Temperatures above 15 °C are required for continuous development (Cox 1974).

The raisin moth shares a principal pheromone component (compound) with the remainder of the stored product moths discussed subsequently, and also with the tobacco moth, *Ephestia elutella* (Hübner) (Brady and Daley 1972; Phelan and Baker 1986). This compound is referred to in the literature by the chemical names (Z,E)-9,12-tetradecadien-1-ol acetate or (Z,E)-9,12- tetradecadien-1-yl acetate, the chemical shorthand Z9,E12-14:Ac, or by the acronyms ZETA or TDA (Kuwahara et al. 1971; Sower et al. 1974; Levinson and Buchelos 1981). However, unlike the other four species, raisin moth males in the study demonstrating this cross-attraction would not fly upwind in a wind tunnel toward calling females (Phelan and Baker 1986), further suggesting that this species lacks necessary pre-adaptation to reproduce in storage.

9.2.2.4 Almond Moth (*Cadra* = *Ephestia cautella*)

Distribution and Host Range

In contrast to the carob moth and the raisin moth, the almond moth is more generally abundant in association with stored products than with orchard or field habitats (Cogburn and Vick 1981). This species has been noted as a widespread pest of dates in the Eastern hemisphere (Jassim et al. 1988; Abo-El-Saad et al. 2011; Aldawood et al. 2013; Marouf et al. 2013). It is, however, more associated with dates later in harvest and post-harvest (Carpenter and Elmer 1978).

While the almond moth is considered cosmopolitan (Ryne et al. 2002), it is also considered more abundant in tropical climates and heated storage (Bowditch and Madden 1996; Olsson et al. 2005a), and in the U.K. and elsewhere this species is referred to as the tropical warehouse moth (Burges and Haskins 1965). In temperate regions, the almond moth is often associated with the confection trade (Olsson et al. 2005b; Ryne et al. 2006; Savoldelli and Süss 2010). In the U.S., the almond moth is apparently of greater abundance in warmer and more humid regions. Studies from the southeastern USA often found the almond moth in similar or greater abundance compared to the Indian meal moth. In the humid Gulf Coast region of Texas (USA), the almond moth was detected in rice storage whereas the Indian meal moth was not (Cogburn and Vick 1981). In North Central Florida, the almond moth and the Indian meal moth were found in similar abundance in trapping of a variety of types of stored products (Vick et al. 1987). Both the Indian meal moth and the almond moth were found associated with stored maize in Southeast South Carolina (USA), but the Indian meal moth was generally more abundant (Arbogast and Chini 2005).

In contrast, the almond moth is relatively rare in more arid regions to the west; e.g., California. Trapping in Central California in the early twentieth century failed to detect the almond moth in bait stations in which raisin moth, Indian meal moth, and Mediterranean flour moth were captured (Donohoe and Barnes 1934). Later in the twentieth century, workers from Central California commented that the almond moth had only appeared there recently, and its abundance was low and variable from year to year (Simmons and Nelson 1975). Subsequent trapping studies with sex pheromone lures were consistent with this conclusion (Soderstrom et al. 1987; Burks et al. 2001). Studies from the date gardens in California's more southerly Coachella Valley also imply that almond moth is not encountered with as much abundance as raisin moth or Indian meal moth (Warner et al. 1990a; Perring and Nay 2015).

The almond moth has a wide host range, like the Indian meal moth and the Mediterranean flour moth. One study found almond moth development nominally faster than Indian meal moth development when moisture content was adjusted to 13 % (LeCato 1976). Another study found that a cereal diet supported good development of both almond moth and Indian meal moth at 70 % RH, but that Indian meal moth survivorship was better than that of almond moth at 30 % RH (Bell 1975). The damage to date fruit from *C. cautella* larvae is exhibited in Fig. 9.1.

Biology and Seasonal Incidence

The almond moth ranges in length from 13–20 mm with wingspread of about 20 mm. Newly emerged adults have greyish-brown forewings with scattered darker patches. Both fore- and hind-wings have short fringes of hairs (Fig. 9.2a). The eggs

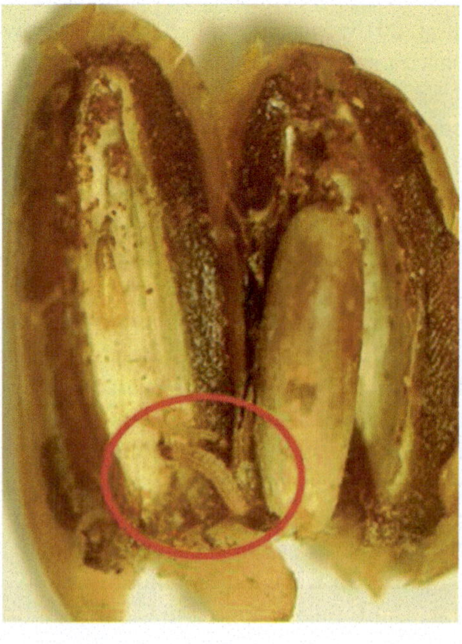

Fig. 9.1 Damage to date fruit from *Cadra cautella* larvae (Photo: Hamadttu A.F. El-Shafie)

Fig. 9.2 Almond moth, *Cadra cautella* life stages. Adult (**a**); egg (**b**); different larval instars (**c**) (size in mm); pupa after removal of silken cocoon (**d**) (Photos: Hamadttu A.F. El-Shafie)

are small round (0.3–0.4 mm) and translucent yellow with a sculptured surface (Fig. 9.2b). They are more oval than typical for Indian meal moth or Mediterranean flour moth (Arbogast et al. 1980).

The larva has white creamy color with brown or purple dots on its back and fine hairs on its body (Fig. 9.2c). The pupa is about 1 cm long with brown color and usually found in silken cocoon (Fig. 9.2d). The female lays eggs on dates bunches, especially those of late-ripening varieties. Eggs can also be laid singly or in clusters on fallen fruits, storehouses, packing houses and factories. The eggs hatch in 3–4 days, and fully-grown larvae can reach 15 mm. There are five larval instars and the larval period lasts for 3 weeks. The pupal period is about 7 days and there are four to five generations in the year depending on the prevailing environmental conditions. The pre-oviposition period is less than 1 day and adults are ready to mate and oviposit immediately after emergence (Marouf et al. 2013). The highest mean fecundity of female almond moth on date varieties was 245.3 eggs on Zahedi. Among varieties studied (Marouf et al. 2013), Deyri was the most favorable for almond moth reproduction (Marouf et al. 2013). *Cadra cautella* has greater moisture requirements than other pyralid moths at similar temperatures (Norris 1934; Bell 1975), and adult moth benefits from access to water (Hagstrum and Tomblin 1975). The almond moth and the Indian meal moth require temperatures between 15 and 35 °C to complete a generation (Bell 1975; Lum 1977), and optimum development occurs between 20 and 30 °C (Bell 1975). Pheromone trapping associated with rice storage in the Texas Gulf Coast and peanut storage in North Central Florida indicated periods of greater abundance in spring and fall, with somewhat reduced abundance during the hotter months of June and July (Cogburn and Vick 1981;

Vick et al. 1987). In grain bins in South Carolina, abundance was greater from the time of harvest in late fall until spring, when the grain in the bins heated into the supra-optimal range (Arbogast and Chini 2005).

Semiochemicals identified in the almond moth include a sex pheromone and food-related kairomones. The sex pheromone of the almond moth includes the aforementioned Z9,E12-14:Ac, in common with other stored product Phycitinae, and also Z-9-tetradecenyl acetate (*Z*9-14:Ac) (Brady 1973; Read and Haines 1976). The almond moth sex pheromone does not, however, contain (Z,E)-9,11-tetradecadien-1-ol (Z9,*E*12-14:OH) (Coffelt and Vick 1987). In contrast, the Mediterranean flour moth sex pheromone contains Z9,E12-14:OH and not (*Z9-14*:Ac) (Kuwahara and Casida 1973); and both Z9,E12-14:OH and Z9-14:Ac are part of the Indian meal moth pheromone (Zhu et al. 1999). Addition of Z9,E12-14:OH to Z9,E12-14:Ac suppresses response of the almond moth to its pheromone (Sower et al. 1974; Read and Haines 1979; Quartey and Coaker 1993). Peak times for courtship in the almond moth occur 5–6 h after dark, and again 1–3 h before daylight. Differences in secondary sex compounds, along with differences in calling time (Krasnoff et al. 1983), are thought to be mechanisms by which these species avoid hybridization. Almond moth adults of both sexes are also attracted to volatile compounds associated with chocolate (Olsson et al. 2005a, b), as is true for the Indian meal moth (Olsson et al. 2005a, b) and the Mediterranean flour moth (Olsson et al. 2006).

9.2.2.5 Indian Meal Moth (*Plodia interpunctella*)

Distribution and Host Range

The Indian meal moth is a small moth (ca. 1.3 cm) with a white and copper wing pattern. It reproduces in storage, and is probably the most widely spread and economically damaging of the stored product moths (Burks and Johnson 2012). Adults are not strong fliers. However, the close association of this species with human food stores and the excellent ability of larvae to find and exploit very small holes in packaging make this species very well adapted for anthropogenic dispersal. The Indian meal moth is a cosmopolitan generalist feeder, found on every continent but Antarctica (Mohandass et al. 2007). At lower temperatures relative humidity has little impact on adults but, at higher temperatures, longevity and fecundity are favored by higher humidity (Mbata 1985). Eggs are white to cream colored, ovoid, 0.3×0.5 mm, with a reticulated pattern (Arbogast et al. 1980). The larvae are variable in color, but generally white to cream. Indian meal moth larvae are distinguishable from related species (e.g., raisin moth, almond moth, Mediterranean flour moth) by their lack of dark-colored pinacula (spots) (Weisman 1987). They grow up to 9–13 mm. Mature larvae in diapause are quite cold tolerant (Read and Haines 1979; Carrillo et al. 2006; Wijayaratne and Fields 2012). Other stages, however, are more susceptible to cold temperature (Bell 1975), and the Indian meal moth is considered to be favored by warmer climates and at a disadvantage to other stored product moths in unheated storage in more northerly climates (Bell et al. 1979; Bell 1975, 1982).

The Indian meal moth is an external feeder that develops on a wide variety of commodities, including grains and cereal products (LeCato 1976; Madrid and Sinha 1982; Mbata 1990), tree and ground nuts (Mbata 1987; Johnson et al. 1992, 2002), dried vegetables (Na and Ryoo 2000; Perez-Mendoza and Aguilera-Peña 2004), and herbs and botanicals (Arbogast et al. 2002; Abdelghany et al. 2010). The Indian meal moth is also an important pest of dried fruit (Simmons and Nelson 1975; Johnson et al. 1995), including dates (Nay et al. 2007; Blumberg 2008). A previous study of stored product moths associated with raisins in marketing channels found that, while relative abundance of species varied with geography, the Indian meal moth was represented across intercontinental marketing channels (Soderstrom et al. 1987). The Indian meal moth has caused up to 30 % economic loss in garlic in Mexico (Perez-Mendoza and Aguilera-Peña 2004), due in part to its rapid growth potential and widespread distribution.

Biology and Seasonal Incidence

The Indian meal moth adult is short-lived, with a mean longevity of 6–8 days at 27–30 °C and 50–65 % RH (Huang and Subramanyam 2003; Arbogast 2007; Razazzian et al. 2015). The importance of watered imbibed by adults varies among stored produce moths of the subfamily Phycitinae, and the available evidence suggests that the Indian meal moth is like the Mediterranean flour moth, *Ephestia kuehniella*, which derives no benefit from access to free water (Norris 1934). The Indian meal moth is able to population maintain population growth between 18 and 34 °C (Howe 1965; Bell 1975; Lum 1977), and larval development requires approximately 40 % RH (or equivalent water activity) (Howe 1965). Laboratory and field studies indicate a flight threshold of 15 °C (Barnes and Kaloostian 1940; Cox et al. 2007).

Sensory cues guiding adult movement for mating and oviposition include visual, audio, chemosensory, and tactile. Males orient to straight edges (Levinson and Hoppe 1983) and show preference for corners of walls with the floor or ceiling (Nansen et al. 2004). Both sexes, and both mated and unmated females, are attracted by blue lights (Cowan and Gries 2009). Sound in the range of 0–80 kHz is produced by wing fanning during courtship, and may have a role in courtship (Trematerra and Pavan 1995). However, sound of this frequency range is also used by bats for echolocation, and artificially-produced sound in this range interferes with male orientation to a female pheromone source (Svensson et al. 2007). Use of such sound reduced reproduction, but did not prevent it (Huang et al. 2003). Chemical cues guiding movement include the female-produced sex pheromone attractive to males (Zhu et al. 1999), water vapor (Nansen et al. 2009), and host-related volatiles (Olsson et al. 2005a). Traps baited with water capture Indian meal moth of both sexes under warmer conditions (Nansen et al. 2009). Males and females (both mated and unmated) fly upwind to and land on laboratory diet or chocolate products, but mated females are more likely than to land on these products compared to unmated females or males (Olsson et al. 2005a).

Oviposition and infestation strategies are varied. The preferred oviposition site is among small grain-sized particles, and the presence of vegetable oils also increased oviposition (Sambaraju and Phillips 2008). In warehouse-like settings, though, eggs are distributed more broadly on and near host commodities (Silhacek et al. 2003; Nansen et al. 2006). Most larvae can find food sources 38 cm from the site of oviposition (Tsuji 2000). Neonate Indian meal moth larvae have poor mandibles but are able to enter packaging through holes as small as 0.293 mm in diameter (Tsuji 1998).

Indian meal moth development time and fecundity, and therefore the overall rate of population increase, are highly dependent on the larval diet (LeCato 1976; Johnson et al. 1992; Subramanyam and Hagstrum 1993; Na and Ryoo 2000). For example (Johnson et al. 1995), studying a strain recently isolated from dried figs, found that the number of progeny of surviving to adulthood was around 75–80 % in pistachios, walnuts, or almonds compared to a laboratory bran diet. Development time was 20–30 % longer. In comparison, the number of adults surviving on commercially processed raisins or prunes was 8–10 % that of the wheat bran diet, and developmental time was 60 % longer. A laboratory colony, also used in that study (Johnson et al. 1995), had markedly better survivorship and quicker development on wheat bran, walnuts, almonds, and pistachios, but slower development in raisins and prunes. Simmons and Nelson (1975) considered the Indian meal moth to be partially a field insect in Central California (N37° latitude), with four overlapping generations and a partial fifth generation surviving overwinter. Automated sampling nets placed outdoors in a raisin drying yard in this area revealed a study increase in abundance of adults from mid-April to September, and a decline in October (Barnes and Kaloostian 1940). Seasonal abundance of Indian meal moth associated with grain bins in South Carolina (N33°) was greatest from fall, through winter, and into spring; and decreased in summer due to excessive heat (Arbogast 2005; Arbogast and Chini 2005). In contrast, Indian meal moth associated with a mill in Kansas, USA (N41°) showed peak abundance between July and September (Campbell and Arbogast 2004), and a phenology study of dried vegetable warehouses in Korea (N36°) found Indian meal moth present from mid-April through October, and suggested a full five generations per year (Na et al. 2000).

9.2.2.6 Mediterranean Flour Moth (*Ephestia* = *Anagasta kuehniella*)

Distribution and Host Range

The Mediterranean flour moth is a cosmopolitan stored product pest, albeit more abundance in temperate climates (Jacob and Cox 1977). Despite lower tolerance for hot conditions, this species has been found in Florida, USA (Vick et al. 1987), and has been noted as a pest of dates in some date growing regions (Jemâa et al. 2012; Ghribi et al. 2012). For example, a 5-year pheromone trapping study in a date garden in Iraq captured more Mediterranean flour moth than almond moth or other related species (Ahmad and Ali 1995). While known principally as a pest of whole and milled grain, the Mediterranean flour moth also infests soy, ground nuts, tree nuts, and other stored products (Sedlacek et al. 1996), in addition to dates.

Biology and Seasonal Incidence

The adult is distinguished from others in this group (almond moth, Indian meal moth, raisin moth) by its larger size (wing span of 2–2.5 cm) and a zigzag pattern over a gray background on the forewings (Jacobs and Calvin 2001; Walker 2007a). The Mediterranean flour moth has a larval mandibular secretion that regulates oviposition, as does the Indianmeal moth (Anderson and Löfqvist 1996). This secretion increases oviposition by conspecific females in both species. In the Mediterranean flour moth, though, high larval abundance reduces oviposition, whereas there is no such effect for the Indian meal moth (Anderson and Löfqvist 1996). Mediterranean flour moth eggs are ellipsoid, 0.5×0.3 mm, and not as reticulated as those of the Indian meal moth (Arbogast et al. 1980).

Mean lifetime fecundity is in the range of 150–250 eggs per female (Cymborowski and Giebułtowicz 1976; Karalius and Buda 1995) when reared on grain. Larvae are light-colored with darker pinacula (Jacobs and Calvin 2001; Walker 2007a). There is a diapause at the end of the last larval instar, as with related moths. But the proportion of individuals entering diapause varied with geographical origin (but not latitude). Diapause was more probable when reared on flour then on an augmented laboratory diet, and more probable when maintained in complete darkness compared to longer photoperiods (Cox et al. 1984). Pupation was more synchronized when larvae were first reared under warmer long-day conditions then transferred to cool short-day conditions than when reared continuously under the latter conditions (Cox 1987). Diapause contributed to cold tolerance in this species. Also larvae were more cold-tolerant than pupae, suggesting that delaying pupation may itself have been adaptive (Cox 1987).

The Mediterranean flour moth has a longer development time than the almond moth or the Indian meal moth when reared on grain at 20–30 °C and 70 % RH (Bell 1975). At 15 °C the Mediterranean flour moth successfully completed a generation whereas the other two species did not, whereas at 30 °C the Indian meal moth and the almond moth successfully reproduced, whereas the Mediterranean flour moth was sterile (Bell 1975). The Mediterranean flour moth and the Indian meal moth were both more tolerant to low humidity compared to the almond moth (Bell 1975). In a date garden in Iraq, the Mediterranean flour moth and related moths were captured in pheromone traps from early April until mid-November (Ahmad and Ali 1995).

9.3 Management of Orchard Pests of Post-harvest Concern

Integrated Pest Management (IPM) is defined as "a comprehensive, systematic approach to commodity protection that emphasized increases information for improved decision making to reduce purchased inputs and minimize social, economic, and environmental consequences" (Hagstrum and Flinn 1996; Trematerra 2012). It is closely related to the concept of sustainability, which required that practices be economically viable, socially supportive, and ecologically sound

(Western SARE 2015). These two concepts thus provide an intellectual framework for selecting pest management tactics for protection of commodities during production, processing, and distribution; in various social, economic, and environmental circumstances. In this section we discuss options for pre-harvest and pre-processing commodity protection, and how these options can be combined with principals of IPM and sustainability in mind. For dates and other dried fruit, it is important to minimize infestation of the fruit in the orchard, and to provide a disinfestation treatment (fumigation or an alternative to fumigation) shortly after harvest to prevent further damage by pests already present in the fruit. Disinfestation is used here in the sense of killing live pests and preventing further population growth (e.g., Bell 2000); not in the sense of causing live pests to emigrate from infested commodity (e.g., Donahaye et al. 1991b). Options for management of pests during processing and storage will be discussed in a subsequent section.

9.3.1 Cultural and Mechanical Control

Physical and cultural methods for protection of dates during production include ground sanitation of orchard gardens, bunch trimming, bunch nets, and early harvest. In the USA, sanitation of the orchard floor has long been suggested for reduction of damage to nitidulid beetles (Stickney et al. 1950). More recent studies have also examined effects of trimming bunches on harvest damage from insect pests (Nay et al. 2006, 2007; Nay and Perring 2009). The purpose of this trimming is to remove immature dates that drop naturally and become stuck in tight bunches. Healthy immature dates are not susceptible to infestation in this mature stage, but decomposition of these dropped dates provides a susceptible host that allows pest populations to increase. Various covers or nets, placed over bunches, have also been used in an attempt to reduce harvest insect pest damage at harvest (Kader and Hussein 2009; Perring and Nay 2015). These covers are placed in August, as the fruit begins to mature. A 2-year study in California in the 1980s compared the effectiveness of treatments of: (1) bunch covers; (2) ground sanitation only; or (3) ground sanitation and bunch thinning with (4) an untreated control against a pest complex comprised of nitidulid beetles, raisin moth, and carob moth. During the summer of the first year of the study, there was an unusually high amount of rain (43 mm in July), whereas in the summer of the subsequent year there were unusually high temperatures and no rain. In the first summer, the sanitation treatments provided greater protection than the bunch covers against pre-harvest drop of insect- or fungus-infested fruit, and all treatments significantly reduced insect and fungal infestation at harvest. During the following dry summer, however, there was less damage and no significant difference among the treatments. Research in Israel also found bunch covers effective against nitidulids, raisin moth and the greater date moth, but indicated that damage by the greater date moth increased rather than decreased if the mesh size was too fine (Blumberg 2008). By the 2000s carob moth has increasingly become the most important part of the pest complex in California, and this pest in particular developed well on abscised (dropped) dates hanging in the bunch (Nay et al. 2006). Use of a tool to get abscised dates out of bunches reduced this problem (Nay et al. 2006, 2007), as did trimming bunches to avoid entrapment of abscised dates (Nay and Perring 2009). Early harvest has also been reported to greatly reduce damage from nitidulid beetles and pyralid moths in certain date cultivars, including Deglet Noor, Medjool, Barhee, and Hayany (Blumberg 2008).

9.3.2 Physical Control

Heat, cold, controlled atmospheres, and ionizing radiation provide alternatives to traditional fumigants (e.g., methyl bromide, phosphine, and sulfuryl fluoride) for disinfestation of dates and other commodities after harvest. Compared to cold, heat has the advantage of killing more quickly (Burks et al. 2000; Fields et al. 2012). It is also often incorporated into processing steps, in which case the disinfestation treatment does not represent additional time or expense (e.g., Johnson et al. 1996). A two-hour heat treatment provided by a pilot-scale forced air system was effective in causing both emigration and mortality of nitidulid beetles infesting Medjool dates (Finkelman et al. 2006), and a subsequent study indicated good results when this system was scaled to a higher capacity and used commercially with organic Deglet Noor dates (Rafaeli et al. 2006). A subsequent study, though, noted quality concerns with the skins when this treatment was used on Deglet Noor dates (Ducom and Ciesla 2009). Heat for disinfestation can be obtained in a variety of ways (Hansen et al. 2011), including electromagnetic energy (Marra et al. 2009). Radio frequency and related means of heating with electromagnetic energy can heat commodities very quickly and can offer differential heating, i.e., the pest is heated more than the commodity. The disadvantage of this method is the need for sophisticated and complex infrastructure. Heating with electromagnetic energy has shown potential for disinfestation of dates (Abo-El-Saad et al. 2011; Manickavasagan et al. 2013).

Low temperature treatments have also been used for disinfestation of dates, and these provide an added benefit of extension of quality (Navarro et al. 2001). Studies examining the effect of cold treatments on pests associated with dates in Israel determined that storage at 0 or −5 °C was an inefficient means of killing nitidulid beetles, but that mortality at −10 to −18 °C was sufficiently rapid that complete control would be assured shortly after the center of a 10 kg container of dates reached the target temperature (Donahaye et al. 1991a). A companion study, in the context of control of pests re-infesting in storage concluded that, for adequate control of the almond moth, red flour beetle, and sawtoothed grain beetle, treatment was required for 10 h at −10 °C, or 3 h at −18 °C (Donahaye et al. 1995). A subsequent study, however, found that 4 h at −12.5 °C was not sufficient to kill all life stages of the Mediterranean flour moth (Andreadis et al. 2012).

Modified and controlled atmosphere treatments encompass a variety of practices (Navarro 2012). Either low oxygen and/or high carbon dioxide environments are created. This may be done by continually purging a tightly sealed structure, or by allowing such an environment to form from respiration of a commodity in a sealed structure. Low pressure (vacuum) also can be used to effectively create a low oxygen environment, and treatments with high carbon dioxide can be shortened with high pressure (Navarro 2012). High carbon dioxide environments have also been used for disinfestation and quality preservation of dates (Al-Redhaiman 2005; Johnson et al. 2009). For example, storage of soft dates under >60 % carbon dioxide – sufficient to kill most stored product pests – resulted in better quality preservation than storage at −18 °C, and achieved this more economically (Navarro et al. 2001).

Treatments with ionizing radiation – also known as irradiation – use high-energy beams or particles (gamma-rays, X-rays, or electrons) generated by an isotope (e.g., cobalt-60 or cesium-137) or linear accelerators to disinfest products. They work by knocking electrons out of their orbits, creating ions and free radicals that are very harmful to living organisms (Hallman 2013). This procedure is safe and does not cause the commodity under treatment to be radioactive, but negative consumer perceptions persist (Hallman 2013). Irradiation is nonetheless accepted by 33 countries (Hallman 2013), and has been used on thousands of tons of commercial fruit in Hawaii (Follett and Weinert 2012). Irradiation treatments for date disinfestations have been reported using gamma rays (Al-Taweel et al. 1990; Zare et al. 1993) and electron beams (Al-Farisi et al. 2013).

9.3.3 Trapping and Monitoring

The lures for monitoring are available for the nitidulid beetles, and for the pyralid moths with the exception of the greater date moth. Light traps have been used to follow the phenology of greater date moth adults (Talhouk 1991).

Containers of fermenting fruit have long been used to monitor nitidulid beetles (Lindgren and Vincent 1953; Coviello et al. 2009). Lures based on the aggregation pheromone are also available (Bartelt et al. 1995). These have been used in California against a five-species complex including the driedfruit beetle, the confused sap beetle, the pineapple beetle, *Carpophilus freemani* Dobson, and the yellow nitidulid *Epuraea luteolus* (Erichson) (Bartelt et al. 1994). In stone fruit in Australia, monitoring with pheromone lures revealed complex seven identified species, including all of the California species plus *Carpophilus davidsoni* Dobson, *C. marginellus* Motschulsky, and *C. davidsoni* Dobson (James et al. 1995). The pheromone requires a fermenting food-source or similar co-attractant and the combination of the two is synergistic; i.e., around 100 times as many beetles are captured compared to either food or pheromone alone (Bartelt et al. 1994). In California, the yellow nitidulid responded only to the co-attractant and not to pheromone (Bartelt et al. 1994). In figs in Central California, response to the food-pheromone combination shows greater numbers than response to the food bait alone, but shows the same phenological trends (Charles S. Burks, unpublished data). Both for the food bait and the food-pheromone combination, the number of beetles caught in traps dropped as the fruit ripens and becomes more attractive (Stickney et al. 1950; Bartelt et al. 1994). These observations suggest that the food bait alone is more practical for monitoring, although the food-pheromone source have been used effectively for mass trapping or attract-and-kill in Australian stone fruit (discussed below).

The study that first elucidated the pheromone for the carob moth found that the complex blend with a trienal major component did not capture as many males as unmated females when used in lures in pheromone traps (Baker et al. 1991). A formate analogue of the major component was tested in an attempt to improve stability. The formate analogue captured more males than the optimized pheromone blend, but still fewer than calling females (Todd et al. 1992). In California, a binomial sampling plan was developed, involving inspection of 150 brown abscised fruit at the time of June drop (Park and Perring 2010). Precision was examined at action thresholds of 7–15 % damage, with the objective of permitting action to be taken against immatures of the generation that would infest marketable fruit. In Iran, some pheromone trapping is done using unmated females as a pheromone source (Mortazavi et al. 2015). In Australia, however, preliminary monitoring guidelines have been developed for the almond industry there based on the pheromone lures (Madge et al. 2012).

Pheromone lures have been used to monitor the raisin moth in date gardens in Israel (Kehat et al. 1992). The number of males captured in pheromone traps correlated well with harvest damage in the range of 13–39 %. It was concluded that, in this circumstance, the pheromone lure provided a useful tool for treatment decisions and treatment timing (Kehat et al. 1992).

9.3.4 Biological Control

The importance of natural control for control of pests of date fruit has been implied or demonstrated in some situations and, in a few cases, either classical or augmentative biological control has proved useful. One study demonstrated the importance of certain ant species in control of the carob moth in California (Nay and Perring 2005). Several parasitoid wasps have consistently been found associated with phyticine moth pest; most consistently, the larval ectoparasitoid *Habrobracon hebetor* (Say) (Donohoe et al. 1949; Simmons and Nelson 1975; Johnson et al. 2000; Blumberg 2008; Hameed et al. 2010). This parasitoid attacks fully mature last instar larvae. It is active at temperatures in which the larvae on which it preys are quiescent. It parasitizes these larvae in winter, although it does not oviposit on them until spring (Donohoe et al. 1949; Simmons and Nelson 1975). In vineyards in Central California, *H. hebetor* was able to sting raisin moth larvae overwintering under the bark in grapes, but not those overwintering under the soil (Donohoe et al. 1949). Despite these limitations, *H. hebetor* and other parasitoid wasps were hypothesized to be responsible for moderating initially very high abundance of raisin moth after its introduction to California early in the twentieth century (Donohoe et al. 1949). A polyembryonic egg-larval parasitoid, *Copidasomopsis* (= *Pentalitomastix*) *plethoricus* (Caltagirone) discovered on the navel orangeworm *Amyelois transitella* (Walker) was imported into Israel for control of the carob moth in almonds and carobs. It established on the former crop, but not the latter (Gothilf 1978). A larval ectoparasitoid, *Goniozus legneri* Gordh, was discovered on carob moth and navel orangeworm in Uruguay (Gordh 1982; Legner and Silveira-Guido 1983). The biology of another larval parasitoid of carob moth, *Apanteles myeloenta* Wilkinson, was characterized in a laboratory study in Iran (Kishani-Farahani et al. 2012a). This was among the numerous parasitoids found associated with the carob moth identified during a survey conducted in three regions of Iran (Kishani-Farahani et al. 2012b). *Habrobracon hebetor* and *A. myeloenta* were also the parasitoids most frequently associated with overwintering carob moth in Iran (Kishani-Farahani et al. 2012b). *Trichogramma* spp. (Hymenoptera: Trichogrammatidae) can be reared economically and are simple to release. They have therefore often been used in inundative releases (Brower et al. 1996). In Egypt, such releases reduced infestation by the spring generation of the greater date moth (Gameel et al. 2015).

Other than insects, biological control agents also include nematodes, fungi, bacteria, and viruses (Flinn and Schöller 2012). In Israel, trials with entomopathogenic nematodes of the genus *Heterorhabditis* had success in a fig orchard, but not in a date garden (Glazer et al. 2007). Spinosyns, a insecticides obtained or derived from the bacteria actinomycete bacterium *Saccharopolyspora spinosa* (Copping and Duke 2007; Hertlein et al. 2011; Sparks et al. 2012), are currently formulated and marketed in the same manner as residual contact insecticides, and will be considered with them; as will insecticide formulations derived from the bacteria *Bacillus thuringiensis* (*Bt*) (Flinn and Schöller 2012).

9.3.5 Reproductive Control

Pheromones have been used for control of nitidulid beetles in stone fruit, and for carob moth in figs. In Australia, a series of studies developed methods using diversion and attract-and-kill (James et al. 2001; Hossain et al. 2006, 2010). Initial attempts at mating disruption of carob moth in dates in California using the formate pheromone analogue yielded modest results (Vetter et al. 2006). Development of this method nonetheless continued, and mating disruption for carob moth remains an established practice in some date gardens in California (Perring et al. 2014; Perring and Nay 2015). Mating disruption of the stored product Phycitinae (almond-moth, Indian meal moth, and Mediterranean flour moth) has received much research and is well established in processing and storage (Trematerra 2012), but not in the field. There is, however, a report that mating disruption of raisin moth was accomplished in Israel using hand-applied reservoir dispensers sold for beet armyworm (which also has ZETA as the principal component of its sex pheromone) (Blumberg 2008).

9.3.6 Chemical Control

Chemical control for insect pests that infest date fruit prior to harvest include residual insecticides used in the field, and fumigants that are often used for disinfestation instead of one of the alternatives discussed in the previous section.

Residual insecticides used to control pests of date fruit in the past have often been organophosphates. In Israel, two applications a year were recommended for control of nitidulid beetles; one mid-summer and one 3–4 weeks before harvest. Malathion was used for this purpose, before being replaced with azinphosmethyl (Blumberg 2008). A similar regime was suggested when raisin moth or carob moth were the primary targets, but with the first application timed to the point where the fruit changes color (Blumberg 2008). A study of chemical control of the carob moth in the USA recommended applications of 5 % malathion dust every 2 weeks after the fruit turned color, with a total of three such applications (Warner et al. 1990a). This study also noted that this treatment regime might be inadequate in taller trees where coverage was problematic, or in wetter years when there was greater pest pressure from nitidulid beetles (Warner et al. 1990a). The practice continued through the 1990s into the 2000s (Nay et al. 2006; Nay and Perring 2006, 2008). A study of residues on dates in Saudi Arabia indicates use of the organochlorines (lindane and dieldrin), and the organophosphates (dimethoate and chlorpyrifos) (El-Saeid and Al-Dosari 2010).

More recently, various factors have resulted in use of a broader range of insecticides for management of date pest fruit. In Israel, the raisin moth showed organophosphate resistance by the 1990s, and pyrethroids (cypermethrin, cyhalothrin) were substituted for control of that species (Blumberg 2008). Trials also examined insect growth regulators, additional pyrethroids (lambda-cyhalothrin and bifenthrin) and neonicotinoids for control of nitidulid beetles (Blumberg 2008). In the U.S., use of malathion for maturing dates was displaced by spinetoram (a semi-synthetic version of compound similar to spinosad, and also from *Saccharopolyspora spinosa*) and methoxyfenozide (a molt-accelerator with greater specificity for Lepidoptera) due to environmental concerns and the availability of these newer chemistries (Perring and Nay 2015). In Egypt, previously used conventional insecticides were banned for use on dates by 2010, and use of biorational or reduced-risk replacements was mandated (Lysandrou et al. 2010). Trials in Egypt examined control of raisin moth using spinetoram in the first week of July, followed by application of methoxyfenozide in the third week of July. This combination significantly reduced damage to Saidi dates at the time of harvest in mid-September (Lysandrou et al. 2010). These Egyptian workers commented on lack of efficacy in previous trials with *Bt* (Lysandrou et al. 2010), and previous studies with *Bt* as a grain protectant found that the Indian meal moth developed resistance rapidly (McGaughey 1985), and that this resistance was stable after exposure to *Bt* was removed (Oppert et al. 2000). These observations suggest caution in using *Bt* on control of other Phycitinae such as the almond moth and the raisin moth. However, another study in Egypt found that winter and spring applications of *Bt* and the fungal product *Beauveria bassiana* (Balsamo) Vuillemin, targeted against overwintering greater date moth, reduced infestation by the first generation (Imam 2012).

Methyl bromide has been the fumigant of choice for disinfestation of dates (Kader and Hussein 2009; Finkelman et al. 2010; Bessi et al. 2015) because it is efficacious and rapid against a wide variety of target pests (Phillips et al. 2012). However, it has been phased out in many places due to environmental restrictions by the Montreal Protocol. In the U.S., the most common alternatives to methyl bromide are phosphine and sulfuryl fluoride (Phillips et al. 2012). These are seen as less favorable due to a longer treatment time (Kader and Hussein 2009), although the treatment time for phosphine can be reduced somewhat by using cylinderized formulations rather than the more common pellets that break down with contact to moisture in the air to generate phosphine gas (Phillips et al. 2012). Different fumigants have different effects on the target pest. For example, there is a linear relationship between concentration and time for methyl bromide and sulfuryl fluoride, so increased concentration reduces the time required for acceptable mortality. This model does not, however, hold for phosphine (Phillips 1998). For that fumigant, maintaining sufficient treatment time will be important for management and prevention of resistance (Opit et al. 2012). Ethyl formate has been registered as a fumigant in Australia and Israel (Finkelman et al. 2010), and has shown promise for disinfesting nitidulid beetles (Finkelman et al. 2010) and carob moth (Bessi et al. 2015) in dates.

Botanically-derived essential oils can be important parts of a push-pull strategy of pest management (Cook et al. 2007). A study in Tunisia examined composition of essential oils of rosemary, thyme and Aleppo pine, and their fumigant and contact toxicity toward carob moth (Amri and Hamrouni 2014).

9.3.7 Integrated Pest Management

Elements of a year-round program of integrated management to control post-harvest pests in date gardens (principally nitidulid beetles, raisin moth, and carob moth) might include: (1) Orchard winter sanitation (floor and crown) to deny overwintering sanctuaries (particularly of carob moth, which does not leave fruit); (2) sanitation of the orchard floor in spring and summer (to reduce nitidulid beetles in wetter years); (3) sanitation of bunches (to reduce summer carry-over of principal pests during the summer); (4) bunch nets to provide a physical barrier as maturing dates in the bunches become more susceptible to infestation; (5) treatment during the maturation period with reduced-risk insecticide; and (6) a fumigation or other disinfestation treatment as soon as possible after harvest to prevent additional crop loss and pest population growth. Local circumstances will dictate economic and practical trade-offs. For example, physical and cultural controls will be particularly important to organic producers, who have fewer insecticide and fumigant options. Relative costs of labor and capital in the growing location also dictate choices among the options outlined above, as do regulations and cultural values (e.g., in killing post-harvest pest vs. treatments causing emigration out of infested fruit).

9.4 Pre-dominantly Post-harvest Insect Pests Infesting Dates

All of the pests listed in this section are beetles with a cosmopolitan distribution. Also, they all fit the generalizations made about the nitidulid beetles in the previous section; i.e., they live long and feed as adults and have male-produced aggregation pheromone (Oehlschlager et al. 1988; Phillips et al. 1993). Propensity for flight varies among these species, but much of the time these beetles walk instead of flying. A previous summary classifying stored product insect species by temperature and humidity categorized all as needing high temperatures for optimal population growth (Howe 1965). All of these beetles were categorized as tolerant of low relative humidity. The red flour beetle was considered cold susceptible, whereas the sawtoothed grain beetle, confused flour beetle, and rusty grain beetle were considered cold hardy (Howe 1965). Illustrated keys for these and other stored product insect pests can be obtained online (Bousquet 1990; Kingsolver 1991), and a previous paper summarizes methods for distinguishing the sex of adult stored products beetles (Halstead 1963).

9.4.1 Sawtoothed Grain Beetle (Oryzaephilus surinamensis)

9.4.1.1 Distribution and Host Range

The sawtoothed grain beetle is a cosmopolitan and polyphagous insect feeding on a wide range of stored products, including stored grain, cereal products, dried fruit, and oilseeds (Back and Cotton 1926; Simmons and Nelson 1975; Rees 1996; Mahroof and Hagstrum 2012; USDA-FGIS, 2015). It is common in tropical areas, but also an important pest of grain and processed grain products in temperate areas (Howe 1956a; Rees 1996). The sawtoothed grain beetles occur in all date-growing areas of the Old World. It is typically a packinghouse and storage insect, and is considered an important storage pest of dates and other dried fruit (Fraenkel and Blewett 1943; Curtis and Clark 1974; Simmons and Nelson 1975; Khairi et al. 2010). It is, however, reported to breed under the bark of trees in tropical and subtropical areas (Linsley 1944), and has been observed under the bark of fig trees in Central California. It prefers fairly dry foods (Simmons and Nelson 1975). It is rarely found in freshly harvested dates (Hussain 1974) or raisins (Simmons and Nelson 1975), but develops high abundance in these dried fruit after long storage. Dates with low moisture content, with the calyx removed, or that are broken or have mechanical damage are likely to be infested by the sawtoothed grain beetle. Adults and larvae feed on dates by making tunnels between the external fruit skin and flesh. In severe infestation, the beetle consumes all date fruit contents leaving the skin or exocarp intact (Fig. 9.3). This is a characteristic damage symptom of sawtoothed grain beetle (Al-Hafidh et al. 1987).

9.4.1.2 Biology and Seasonal Incidence

This beetle is flattened, reddish-brown, and about 2.5–3.5 mm long (USDA-FGIS, 2015). It has well-developed wings, but rarely flies (Cox and Dolder 1995). A rotary net operated continuously at a raisin yard in Fresno, California captured some sawtoothed grain beetle, but far fewer compared to the red flour beetle and rusty grain

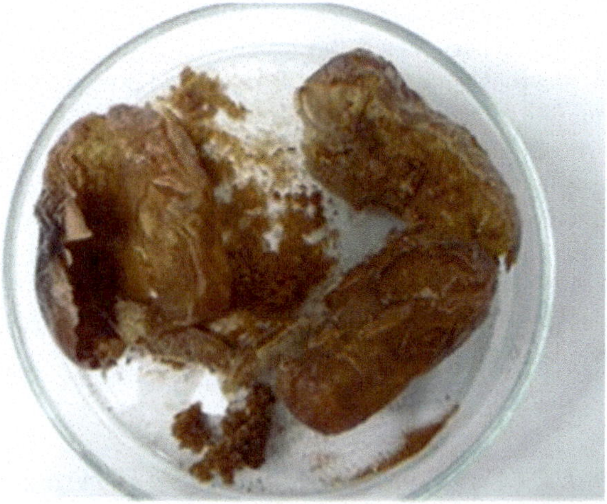

Fig. 9.3 Date fruit damaged by the sawtoothed grain beetle, *Oryzaephilus surinamensis* (Photo: Hamadttu A.F. El-Shafie)

beetle (Barnes and Kaloostian 1940). The genus can be easily identified by six saw-toothed like projections on the sides of the adult thorax (Fig. 9.4a). The males have pointed out posterior femora, each armed with a tooth, which serves as an important character for distinguishing them from females, the femora of which are unarmed (Back and Cotton 1926). Eggs are laid singly, in small clusters and are usually put in hidden places in some crevices in food supply, or laid loosely in finely ground food such as flour or meal. The female has an elaborate egg laying behavior, consisting of maneuvering around a broken grain seeking a crevice in which to deposit eggs. The female usually thrust her ovipositor in small crevice and tuck away egg in it; the whole process may take 3 min. The pre-oviposition period is 5–7 days after emergence. The oviposition period is quite long ranging from about 2–5 months. The total number of eggs laid by individual females ranged from 45 to 285 eggs. The egg is white, shiny and elongate-oval in form, it measures 0.83–0.88 mm in length and 0.25 mm in width (Fig. 9.4b). The young embryo when fully developed nearly fills the eggshell, which becomes wrinkled and undulated with the movement of the embryo. The egg usually hatches in 3–5 days depending on temperature. When first hatched, the larva is about 0.80–0.90 mm in length, and the width of head capsule is 0.24 mm (Back and Cotton 1926). The fully-grown larva attains a length of 2.5–2.8 mm with a head capsule ranged from 0.46–0.54 mm (Fig. 9.4c). The larva is active and free living and moves frequently while feeding (Back and Cotton 1926). The larval period can be 12 days or 2 weeks in midsummer, however, it can last for 4–7 weeks during spring. The larva molts three to four times before pupation depending on temperature (Back and Cotton 1926). The pupa is formed in a crude pupal cell or cocoon made of particles of seed or other foodstuffs. The larva attaches itself by the anal end to some solid objects. The pre-pupal period ranges from 1 to 7 days and the pupal period from 6 to 12 days. Pupation can also occur without cocoon, just between food particles. The pupa has the same six projections on each side of the thorax (Fig. 9.4d).

Fig. 9.4 Sawtoothed grain beetle, *Oryzaephilus surinamensis* life stages; adult (**a**); egg (**b**); larva (**c**); pupa (**d**) (Photos: Hamadttu A.F. El-Shafie)

The life cycle from egg to adult of the sawtoothed grain beetle may range from 27–315 days (Simmons and Nelson 1975). It is capable of completing development between 17.5 and 37.5 °C, with fastest development at 32.5–35 °C and 90 % RH (Howe 1956a). The beetle may have 4–5 generation annually in temperate zones while breeding continues in subtropical and tropical climates (Back and Cotton 1926). Most developmental studies focused on grain and the sawtoothed grain beetle develops somewhat more slowly on dried fruit (Fraenkel and Blewett 1943; Curtis and Clark 1974), but may be aided by fungal growth on the host product (Sinha 1965).

The merchant grain beetle, *Oryzaephilus mercator* (Fauvel) is closely related to the sawtoothed grain beetle. It is difficult to separate these two beetles morphologically; however, under magnification, the sawtoothed grain beetle has exposed eyes, whereas the merchant grain beetle has the eyes more protected by small knobs behind the eyes (Kingsolver 1991). The merchant grain beetle has a slightly lower optimum temperature (31–34 °C), but is also less chill tolerant and less tolerant of low humidity (Howe 1965). The sawtoothed grain beetle usually passes the winter in the adult stage, breeding normally ceases late in the fall and commences again in the spring (Back and Cotton 1926).

9.4.2 Flour Beetles (Tribolium spp.)

9.4.2.1 Distribution and Host Range

The flour beetles, red flour beetle *Tribolium castaneum* (Herbst) and confused flour beetle *T. confusum* (Jacquelin du Val) are cosmopolitan in nature (Good 1936). The red flour beetle is most common in tropical to warm-temperate areas, whereas the confused flour beetle is uncommon in tropical areas but widespread in temperate areas (Good 1936; Rees 1996). Both are polyphagous general feeders and infest variety of food including grain cereals, flour, pasta, crackers, cake mix, meal, beans, dried pet food, chocolate, dried flowers, spices, nuts, seeds, dried museum specimens (Simmons and Nelson 1975; Rees 1996) and stored dates (Al-Zadjali et al. 2006; Kader and Hussein 2009; Hassan et al. 2013). Experiments with different types of grains and flours characterized interactions of these two species as mutual predation. It was concluded that the two species would not coexist indefinitely, and, for a given temperature, which species remained would depend upon the nutritional resources (Sokoloff and Lerner 1967). In a processing environment, that probably means that the two species can partition resources in a facility at lower abundance, but will not co-exist in higher abundance (Rees 1996). For both species, heavy feeding leaves chemicals (benzoquinones) with a disagreeable odor which is not removed by cooking (Hodges et al. 1996).

9.4.2.2 Biology and Seasonal Incidence

Both species have same size and appearance along with the similar life histories and habits, but some differences in their morphology and biology are presented in Table 9.3.

The adults of both species are elongated, oval in shape, small, shiny, reddish brown and measure about 3.5 mm in size. During 5–8 month viable period of life, a single female deposits about 300–400 white, microscopic and sticky eggs in or on the substrate. The incubation period may last for 5–12 days after which small, creamy yellow or brownish white elateriform larvae hatch out. The larvae typically go through six to seven instars depending upon the food source and prevailing environmental conditions (Mahroof and Hagstrum 2012). The range of temperature permitting complete development is 22–40 °C for the red flour beetle, with an optimum temperature range of 32–35 °C (Howe 1956b). For the confused flour beetle the

Table 9.3 Differences between the red flour beetle and the confused flour beetle

Red flour beetle, *T. castaneum*	Confused flour beetle, *T. confusum*
Originated from Asian subcontinent and Australia and found throughout tropical and warmer temperate regions of the world	Originated from Africa and found in higher latitudes and cooler areas throughout the world
Last three segments of antennae club shaped	Last four segments of antennae gradually clubbed towards the apex
Head is without beak and visible from above, and side of the thorax are slightly curved	Sides of the thorax are more parallel
The eyes are separated by 33 % of head width	Eyes are separated by 50 % of head width
Capable of flight; fly under certain conditions (evening or changing weather)	Not known to fly
No ridge is present at head capsule	Sides of head capsule are notched at the eyes forming a ridge

Simmons and Nelson (1975), Bousquet (1990), Baldwin and Fasulo (2010), and USDA-FGIS (2015)

permissible temperature range is 21–37.5 and the optimum temperature range is 30–33 °C (Howe 1960, 1962). Both species survive low humidity conditions, but reproduce better under high humidity (Howe 1965).

The mature larvae transform into white pupae which transform into the adult beetle after approximately 8 days. The period of egg to adult emergence may vary from 40 to 90 days, and is shorter in *T. castaneum* than in *T. confusum.* Adults live 130–200 days at 18–29 °C (Mahroof and Hagstrum 2012; USDA-FGIS, 2015). A study on the temporal distribution of coleopteran beetles including *Tribolium* spp. in a flour mill in central Italy over 1 year revealed that the greatest abundance was found during July and August (Trematerra and Sciarretta 2004). The rotary flight net study in the Fresno, California raisin yard captured moderately high numbers of red flour beetle in flight, but only from August to October (peak captures in September), and only near sunset (Barnes and Kaloostian 1940).

9.4.3 Rusty Grain Beetle (*Cryptolestes ferrugineus*)

9.4.3.1 Distribution and Host Range

The rusty grain beetle is a cosmopolitan stored product pest (Howe and Lefkovitch 1957; Bishop 1959) assumed to have evolved from a habitat of the underside the bark of shrubs and trees (Suresh et al. 2001) to stored commodities like rice, corn, copra, sorghum, barley, coffee berries, cassava root, cotton seed, wheat, nutmeg, flour, peanut, oilseeds, dried fruits and stored dates (Simmons and Nelson 1975; Suresh et al. 2001; Mahroof and Hagstrum 2012).

Despite the presence of the rusty grain beetle in many tropical climates (Bishop 1959) and its ability to thrive in high heat (Howe 1965), it also is tolerant of cold and does will in the protected environment of grain bins in colder temperate regions. For instance, in Canada it is the most serious pest of grain in farm bin storage (Bousquet 1990). It benefits from the presence of fungus (Sinha 1965) and is common in raisins that have been damaged by rain and become moldy. Raisins are infested both in the field and in storage (Simmons and Nelson 1975).

9.4.3.2 Biology and Seasonal Incidence

The rusty grain beetle is a small light to dark reddish brown beetle measuring 1.5–2.0 mm in length, with hair-like antennae. A single female can lay 200–500 white eggs during their life span (Mason and McDonough 2012) 130–180 days at 21–30 °C (Bishop 1959). Eggs hatch in 3–5 days at 30 °C and >65 % RH (Bishop 1959). The larvae measure 3 mm in length, and are cylindrical, cigar shaped with yellowish white body having black head. Two reddish brown spots are present at its anal segment. The larval development completes in four instars within 32–37 days (Bishop 1959).

Before going into pupation, the last larval instar yields a minute quantity of silk for cocoon formation that is fastened with the host commodity. Five days after pupation, adult emerge out from the cocoon (Mason and McDonough 2012). The complete development from egg to adult requires a temperature range of 20–40 °C (Bishop 1959; Mason and McDonough 2012).

The rusty grain beetle is a strong flyer (Mason and McDonough 2012). A laboratory study of general biology revealed a tendency to fly up and out of their medium at temperatures between 27 and 38 °C (Bishop 1959). This observation was later confirmed in experiments in a purpose-built flight chamber (Cox and Dolder 1995). In the field, rusty grain beetle was captured in the rotary net in June through August. Like the red flour beetle, it flew primarily at dusk (Barnes and Kaloostian 1940). In Kansas, traps in the tops of grain bins indicated inward movement of the rusty grain beetle from the time the wheat was harvested and binned in early July until the grain and the ambient temperature cooled in September, some 60–80 days later (Hagstrum 2000).

9.5 Management of Date Pests in Processing and Storage

Compared to the orchard, management of pests in a post-harvest environment takes place in the context of different standards and constraints. There is a much lower tolerance for pests, and retaining biological control services from an existing natural ecosystem is not relevant. Maintaining sanitation and commodity quality and excluding pests became the first lines of defense against reinfestation.

9.5.1 Mechanical Control

The post-harvest equivalent of cultural control is good facility design and management (Heaps 2012). A Master Sanitation Plan should be used, with written checklists to ensure and document regular and timely sanitation (Heaps 2012).

Best practices include a number of elements outside and inside the facility (Heaps 2012). Examples on the exterior of the plant include keeping trash areas clean. The interiors of cleaned dumpsters should be treated with insecticide to keep flies down during the summer. Landscape plants with flowers or fruit that might attract pests should be minimized, as should hedges or other plantings that might provide a harborage to rodents and other pests. Outside lighting directly on building exteriors or above doors and any necessary security lights at least 5 m from exterior doors should be avoided. The entrance, exit, and dock doors should not open directly into manufacturing areas. Doors should be tightly sealed along the bottom.

Inside the plant, these best practices include sealing wall openings (Mullen and Pedersen 2000), and using insect light and pheromone traps should be placed by doors. Warehouses are an area of particular pest vulnerability; monitoring and control efforts should take this into account (Heaps 2012). Windows are an area of vulnerability; there should not be more windows than necessary. If windows are necessary for ventilation, then they should have screens that are sufficiently fine and well repaired in order to exclude insects. False ceilings should be avoided because they are hard to clean, and pests can congregate there. Equipment and electrical conduits should be planned and placed to avoid inaccessible places, and clean motor regularly to avoid harborages for stored product insects (Heaps 2012).

Application of best management practices for dates have been described succinctly in an online publication of the International Center for Agricultural Research in Dry Areas (ICARDA) (Kader and Hussein 2009). Sometimes processing steps can also provide disinfestation, as was shown for the navel orangeworm on pistachios in California (Johnson et al. 1996). In dates, steps such as freezing to accelerate ripening, solar heating for drying, or pasteurization (Kader and Hussein 2009) might achieve such disinfestation. More generally, given the previous description of post-harvest beetle pests as opportunistic fungivores seeking broken and infested commodity, the benefit of best storage practices (Kader and Hussein 2009) for preventing insect infestation of stored dates should be evident.

9.5.2 Physical Control

Heating for disinfestation of insect pests from inaccessible parts of plants is an established practice in the U.S. milling industry (Burks et al. 2000; Fields et al. 2012). A typical heat treatment last 24–36 h, with a heating rate of 3–5 °C per hour, the target temperature of 50 °C, and a total treatment time of 24–36 h. The two basic approaches are recirculation of air using heaters inside the facility, or using forced air to circulate heat from heaters outside the facility. The recirculation approach is more energy-efficient but results in more difficulty in obtaining even heating (particularly vertically), whereas the forced air approach has the converse advantage and disadvantage (Fields et al. 2012).

9.5.3 *Trapping and Monitoring*

Most orchard-based Integrated Pest Management which involve explicit action thresholds and implicit acceptable populations. In contrast, IPM in the food industry has a goal of zero pest abundance (Campbell et al. 2012), and the trade-off is between the expense of the detection effort to support this goal and the financial risk of not making sufficient effort. Monitoring programs in food facilities attempt to provide detection in time and space (Campbell et al. 2012). In space, tools such as contour analysis can be used to determine where within a facility a problem is occurring. In time, trapping records are compared to previous trend to determine when there is an increased problem or (in the case of traps near doors or intake areas) increased pressure.

Different types of traps are used for the pests discussed in this chapter. The almond moth, Indian meal moth, and Mediterranean flour moth all occur in processing and storage areas, and (in the absence of mating disruption) are efficiently detected at some distance by sticky traps with ZETA lures (Toews and Nansen 2012; Trematerra 2012). In contrast, the sawtoothed grain beetle, flour beetles, and rusty grain beetle are typically monitored with floor traps containing food bait and, in some cases, a pheromone lure (Toews and Nansen 2012). These traps attract at a more local distance (Toews and Nansen 2012), and share the problem of being outcompeted by competing food supplies as previously described for nitidulid aggregation pheromones. The latter problem illustrates the importance of sanitation for effective monitoring for these pests.

The more local natural captures in beetle traps can be an advantage for spatial detection (Toews and Nansen 2012). In monitoring, detection can be used in a variety of ways. Often the most appropriate immediate step is sanitation to remove a previously undetected patch of waste material serving as refuge, or a lot of infested commodity; or perhaps sealing a previously undetected entrance route from outside the facility. Other possible responses include the biorational and chemical responses described below.

9.5.4 *Biological Control*

Parasitoids that have been used for control of stored product moths (almond moth, Indian meal moth, and Mediterranean flour moth) include *Trichogramma evanescens* (Westwood) (Brower 1983), *Habrobracon hebetor* (Ghimire and Phillips

2014), and *Venturia canescens* (Gravenhorst) (Brower et al. 1996). *Trichogramma evanescens* is a polyphagous egg parasite (Brower et al. 1996). *Trichogramma* spp. can be reared economically and are simple to release. They have therefore often been used in inundative releases (Brower et al. 1996), and in Germany, have been used for over a decade for commercial control of Indian meal moth and other stored product moths (Adarkwah and Schöller 2012). *Habrobracon hebetor* is a larval ectoparasite that has also been used for control of stored product pests on a commercial scale (Schöller and Flinn 2000).

Trials of *H. hebetor* in combination with another *Trichogramma* spp., *T. deion*, found that *H. hebetor* suppressed Indian meal moth in both bulk and bagged cornmeal (Grieshop et al. 2006). Releases with both *H. hebetor* and *T. deion* resulted in further suppression in bagged commodity, but not in bulk commodity (Grieshop et al. 2006). It was concluded that dual release of both the larval and egg parasitoids would be useful in situations where sanitation difficulties created refugia (Grieshop et al. 2006). *Venturia canescens* is a polyphagous and widely distributed larval endoparasite of Indian meal moth and related stored product moths. It has been widely used as a model organism in basic studies, but less frequently used as a biological control agent. Earlier studies indicated that *H. hebetor* attacks host larvae previously parasitized by *V. canescens*, killing the latter (Brower et al. 1996). A more recent study, however, suggested that, for inundative release, these two species together could be more effective than either alone (Adarkwah and Schöller 2012).

9.5.5 Reproductive Control

Mating disruption for almond moth, Indian meal moth, and Mediterranean flour moth inside commercial processing and storage facilities is a well-established commercial practice. Research has compared mating disruption, mass trapping, and attract-and-kill (also known as attracticide). Mating disruption can work by various mechanisms, depending on the species and the formulation involved (Cardé and Minks 1995; Miller et al. 2006). Aerosol devices releasing relatively large amounts of pheromone in pulses are used increasingly commonly for mating disruption for orchard pests (McGhee and Gut 2014; Reinke et al. 2014). Such aerosol formulations have been tested experimentally for Indian male moth (Burks et al. 2011), but commercially available mating disruption formulations involve passive release of pheromone from a greater number of point sources (Burks and Kuenen 2012), similar to the formulations that have historically been more common for use against other moth pests in orchards.

Various studies have shown evidence for control of Indian meal moth and other stored product Phycitinae using mating disruption (Ryne et al. 2007; Burks et al. 2011; Trematerra et al. 2011), mass trapping (Trematerra and Gentile 2010), and attract-and-kill (Trematerra and Capizzi 1991; Nansen and Phillips 2004). One study compared these methods in small, commercial facilities in Dallas, Texas (USA). Treatments were assessed by communication disruption (i.e., suppression of the number of males captured by monitoring traps with pheromone) and reduction of fertility (as determined by the number of larvae recovered from food cups in the experimental facilities) (Campos and Phillips 2014). Mating disruption was provided by using a density of lures equal to that used in conjunction with the mass trapping and attract-and-kill formulations. Compared to untreated facilities, all three pheromonal control methods (mating disruption, mass trapping, attract-and-kill) resulted in a reduction of sexual communication and fertility (Campos and Phillips 2014). There were, however, no significant or consistent differences in the males captured or larvae recovered between the pheromonal control methods. This latter observation suggests that the mating disruption effect provided by the lures associated with the mass trapping and attract-and-kill formulations were sufficient to reduce fertility. Under the circumstances of the study, trapping or intoxicating males responding to these lures provided no practical additional benefit.

9.5.6 Chemical Control

Chemical control, as it pertains to food processing and storage facilities, includes: (1) general surface treatments with residual insecticides, (2) insecticide aerosol treatments, and (3) structural fumigation (Arthur and Subramanyam 2012; Phillips et al. 2012). Insecticides used for surface treatments in the U.S. include the pyrethroid cyfluthrin, the neonicotinoid chlorfenapyr, and the growth regulators hydroprene and pyriproxyfen. The presence of food or other waste reduces the effectiveness of these surface treatments (Arthur and Subramanyam 2012).

Some insecticides can also be applied as 5- to 50-μm aerosols, which are applied by automated systems. These give wide coverage, but do not penetrate like fumigants. Insecticides applied in this manner include pyrethrins and methoprene (Arthur and Subramanyam 2012).

Structural fumigation, like heat treatment, provides a way to disinfest very inaccessible parts of plants (Phillips et al. 2012). As with heat treatments, the plant must be shut down and carefully prepared. For this purpose, phosphine has the disadvantage of reacting with and damaging copper and integrated circuits. In the past methyl bromide was more often used for structural fumigation. Currently in the USA, sulfuryl fluoride is frequently used for structural fumigation (Phillips et al. 2012).

9.5.7 Integrated Pest Management

As previously indicated, IPM for food plants uses detection for the goal of maintaining a zero-abundance of insect pests inside of food processing and storage plants. Elements of such an IPM plan include: (1) an appropriately designed and well maintained facilities for storage and process; (2) a Master Sanitation Plan to provide attention to and documentation of sanitation in order to minimize ability of secondary pests to become established; (3) maintenance of a monitoring program based in the context of good manufacturing practices; and (4) implementation of appropriate response if monitoring indicates problem. These responses might include removing infested commodity; commodity re-treatment; or facility treatment by heat, surface treatment, aerosol, or structural fumigation. Inundative biological control and mating disruption or other pheromone control techniques can provide additional layers of protection to assure a minimal pest population within the plant.

9.6 Future Research

A pest complex of nitidulid beetles and pyralid moths continues to pose a formidable challenge to production of high quality date fruit in many parts of the world, ongoing advances in pest management research and development notwithstanding. Additional post-harvest losses to key stored product moths and beetle pests speak to challenges in date processing and storage. Future research directions for control of orchard pests of date fruit might include improvement of semiochemical management of nitidulid beetles and carob moth. The diversion and attract-and-kill system used in Australia, as described previously, might be adapted and used as part of a push-pull system (Cook et al. 2007) in date gardens. Obtaining a more attractive carob moth pheromone is still to be desired. Alternatively, if this goal continues to be elusive, examining ways to obtain non-competitive mating disruption (Miller and Gut 2015) with known components or analogues might prove useful. Finally, continuing losses to stored product pests, particularly beetles, speaks to a need for date storage and technologies that are both effective and useable.

9.7 Summary

A variety of moth and beetle pests infest date fruit before and after harvest. The nitidulid beetles, carob moth, and the raisin moth are the pre-harvest date pests across wide geographical areas of the world. However, the almond moth, Indian meal moth, and Mediterranean flour moth may infest dates before as well as after harvest. The fruit infestations in the orchard can be reduced with an integrated management program including sanitation of the orchard floor and bunches, use of bunch protectors, well-timed insecticide treatments with reduced risk compounds, and prompt disinfestation following harvest. After harvest, dates during processing and storage remain at risk from the almond moth, Indian meal moth, and Mediterranean flour moth. As dates get dried, they are at greater risk from the sawtoothed grain beetle, while dates with spoilage problems are more susceptible to Mediterranean flour moth and rusty grain beetle infestation. These post-harvest pests can be treated with additional disinfestation treatments. But the best way to eliminate these pests is to store dates in a manner that best retains their quality and avoids spoilage. Research is needed regarding improved effectiveness of pheromone formulations for carob moth, improved implementation of pheromone-based control methods for nitidulid beetles, and ongoing development in effective storage technologies.

Acknowledgments The authors would like to thank and acknowledge Agric. Engineer Ibrahim Bou-Khowh, Date Palm Research Center of Excellence, King Faisal University for his kind assistance in photographing the insects displayed in this chapter. Mention of trade names or commercial products in this publication is solely for the purpose of providing specific information and does not imply recommendation or endorsement by the United States Department of Agriculture. USDA is an equal opportunity provider and employer.

References

Abdelghany, A. Y., Awadalla, S. S., Abdel-Baky, N. F., El-Syrafi, H. A., & Fields, P. G. (2010). Stored-product insects in botanical warehouses. *Journal of Stored Products Research, 46*, 93–97.

Abo-El-Saad, M. M., El Shafie, H. A., Al Ajlan, A. M., & Bou-Khowh, I. A. (2011). Non-chemical alternatives to methyl bromide against *Ephestia cautella* (Lepidoptera: Pyralidae): Microwave and ozone. *Agriculture and Biology Journal of North America, 2*, 1222–1231.

Adarkwah, C., & Schöller, M. (2012). Biological control of *Plodia interpunctella* (Lepidoptera: Pyralidae) by single and double releases of two larval parasitoids in bulk stored wheat. *Journal of Stored Products Research, 51*, 1–5.

Ahmad, T. R., & Ali, M. A. (1995). Forecasting emergence and flight of some *Ephestia* spp. (Lep., Pyralidae) based on pheromone trapping and degree-day accumulations. *Journal of Applied Entomology, 119*, 611–614.

Al-Azawi, A. S. (1986). A survey of insect pests of date palm in Qatar. *Date Palm Journal, 4*, 247–266.

Aldawood, A. S., Rasool, K. G., Alrukban, A. H., Alan, S., Mureed, H., Sutanto, K. D., & Muhammad, T. (2013). Effects of temperature on the development of *Ephestia cautella* (Walker) (Pyralidae: Lepidoptera): A case study for its possible control under storage conditions. *Pakistan Journal of Zoology, 45*, 1573–1578.

Aldryhim, Y. (2008). Greater date moth, *Arenipses sabella* Hmps. (Lepidoptera: Pyralidae). In J. L. Capinera (Ed.), *Encyclopedia of entomology* (1716–1717 pp). Dordrecht: Springer.

Al-Farisi, M., Abuagla, A., Mohamed, E., & Gohs, U. (2013). The effect of electron beam on dates infestation. *Food Control, 33*, 157–161.

Al-Hafidh, E. M. T., Al-Kawaga, A. A., & Al-Ahad, I. A. (1987). Infestation of date palm varieties by stored product insects in orchard. *Date Palm Journal, 5*, 233–237.

Al-Izzi, M. J., Al-Maliky, S. K., Younis, M. A., & Jabbo, N. F. (1985). Bionomics of *Ectomyelois ceratoniae* (Lepidoptera: Pyralidae) on pomegranates in Iraq. *Environmental Entomology, 14*, 149–153.

Al-Redhaiman, K. N. (2005). Chemical changes during storage of 'Barhi' dates under controlled atmosphere conditions. *HortScience, 40*, 1413–1415.

Alrubeai, H. F. (1987). Growth and development of *Ectomyelois ceratoniae* (Lepidoptera: Pyralidae) under laboratory mass-rearing conditions. *Journal of Stored Products Research, 23*, 133–135.

Al-Taweel, A. A., Ahmed, M. S. H., Kadhum, S. S., & Hameed, A. A. (1990). Effects of gamma radiation on the progeny of irradiated *Ephestia cautella* (Walker) (Lepidoptera: Pyralidae) males. *Journal of Stored Products Research, 26*, 233–236.

Al-Zadjali, T. S., Abdallah, F. F., & El-Haidari, H. S. (2006). Insect pests attacking date palms and dates in Sultanate of Oman. *Egyptian Journal of Agricultural Research, 84*, 51–59.

Amri, I., & Hamrouni, L. (2014). Essential oils as biological alternatives to protect date palm (*Phoenix dactylifera* L.) against *Ectomyelois ceratoniae* Zeller (Lepidoptera: Pyralidae). *Chilean Journal of Agricultural Research, 74*, 273–279.

Anderson, P., & Löfqvist, J. (1996). Asymmetric oviposition behaviour and the influence of larval competition in the two pyralid moths *Ephestia kuehniella* and *Plodia interpunctella. Oikos, 76*, 47–56.

Andreadis, S. S., Eliopoulos, P. A., & Savopoulou-Soultani, M. (2012). Cold hardiness of immature and adult stages of the Mediterranean flour moth, *Ephestia kuehniella. Journal of Stored Products Research, 48*, 132–136.

Arbogast, R. T. (2005). Seasonal flight activity of stored-product moths (Lepidoptera: Pyralidae, Gelechiidae) in South Carolina, USA. *Entomological News, 116*, 197–208.

Arbogast, R. T. (2007). A wild strain of *Plodia interpunctella* (Hübner) (Lepidoptera: Pyralidae) from farm-stored maize in South Carolina: Effect of temperature on mating, survival, and fecundity. *Journal of Stored Products Research, 43*, 503–507.

Arbogast, R. T., & Chini, S. R. (2005). Abundance of *Plodia interpunctella* (Hübner) and *Cadra cautella* (Walker) infesting maize stored on South Carolina farms: Seasonal and non-seasonal variation. *Journal of Stored Products Research, 41*, 528–543.

Arbogast, R. T., Leonard Lecato, G., & Van Byrd, R. (1980). External morphology of some eggs of stored-product moths (Lepidoptera: Pyralidae, Gelechiidae, Tineidae). *International Journal of Insect Morphology and Embryology, 9*, 165–177.

Arbogast, R. T., Kendra, P. E., Mankin, R. W., & McDonald, R. C. (2002). Insect infestation of a botanicals warehouse in north-central Florida. *Journal of Stored Products Research, 38*, 349–363.

Arthur, F. H., & Subramanyam, B. (2012). Chemical control in stored products. In D. W. Hagstrum, T. W. Phillips, & G. Cuperus (Eds.), *Stored product protection, S156* (95–100 pp). Manhattan: Kansas State University.

Assari, M. J., & Khajepour, S. (2013). Population fluctuation of *Ephestia figulilella* in Kerman province of Iran. *Archives of Phytopathology and Plant Protection, 46*, 862–867.

Avidov, Z., & Harpaz, I. (1969). *Plant pests of Israel.* Jerusalem: Israel Universities Press.

Bachrouch, O., Jemâa, J. M.-B., Wissem, A. W., Talou, T., Marzouk, B., & Abderraba, M. (2010). Composition and insecticidal activity of essential oil from *Pistacia lentiscus* L. against *Ectomyelois ceratoniae* Zeller and *Ephestia kuehniella* Zeller (Lepidoptera: Pyralidae). *Journal of Stored Products Research, 46*, 242–247.

Back, E. A., & Cotton, R. T. (1926). Biology of the saw-toothed grain beetle, *Oryzaephilus surinamensis* L. *Journal of Agricultural Research, 33*, 435–452.

Baker, T. C., Francke, W., Millar, J. G., Löfstedt, C., Hansson, B., Du, J. W., Phelan, P. L., Vetter, R. S., Youngman, R., & Todd, J. L. (1991). Identification and bioassay of sex pheromone components of carob moth, *Ectomyelois ceratoniae* (Zeller). *Journal of Chemical Ecology, 17*, 1973–1988.

Baldwin, R., & Fasulo, T. R. (2010). *Confused flour beetle, Tribolium confusum Jacquelin du Val (Insecta: Coleoptera: Tenebrionidae) and red flour beetle, Tribolium castaneum (Herbst) (Insecta: Coleoptera: Tenebrionidae)* (University of Florida Institute of Food and Agricultural Sciences Publication No. EENY289). Gainesville: University of Florida Institute of Food and Agricultural Sciences.

Barnes, D. F., & Kaloostian, G. H. (1940). Flight habits and seasonal abundance of dried-fruit insects. *Journal of Economic Entomology, 33*, 115–119.

Bartelt, R. J., Dowd, P. F., Vetter, R. S., Shorey, H. H., & Baker, T. C. (1992). Responses of *Carpophilus hemipterus* (Coleoptera: Nitidulidae) and other sap beetles to the pheromone of *C. hemipterus* and host-related coattractants in California field tests. *Environmental Entomology, 21*, 1143–1153.

Bartelt, R. J., Vetter, R. S., Carlson, D. G., & Baker, T. C. (1994). Responses to aggregation pheromones for five *Carpophilus* species (Coleoptera: Nitidulidae) in a California date garden. *Environmental Entomology, 23*, 1534–1543.

Bartelt, R. J., Vetter, R. S., Carlson, D. G., Petroski, R. J., & Baker, T. C. (1995). Pheromone combination lures for *Carpophilus* (Coleoptera: Nitidulidae) species. *Journal of Economic Entomology, 88*, 864–869.

Bell, C. H. (1975). Effects of temperature and humidity on development of four pyralid moth pests of stored products. *Journal of Stored Products Research, 11*, 167–175.

Bell, C. H. (1982). Observations on the intensity of diapause and cold tolerance in larvae from twelve populations and two reciprocal crosses of the Indian meal moth, *Plodia interpunctella. Physiological Entomology, 7*, 371–377.

Bell, C. H. (2000). Fumigation in the 21st century. *Crop Protection, 19*, 563–569.

Bell, C. H., Bowley, C. R., Cogan, P. M., & Sharma, S. (1979). Diapause in twenty-three populations of *Plodia interpunctella* (Hübner) (Lep., Pyralidae) from different parts of the world. *Ecological Entomology, 4*, 193–197.

Bessi, H., Bellagha, S., Lebdi, K. G., Bikoba, V., & Mitcham, E. J. (2015). Ethyl formate fumigation of dry and semidry date fruits: Experimental kinetics, modeling, and lethal effect on carob moth. *Journal of Economic Entomology, 108*, 993–999.

Bishop, G. W. (1959). The comparative bionomics of American *Cryptolestes* (Coleoptera-Cucujidae) that infest stored grain. *Annals of the Entomological Society of America, 52*, 657–665.

Blumberg, D. (2008). Date palm arthropod pests and their management in Israel. *Phytoparasitica, 36*, 411–448.

Boshra, S. A., & Mikhaiel, A. A. (2006). Effect of gamma irradiation on pupal stage of *Ephestia calidella* (Guenee). *Journal of Stored Products Research, 42*, 457–467.

Bousquet, Y. (1990). *Beetles associated with stored products in Canada: An identification guide* (Agriculture Canada Publication No. 1837). Ottawa: Agriculture Canada.

Bowditch, T. G., & Madden, J. L. (1996). Spatial and temporal distribution of *Ephestia cautella* (Walker) (Lepidoptera: Pyralidae) in a confectionary factory: Causal factors and management implications. *Journal of Stored Products Research, 32*, 123–130.

Brady, U. E. (1973). Isolation, identification and stimulatory activity of a second component of the sex pheremone system (complex) of the female almond moth, *Cadra cautella* (Walker). *Life Sciences, 13*, 227–235.

Brady, U. E., & Daley, R. C. (1972). Identification of a sex pheromone from the female raisin moth, *Cadra figulilella. Annals of the Entomological Society of America, 65*, 1356–1358.

Brower, J. H. (1983). Eggs of stored-product lepidoptera as hosts for *Trichogramma evanescens* [Hym.: Trichogrammatidae]. *Entomophaga, 28*, 355–361.

Brower, J. H., Smith, L., Vail, P. V., & Flinn, P. W. (1996). Biological control. In B. Subramanyam & D. W. Hagstrum (Eds.), *Integrated management of insects in stored products* (223–268 pp). New York: Marcel Dekker.

Burges, H. D., & Haskins, P. F. (1965). Life cycle of the tropical warehouse moth *Cadra cautella* (Wlk.) at controlled temperatures and humidities. *Bulletin of Entomological Research, 55*, 775–779.

Burks, C. S., & Johnson, J. J. (2012). Biology, behavior, and ecology of stored grain and legume insects. In D. W. Hagstrum, T. W. Phillips, & G. W. Cuperus (Eds.), *Stored product protection* (Publication No. S156, 7–20 pp). Manhattan: Kansas State University.

Burks, C. S., & Kuenen, L. P. S. (2012). Effect of mating disruption and lure load on the number of *Plodia interpunctella* (Hübner) (Lepidoptera: Pyralidae) males captured in pheromone traps. *Journal of Stored Products Research, 49*, 189–195.

Burks, C. S., Johnson, J. A., Maier, D. E., & Heaps, J. W. (2000). Temperature. In B. Subramanyam & D. W. Hagstrum (Eds.), *Alternatives to pesticides in stored-product IPM* (73–104 pp). Boston: Kluwer.

Burks, C. S., McLaughlin, J. R., Miller, J. R., & Brandl, D. G. (2001). Mating disruption for control of the Indianmeal moth in a warehouse. In G. Obenauf & R. Obenauf (Eds.), *2001 Annual international research conference on methyl bromide alternatives and emissions reductions*(3 p.). San Diego: Methyl Bromide Alternatives Outreach.
Burks, C. S., McLaughlin, J. R., Miller, J. R., & Brandl, D. G. (2011). Mating disruption for control of *Plodia interpunctella* (Hübner) (Lepidoptera: Pyralidae) in dried beans. *Journal of Stored Products Research, 47*, 216–221.
Buxton, P. A. (1920). Insect pests of dates and the date palm in Mesopotamia and elsewhere. *Bulletin of Entomological Research, 11*, 287–304.
Calderon, M., Navarro, S., & Donahaye, E. (1969). *Ectomyelois ceratoniae* (Zell.) (Lep., Phycitidae), a major pest of stored almonds in Israel. *Journal of Stored Products Research, 5*, 427–428.
Campbell, J. F., & Arbogast, R. T. (2004). Stored-product insects in a flour mill: Population dynamics and response to fumigation treatments. *Entomologia Experimentalis et Applicata, 112*, 217–225.
Campbell, J. F., Perez-Mendoza, J., & Weier, J. (2012). Insect pest management decisions in food processing facilities. In D. W. Hagstrum, T. W. Phillips, & G. W. Cuperus (Eds.), *Stored product protection no. S156* (219–232 pp). Manhattan: Kansas State University.
Campos, M., & Phillips, T. W. (2014). Attract-and-kill and other pheromone-based methods to suppress populations of the Indianmeal moth (Lepidoptera: Pyralidae). *Journal of Economic Entomology, 107*, 473–480.
Cardé, R. T., & Minks, A. K. (1995). Control of moth pests by mating disruption: Successes and constraints. *Annual Review of Entomology, 40*, 559–585.
Carpenter, J. B., & Elmer, H. S. (1978). *Pests and diseases of the date palm* (United States Department of Agriculture Handbook No. 527). Washington, DC: United States Department of Agriculture.
Carrillo, M. A., Moon, R. D., Wilcke, W. F., Morey, R. V., Kaliyan, N., & Hutchison, W. D. (2006). Overwintering mortality of Indianmeal moth (Lepidoptera: Pyralidae) in Southern Minnesota. *Environmental Entomology, 35*, 843–855.
Coffelt, J. A., & Vick, K. W. (1987). Sex pheromones of *Ephestia cautella* (Walker) (Lepidoptera: Pyralidae): Influence of mating on pheromone titer and release rate. *Journal of Stored Products Research, 23*, 119–123.
Cogburn, R. R., & Vick, K. W. (1981). Distribution of angoumois grain moth, almond moth, and Indian meal moth in rice fields and rice storages in Texas as indicated by pheromone-baited adhesive traps. *Environmental Entomology, 10*, 1003–1007.
Cook, S. M., Khan, Z. R., & Pickett, J. A. (2007). The use of push-pull strategies in integrated pest management. *Annual Review of Entomology, 52*, 375–400.
Copping, L. G., & Duke, S. O. (2007). Natural products that have been used commercially as crop protection agents. *Pest Management Science, 63*, 524–554.
Cossé, A. A., Endris, J. J., Millar, J. G., & Baker, T. C. (1994). Identification of volatile compounds from fungus-infected date fruit that stimulate upwind flight in female *Ectomyelois ceratoniae. Entomologia Experimentalis et Applicata, 72*, 233–238.
Coviello, R. L., Bentley, W. J., Michailides, T. J., Ferguson, L., & Westerdahl, B. B. (2009). *UC IPM pest management guidelines: Fig. Publication No. 447*. Oakland: University of California Statewide Integrated Pest Management Program.
Cowan, T., & Gries, G. (2009). Ultraviolet and violet light: Attractive orientation cues for the Indian meal moth, *Plodia interpunctella. Entomologia Experimentalis et Applicata, 131*, 148–158.
Cox, P. D. (1974). The influence of temperature and humidity on the life-cycles of *Ephestia figulilella* Gregson and *Ephestia calidella* (Guenee) (Lepidoptera: Phycitidae). *Journal of Stored Products Research, 10*, 43–55.

Cox, P. D. (1975a). The influence of photoperiod on the life-cycles of *Ephestia calidella* (Guenee) and *Ephestia figulilella* Gregson (Lepidoptera: Phycitidae). *Journal of Stored Products Research, 11*, 75–78.

Cox, P. D. (1975b). The suitability of dried fruits, almonds, and carobs for the development of *Ephestia figulilella* Gregson, *E. calidella* (Guenee) and *E. cautella* (Walker) (Lepidoptera: Phycitidae). *Journal of Stored Products Research, 11*, 229–233.

Cox, P. D. (1976). The influence of temperature and humidity on the life-cycle of *Ectomyelois ceratoniae* (Zeller) (Lepidoptera: Phycitidae). *Journal of Stored Products Research, 12*, 111–117.

Cox, P. D. (1979). The influence of photoperiod on the life-cycle of *Ectomyelois ceratoniae* (Zeller) (Lepidoptera: Pyralidae). *Journal of Stored Products Research, 15*, 111–115.

Cox, P. D. (1987). Cold tolerance and factors affecting the duration of diapause in *Ephestia kuehniella* Zeller (Lepidoptera: Pyralidae). *Journal of Stored Products Research, 23*, 163–168.

Cox, P. D., & Dolder, H. S. (1995). A simple flight chamber to determine flight activity in small insects. *Journal of Stored Products Research, 31*, 311–316.

Cox, P. D., Allen, L. P., Pearson, J., & Beirne, M. A. (1984). The incidence of diapause in seventeen populations of the flour moth, *Ephestia kuehniella* Zeller (Lepidoptera: Pyralidae). *Journal of Stored Products Research, 20*, 139–143.

Cox, P. D., Wakefield, M. E., & Jacob, T. A. (2007). The effects of temperature on flight initiation in a range of moths, beetles and parasitoids associated with stored products. *Journal of Stored Products Research, 43*, 111–117.

Curtis, C. E., & Clark, J. D. (1974). *Comparative Biologies of Oryzaephilus surinamensis and O. mercator (Coleoptera: Cucujidae) on dried fruits and nuts* (Technical Bulletin No. 1488). Washington, DC: United States Department of Agriculture.

Cymborowski, B., & Giebułtowicz, J. M. (1976). Effect of photoperiod on development and fecundity in the flour moth, *Ephestia kuehniella*. *Journal of Insect Physiology, 22*, 1213–1217.

Donahaye, E., Navarro, S., & Rindner, M. (1991a). The influence of low temperatures on two species of *Carpophilus* (Coleoptera; Nitidulidae). *Journal of Applied Entomology, 111*, 297–302.

Donahaye, E., Navarro, S., Rindner, M., & Dias, R. (1991b). The influence of different treatments causing emigration of nitidulid beetles. *Phytoparasitica, 19*, 273–282.

Donahaye, E. J., Navarro, S., & Rinder, M. (1995). Low temperature as an alternative to fumigation for disinfecting dried fruit from three insect species. *Journal of Stored Products Research, 31*, 63–70.

Donohoe, H. C., & Barnes, D. F. (1934). Notes on field trapping of Lepidoptera attacking dried fruits. *Journal of Economic Entomology, 27*, 1067–1072.

Donohoe, H. C., Simmons, P., Barnes, D. F., Kaloostian, G. H., & Heinrich, C. (1949). *Biology of the raisin moth* (Technical Bulletin No. 994, 23 p). Washington, DC: United States Department of Agriculture.

Ducom, P., & Ciesla, Y. (2009). Preliminary tests on quality of high moisture dates with five alternatives to methyl bromide. In G. L. Obenauf (Ed.), *Annual international research conference on methyl bromide alternatives and emissions reductions* (71–74 pp). San Diego: Methyl Bromide Alternatives Outreach.

El-Saeid, M. H., & Al-Dosari, S. A. (2010). Monitoring of pesticide residues in Riyadh dates by SFE, MSE, SFC, and GC techniques. *Arabian Journal of Chemistry, 3*, 179–186.

El-Shafie, H. A. F. (2012). Review: List of arthropod pests and their natural enemies identified worldwide on date palm, *Phoenix dactylifera* L. *Agriculture and Biology Journal of North America, 3*, 516–524.

El-Sherif, S. I., Elwan, E. A., & Abd-El-Razik, M. I. E. (1996). Insect pests of date palm trees in northern Sinai, Egypt. *Libya Arab Journal of Plant Protection, 8*, 72–76.

ESA (Entomological Society of America). (2015). *Common names of insects and related organisms*. Annapolis: Entomological Society of America.

Ewing, C. P., & Cline, A. R. (2005). Key to adventive sap beetles (Coleoptera: Nitidulidae) in Hawaii, with notes on records and habits. *Coleopterists Bulletin, 59*, 167–183.

FAO (Food and Agriculture Organization). (2015). *Crop production and trade data, food and agriculture organization of the United Nations*. Available at: http://faostat3.fao.org/. Accessed on 19 June 2015.

Ferguson, D. C. (1987). Adult moths (Lepidoptera). In J. R. Gorham (Ed.), *Insect and mite pests in food: An illustrated key* (231–244 pp). Washington, DC: United States Government Printing Office.

Fields, P. G., Subramanyam, B., & Hulasare, R. (2012). Extreme temperatures. In D. W. Hagstrum, T. W. Phillips, & G. W. Cuperus (Eds.), *Stored product protection no. S156* (179–190 pp). Manhattan: Kansas State University.

Finkelman, S., Navarro, S., Rindner, M., & Dias, R. (2006). Use of heat for disinfestation and control of insects in dates: Laboratory and field trials. *Phytoparasitica, 34*, 37–48.

Finkelman, S., Lendler, E., Navarro, S., Navarro, H., & Ashbell, G. (2010). New prospects for ethyl formate as a fumigant for the date industry. In: M. O. Carvalho, P. G. Fields, C. S. Adler, F. H. Arthur, C. G. Athanassiou, J. F. Campbell, F. Fleurat-Lessard, P. W. Flinn, R. J. Hodges, A. A. Isikber, S. Navarro, R. T. Noyes, J. Riudavets, K. K. Sinha, G. R. Thorpe, B. H. Timlick, P. Trematerra, & N. D. G. White (Eds.), *Proceedings of the 10th international working conference for stored product protection* (359–364 pp). Estoril: Julius-Kühn-Archiv.

Flinn, P. W., & Schöller, M. (2012). Biological control: Insect pathogens, parasitoids, and predators. In D. W. Hagstrum, T. W. Phillips, & G. W. Cuperus (Eds.), *Stored product protection no. S156* (203–212 pp). Manhattan: Kansas State University.

Follett, P. A., & Weinert, E. D. (2012). Phytosanitary irradiation of fresh tropical commodities in Hawaii: Generic treatments, commercial adoption, and current issues. *Radiation Physics and Chemistry, 81*, 1064–1067.

Fraenkel, G., & Blewett, M. (1943). The natural foods and the food requirements of several species of stored products insects. *Transactions of the Royal Entomological Society of London, 93*, 457–490.

Gameel, S. M. M., Ewais, M. A., & Sayed, A. A. (2015). Using of *Trichogramma evanescens* West (Hymenoptera: Trichogrammatidae) for controlling *Arenipses sabella* hampson and *Batrachedra amydraula* Meyrick in the date palm fields at the New Valley-Egypt. *Egyptian Journal of Biological Sciences, 61*, 35–41.

Gautam, S. G., Opit, G. P., Margosan, D., Tebbets, J. S., & Walse, S. (2014). Egg morphology of key stored-product insect pests of the United States. *Annals of the Entomological Society of America, 107*, 1–10.

Ghimire, M. N., & Phillips, T. W. (2014). Oviposition and reproductive performance of *Habrobracon hebetor* (Hymenoptera: Braconidae) on six different pyralid host species. *Annals of the Entomological Society of America, 107*, 809–817.

Ghribi, D., Elleuch, M., Abdelkefi, L., & Ellouze-Chaabouni, S. (2012). Evaluation of larvicidal potency of *Bacillus subtilis* SPB1 biosurfactant against *Ephestia kuehniella* (Lepidoptera: Pyralidae) larvae and influence of abiotic factors on its insecticidal activity. *Journal of Stored Products Research, 48*, 68–72.

Glazer, I., Eliyau, M., Salame, L., Nakash, Y., & Blumberg, D. (2007). Evaluation of the efficacy of the entomopathogenic nematodes *Heterorhabditis* sp. against sap beetles (Coleoptera: Nitidulidae). *BioControl, 52*, 259–270.

Good, N. E. (1936). *The flour beetles of the genus Tribolium* (Technical Bulletin No. 498). Washington, DC: United States Department of Agriculture.

Gordh, G. (1982). A new species of *Goniozus* imported into California for the biological control of navel orangeworm [Hymenoptera: Bethylidae; Lepidoptera: Pyralidae]. *Entomological News, 93*, 136–138.

Gothilf, S. (1970). The biology of the carob moth *Ectomyelois ceratoniae* (Zell.) in Israel. III Phenology on various hosts. *Israel Journal of Entomology, 5*, 161–175.

Gothilf, S. (1978). Establishment of the imported parasite *Pentalitomastix plethoricus* [Hym.: Encyrtidae], on *Ectomyelois ceratoniae* [Lep.: Phycitidae] in Israel. *Entomophaga, 23*, 299–302.

Gothilf, S. (1984). Biology of *Spectrobates ceratoniae* on almonds in Israel. *Phytoparasitica, 12*, 77–87.

Grieshop, M. J., Flinn, P. W., & Nechols, J. R. (2006). Biological control of Indianmeal moth (Lepidoptera: Pyralidae) on finished stored products using egg and larval parasitoids. *Journal of Economic Entomology, 99*, 1080–1084.

Hagstrum, D. W. (2000). Using five sampling methods to measure insect distribution and abundance in bins storing wheat. *Journal of Stored Products Research, 36*, 253–262.

Hagstrum, D. W., & Flinn, P. W. (1996). Integrated pest management. In B. Subramanyam & D. W. Hagstrum (Eds.), *Integrated management of insect in stored products* (399–408 pp). New York: Marcel Dekker.

Hagstrum, D. W., & Tomblin, C. F. (1975). Relationship between water consumption and oviposition by *Cadra cautella* (Lepidoptera: Phycitidae). *Journal of the Georgia Entomological Society, 58*, 561–566.

Hallman, G. J. (2013). Control of stored product pests by ionizing radiation. *Journal of Stored Products Research, 52*, 36–41.

Halstead, D. G. H. (1963). External sex differences in stored-products Coleoptera. *Bulletin of Entomological Research, 54*, 119–134.

Hameed, A. A., Al-Taweel, A. A., Eass, A. J., Shalal, R. N., & Al-Garbawi, Z. A. (2010). Surveying *Ephestia* spp. and the parasitoid *Bracon hebetor* in date palm orchards and date stores of five governorates in Iraq. *Acta Horticulturae, 882*, 985–993.

Hansen, J. D., Johnson, J. A., & Winter, D. A. (2011). History and use of heat in pest control: A review. *International Journal of Pest Management, 57*, 267–289.

Hassan, M. W., Jamil, M., Iqbal, J., Hussain, T., Iqbal, M., Khan, M. A., Ahmed, I., & Ali, L. (2013). Comparative infestation of some date fruit varieties by red flour beetle *Tribolium castaneum* (Coleoptera: Tenebrionidae) as a storage pest. *World Applied Sciences Journal, 25*, 1130–1133.

Heaps, J. W. (2012). Food plant sanitation, pest exclusion, and facility design. In D. W. Hagstrum, T. Phillips, & G. W. Cuperus (Eds.), *Stored product protection no. S156* (85–93 pp). Manhattan: Kansas State University.

Hertlein, M. B., Thompson, G. D., Subramanyam, B., & Athanassiou, C. G. (2011). Spinosad: A new natural product for stored grain protection. *Journal of Stored Products Research, 47*, 131–146.

Hodges, R. J., Robinson, R., & Hall, D. R. (1996). Quinone contamination of dehusked rice by *Tribolium castaneum* (Herbst) (Coleoptera: Tenebrionidae). *Journal of Stored Products Research, 32*, 31–37.

Horak, M. (1994). A review of *Cadra* Walker in Australia: Five new native species and the two introduced pest species (Lepidoptera: Pyralidae: Phycitinae). *Australian Journal of Entomology, 33*, 245–262.

Hossain, M. S., Williams, D. G., Mansfield, C., Bartelt, R. J., Callinan, L., & Il'ichev, A. L. (2006). An attract-and-kill system to control *Carpophilus* spp. in Australian stone fruit orchards. *Entomologia Experimentalis et Applicata, 118*, 11–19.

Hossain, M. S., Hossain, M. A. B. M., Williams, D. G., & Chandra, S. (2010). Potential to reduce the spatial density of attract and kill traps required for effective control of *Carpophilus* spp. (Coleoptera: Nitidulidae) in stone fruit in Australia. *Australian Journal of Entomology, 49*, 170–174.

Howe, R. W. (1956a). The biology of the two common storage species of *Oryzaephilus* (Coleoptera, Cucujidae). *Annals of Applied Biology, 44*, 341–355.

Howe, R. W. (1956b). The effect of temperature and humidity on the rate of development and mortality of *Tribolium castaneum* (Herbst) (Coleoptera, Tenebrionidae). *Annals of Applied Biology, 44*, 356–368.

Howe, R. W. (1960). The effects of temperature and humidity on the rate of development and the mortality of *Tribolium confusum* Duval (Coleoptera, Tenebrionidae). *Annals of Applied Biology, 48*, 363–376.

Howe, R. W. (1962). The effects of temperature and humidity on the oviposition rate of *Tribolium castaneum* (Hbst.) (Coleoptera, Tenebrionidae). *Bulletin of Entomological Research, 53*, 301–310.

Howe, R. W. (1965). A summary of estimates of optimal and minimal conditions for population increase of some stored products insects. *Journal of Stored Products Research, 1*, 177–184.

Howe, R. W., & Lefkovitch, L. P. (1957). The distribution of the storage species of *Cryptolestes* (Col., Cucujidae). *Bulletin of Entomological Research, 48*, 795–809.

Huang, F., & Subramanyam, B. (2003). Effects of delayed mating on reproductive performance of *Plodia interpunctella* (Hübner) (Lepidoptera: Pyralidae). *Journal of Stored Products Research, 39*, 53–63.

Huang, F., Subramanyam, B., & Taylor, R. (2003). Ultrasound affects spermatophore transfer, larval numbers, and larval weight of *Plodia interpunctella* (Hübner) (Lepidoptera: Pyralidae). *Journal of Stored Products Research, 39*, 413–422.

Hussain, A. A. (1963). Notes on borers of date palms in Iraq. *Bulletin of Entomological Research, 54*, 345–348.

Hussain, A. A. (1974). *Date palms and dates with their pests in Iraq*. Baghdad: Ministry of Higher Education and Scientific Research, University of Baghdad.

Imam, A. I. (2012). Evaluation of some integrated managements to combat the greater date moth, *Arenipses sabella* hmps. (Lepidoptera: Pyralidae), on date palm trees under siwa oasis conditions in Egypt. *Egyptian Journal of Biological Pest Control, 22*, 157–160.

Jacob, T. A., & Cox, P. D. (1977). The influence of temperature and humidity on the life-cycle of *Ephestia kuehniella* Zeller (Lepidoptera: Pyralidae). *Journal of Stored Products Research, 13*, 107–118.

Jacobs, S. B., & Calvin, D. D. (2001). *Mediterranean flour moth, entomological notes*. Pennsylvania: The Pennsylvania State University, University Park, State College.

James, D. G., & Vogele, B. (2000). Development and survivorship of *Carpophilus hemipterus* (L.), *Carpophilus mutilatus* Erichson and *Carpophilus humeralis* (F.) (Coleoptera: Nitidulidae) over a range of constant temperatures. *Australian Journal of Entomology, 39*, 180–184.

James, D. G., Bartelt, R. J., & Faulder, R. J. (1994). Attraction of *Carpophilus* spp. (Coleoptera: Nitidulidae) to synthetic aggregation pheromones and host-related coattractants in Australian stone fruit orchards: Beetle phenology and pheromone dose studies. *Journal of Chemical Ecology, 20*, 2805–2819.

James, D. G., Faulder, R. J., & Bartelt, R. J. (1995). Fauna and seasonal abundance of *Carpophilus* spp. (Coleoptera: Nitidulidae) in four stone fruit growing regions of Southeastern Australia as determined by pheromone-trapping. *Australian Journal of Entomology, 34*, 327–333.

James, D. G., Vogele, B., Faulder, R. J., Bartelt, R. J., & Moore, C. J. (2001). Pheromone-mediated mass trapping and population diversion as strategies for suppressing *Carpophilus* spp. (Coleoptera: Nitidulidae) in Australian stone fruit orchards. *Agricultural and Forest Entomology, 3*, 41–47.

Jassim, H. K., Abdulla, L. M., & Abd-Al-Ahad, I. (1988). Determination of the exact concentration of *Beauveria bassiana* (Vuill.) to control the larvae of the fig moth, *Ephestia cautella* (Walk.) on stored dates in Iraq. *Arab Journal of Plant Protection, 6*, 44–45.

Jemâa, J. M.-B., Haouel, S., Bouaziz, M., & Khouja, M. L. (2012). Seasonal variations in chemical composition and fumigant activity of five *Eucalyptus* essential oils against three moth pests of stored dates in Tunisia. *Journal of Stored Products Research, 48*, 61–67.

Johnson, J. A., Wofford, P. L., & Whitehand, L. C. (1992). Effect of diet and temperature on development rates, survival, and reproduction of the Indianmeal moth (Lepidoptera: Pyralidae). *Journal of Economic Entomology, 85*, 561–566.

Johnson, J. A., Wofford, P. L., & Gill, R. F. (1995). Developmental thresholds and degree-day accumulations of Indianmeal moth (Lepidoptera: Pyralidae) on dried fruits and nuts. *Journal of Economic Entomology, 88*, 734–742.

Johnson, J. A., Gill, R. F., Valero, K. A., & May, S. A. (1996). Survival of navel orangeworm (Lepidoptera: Pyralidae) during pistachio processing. *Journal of Economic Entomology, 89*, 197–203.

Johnson, J. A., Valero, K. A., Hannel, M. M., & Gill, R. F. (2000). Seasonal occurrence of postharvest dried fruit insects and their parasitoids in a culled fig warehouse. *Journal of Economic Entomology, 93*, 1380–1390.

Johnson, J. A., Vail, P. V., Brandl, D. G., Tebbets, J. S., & Valero, K. A. (2002). Integration of nonchemical treatments for control of postharvest pyralid moths in almonds and raisins. *Journal of Economic Entomology, 95*, 190–199.

Johnson, J. A., Yahia, E. M., & Brandl, D. G. (2009). Dried fruits and tree nuts. In E. M. Yahia (Ed.), *Modified and controlled atmospheres for the storage, transportation, and packaging of horticultural commodities* (507–526 pp). Boca Raton: CRC Press/Taylor & Francis.

Kader, A. A., & Hussein, A. M. (2009). *Harvesting and postharvest handling of dates*. Aleppo: ICARDA.

Karalius, V., & Buda, V. (1995). Mating delay effect on moths' reproduction: Correlation between reproduction success and calling activity in females *Ephestia kuehniella, Cydia pomonella, Yponomeuta cognagellus* (Lepidoptera: Pyralidae, Tortricidae, Yponomeutidae). *Pheromones, 5*, 169.

Kehat, M., & Greenberg, S. (1969). The biology and phenology of *Arenipses sabella* Hmps. and *Cadra figulilella* (Gregson) (Lepidoptera, Pyralidae) on dates in Israel. *Bulletin of Entomological Research, 58*, 411–420.

Kehat, M., Blumberg, D., & Williams, R. N. (1983). Seasonal abundance of sap beetles (Coleoptera: Nitidulidae) in date plantations in Israel. *Phytoparasitica, 11*, 109–112.

Kehat, M., Eitam, A., Blumberg, D., Dunkelblum, E., Anshelevich, L., & Gordon, D. (1992). Sex pheromone traps for detecting and monitoring the raisin moth, *Cadra figulilella*, in date palm plantations. *Phytoparasitica, 20*, 99–106.

Khairi, M. M. A., Elhassan, M. I., & Bashab, F. A. (2010). The status of date palm cultivation and date production in Sudan. *Acta Horticulturae, 882*, 37–42.

Khajepour, S., Izadi, H., & Asari, M. J. (2012). Evaluation of two formulated chitin synthesis inhibitors, hexaflumuron and lufenuron against the raisin moth, *Ephestia figulilella. Journal of Insect Science, 12*, 102.

Kingsolver, J. M. (1991). Adult beetles (Coleoptera). In J. R. Gorham (Ed.), *Insect and mite pests in food: An illustrated key* (Agricultural Handbook No. 655). Washington, DC: United States Department of Agriculture.

Kishani-Farahani, H., Bell, H., & Goldansaz, S. H. (2012a). Biology of *Apanteles myeloenta* (Hymenoptera: Braconidae), a larval parasitoid of carob moth *Ectomyelais ceratoniae* (Lepidoptera: Pyralidae). *Journal of Asia-Pacific Entomology, 15*, 607–610.

Kishani-Farahani, H., Goldansaz, S. H., & Sabahi, Q. (2012b). A survey on the overwintering larval parasitoids of *Ectomyelois ceratoniae* in three regions in Iran. *Crop Protection, 36*, 52–57.

Krasnoff, S. B., Vick, K. W., & Mankin, R. W. (1983). Female calling behavior in *Ephestia elutella* and *E. figulilella* (Lepidoptera: Pyralidae). *Florida Entomologist, 66*, 249–254.

Kuwahara, Y., & Casida, J. E. (1973). Quantitative analysis of the sex pheromone of several phycitid moths by electron capture gas chromatography. *Agricultural and Biological Chemistry, 37*, 681–684.

Kuwahara, Y., Kitamura, C., Takahashi, S., Hara, H., Ishii, S., & Fukami, H. (1971). Sex pheromone of the almond moth and the Indian meal moth: *Cis-9,trans-12-tetradecadienyl* acetate. *Science (Washington, DC), 171*, 801–802.

LeCato, G. L. (1976). Yield, development, and weight of *Cadra cautella* (Walker) and *Plodia interpunctella* (Hübner) on twenty-one diets derived from natural products. *Journal of Stored Products Research, 12*, 43–47.

Legner, E. F., & Silveira-Guido, A. (1983). Establishment of *Goniozus emigratus* and *Goniozus legneri* [Hym: Bethylidae] on navel orangeworm, *Amyelois transitella* [Lep: Phycitidae] in California and biological control potential. *Entomophaga, 28*, 97–106.

Levinson, H. Z., & Buchelos, C. T. (1981). Surveillance of storage moth species (Pyralidae, Gelechiidae) in a flour mill by adhesive traps with notes on the pheromone—Mediated flight behaviour of male moths. *Zeitschrift für Angewandte Entomologie, 92*, 233–251.

Levinson, H. Z., & Hoppe, T. (1983). Preferential flight of *Plodia interpunctella* and *Cadra cautella* (Phycitinae) toward figure of definite shape and position with notes on the interaction between optical and pheromone stimuli. *Zeitschrift für Angewandte Entomologie, 96*, 491–500.

Levi-Zada, A., David, M., Fefer, D., Seplyarsky, V., Sadowsky, A., Dobrinin, S., Ticuchinski, T., Harari, D., Blumberg, D., & Dunkelblum, E. (2014). Circadian release of male-specific components of the greater date moth, *Aphomia (Arenipses) sabella*, using sequential SPME/GC/MS analysis. *Journal of Chemical Ecology, 40*, 236–243.

Lindgren, D. L., & Vincent, L. E. (1953). Nitidulid beetles infesting California dates. *Hilgardia, 22*, 97–118.

Linsley, E. G. (1944). Natural sources, habitats, and reservoirs of insects associated with stored food products. *Hilgardia, 16*, 187–224.

Lum, P. T. M. (1977). High temperature inhibition of development of eupyrene sperm and reproduction in *Plodia interpunctella* and *Ephestia cautella. Journal of the Georgia Entomological Society, 12*, 199–203.

Lysandrou, M., Temerak, S. A., & Sayed, A. A. (2010). The use of different insect control regimes using three green chemicals to combat *Batrachedra amydraula* Meyrick and *Cadra* spp. on date palm fruit in Egypt. *Acta Horticulturae, 882*, 481–489.

Madge, D., Taylor, C., & Williams, D. (2012). *Carob moth in almonds: Preliminary monitoring guidelines*. Berri: Almond Board of Australia.

Madrid, F. J., & Sinha, R. N. (1982). Feeding damage of three stored-product moths (Lepidoptera: Pyralidae) on wheat. *Journal of Economic Entomology, 75*, 1017–1020.

Mahroof, R., & Hagstrum, D. W. (2012). Biology, behavior, and ecology of insects in processed commodities. In D. W. Hagstrum, T. W. Phillips, & G. W. Cuperus (Eds.), *Stored product protection* (Publication S156, 33–44 pp). Manhattan: Kansas State University.

Manickavasagan, A., Alahakoon, P. M. K., Al-Busaidi, T. K., Al-Adawi, S., Al-Wahaibi, A. K., Al-Raeesi, A. A., Al-Yahyai, R., & Jayas, D. S. (2013). Disinfestation of stored dates using microwave energy. *Journal of Stored Products Research, 55*, 1–5.

Marouf, A., Amir-Maafi, M., & Shayesteh, N. (2013). Two-sex life table analysis of population characteristics of almond moth, *Cadra cautella* (Lepidoptera: Pyralidae) on dry and semi-dry date palm varieties. *Journal of Crop Protection, 2*, 171–181.

Marra, F., Zhang, L., & Lyng, J. G. (2009). Radio frequency treatment of foods: Review of recent advances. *Journal of Food Engineering, 91*, 497–508.

Mason, L. J. (2010). *Dried fruit beetle Carpophilus hemipterus (L.) and corn sap beetle Carpophilus dimidiatus (L.)* (Purdue Extension No. E-229-W). West Lafayette: Purdue University.

Mason, L. J., & McDonough, M. (2012). Biology, behavior, and ecology of stored grain and legume insects. In D. W. Hagstrum, T. W. Phillips, & G. W. Cuperus (Eds.), *Stored product protection. S156* (7–20 pp). Manhattan: Kansas State University.

Mbata, G. N. (1985). Some physical and biological factors affecting oviposition by *Plodia interpunctella* (Hubner) (Lepidoptera: Phycitidae). *Insect Science and Its Applications, 6*, 597–604.

Mbata, G. N. (1987). Studies on induction of larval diapause in a Nigerian strain of *Plodia interpunctella* (Hübner) (Lepidoptera: Pyralidae). *Insect Science and Its Applications, 8*, 317–322.

Mbata, G. N. (1990). Suitability of maize varieties for the oviposition and development of *Plodia interpunctella* (Hubner) (Lepidoptera: Pyralidae). *Tropical Pest Management, 36*, 122–127.

McGaughey, W. H. (1985). Insect resistance to the biological insecticide *Bacillus thuringiensis. Science, 229*, 193–195.

McGhee, P., & Gut, L. (2014). Aerosol emitters disrupt codling moth, *Cydia pomonella*, competitively. *Pest Management Science, 70*, 1859–1862.

Mehrnejad, M. R. (1994). Carob moth, a pest of pistachio nut in Iran. *Acta Horticulturae, 419*, 365–372.

Michael, P. (1968). The carob moth. *Journal of Agriculture, Western Australia, 9*, 81–82.

Miller, J. R., & Gut, L. J. (2015). Mating disruption for the 21st century: Matching technology with mechanism. *Environmental Entomology, 44*(3), 427–453. doi:http://dx.doi.org/10.1093/ee/nvv052.

Miller, J. R., Gut, L., de Lame, F. M., & Stelinski, L. L. (2006). Disruption of moth sexual communication by point sources of sex pheromone (Part I): Theory. *Journal of Chemical Ecology, 32*, 2089–2114.

Mohandass, S., Arthur, F. H., Zhu, K. Y., & Throne, J. E. (2007). Biology and management of *Plodia interpunctella* (Lepidoptera: Pyralidae) in stored products. *Journal of Stored Products Research, 43*, 302–311.

Mohseni, A. (2009). The situation of pomegranate orchards in Iran. *Acta Horticulturae, 818*, 35–42.

Mortazavi, S., Samih, M. A., Ghaharieh, H., & Jafari, A. (2015). Effect of some diets on demographic parameters of *Ectomyelois ceratoniae* (Zeller) (Lepidoptera: Pyralidae) in vitro. *Journal of Plant Protection Research, 55*, 212–219.

Mozaffarian, F., Mardi, M., Sarafrazi, A., & Nouri Ganbalani, G. (2008). Assessment of geographic and host-associated population variations of the carob moth, *Ectomyelois ceratoniae*, on pomegranate, fig, pistachio, and walnut, using AFLP markers. *Journal of Insect Science, 8*(6), 1–9.

Mullen, M. A., & Pedersen, J. R. (2000). Sanitation and exclusion. In B. Subramanyam & D. W. Hagstrum (Eds.), *Alternatives to pesticides in stored-product IPM* (29–50 pp). Boston: Kluwer.

Na, J. H., & Ryoo, M. I. (2000). The influence of temperature on development of *Plodia interpunctella* (Lepidoptera: Pyralidae) on dried vegetable commodities. *Journal of Stored Products Research, 36*, 125–129.

Na, J. H., Ryoo, M. I., & Sone, J. (2000). A phenology model for seasonal occurrence of *Plodia interpunctella* (Lepidoptera: Pyralidae) in stored product storage. *Journal of Asia-Pacific Entomology, 3*, 77–81.

Nansen, C., & Phillips, T. W. (2004). Attractancy and toxicity of an attracticide for Indianmeal moth, *Plodia interpunctella* (Lepidoptera: Pyralidae). *Journal of Economic Entomology, 97*, 703–710.

Nansen, C., Phillips, T. W., & Sanders, S. (2004). Effects of height and adjacent surfaces on captures of Indianmeal moth (Lepidoptera: Pyralidae) in pheromone-baited traps. *Journal of Economic Entomology, 97*, 1284–1290.

Nansen, C., Meikle, W. G., & Phillips, T. W. (2006). Ovipositional response of Indianmeal moth (Lepidoptera: Pyralidae) to size, quality, and number of food patches. *Annals of the Entomological Society of America, 99*, 253–260.

Nansen, C., Davidson, D., & Porter, P. (2009). Using water bottles for trapping of Indianmeal moths in stored peanuts. *Entomologia Experimentalis et Applicata, 133*, 251–259.

Navarro, S. (2012). The use of modified and controlled atmospheres for the disinfestation of stored products. *Journal of Pest Science, 85*, 301–322.

Navarro, S., Donahaye, E., & Calderon, M. (1986). Development of the carob moth, *Spectrobates ceratoniae*, on stored almonds. *Phytoparasitica, 14*, 177–186.

Navarro, S., Donahaye, J. E., Rindner, M., & Azrieli, A. (2001). Storage of dates under carbon dioxide atmosphere for quality preservation. In: *Proceedings of international conference on controlled atmosphere and fumigation in stored products* (231–239 pp). Fresno: Executive Printing Services.

Nay, J. E., & Perring, T. M. (2005). Impact of ant predation and heat on carob moth (Lepidoptera: Pyralidae) mortality in California date gardens. *Journal of Economic Entomology, 98*, 725–731.

Nay, J. E., & Perring, T. M. (2006). Effect of fruit moisture content on mortality, development, and fitness of the carob moth (Lepidoptera: Pyralidae). *Environmental Entomology, 35*, 237–244.

Nay, J. E., & Perring, T. M. (2008). Influence of host plant stages on carob moth (Lepidoptera: Pyralidae) development and fitness. *Environmental Entomology, 37*, 568–574.

Nay, J. E., & Perring, T. M. (2009). Effect of center cut strand thinning on fruit abscission and *Ectomyelois ceratoniae* (Lepidoptera: Pyralidae) infestation in California date gardens. *Journal of Economic Entomology, 102*, 948–953.

Nay, J. E., Boyd, E. A., & Perring, T. M. (2006). Reduction of carob moth in 'Deglet Noor' dates using a bunch cleaning tool. *Crop Protection, 25*, 758–765.
Nay, J. E., Park, Y.-L., & Perring, T. M. (2007). Effect of bunch sanitation on spatial distributions of abscised fruit and phycitine moths (Lepidoptera: Pyralidae) in California date gardens. *Journal of Economic Entomology, 100*, 1773–1780.
Neunzig, H. H. (1979). *Systematics of immature phycitines (Lepidoptera: Pyralidae) associated with leguminous plants in the Southern United States* (Technical Bulletin Number 1859). Washington, DC: United States Department of Agriculture.
Norris, M. J. (1934). Contributions towards the study of insect fertility. III. Adult nutrition, fecundity, and longevity in the genus *Ephestia. Proceedings of the Zoological Society of London, UK, 104*, 333–360.
Oehlschlager, A. C., Pierce, A. M., Pierce, H. D., Jr., & Borden, J. H. (1988). Chemical communication in cucujid grain beetles. *Journal of Chemical Ecology, 14*, 2071–2098.
Olsson, P.-O., Anderbrant, O., & Löfstedt, C. (2005a). Flight and oviposition behavior of *Ephestia cautella* and *Plodia interpunctella* in response to odors of different chocolate products. *Journal of Insect Behavior, 18*, 363–380.
Olsson, P. O. C., Anderbrant, O., Löfstedt, C., Borg-Karlson, A.-K., & Liblikas, I. (2005b). Electrophysiological and behavioral responses to chocolate volatiles in both sexes of the Pyralid moths *Ephestia cautella* and *Plodia interpunctella. Journal of Chemical Ecology, 31*, 2947–2961.
Olsson, P.-O., Anderbrant, O., & Lofstedt, C. (2006). Attraction and oviposition of *Ephestia kuehniella* induced by volatiles identified from chocolate products. *Entomologia Experimentalis et Applicata, 119*, 137–144.
Opit, G., Collins, P. J., & Daglish, G. J. (2012). Resistance management. In D. W. Hagstrum, T. W. Phillips, & G. W. Cuperus (Eds.), *Stored product protection* (143–155 pp). Manhattan: Kansas State University.
Oppert, B., Hammel, R., Throne, J. E., & Kramer, K. J. (2000). Fitness costs of resistance to *Bacillus thuringiensis* in the Indianmeal moth, *Plodia interpunctella. Entomologia Experimentalis et Applicata, 96*, 281–287.
Park, J. J., & Perring, T. M. (2010). Development of a binomial sampling plan for the carob moth (Lepidoptera: Pyralidae), a pest of California dates. *Journal of Economic Entomology, 103*, 1474–1482.
Parsons, C. T. (1943). A revision of nearctic Nitidulidae (Coleoptera). *Bulletin of the Museum of Comparative Zoology, 92*, 119–278.
Perez-Mendoza, J., & Aguilera-Peña, M. (2004). Development, reproduction, and control of the Indian meal moth, *Plodia interpunctella* (Hübner) (Lepidoptera: Pyralidae), in stored seed garlic in Mexico. *Journal of Stored Products Research, 40*, 409–421.
Perring, T. M., & Nay, J. E. (2015). Evaluation of bunch protectors for preventing insect infestation and preserving yield and fruit quality of dates, *Phoenix dactylifera* L. *Journal of Economic Entomology, 108*, 654–661.
Perring, T., Braverman, M., & Mafra-Neto, A. (2014). *Pheromone-based mating disruption reduces pest attack on, the IR4 project newsletter* (The IR-4 project, 3 p). New Brunswick: Rutgers.
Phelan, P. L., & Baker, T. C. (1986). Cross-attraction of five species of stored-product Phycitinae (Lepidoptera: Pyralidae) in a wind tunnel. *Environmental Entomology, 15*, 369–372.
Phillips, T. W. (1998). Cylinder-based phosphine for control of postharvest insect pests. In *Annual international conference on methyl bromide alternatives and emissions reductions* (Paper No. 86). Fresno: Methyl Bromide Alternatives Outreach.
Phillips, T. W., Jiang, X. L., Burkholder, W. E., Phillips, J. K., & Tran, H. Q. (1993). Behavioral responses to food volatiles by two species of stored-product Coleoptera, *Sitophilus oryzae* (Curculionidae) and *Tribolium castaneum* (Tenebrionidae). *Journal of Chemical Ecology, 19*, 723–734.

Phillips, T. W., Thoms, E. M., DeMark, J., & Walse, S. (2012). Fumigation. In D. W. Hagstrum, T. W. Phillips, & G. W. Cuperus (Eds.), *Stored product protection* (157–178 pp). Manhattan: Kansas State University.

Quartey, G. K., & Coaker, T. H. (1993). Role of sex pheromone components in the orientation behaviour of *Ephestia cautella. Entomologia Experimentalis et Applicata, 66*, 237–245.

Rafaeli, A., Kostukovsky, M., & Carmeli, D. (2006). Successful disinfestations of sap-beetle contaminations from organically grown dates using heat treatment: A case study. *Phytoparasitica, 34*, 204–212.

Razazzian, S., Hassani, M. R., Imani, S., & Shojai, M. (2015). Life table parameters of *Plodia interpunctella* (Lepidoptera: Pyralidae) on four commercial pistachio cultivars. *Journal of Asia-Pacific Entomology, 18*, 55–59.

Read, J. S., & Haines, C. P. (1976). The functions of the female sex pheromones of *Ephestia cautella* (Walker) (Lepidoptera, Phycitidae). *Journal of Stored Products Research, 12*, 49–53.

Read, J. S., & Haines, C. P. (1979). Secondary pheromone components and synergism in stored-products phycitinae. *Journal of Chemical Ecology, 5*, 251–257.

Rees, D. P. (1996). Coleoptera. In B. H. Subramanyam & W. David (Eds.), *Integrated management of insect in stored products* (1–39 pp). New York: Marcel Dekker.

Reinke, M. D., Siegert, P. Y., McGhee, P. S., Gut, L. J., & Miller, J. R. (2014). Pheromone release rate determines whether sexual communication of oriental fruit moth is disrupted competitively vs. non-competitively. *Entomologia Experimentalis et Applicata, 150*, 1–6.

Ryne, C., Ekeberg, M., Olsson, P.-O. C., Valeur, P. G., & Löfstedt, C. (2002). Water revisited: A powerful attractant for certain stored-product moths. *Entomologia Experimentalis et Applicata, 103*, 99–103.

Ryne, C., Ekeberg, M., Jonzén, N., Oehlschlager, C., Löfstedt, C., & Anderbrant, O. (2006). Reduction in an almond moth *Ephestia cautella* (Lepidoptera: Pyralidae) population by means of mating disruption. *Pest Management Science, 62*, 912–918.

Ryne, C., Svensson, G. P., Anderbrant, O., & Löfstedt, C. (2007). Evaluation of long-term mating disruption of *Ephestia kuehniella* and *Plodia interpunctella* (Lepidoptera: Pyralidae) in indoor storage facilities by pheromone traps and monitoring of relative aerial concentrations of pheromone. *Journal of Economic Entomology, 100*, 1017–1025.

Sambaraju, K. R., & Phillips, T. W. (2008). Effects of physical and chemical factors on oviposition by *Plodia interpunctella* (Lepidoptera: Pyralidae). *Annals of the Entomological Society of America, 101*, 955–963.

Savoldelli, S., & Süss, L. (2010). Integrated control of *Ephestia cautella* (Walker) in a confectionary factory. In: M. O. Carvalho, P. G. Fields, C. S. Adler, F. H. Arthur, C. G. Athanassiou, J. F. Campbell, F. Fleurat-Lessard, P. W. Flinn, R. J. Hodges, A. A. Isikber, S. Navarro, R. T. Noyes, J. Riudavets, K. K. Sinha, G. R. Thorpe, B. H. Timlick, P. Trematerra, & N. D. G. White (Eds.), *Proceedings of the 10th international working conference on stored product protection* (991–992 pp). Estoril: Julius-Kühn-Archiv.

Schmidt, C. T. (1935). Biological studies on the Nitidulid beetles found in pineapple fields (Nitidulidae, Coleoptera). *Annals of the Entomological Society of America, 28*, 475–511.

Schöller, M., & Flinn, P. W. (2000). Parasitoids and predators. In B. Subramanyam & D. W. Hagstrum (Eds.), *Alternatives to pesticides in stored-product IPM* (229–271 pp). Norwell: Kluwer.

Sedlacek, J. D., Weston, P. A., & Barney, R. J. (1996). Lepidoptera and psocoptera. In B. Subramanym & D. W. Hagstrum (Eds.), *Integrated management of insects in stored products* (41–70 pp). New York: Marcel Dekker.

Silhacek, D., Murphy, C., & Arbogast, R. T. (2003). Behavior and movements of Indian meal moths (*Plodia interpunctella* Hübner) during commodity infestation. *Journal of Stored Products Research, 39*, 171–184.

Simmons, P., & Nelson, H. D. (1975). *Insects on dried fruits* (Agriculture Handbook 464). Washington, DC: Agricultural Research Service United States Department of Agriculture.

Simmons, P., Reed, W. D., & McGregor, E. A. (1931). Circular No. 157: *Fig insects in California*. Washington, DC: United States Department of Agriculture, Bureau of Entomology.

Sinha, R. N. (1965). Development of *Cryptolestes ferrugineus* (Stephens) and *Oryzaephilus mercator* (Fauvel.) on seed-borne fungi. *Entomologia Experimentalis et Applicata, 8*, 309–313.

Soderstrom, E., Hinsch, R., Bongers, A., Brandl, D., & Hoogendorn, H. (1987). Detecting adult Phycitinae (Lepidoptera: Pyralidae) infestations in a raisin-marketing channel. *Journal of Economic Entomology, 80*, 1229–1232.

Sokoloff, A., & Lerner, M. (1967). Laboratory ecology and mutual predation of *Tribolium* species. *The American Naturalist, 101*, 261–276.

Solis, A. M. (2006). *Key to selected Pyraloidea (Lepidoptera) larvae intercepted at U.S. ports of entry: Revision of Pyraloidea in "keys to some frequently intercepted lepidopterous larvae" by Weisman 1986*. Washington, DC: United States Department of Agriculture, Agricultural Research Service.

Sower, L. L., Vick, K. W., & Tumlison, J. H. (1974). (Z, E)-9,12-tetradecadien-1-ol: A chemical released by female Plodia interpunctella that inhibits the sex pheromone response of male Cadra cautella. *Environmental Entomology, 3*, 120–122.

Sparks, T. C., Dripps, J. E., Watson, G. B., & Paroonagian, D. (2012). Resistance and cross-resistance to the spinosyns – A review and analysis. *Pesticide Biochemistry and Physiology, 102*, 1–10.

Stickney, F. S., Barnes, D. F., & Simmons, P. (1950). *Date palm insects in the United States* (USDA Circular No. 846). Washington, DC: United States Department of Agriculture.

Subramanyam, B., & Hagstrum, D. W. (1993). Predicting development times of six stored-product moth species (Lepidoptera: Pyralidae) in relation to temperature, relative humidity, and diet. *European Journal of Entomology, 90*, 51–64.

Suresh, S., White, N. D. G., Jayas, D. S., & Hulasare, R. B. (2001). Mortality resulting from interaction between the red flour beetle and the rusty grain beetle. *Proceedings of the Entomological Society of Manitoba, Canada, 57*, 11–18.

Svensson, G. P., Löfstedt, C., & Skals, N. (2007). Listening in pheromone plumes: Disruption of olfactory-guided mate attraction in a moth by a bat-like ultrasound. *Journal of Insect Science, 7*, 59.

Talhouk, A. S. (1991). On the management of the date palm and its arthropod enemies in the Arabian Peninsula. *Journal of Applied Entomology, 111*, 514–520.

Todd, J. L., Millar, J. G., Vetter, R. L., & Baker, T. C. (1992). Behavior and physiological activity of (Z, E)-7,9,11-dodecatrienyl formate, a mimic of the major sex pheromone component of carob moth, Ectomyelois ceratoniae. *Journal of Chemical Ecology, 18*, 2331.

Toews, M. D., & Nansen, C. (2012). Trapping and interpreting captures of stored grain insects. In D. W. Hagstrum, T. W. Phillips, & G. W. Cuperus (Eds.), *Stored product protection. Publication S156* (243–261 pp). Manhattan: Kansas State University.

Trematerra, P. (2012). Advances in the use of pheromones for stored-product protection. *Journal of Pest Science, 85*, 285–299.

Trematerra, P., & Capizzi, A. (1991). Attracticide method in the control of *Ephestia kuehniella* Zeller – Studies on effectiveness. *Zeitschrift für Angewandte Entomologie, 111*, 451–456.

Trematerra, P., & Gentile, P. (2010). Five years of mass trapping of *Ephestia kuehniella* Zeller: A component of IPM in a flour mill. *Journal of Applied Entomology, 134*, 149–156.

Trematerra, P., & Pavan, G. (1995). Ultrasound production in the courtship behaviour of *Ephestia cautella* (Walk.), *E. Kuehniella* Z. and *Plodia interpunctella* (Hb.) (Lepidoptera: Pyralidae). *Journal of Stored Products Research, 31*, 43–48.

Trematerra, P., & Sciarretta, A. (2004). Spatial distribution of some beetles infesting a feed mill with spatio-temporal dynamics of *Oryzaephilus surinamensis*, *Tribolium castaneum* and *Tribolium confusum*. *Journal of Stored Products Research, 40*, 363–377.

Trematerra, P., Athanassiou, C., Stejskal, V., Sciarretta, A., Kavallieratos, N., & Palyvos, N. (2011). Large-scale mating disruption of *Ephestia* spp. and *Plodia interpunctella* in Czech Republic, Greece and Italy. *Journal of Applied Entomology, 135*, 749–762.

Tsuji, H. (1998). Experimental invasion of a food container by first-instar larvae of the Indian meal moth, *Plodia interpunctella* Hübner, through pinholes. *Medical Entomology and Zoology, 49*, 99–104.

Tsuji, H. (2000). Ability of first instar larvae of Indian meal moth, *Plodia interpunctella* Hübner, to reach their food. *Medical Entomology and Zoology, 51*, 283–287.

USDA-AMS. (2013). *Grading, certification, and verification*. Washington, DC: United States Department of Agriculture.

USDA-FGIS. (2015). *Stored-grain insect reference*. Washington, DC: United States Department of Agriculture, Federal Grain Inspection Service [USDA-FGIS].

Vetter, R. S., Tatevossian, S., & Baker, T. C. (1997). Reproductive behavior of the female carob moth, (Lepidoptera: Pyralidae). *Pan-Pacific Entomologist, 73*, 28–35.

Vetter, R. S., Millar, J. G., Vickers, N. J., & Baker, T. C. (2006). Mating disruption of carob moth, *Ectomyelois ceratoniae*, with a sex pheromone analog. *Southwestern Entomologist, 33*, 33–47.

Vick, K. W., Coffelt, J. A., & Weaver, W. A. (1987). Presence of four species of stored-product moths in storage and field situations in north-central Florida as determined with sex pheromone-baited traps. *Florida Entomologist, 70*, 488–492.

Walker, K. R. (2007a). *Mediterranean flour moth (Ephestia kuehniella)*. Updated on 18 Sept 2007 1:01:25 PM. Canberra: PaDIL.

Walker, K. R. (2007b). PaDIL. Confused sap beetle (*Carpophilus mutilatus*) Updated on 14 June 2007 10:55:09 PM. Victoria: Australian Biosecurity.

Warner, R. L., Barnes, M. M., & Laird, E. F. (1990a). Chemical control of a carob moth, *Ectomyelois ceratoniae* (Lepidoptera: Pyralidae), and various nitidulid beetles (Coleoptera) on Deglet Noor dates in California. *Journal of Economic Entomology, 83*, 2357–2361.

Warner, R. L., Barnes, M. M., & Laird, E. F. (1990b). Reduction of insect infestation and fungal infection by cultural practice in date gardens. *Environmental Entomology, 19*, 1618–1623.

Weisman, D. M. (1987). Larval moths (Lepidoptera). In J. R. Gorham (Ed.), *Insect and mite pests in food: An illustrated key* (245–268 pp). Washington, DC: United States Government Printing Office.

Western SARE. (2015). *What is sustainable agriculture?* Logan: Western SARE, United States Department of Agriculture.

Wijayaratne, L. K., & Fields, P. G. (2012). Effects of rearing conditions, geographical origin, and selection on larval diapause in the Indianmeal moth, *Plodia interpunctella*. *Journal of Insect Science, 12*, 119.

Williams, R. N., Fickle, D. S., Kehat, M., Blumberg, D., & Klein, M. G. (1983). *Research circular 278: Bibliography of the genus Carpophilus Stephens (Coleoptera: Nitidulidae)*. Wooster: The Ohio State University, Ohio Agricultural Research and Development Center.

Wright, G. C. (2007). An overview of the changing date industry in the United States. *Acta Horticulturae, 736*, 47–52.

Zare, Z., Sayhoon, M., & Maghsoudi, V. (1993). Irradiation disinfestation and decontamination of Iranian dates and pistachio nuts. *Radiation Physics and Chemistry, 42*, 301–305.

Zaviezo, T., Romero, A., Castro, D., & Wagner, A. (2007). First record of *Goniozus legneri* (Hymenoptera: Bethylidae) in Chile. *Ciencia e Invetigación Agraria, 34*, 49–52.

Zhu, J., Ryne, C., Unelius, C. R., Valeur, P. G., & Löfstedt, C. (1999). Reidentification of the female sex pheromone of the Indian meal moth, *Plodia interpunctella*: Evidence for a four-component pheromone blend. *Entomologia Experimentalis et Applicata, 92*, 137–146.

Chapter 10
Phytoplasmas and Their Insect Vectors: Implications for Date Palm

Geoff M. Gurr, Assunta Bertaccini, David Gopurenko, Robert R. Krueger, Khalid A. Alhudaib, Jian Liu, and Murray J. Fletcher

Abstract Phytoplasmas are transmitted chiefly by insects, most commonly planthoppers and leafhoppers. Molecular genetic analyses have improved the understanding of phytoplasma taxonomy, and also enhanced the ability to identify phytoplasmas detected in hosts and insect vectors. Date palm is affected by Al-Wijam disease in Saudi Arabia and molecularly indistinguishable phytoplasmas

---

G.M. Gurr (✉)
Fujian Agriculture and Forestry University, Fuzhou, China

Graham Centre, Charles Sturt University and NSW Department of Primary Industries, Orange, NSW, Australia
e-mail: ggurr@csu.edu.au

A. Bertaccini
Alma Mater Studiorum, University of Bologna, Bologna, Italy
e-mail: assunta.bertaccini@unibo.it

D. Gopurenko
NSW Department of Primary Industries, Wagga Wagga, NSW, Australia
e-mail: david.gopurenko@dpi.nsw.gov.au

R.R. Krueger
USDA-ARS, National Clonal Germplasm Repository for Citrus and Dates, Riverside, CA, USA
e-mail: robert.krueger@ars.usda.gov

K.A. Alhudaib
Plant Protection Department, College of Agricultural and Food Sciences, King Faisal University, Hofuf, Alhasa, Saudi Arabia
e-mail: kalhudaib@kfu.edu.sa

J. Liu
Fujian Agriculture and Forestry University, Fuzhou, China
e-mail: liujianwh@gmail.com

M.J. Fletcher
New South Wales Department of Primary Industries, Charles Sturt University, Orange, Australia
e-mail: murray.fletcher@dpi.nsw.gov.au

W. Wakil et al. (eds.), *Sustainable Pest Management in Date Palm: Current Status and Emerging Challenges*, Sustainability in Plant and Crop Protection,
DOI 10.1007/978-3-319-24397-9_10

(16SrI group) were recovered from affected palms and from *Cicadulina bipunctata*, an insect commonly found on the palms. The phytoplasma that is associated with the lethal yellowing disease in coconut palm (16SrIV-A) can also infect date palm. In the Americas lethal yellowing is likely to be transmitted by *Haplaxius crudus* (formerly *Myndus crudus*). Texas phoenix decline is reported from warm regions of South-East USA and may be transmitted by two species of Derbidae. Phytoplasmas belonging to the 16SrIV-F and 16SrXIV groups have also been identified from date palm growing in the USA and Africa respectively, though vectors have not been identified. Preventing spread in infected vegetative planting material and of vectors is key to limiting the impact of phytoplasma diseases. Management in affected areas can use antibiotics on high value trees, but this is not economical for extensive crops. In these situations, vector control by insecticide use or habitat management can be useful, but the long lifespan of individual palms means that even low vector pressure can lead to infection over successive years. The development of resistant varieties and replanting is the most effective long-term approach developed so far for phytoplasma disease management in this plant species.

10.1 Introduction

Date palm (*Phoenix dactylifera* L.) is an important crop in many countries that have a climate suitable for its cultivation (Chao and Krueger 2007). As with all cultivated crops, various pests and pathogens compromise the production and might also threaten the life of the tree. Most attention has been given to the diseases of date palm caused by fungal pathogens (Carpenter and Elmer 1978; Djerbi 1983; Ploetz et al. 2003). Some of these, such as Bayoud disease caused by *Fusarium oxysporum* Schlecht. f. sp. *albedinis* (Kill & Maire) Malençon, are devastating and can destroy or cripple entire industries (Djerbi 1983). To date, there have been no reports of virus or viroid diseases of date palm. There have been several reports of date palm diseases associated with the type of bacteria known as phytoplasmas.

Phytoplasmas belong to the taxonomic domain Bacteria but unlike most bacteria they lack a cell wall (Fig. 10.1) and have a reduced genome (Oshima et al. 2004). They exist as intracellular parasites in plant hosts and, as they lack many of the genes required for metabolism (Oshima et al. 2004) are thought to cause disease symptoms in hosts by aggressively sequestering metabolites produced by host cells and altering the expression of plant homeotic genes (Himeno et al. 2011). Phytoplasmas have also been shown to alter the morphology and metabolism of the plant to benefit the longevity of vector insects (Sugio et al. 2011). The longest-known phytoplasma-associated disease affecting date palms is Lethal Yellows (LY), first described in coconut (*Cocos nucifera* L.) in 1891 and associated with phytoplasmas in 1972 (Plavsic-Banjac et al. 1972; Tsai and Harrison 2003). Although date palm is highly susceptible to LY, the disease has had little economic impact on the date industry since it has only been reported in areas outside of primary date

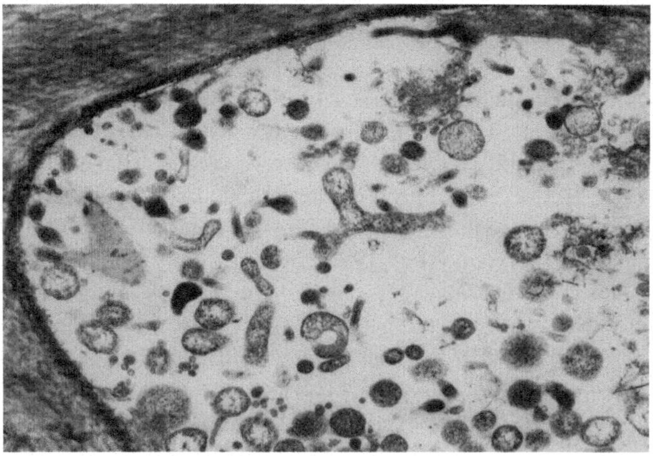

Fig. 10.1 Phytoplasmas showing different shape and sizes in a sieve tube (X 10,000) (Photo: Assunta Bertaccini)

production. In the USA, a decline of *Phoenix* spp. was first reported in the 1970s from the Rio Grande Valley of Texas, USA (McCoy et al. 1980a, b; Miller et al. 1980). Although at this time it cannot be stated with certainty due to the time lapse, it is quite probable that this was the same decline (Texas Phoenix Decline or TPD) later reported to be associated with a LY-group phytoplasma (Harrison et al. 2002a). Similar symptoms were reported from ornamental *Phoenix* spp. in Florida, USA (Elliott 2009) and associated with the TPD strain phytoplasma, the LY strain phytoplasma and a previously unreported phytoplasma (Harrison et al. 2008). This was the first report of TPD outside of Texas and its presence in Florida is apparently increasing but not at a rapid rate (Florida Department of Agriculture and Consumer Services 2008). In the USA reports of LY and TPD on date palms are confined to areas where date palms are grown as ornamentals or specimen trees (Ploetz et al. 2003; Elliott 2009). No phytoplasma-associated declines have been reported from the USA date producing areas in California and Arizona (USA).

In 2000, two new declines of date palm associated with phytoplasmas were reported from Sudan. White tip disease (Cronjé et al. 2000a) and slow decline (Cronjé et al. 2000b) were associated with closely related phytoplasmas that were also similar to Bermuda grass *Cynodon dactylon* (L.) white leaf-associated phytoplasmas, but not to LY phytoplasmas. Al-Wijam disease was first reported from the Al-Ahsa area of Saudi Arabia in the 1950s (Carpenter and Elmer 1978) and in 2007, phytoplasmas related to '*Candidatus* Phytoplasma asteris' were associated with the disease (Alhudaib et al. 2007a, b).

Thus, in the last 10–15 years, reports of phytoplasmas occurring and being associated with date palms have appeared indicating that these pathogens have growing importance also in this crop species. This chapter will examine in depth general phytoplasma taxonomy, detection methods, vector relationships, and control. More information on the phytoplasmas associated with date palms will also be presented.

10.2 Phytoplasma Taxonomy

Phytoplasmas are classified as '*Candidatus* Phytoplasma' genus, in the bacterial domain. Previously called mycoplasma-like organisms (MLOs) classification of phytoplasmas is historically based on 16S ribosomal gene sequences and was first established using 16S ribosomal groups determined after PCR/RFLP analyses with informative restriction enzymes that allows phytoplasma differentiation into groups and subgroups (Fig. 10.2) (Lee et al. 1998). More recently an automated computer-simulated virtual gel RFLP analysis system named iPhyClassifier was developed (Zhao et al. 2009) which uses sequence data available from GenBank to expand and update the ribosomal-based classification. "Currently thirty three 16Sr groups with more than 100 subgroups are recognized (revised in Bertaccini et al., 2014)."

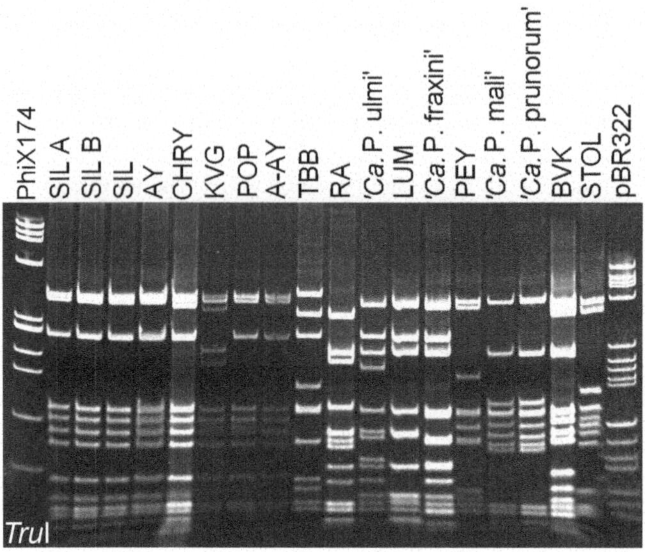

Fig. 10.2 Polyacrylamide gel 5 % shows the RFLP patterns with TruI obtained with the primer pair P1/P7 in direct PCR from field collected phytoplasma infected samples (SIL A, SIL B, SIL, 16SrI-B); AY, aster yellows (16SrI-B); CHRY, Chrysanthemum yellows (16SrI-A); KVG, clover phyllody (16SrI-C) from Germany; POP, populus yellows from Croatia (16SrI-O); A-AY, aster yellows from apricot from Spain (16SrI-F); TBB, tomato big bud, from Australia (16SrII-D); RA, Ranunculus virescence from Italy (16SrIII-B); '*Ca.* P. ulmi' (elm yellows, 16SrV-A); LUM, lucerne virescence from France (16SrVI); '*Ca.* P. fraxini' (ash yellows, 16SrVII-A); PEY, Pichris echioides yellows (16SrIX); '*Ca.* P. mali' (apple proliferation, 16SrX-A); '*Ca.* P. prunorum' (European stone fruit yellows, 16SrX-B); BKV, leafhopper transmitted from Germany (16SrXI); STOL, stolbur from Serbia ('*Ca.* P. pruni', 16SrXII-A); phiX174 is the marker ΦX174 HaeIII digested. The fragment sizes in base pairs from top to bottom are: 1353, 1078, 872, 603, 310, 281, 271, 234, 194, 118, and 72; pBR322 is the marker pBR322 BsuRI digest and the fragment sizes in base pairs from *top* to *bottom* are: 587, 540, 502, 458, 434, 267, 234, 213, 192, 184, 124, 123, 104, 89, 80, 64, 57, and 51 (Photo: Assunta Bertaccini)

The '*Candidatus* Phytoplasma' species description has a threshold of <97.5 % similarity on the 16S rDNA gene to any previously described '*Ca.* Phytoplasma' species is also employed and officially approved for description of new taxa (IRPCM 2004). In Table 10.1 these two classifications are summarized for strains having received officially both designations. Several important phytoplasma taxa are still not reported as '*Candidatus* species', mainly since they fail the 97.5 % similarity criteria on 16S rDNA and biological properties are not available to support their classification.

The only way to get the '*Candidatus*' status in cases in which the threshold homology on 16SrDNA is above 97.5 % is to meet biological differences such as specific insect host or geographic separation. The '*Ca.* Phytoplasma' spp. infecting pome and stone fruit differ in being vectored by different *Cacopsylla* species or '*Ca.* P. balanitae' infects a plant species, *Balanites triflora* Tiegh., only reported to grow in Myanmar (Seemüller and Schneider 2004; Win et al. 2013).

However a number of phytoplasma ribosomal groups still cannot meet the taxonomic requirements for '*Candidatus*' designation. This is the situation for phytoplasmas associated with some very important and epidemic diseases such as palm lethal yellows (LY) in America and grapevine 'flavescence dorée' in Europe (Martini et al. 2002; Ntushelo et al. 2013a). The phytoplasmas associated with devastating outbreaks of LY have been classified as members of group 16SrIV subgroup A, while related phytoplasmas classified as subgroups 16SrIV-B, -D, -E and -F strains are associated with localized outbreaks of disease often involving coconut and other palms, including several species with no prior history of susceptibility to subgroup 16SrIV-A strains. Very recently the phytoplasmas associated with similar LY diseases in Africa were assigned to the newly described '*Ca.* P. palmicola' taxon (Table 10.1) (Harrison et al. 2014).

In the last decade, the number of phytoplasma strains reported worldwide has increased exponentially which made updating the classification scheme difficult. Accordingly, a DNA barcode system based on highly conserved genes such as tuf and the 600 bp of the 3′ region of 16S rDNA was developed and allows quick and specific identification (Makarova et al. 2012). Moreover, the highly conserved double copy 16S rDNA is affected by the not uncommon presence of 16S rDNA interoperon sequence heterogeneity, and is also problematic for differentiation of closely related strains (Mitrović et al. 2011).

An accurate molecular distinction is therefore necessary for phytoplasma strain characterization and epidemiologic studies. Multilocus typing using other genes such as ribosomal proteins, secY, secA, groEL, nitroreductase and rhodonase-like are now currently employed as supplementary tools and provide reliable information for phytoplasma strains differentiation when used in combination with 16SrDNA (Martini et al. 2007; Hodgetts et al. 2008; Mitrović et al. 2011; Lee et al. 2012; Fránová et al. 2013). Very recently lethal wilt of oil palm in Colombia was associated with related to '*Ca.* P. asteris' strain. The characterization of the *groEL*, *amp*, and *rp* genes by RFLP and sequence data, shows that this strain could be differentiated from other aster yellows phytoplasmas including a strain in corn in Colombia (Alvarez et al. 2014).

Table 10.1 Phytoplasmas classified based on 16S rDNA sequencing and corresponding provisional scientific names (see text for detail)

16Sr subgroup	Phytoplasma strain	'*Candidatus* species'	GenBank accession
16SrI-B	Aster yellows	'*Ca.* P. asteris'	M30790
16SrI-Y	"Brote grande" of tomato	'*Ca.* P. lycopersici'	EF199549
16SrII-B	Lime witches' broom	'*Ca.* P. aurantifolia'	U15442
16SrII-D	Papaya mosaic	'*Ca.* P. australasia'	Y10096
16SrIII-A	Peach X disease	'*Ca.* P. pruni'	JQ044392/JQ044393
16SrV-A	Elm yellows	'*Ca.* P. ulmi'	AY197655
16SrV-B	Jujube witches' broom	'*Ca.* P. ziziphi'	AB052876
16SrV-E	Rubus stunt	'*Ca.* P. rubi'	AY197648
16SrV-F	Balanite witches' broom	'*Ca.* P. balanitae'	AB689678
16SrVI-A	Clover proliferation	'*Ca.* P. trifolii'	AY390261
16SrVI-I	Passionfruit	'*Ca.* P. sudamericanum'	GU292081
16SrVII-A	Ash yellows	'*Ca.* P. fraxini'	AF092209
16SrIX-B	Almond witches' broom	'*Ca.* P. phoenicium'	AF515636
16SrX-A	Apple proliferation	'*Ca.* P. mali'	AJ542541
16SrX-B	European stone fruit yellows	'*Ca.* P. prunorum'	AJ542544
16SrX-C	Pear decline	'*Ca.* P. pyri'	AJ542543
16SrX-D	Spartium witches' broom	'*Ca.* P. spartii'	X92869
16SrXI-A	Rice yellow dwarf	'*Ca.* P. oryzae'	AB052873
16SrXII-A	Stolbur	'*Ca.* P. solani'	AF248959
16SrXII-B	Australian grapevine yellows	'*Ca.* P. australiense'	L76865
16SrXII-D	Japanese hydrangea phyllody	'*Ca.* P. japonicum'	AB010425
16SrXII-E	Yellows diseased strawberry	'*Ca.* P. fragariae'	DQ086423
16SrXII-H	Bindweed yellows	'*Ca.* P. convolvuli'	JN833705
16SrXIV-A	Bermudagrass white leaf	'*Ca.* P. cynodontis'	AJ550984
16SrXV-A	Hibiscus witches' broom	'*Ca.* P. brasiliense'	AF147708
16SrXVI-A	Sugarcane yellow leaf syndrome	'*Ca.* P. graminis'	AY725228
16SrXVII-A	Papaya bunchy top	'*Ca.* P. caricae'	AY725234
16SrXVIII-A	American potato purple top wilt	'*Ca.* P. americanum'	DQ174122
16SrXIX-A	Chestnut witches' broom	'*Ca.* P. castaneae'	AB054986

(continued)

Table 10.1 (continued)

16Sr subgroup	Phytoplasma strain	'*Candidatus* species'	GenBank accession
16SrXX-A	Rhamnus witches' broom	'*Ca*. P. rhamni'	AJ583009
16SrXXI-A	Pinus phytoplasma	'*Ca*. P. pini'	AJ310849
16SrXXII-A	Mozambique coconut yellows	'*Ca*. P. palmicola'	KF751388
16SrXXIX-A	Cassia witches' broom	'*Ca*. P. omanense'	EF666051
16SrXXX-A	Salt cedar witches' broom	'*Ca*. P. tamaricis'	FJ432664
16SrXXXI-A	Soybean stunt	'*Ca*. P. costaricanum'	HQ225630
16SrXXXII-A	Malaysian periwinkle virescence	'*Ca*. P. malaysianum'	EU371934
16SrXXXIII-A	Allocasuarina phytoplasma	'*Ca*. P. allocasuarinae'	AY135523

Host range, geographic distribution and vector transmission specificity are biological properties that need to be included for unique phytoplasma species delineation, especially if in the near future the dropping of the '*Candidatus*' classification will be achieved by acquisition of the phenotypic data necessary for this (Contaldo et al. 2012).

As for most microorganisms, identification of phytoplasmas in field collected material is carried out on populations rather than on individual clones. When the diseases are endemic PCR detection and identification is usually able to detect only the most abundant phytoplasma present in the sample at the time of collection. Multiple phytoplasmas were found in grapevine plants after examination by group specific nested-PCR using deep amplicon sequencing, which was able to confirm the presence of several 16Sr groups/'*Ca*. P. species' in single grapevine samples (Nicolaisen et al. 2011).

10.3 Phytoplasma Detection and Study

Since the first discovery of phytoplasmas over four decades ago (Doi et al. 1967), there has been considerable focus on understanding the tritrophic relationships among phytoplasmas, arthropod vectors and host plants. Central to this understanding has been the ongoing development of accurate methods of phytoplasma detection and identification (Firrao et al. 2007; Lee et al. 2007). Because phytoplasmas could not be readily cultured in vitro until recently (Bertaccini et al. 2014) the group has proven more challenging to isolate and study than other bacteria or pathogens such as fungi. Transmission electron microscopy was widely used in early studies to visualize the pathogen directly within phloem (Fig. 10.1). This laborious approach has been superseded by the advent and availability of molecular approaches.

Disorders and outward visual symptoms of phytoplasma infection vary considerably among plant hosts according to a range of factors such as concentration and localization of phytoplasma in host tissues, seasonality of infection and ultimately the metabolic interactions occurring between the actual phytoplasma and host species involved (see references in Bertaccini (2007), Firrao et al. (2007), and Bertaccini et al. (2014)). In some perennial woody plant hosts, phytoplasmas may lay seasonally dormant (Seemüller et al. 1984; Jarausch et al. 1999), or accumulate in asymptomatic host, species that act as reservoirs for spread to less tolerant hosts (Carraro et al. 1998). Finally, a particular phytoplasma can induce different symptoms among multiple hosts, and some shared symptoms among infected hosts can occur through infection by different phytoplasmas and or other unrelated causes (see references in Bertaccini et al. 2014). Accordingly, rapid laboratory based methods of phytoplasma detection and identification are pre-requisites for early control of infected hosts, for phytosanitary screening processes and for biosecurity concerns regarding cross-border disease outbreaks resulting from introduction of phytoplasma infected vectors and or hosts (Schaad et al. 2003; Bertin et al. 2007).

Early attempts at phytoplasma detection and identification generally relied on broad biological descriptions of host plant specificity and disease symptomatology (McCoy et al. 1989). These descriptive properties of phytoplasma infection remain integral to categorizations of the group, and are supported by DNA based methods of phytoplasma detection and classification. Initial DNA based methods of phytoplasma detection involved use of molecular probes in serological or DNA hybridization assays (see references in Firrao et al. 2007). Though useful for providing primary attempts at classification, these initial methods were not sensitive enough to be used in routine diagnostics, mainly due to low concentrations of phytoplasmas generally present in woody infected plant hosts (Kollar et al. 1990). These methods were superseded by the rapid development of polymerase chain reaction (PCR) based approaches (see references in Lee et al. 2000) which allowed targeted amplification of diagnostic gene regions from trace amounts of phytoplasma present in background host tissues. The information gained from targeted PCR not only provided a rapid and reliable method for screening hosts for phytoplasma presence but also provided a means to allow subsequent analyses used for identification of phytoplasma groups, subgroups and species.

As discussed in the previous section, phytoplasma taxonomy is based on assessment of the 16S rDNA gene. Conserved nucleotide regions within the 16S rDNA gene among phytoplasmas allows for design of PCR primers that can be tailored for exclusive amplification of phytoplasmas present in host tissues and can be used as a general PCR diagnostic for phytoplasma presence (Lee et al. 1993; Namba et al. 1993; Schneider et al. 1993). Since the development of PCR for phytoplasma detection (Deng and Hiruki 1990, 1991), a plethora of studies has emerged detailing PCR primers and conditions optimized for both universal and specific amplification of 16S rDNA from phytoplasmas (see references in Lee et al. 1998). The vast majority of these studies have employed a nested-PCR procedure to overcome problems faced in amplifying phytoplasma targets present at very low concentrations within host tissues (Lee et al. 1994, 1995). Nested-PCR, which uses diluted products of a direct PCR for the template to start a second round of PCR (using internal nested primers in the second round), is an efficient means of obtaining high concentration phytoplasma specific PCR products used in downstream diagnostics. Diagnostics for phytoplasma presence/absence among samples is then determined by staining of phytoplasma positive PCR products after gel electrophoresis run alongside positive/negative PCR controls and size markers.

Although nested-PCR assays provide an inexpensive and sensitive means for detection of phytoplasma presence in host tissue, additional analyses are required to determine the identity of detected phytoplasmas. Restriction fragment length polymorphism (RFLP) analysis has been the historically dominant method of identifying phytoplasma groups/subgroups present at individual samples in PCR surveys (Lee et al. 1993, 1998, 2007) including studies of infected date palm hosts (Harrison et al. 2002a; Alhudaib 2007a; Harrison et al. 2008; Ong and Mcbride 2009). RFLP uses restriction endonucleases to cleave double stranded DNA molecules at recognition-sites containing a specific nucleotide sequence. After cleavage of recognition sites with one or more endonucleases, different sized DNA fragments are separated by electrophoresis and band pattern profiles specific to phytoplasma groups identified (Lee et al. 1998). RFLP is inexpensive and allows researchers to rapidly process and identify large numbers of samples to phytoplasma ribosomal group and, in many cases, to subgroups. Similar to RFLP analysis, quantitative PCR (qPCR), involving real-time amplification, can detect phytoplasmas to the level of ribosomal group (Hren et al. 2007; Hodgetts et al. 2009). The approach is based on the annealing of fluorescently labeled oligonucleotide probes to sequence specific targets and their detection and quantification using an optical system involving PCR mediated light emission and capillary electrophoresis. Quantitative-PCR is rapid and can determine the quantity of a target DNA molecule relative to background material. This is a useful attribute for phytoplasma research as it allows the investigator to quantify differences in phytoplasma concentration among different host tissue samples (Christensen et al. 2004; Jarausch et al. 2004). By incorporating internal analytical amplification controls, qPCR also allows an empirical distinction to separate uninfected plant material and false-negative results that may be due to PCR inhibitors present in host plant tissues (Baric and Dalla-Via 2004). Furthermore, because qPCR amplification and detection methods are fully integrated in real-time PCR platforms, risks of false-positive presence caused by cross contamination among PCR amplicons are greatly reduced.

Loop-mediated isothermal amplification (LAMP) was developed by Notomi et al. (2000) and has great potential as an improved technology over the nested-PCR based methods for detection of phytoplasmas in host tissues (Tomlinson et al. 2010; Bekele et al. 2011; Obura et al. 2011). LAMP is an autocycling, strand-displacement DNA synthesis tool used for generating secondary-folding DNA structures (<300 base pairs) able to detect phytoplasma presence. The method also allows for rapid amplification of target DNA under isothermal conditions, with high target specificity and minimal laboratory processing. Furthermore, products of LAMP are detectable by several means including gel electrophoresis, optical scanning in real-time PCR and potentially by simple visual inspection aided by use of chemical precipitation

(Fukuta et al. 2003) or intercalating dyes (Iwamoto et al. 2003). LAMP is reported to have an almost tenfold decrease in assay time and 20-fold increase in sensitivity to phytoplasma titer compared to nested-PCR assays (Obura et al. 2011). Currently, LAMP primers for phytoplasma detection have been identified which are specific to some 16Sr groups (Hodgetts et al. 2011) or assemblies of related groups to the exclusion of others (Obura et al. 2011). It is conceivable that primers will be further identified to allow broader LAMP detection of all 16Sr phytoplasma groups. It is also likely that LAMP applications will be developed to allow in situ field detections of phytoplasmas, though incorporation of a rapid and necessary first step DNA extraction procedure is currently a limiting factor in the development of such a methodology (Hodgetts et al. 2011).

Although expedient, the above diagnostic molecular methods (PCR/RFLP, qPCR, LAMP) only screen sequence-specific sites that are generally conserved at, and diagnostic for, particular phytoplasma groups. Alteration of these sites by novel mutations can cause phytoplasma misidentifications. Further, the methods are insensitive to nucleotide mutations that occur outside of the targeted sites. Effectively these mutations remain hidden from the analyses and represent undetected polymorphisms potentially useful for identifying additional phytoplasma diversity. Ultimately, complete detection of all nucleotide polymorphisms at a sampled gene region is directly achieved by Sanger sequencing, used to identify the chain of nucleotides present along DNA molecules. Aligned sequence data is used to provide confirmation of species identity by comparative genetic distance based and/or phylogenetic analyses. Sequence analysis also provides a means of detecting nucleotide variation among samples that may signal the presence of taxonomic diversity. Although only marginally more expensive than RFLP and qPCR analyses, sequencing is confounded by PCR heterogeneity caused by co-amplification of different phytoplasmas present in a sample or from heterogeneity between the two ribosomal operons present in phytoplasmas (Schneider and Seemüller 1994). Cloning multiple replicates of PCR products prior to sequencing using commercially available recombinant vector kits is an expedient means of dealing with sequence heterogeneity problems but is an added cost to the analyses and requires sufficient replication to improve probability of detecting low-copy phytoplasma variants present in an amplicon.

10.4 Transmission of Phytoplasmas

Phytoplasmas are intracellular parasites characterized by a greatly reduced genome (Oshima et al. 2004) and lack of a cell wall making their survival outside of a living host very difficult. Transmission from plant to plant is, therefore, by insect vectors, parasitic weeds such as dodder (*Cuscuta* spp.) or cultural practices such as grafting, cutting and propagation. Of these mechanisms, insect vectoring seems to be the most important. Determining vector status is challenging due to of the nature of phytoplasma pathosystems and the type of pathogen. A study aiming to identify the vector species in a given pathosystem clearly needs to sample potential insect vectors from the field. Methods used include yellow sticky traps that can be positioned in the canopy of palm trees, hand collection into vials of adults on foliage and nymphs from grassy host roots, sweep-netting of low-lying vegetation and Malaise trapping. The latter is particularly useful if studying the movement of suspected vector species between different types of habitat as this form of trap allows trapping from either or both directions (Weintraub and Beanland 2006). Some phytoplasma vectors such as those in the family Cixiidae develop as nymphs on the roots of grasses whilst the adults move to trees and are responsible for vectoring phytoplasmas. Accordingly, sampling nymphs by hand searching in soil litter (such as mulch) or within the soil in the proximity of roots can be an important complement to sampling adults. A combination of methods is necessary in the early stages of a study so that a wide range of possible insect vector species is sampled for testing, usually of the head by molecular assays.

A complication in seeking to establish the identity of an insect vector is that a given plant species may be fed upon by multiple species, even within the subfamilies often implicated as vectors of other phytoplasmas (Table 10.1). A phytoplasma may also be introduced to plants by non-feeding insects through their habit of explorative probing of plant tissue. Accordingly, multiple taxa may need to be considered and accurate taxonomic identification of the insects is important. Further, there are instances of phytoplasma vectors feeding on more than one host plant. A particularly important type of this effect is vectors having a life cycle in which the immature nymphs feed on a different plant species (usually grass roots) to the adults. An important example is the cixiid *Haplaxius crudus* (Van Duzee) (formerly *Myndus crudus*) which is a vector of Palm Lethal Yellowing Disease in Florida (Harrison et al. 2008), but feeds as a nymph on the roots of various grass species including Bermuda grass. Information on host range for vectors is important because it can lead to possible disease management strategies based on removal and replacement of such alternative host plants.

The need for sampling multiple taxa of suspected insect vectors is compounded by the need to sample adults as well as nymphs and both sexes. Vector status can be affected by these parameters. Furthermore, not all vectors in a population will be infected with the phytoplasmas and this necessitates the capture and processing of multiple individuals if a reliable indication of the level of association between a phytoplasma and an insect is to be achieved. A final series of complications revolves around the fact that a given insect species may be a vector for more than one phytoplasma species. Accordingly, it is important to consider the range of possible relationship permutations in a given setting as a three dimensional matrix with the possibility of multiple species of vector, multiple phytoplasma species and multiple host plants.

10.4.1 Association and Causality

The section on methods for detection and identification of phytoplasmas makes it clear that this is a methodologically challenging group of pathogens. Accordingly, identifying the vectors of a given phytoplasma involve two important steps. The first is to consider association. This involves assays – which usually imply molecular approaches – to determine which vector species are positive for presence of phytoplasma DNA. In step two, work is designed to establish causality. A restricted range of species is used in transmission tests in which insects are taken from a field location in which symptomatic plants are common (Howard et al. 1983) or from a colony held on symptomatic plants. The insects are then placed on phytoplasma free plants. To ensure these plants are free from phytoplasma, plants are grown from seed inside vector proof cages and are tested for phytoplasma DNA at commencement. The plants are then monitored for symptoms and eventually sampled for molecular assays for phytoplasma DNA presence. Important methodological considerations in these transmission studies is the acquisition access period (AAP). The AAP is the time for which a vector individual needs to feed on a phytoplasma-infected plant in order to acquire the phytoplasma. This period is relatively short, typically ranging from several minutes to several hours (Weintraub and Beanland 2006). It can be affected by the titer (density) of phytoplasmas in the host phloem and any difference between the plant part fed upon and the plant part in which phytoplasma may be concentrated. As important is the latency period, the time between phytoplasma acquisition and when the individual vector is competent to transmit the microbe to a host plant. Latency periods range from several to 80 days (Weintraub and Beanland 2006). This is considerably longer than the AAP because vectors are able to transmit the phytoplasma only after the microbe has crossed from the alimentary canal to the haemolymph and then to the salivary glands. In other words, the vector has to become sufficiently infected itself by the phytoplasma to be capable of transmitting it to its host plant. Maixner noted at a conference (Michael Maixner – pers. comm., 11th International Auchenorrhyncha Congress, Potsdam, Germany, August 5–9, 2002) that phytoplasmas could be considered agents of diseases of the vector rather than of the plant. A fundamental limitation of even the most modern approaches for phytoplasma study is that the lack of routine methods to culture them in vitro. Accordingly, transmission tests need to use infective vectors rather than cultured microbes and cannot culminate in the isolation of the microbe and aseptic culturing that are normally used to satisfy Koch's postulates. Recent advances in rearing methods for phytoplasmas (Contaldo et al. 2012). Phytopathologia mediterranea do, however, look promising so more studies involving culturing are likely to be reported in coming years.

Whilst many studies are able to rapidly obtain data on association between a suspected vector and a phytoplasma, moving from this stage to establishing causality is less often accomplished as it is logistically demanding. The work requires known phytoplasma-free host plants, known infective vectors and an appropriate time period for test plants to acquire the phytoplasma and develop symptoms, typically a number of months. Plants need to be kept free from other possible vector species during this extended period and the exclusion cages need to be large enough to guard against the possibility of a 'wild' vector feeding on the test plant foliage through the fabric of the cage. Despite the demands of transmission testing, such work is critical because data on association alone is unreliable since a given insect may be positive for phytoplasma DNA simply from feeding on the phloem of an infected plant and carrying the phytoplasma in its gut. Dissection of the salivary glands of insects and subjecting only this tissue to PCR amplification is possible but the risk remains of this tissue being contaminated by the contents of the insect's feeding apparatus and alimentary canal.

A relatively new method has sought to provide more rapid confirmation of vector status (Zhang et al. 1998; Tanne et al. 2001). This relies on the fact that vectors maintain phytoplasmas in their saliva during probing and feeding. Accordingly, individual insects are held in a small sample tube that also contains a small volume of sucrose solution behind a parafilm barrier. Hemipterans will typically probe this material and feed from the solution, contaminating it with phytoplasmas and allowing subsequent PCR analysis of the solution. An advantage of this method is its speed. Used alone, however, this method does not establish beyond doubt that a given insect can vector a given phytoplasma because the test conditions are artificial and do not include host plant material. Potentially a given host species may not be attractive to a given insect nor susceptible to a given phytoplasma. In some cases a plant may become infected by a phytoplasma but the vector is not able to acquire the phytoplasma from such 'dead end' host plants (Weintraub and Beanland 2006). The saliva assay method can, however, be a useful intermediate step between studies of association that can narrow the number of candidate species and more expensive definitive tests of causality. Species that are observed to feed though parafilm barriers yet do not lead to detectable phytoplasma DNA in the sucrose solution can be excluded from costly transmission tests of causality. Further, because the test insects usually remain alive during the 1–2 day periods of access to sucrose solution, an individual insect that is known to be infective is then available for follow up studies of causality or behavior.

10.4.2 Known Phytoplasma Vector Taxa

The complications inherent in phytoplasma studies detailed in the preceding section mean that vectors are not known for most phytoplasma diseases. The vector of a given phytoplasma may be well known in one continent but absent from another part of the world in which the same phytoplasma is associated with disease in the same host plant so must be transmitted by another vector species. For example, Lethal Yellowing of palm is vectored by *H. crudus* in the Caribbean while in the South Pacific island of Vanuatu, it is vectored by the congeneric *Haplaxius taffini* Bonfils (Wilson 1987).

Notwithstanding the challenges of identifying phytoplasma vectors, Weintraub and Beanland (2006) provide a list of 113 insect species of vector in their comprehensive review of phytoplasma vector taxa and biology. An additional four species are listed in an update (Weintraub 2007). Most of the vectors are Auchenorrhyncha, a suborder of Hemiptera, the true bugs. The Auchenorrhyncha contains many of the insects formerly, but erroneously, known as Homoptera including the leafhoppers, planthoppers, cicadas and spittle-bugs. All members of this suborder have wings of uniform texture and many feed on plant phloem. Notably, however, cicadas and spittlebugs are xylem feeding as are the members of the leafhopper subfamily Cicadellinae. These insects are generally larger than the phloem feeders and have a swollen frontoclypeus which contains a powerful suctorial pump to move sufficient quantity of xylem liquid so the insects can extract sufficient nutrient. In contrast, some members of the leafhopper subfamily Typhlocybinae feed on cell contents, a process termed cell-rupture feeding. By far the largest number of vector of species (57, Table 10.2) is found in the subfamily Deltocephalinae. Phytoplasmas are vectored by insects in eight other subfamilies of the Cicadellidae (leafhopper) family. The superfamily Fulgoroidea (planthoppers) contains seven subfamilies of known phytoplasma vectors of which the Cixiidae and Delphacidae are the most important. In addition to the many Auchenorrhyncha vectors of phytoplasmas, seven species of plant louse (suborder Sternorrhyncha, superfamily Psylloidea), one species of stink bug (suborder Heteroptera, family Pentatomidae) and one species of lacebug (suborder Heteroptera, family Tingidae) are known phytoplasma vectors (Table 10.2).

Very little information is available on the vectors of date palm phytoplasmas. A study in Saudi Arabia of Al-Wijam disease of date palm included PCR assays using 42 *Cicadulina bipunctata* (Melichar) and 18 *Assymetrasca decedens* (Paoli) (Alhudaib et al. 2007a). Phytoplasma DNA was amplified from 12 of the *C. bipunctata* and found to be 16SrI group ('*Ca.* P. asteris') and identical to the amplification products from symptomatic date palm. In contrast, the DNA from *A. decedens* was of the 16SrXII-A ("stolbur") group that was not found in date palm samples. Alhudaib et al. (2007a) identified *C. bipunctata* (Fig. 10.3) as the potential vector of Al-Wijam disease following the identification of the phytoplasma with 100 % identity to the 16S rDNA of phytoplasma in the date palm in *C. bipunctata* captured in Al-Ahsa oasis Saudi Arabia on Al-Wijam-affected date palms.

Though we are not aware of any studies on date palm in which *H. crudus* has been definitively identified as the vector, this species is known as a vector of Lethal Yellowing in Florida (Harrison et al. 2008). *H. crudus* is widely distributed in the Caribbean and Southeastern USA and transmits the disease to coconut, *C. nucifera*. Its host range also includes various grass species on the roots of which the nymphs feed and several palm species including date palm (Elliott and Harrison 2007; Elliott 2009).

The insect vectors of other phytoplasmas detected in date palm (see following section) are not known.

10.5 Phytoplasma-Palm Pathosystems

10.5.1 Phytoplasma-Associated Date Palm Diseases in the Middle East

The first published of Al-Wijam disease of date palm reported it as widespread in Al-Ahsa, Saudi Arabia. The symptoms include yellow streaking and stunting of leaves and fruits and fruit stalk size reduced by 30 % (Figs. 10.4 and 10.5). The extent of symptoms was dependant on date palm variety (Alhudaib et al. 2007a, b). The association of viral, fungal and nematode pathogens with the disease is not consistent (El-Arosi et al. 1982; Abdulsalam et al. 1992, 1993). A phytoplasma was suspected to be associated with Al-Wijam-affected palms, following histopathology

Table 10.2 Summary of Hemiptera taxa with known vectors of phytoplasmas

Suborder	Family/ super-family	Subfamily	Number of known vector species	Possible date palm phytoplasma vector species (phytoplasma disease, 16Sr subgroup, if known)
Auchenorrhyncha	Cicadellidae	Megophthalminae: Agalliini	1	
		Aphrodinae	2	
		Cicadellinae	1	
		Coelidiinae	5	
		Deltocephalinae	57	*Cicadulina bipunctata* (Al-Wijam disease, 16SrI) (Alhudaib et al. 2007b)
		Iassinae	1	
		Idiocerinae	2	
		Macropsinae	5	
		Scarinae	1	
		Typhlocybinae	3	*Asymmetrasca decedens* (16SrXII-A), (Alhudaib et al. 2007a)
	Fulgoroidea: Cixiidae		7	*Haplaxius crudus* (16SrIV-A), (Harrison et al. 2008), see text.
	Delphacidae		4	
	Derbidae		3	*Patara albida*, *Omolicna joi* (Texas Phoenix Decline, 16SrIV-D), (Halbert et al. 2014)
	Flatidae		1	
Sternorrhyncha	Psylloidea		7	
Heteroptera	Tingidae		1	
	Pentatomidae		1	

Columns 1–4 based on data from Weintraub (2007) and Weintraub and Beanland (2006)

Fig. 10.3 *Cicadulina bipunctata*, the likely vector insect of Saudi Arabian Al-Wijam disease of date palm (Photo: Khalid A. Alhudaib)

Fig. 10.4 Symptoms of Al-Wijam disease of date palm in Saudi Arabia (clockwise from *top left*: fruits, frond rachis and inflorescence) (Photos: Khalid A. Alhudaib)

and antibiotic therapy studies (Abdulsalam et al. 1993). A phytoplasma found in Saudi Arabia had similar symptoms to LY of coconut palm found in Florida; however they found that the identity on 16S rDNA of this date palm phytoplasma with lethal yellowing was only 87 % (El-Zayat et al. 2000). Other diseases of date palm including white tip die-back of date palm and slow decline of date palm found in North Africa (Cronjé et al. 2000a, b), yellowing in Kuwait (Al-Awadhi et al. 2002), and lethal decline in America (see below) (Harrison et al. 2002a) have been associated with phytoplasmas. Lethal decline (LD) of date palm is associated with a phytoplasma that has been classified into the 16SrIV group, subgroup D. The diverse 16SrIV group of phytoplasmas is associated with diseases of coconut and other palms in Central America, the Caribbean and East and West Africa (Harrison and Jones 2003). The phytoplasmas associated with Al-Wijam in Al-Ahsa and possible vectors have been characterized using nested PCR assays (Table 10.2) (Alhudaib et al. 2007b). Two different groups of phytoplasma infecting date palm were

Fig. 10.5 Date palm affected by Al-Wijam disease in Saudi Arabia (Photo: Khalid A. Alhudaib)

identified by comparing the results with phytoplasmas in the GenBank: 16SrI in Al-Ahsa and 16SrII in other locations of Saudi Arabia.

10.5.2 Phytoplasma-Associated Date Palm Diseases in the Western Hemisphere

Lethal yellows (LY) was first reported as a disease of coconut in Jamaica in 1891 and was associated with a phytoplasma in 1972 (Tsai and Harrison 2003). Various similar declines of coconut later determined to be LY were reported from Caribbean and African countries over the years (Howard 1983; McCoy et al. 1983). An epidemic outbreak of LY occurred in Southern Florida starting in the late 1960s and early 1970s (Tsai and Harrison 2003). This epidemic was confined to the southernmost counties in Florida, including the Miami area (Tsai and Harrison 2003). Although recently date palms have been imported from California into Florida and planted as ornamentals or specimen trees, during the 1970s date palms were not common in Florida. However, various *Phoenix* spp. were planted as ornamentals in South Florida and in addition were present in germplasm collections such as the United States Department of Agriculture (USDA) Subtropical Research Unit and Fairchild Tropical Botanical Garden, both located near Miami. Thomas (1974, 1979) and Howard et al. (1979) reported *P. dactylifera*, *P. reclinata* Jacq., *P. sylvestris* L. and *P. canariensis* Chabaud, as well as various other palm genera and species, to be susceptible to a lethal decline associated with mycoplasma-like bodies observed by electron microscopy. *Phoenix dactylifera* appeared to be more susceptible than other *Phoenix* spp., being equivalent in this respect to *C. nucifera* (Howard and Barrant 1989). Symptom expression in *Phoenix* spp. differed somewhat from the classical symptoms shown in *C. nucifera*. Leaf discoloration in *Phoenix* spp. was dusty gray compared to the yellow color exhibited by *C. nucifera*; the first symptom expressed in *Phoenix* spp. was spear leaf necrosis, as compared to immature fruit drop in *C. nucifera* and immature fruit drop and inflorescence necrosis were early symptoms in *Phoenix* spp., compared to being the first and second symptoms, respectively, in *C. nucifera* (McCoy et al. 1983). After a long search for a vector, LY was shown to be vectored by *H. crudus* (Howard et al. 1983, 1984). *Haplaxius crudus* was capable of transmitting LY to *P. canariensis*, although the rate of disease development was slow compared to most other palm species surveyed (Howard et al. 1984). Populations of *H. crudus* on mature *P. dactylifera* were not statistically different from those on mature *C. nucifera* (Howard 1980). However, there are apparently no reports actually confirming *H. crudus* as a vector of LY in *P. dactylifera*.

In 1978, *P. dactylifera* and *P. canariensis* specimen trees in the lower Rio Grande Valley in Texas were observed to be declining (Miller et al. 1980; McCoy et al. 1980a, b). Symptoms were similar to decline of *Phoenix* spp. in Florida associated with phytoplasmas in that the initial symptom consisted of gray discoloration of the spear leaf, followed by fruit drop and inflorescence necrosis. Although no definitive link between the declining palms in Texas and LY could be made, it was assumed that the causal agent was probably the same due to the similarities in symptomatology and etiology (McCoy et al. 1980b). Little additional research was done in this area for a number of years and the disease remained quiescent apparently due to the relatively low population densities and species diversity of palms in the lower Rio Grande Valley. However, in 2001 a decline disease with similar symptomatology was observed on *P. canariensis* in the Corpus Christi area of Texas (Harrison et al. 2002a). Symptomatic *P. canariensis* were associated with LY-group (16SrIV) phytoplasmas. This strain was designated Texas Phoenix decline (TPD) and was designated a new subgroup (16SrIV-D) of the LY phytoplasmas (Harrison et al. 2002a). The TPD phytoplasma was very similar to the previously published *Carludovica palmata* yellows (CPY) (Cordova et al. 2000), which was also included in 16SrIV-D. Harrison et al. (2002a) stated that it is not possible to determine whether or not this outbreak was related to the previous outbreaks in the Rio Grande Valley (approximately 250 km south of Corpus Christi) or another source (such as Yucatán, Mexico, source of the CPY strain report). They note that, at that time, losses of *P. canariensis* in the Corpus Christi area were larger in scope than "minimal" losses of this species due to LY in South Florida as reported by Howard and Barrant (1989). Elliott and Harrison (2007) referenced an undated report from the Texas Department of Agriculture (USA) that *H. crudus* was rarely found in areas with TPD whereas several other planthoppers that could potentially vector TPD were abundant. They concluded that it was unlikely that *H. crudus* was the vector of TPD in Texas. Additional details on this report were unavailable at the time of writing. Currently, TPD occurs sporadically in Texas, being contained as delimited by quarantine since it "does not seem to persist at any location over a prolonged period of time after the removal of the infested tree" (Awinash Bhaktar – personal communication, 2014).

In 2007, *Phoenix* spp. growing in Hillsborough County, Florida were observed to be declining with symptoms similar to those associated with LY and TPD (Harrison et al. 2008; Elliott 2009). Symptomatic trees were assayed and most were found to have TPD strain phytoplasmas associated with them. The TPD phytoplasma was also present in non-*Phoenix* spp., as was a novel phytoplasma tentatively classified as 16SrIV-F; this latter was also present in mixed phytoplasma infections in *P. dactylifera* (Harrison et al. 2008). In Florida, TPD has also been reported to cause decline in Florida's native *Sabal palmetto* (Walt.) Lodd. (Harrison et al. 2009) and a few other non-*Phoenix* genera (Ntushelo et al. 2013b). Additional symptomatic and positive trees were found in Manatee and Sarasota Counties and, later, even farther north in Lake and Duval counties (Florida Department of Agriculture and Consumer Services 2010). These counties are further north than those in Southern Florida in which LY and related palm declines had been reported (Harrison et al. 2008; Elliott 2009). In Florida, TPD is considered a pest of limited distribution. To date (2014), there have been 37 confirmed reports, of which 21 were in 2008–2009 (Gregory Hodges – personal communication, 2014). Of the 37 confirmed reports, 18 have been from *Phoenix* spp. and 16 from *S. palmetto* (Gregory Hodges – pers. comm., 2014). Harrison et al. (2008) stated that it is probable that the pathogen, or an infected vector, was introduced from South Texas. This is in contrast with the viewpoint in Texas that TPD came from Florida (Awinash Bhaktar – personal communication, 2014). Previously, Harrison et al. (2002b) had reported that LY was associated with the 16SrIV-A phytoplasma, which was assumed to be associated with the decline of various palm species in South Florida. Subgroup 16SrIV-A phytoplasmas have been detected in *P. canariensis*, *P. dactylifera*, *P. reclinata*, *P. rupicola* T. Anders. and *P. sylvestris* in South Florida (Nigel Harrison – personal communication, 2014). Harrison et al. (2008) reported two additional phytoplasma strains affecting various palm species, in an area further north than previously reported, which increased both the range of phytoplasma-associated palm declines and potentially the vectors responsible for their dispersal. Although the TPD 16SrIV-D phytoplasma and the LY 16SrIV-A phytoplasma are distinct, they are very closely related (Ntushelo et al. 2012). Both phytoplasmas are present in the Yucatán Peninsula of Mexico (Vázquez-Euán et al. 2011) and Southern Florida but not in Texas (Nigel Harrison – personal communication, 2014). It is not currently known whether *H. crudus* can vector TPD as well as LY, or indeed what the vector for TPD actually is. It was thought that *H. crudus* was restricted to Southern Florida and therefore unlikely to be a TPD vector in areas further north, however, Halbert et al. (2014) reported that *H. crudus* can overwinter as far north in Florida as Gainesville (29.63380 N, 82.37200 W) and speculated that the northern limit of LY might be due to higher latitudinal intolerance of the pathogen rather than the putative vector. In addition, Halbert et al. (2014) identified *Patara albida* Westwood and a new species of planthopper, *Omolicna joi* Wilson, Halbert and Bextine, as potential vectors of TPD based upon their presence on susceptible palm species.

In the USA, dates are produced only in a limited geographic area in South California and Arizona (Krueger 2015). This region is geographically remote from Florida and Texas; however, the potential threat of LY to date production in California and Arizona has long been recognized (McCoy et al. 1976, 1982). Howard et al. (1985) experimentally established a small plantation of commercial date palm varieties in Southern Florida. All were found to be hosts of *H. crudus* although it was present in lower populations than on coconut palms nearby. Howard (1992) reported that 94.1 % of the five cultivars (Deglet Noor, Zahidi, Thoory, Medjool, Halawy) developed symptoms of LY and phytoplasmas were observed in the phloem of all symptomatic palms. Halawy palms survived infection longer than the other cultivars. As *P. dactylifera* is also susceptible to TPD, which may also be either more widely established further north in the Western Hemisphere or be vectored by more northerly established insects, TPD (and possibly other phytoplasmas) can also be considered a potential threat to the U.S. date industry. Date palms with symptoms similar to those associated with LY or TPD have not been observed in US date production. Consequently, there has been no testing for phytoplasmas and they are apparently not present in the US date production area. The climate of U.S. in date production areas is much different from that of Florida or Texas and so might not be conducive to the pathogen or potential vectors. *Haplaxius crudus* and other potential phytoplasma vectors as reported by Weintraub and Beanland (2006) and Weintraub (2007) are not known to occur in California. One planthopper species, *Asarcopus palmarum* Horváth is present at very low incidence and population in California; however, its status as a vector is not known (Thomas M. Perring – personal communication, 2014). The production area in California is currently protected by a state internal quarantine that prohibits the introduction of palm species into the production area.

10.6 Management of Phytoplasmas and Insect Vectors

In management of phytoplasma diseases, the primary concern is prevention and containment rather than treatment. Commercial movement of living plants from locations affected by a phytoplasma to disease-free areas is generally not permitted. However, such diseases may not be recognized until after significant distribution of infected material has taken place. Further, quarantine rigor varies a lot according to the specific geographic areas involved. Effective quarantine can also help minimize the impact of phytoplasmas by restricting the dispersal and establishment of insect vector species.

In affected areas, disease management methods include control of insect vectors and host plant reservoirs in weeds (which may be latent hosts not exhibiting symptoms), roguing of symptomatic plants and avoiding planting susceptible crops next to crops harboring phytoplasmas (Lee et al. 2000; Eziashi and Omamor 2010). Phytoplasma diseases can be prevented or controlled in individual host plants by injection of antibiotic tetracycline-type, oxytetracycline HCl (Terramycin) into the trunk. As a therapeutic measure, biweekly systemic treatment on a four monthly schedule should be applied once the first symptoms have appeared. Whilst this is practicable and economic for ornamental or valuable palms such as in tourist sites or hotels, it is less appropriate for treating large scale crop palm plantations. In these circumstances, the use of genetically resistant ecotypes and hybrids offers the only practical long-term solution and has proven to be effective in some areas affected by LY of coconut.

For phytoplasma vector management, screening to achieve physical exclusion is the most effective method to attain the high levels of vector control that are necessary to prevent a host plant becoming infected. In the case of palms that live for decades, the need for stringent vector control is greater than in shorter-lived field crops that are vulnerable to infection for only months. Though exclusion of vectors can be attained for greenhouse crops, it is not practicable for most other types of crops (Weintraub 2007) including palms. One practical advantage of knowing the vectors of phytoplasma in a given pathosystem is that it allows the targeted use of insecticides; spraying only when and where monitoring shows the vector to be present. This lessens the material and labor costs as well as environmental impact and can give useful levels of disease management (Munyaneza et al. 2010).

Habitat management and the use of genetically modified crops are now replacing the use of chemical control which was widely used for decades (Weintraub and Beanland 2006) though spraying remains useful during an outbreak of phytoplasma disease if the application is done at the right time. As an illustration of the scope for habitat management in the Columbia Basin and Yakima Valley of the USA, outbreaks of potato purple top disease were caused by the beet leafhopper-transmitted virescence agent (BLTVA) phytoplasma. This phytoplasma was known to be transmitted by the beet leafhopper, *Circulifer tenellus* (Baker). Commonly found in crops and adjoining weedy areas, when collected near potato fields 29.6 % of overwintering leafhoppers carried the phytoplasma. In situations such as this, the manipulation of weedy habitats near vulnerable crops is important for control of vectors of phytoplasma diseases. In the case of palm-phytoplasma pathosystems, LY of coconut is promoted by the presence of certain grass species on which the nymphs develop. This favors disease incidence to the extent that palms growing on sandy beaches are often free of disease whilst nearby palms are affected. This has promoted the practice of host grass removal and use of mulches to suppress vector pressure (Howard and Oropeza 1998).

Sterile-male release methods offer scope for vector control in instances where the vector can be bred and females mate only once. This approach is particularly applicable on islands where the isolation facilitates over flooding of the wild population with released sterile males. Whilst this approach has been considered for phytoplasma pathosystems (Ogle and Harries 2005) we are unaware of any reported success; probably because the females are able to mate more than once. A potentially powerful tool for controlling phytoplasma transmission is through the manipulation of symbiotic bacteria to reduce vector competence (Weintraub 2007). However, the major challenge is the delivery of the transgenic bacteria to the target vectors without negative effect to the environment or other insect populations.

Overall, preventing a phytoplasma-associated disease from becoming established in a given area is central to effective management of this threat to palms. Growing awareness of phytoplasma and the advent of molecular methods for their detection should help this to be achieved more effectively in future years. Whilst there is a range of methods available for vector control and antibiotics can be used for direct control of phytoplasmas, the development of resistant cultivars and gradual replanting of affected areas with these is the best long-term strategy. Clearly, however, pathogen populations are able to adapt to new host genotypes and resistance "breakdown" has been reported in many pathosystems. The risk of this occurring in phytoplasma-palm interactions is particularly high because the generation time of phytoplasmas is very short compared with the lifespan of a given stand of palms. Ultimately this will demand the use of resistance management strategies such as "stacking" of resistance genes in a given cultivar, using mixed cultivar plantings rather than the sole use of a given cultivar over wide areas, and combining host plant resistance with other methods including vector control. Presently, however, the development of such advanced integrated disease management systems for phytoplasmas is in its infancy.

10.7 Outlook and Conclusion

Despite the economic importance of date palm, it is a relatively under-researched crop species. The lack of information on phytoplasma diseases in this crop is compounded by the general difficulties of working with this group of bacteria. The high and growing level of availability of molecular methods, technicians and adequately equipped laboratories is leading to a rapid growth in research on phytoplasma pathosystems and date palm will benefit from this general trend. The available information on date palm phytoplasmas is sufficient to conclude that this group of pathogens constitutes a significant threat to the crop but that there are important knowledge gaps, particularly in relation to insect vectors of the several phytoplasmas that have been reported from date palm. Key priorities for minimizing the impact of phytoplasmas on this crop are twofold. First, quarantine and phytosanitary measures need to be rigorously applied to minimise the international spread of infected planting material and insect vectors. Second, disease resistant varieties need to be developed for the prevalent phytoplasmas. Sustainable management of date palm phytoplasmas will also require accurate knowledge of insect vector species and methods for their suppression.

Acknowledgments We thank Mrs. Anne Johnson of Charles Sturt University, Australia for assistance with manuscript preparation. Geoff Gurr is supported by the Australian Research Council, Australia and the Chinese Thousand Talents Program, China.

References

Abdulsalam, K., Nageeb, M., Rezk, M., & Abdel-Megeed, M. (1992). Survey of certain fungi associated with Wijamed date palm trees in Al-Hassa Oasis of Saudi Arabia. *Annals of Agricultural Science (Ain Shams University, Egypt), 37*, 603–611.

Abdulsalam, K., Abdel-Megeed, M., Rezk, M., & Nageeb, M. (1993). The influence of oxytetracycline on Wijamed date palm trees. *Annals of Agricultural Science (Ain Shams University, Egypt), 38*, 301–309.

Al-Awadhi, H., Hanif, A., Suleman, P., & Montasser, M. (2002). Molecular and microscopical detection phytoplasma associated with yellowing disease of date palm *Phoenix dactylifera* L in Kuwait. *Kuwait Journal of Science and Engineering, 29*(2), 87–109.

Alhudaib, K., Arocha, Y., Wilson, M., & Jones, P. (2007a). "Al-Wijam", a new phytoplasma disease of date palm in Saudi Arabia. *Bulletin of Insectology, 60*(2), 285–286.

Alhudaib, K., Arocha, Y., Wilson, M., & Jones, P. (2007b). Identification and molecular characterization of a phytoplasma associated with Al-Wijam disease of date palm in Saudi Arabia. *Arab Journal of Plant Protection, 25*, 116–122.

Alvarez, E., Mejía, J. F., Contaldo, N., Paltrinieri, S., Duduk, B., & Bertaccini, A. (2014). '*Candidatus* Phytoplasma asteris' strains associated with oil palm lethal wilt in Colombia. *Plant Disease, 98*(3), 311–318.

Baric, S., & Dalla-Via, J. (2004). A new approach to apple proliferation detection: A highly sensitive real-time PCR assay. *Journal of Microbiological Methods, 57*(1), 135–145.

Bekele, B., Hodgetts, J., Tomlinson, J., Boonham, N., Nikolić, P., Swarbrick, P., & Dickinson, M. (2011). Use of a realtime LAMP isothermal assay for detecting 16SrII and XII phytoplasmas in fruit and weeds of the Ethiopian Rift Valley. *Plant Pathology, 60*(2), 345–355.

Bertaccini, A. (2007). Phytoplasmas: Diversity, taxonomy, and epidemiology. *Frontiers in Bioscience, 12*, 673–689.

Bertaccini, A., Duduk, B., Paltrinieri, S., & Contaldo, N. (2014). Phytoplasmas and phytoplasma diseases: A severe threat to agriculture. *American Journal of Plant Sciences, 5*(12), 1763–1788.

Bertin, S., Guglielmino, C., Karam, N., Gomulski, L., Malacrida, A., & Gasperi, G. (2007). Diffusion of the Nearctic leafhopper *Scaphoideus titanus* Ball in Europe: A consequence of human trading activity. *Genetica, 131*(3), 275–285.

Carpenter, J. B., & Elmer, H. S. (1978). *Pests and diseases of the date palm* (United States Department of Agriculture – Agricultural Research Service Handbook No 527, 42 pp). Washington, DC: Department of Agriculture, Science and Education Administration.

Carraro, L., Loi, N., Ermacora, P., & Osler, R. (1998). High tolerance of European plum varieties to plum leptonecrosis. *European Journal of Plant Pathology, 104*(2), 141–145.

Chao, C. T., & Krueger, R. R. (2007). The date palm (*Phoenix dactylifera* L.): Overview of biology, uses, and cultivation. *HortScience, 42*(5), 1077–1082.

Christensen, N. M., Nicolaisen, M., Hansen, M., & Schulz, A. (2004). Distribution of phytoplasmas in infected plants as revealed by real-time PCR and bioimaging. *Molecular Plant-Microbe Interactions, 17*(11), 1175–1184.

Contaldo, N., Bertaccini, A., Paltrinieri, S., Windsor, H., & Windsor, G. (2012). Axenic culture of plant pathogenic phytoplasmas. *Phytopathologia Mediterranea, 51*(3), 607–617.

Cordova, I., Oropeza, C., Almeyda, H., & Harrison, N. A. (2000). First report of a phytoplasma-associated leaf yellowing syndrome of palma jipi plants in Southern México. *Plant Disease, 84*(7), 807.

Cronjé, P., Dabek, A. J., Jones, P., & Tymon, A. M. (2000a). First report of a phytoplasma associated with a disease of date palms in North Africa. *Plant Pathology, 49*(6), 801.

Cronjé, P., Dabek, A. J., Jones, P., & Tymon, A. M. (2000b). Slow decline: A new disease of mature date palms in North Africa associated with a phytoplasma. *Plant Pathology, 49*(6), 804.

Deng, S., & Hiruki, C. (1990). Enhanced detection of a plant pathogenic mycoplasma-like organism by polymerase chain reaction. *Proceedings of the Japan Academy. Series B, Physical and Biological Sciences, 66*(7), 140–144.

Deng, S., & Hiruki, C. (1991). Amplification of 16S rRNA genes from culturable and nonculturable mollicutes. *Journal of Microbiological Methods, 14*(1), 53–61.

Djerbi, M. (1983). Diseases of the date palm. *FAO Regional Project for Palms and Dates. Baghdad, Iraq* (112 p).

Doi, Y., Teranaka, M., Yora, K., & Asuyama, H. (1967). Mycoplasma or PLT grouplike microorganisms found in the phloem elements of plants infected with mulberry dwarf, potato witches' broom, aster yellows or paulownia witches' broom. *Annals of Phytopathological Society, Japan, 33*, 259–266.

El-Arosi, H., El-Said, H., Najieb, M., & Jabeen, N. (1982). Al-Wijam, decline date palm disease. In Y. M. Makki (Ed.), *Proceedings of the first symposium on date palm*, 23–25 March 1982, Al-Hassa (388–403 pp).

Elliott, M. L. (2009). Emerging palm diseases in Florida. *HortTechnology, 19*(4), 717–718.

Elliott, M. L., & Harrison, N. A. (2007). *Palm diseases caused by phytoplasmas*. Available at: http://www.freshfromflorida.com/content/download/9785/134978/Palm%20Diseases%20Caused%20by%20Phytoplasmas.pdf. Accessed on 18 July, 2014.

El-Zayat, M., Abdusalm, K., Shamlool, A., Djerbi, M., & Hadidi, A. (2000). Phytoplasma detected in date palm trees infected by Al-Wijam in Kingdom of Saudi Arabia. In *Proceedings of the date palm international symposium*, 22–25 February 2000, Windhoek (230–236 pp).

Eziashi, E., & Omamor, I. (2010). Lethal yellowing disease of the coconut palms (*Cocos nucifera* L.): An overview of the crises. *African Journal of Biotechnology, 9*(54), 9122–9127.

Firrao, G., Garcia-Chapa, M., & Marzachì, C. (2007). Phytoplasmas: Genetics, diagnosis and relationships with the plant and insect host. *Frontiers in Bioscience, 12*, 1353–1375.

Florida Department of Agriculture and Consumer Services. (2008). *Pest alert: Texas phoenix palm decline*. Available at: http://www.freshfromflorida.com/Divisions-Offices/Plant-Industry/Plant-Industry-Publications/Pest-Alerts/Pest-Alerts-Texas-Phoenix-Palm-Decline. Accessed on 18 July 2014.

Florida Department of Agriculture and Consumer Services. (2010). *Texas phoenix palm decline sites*. Available at: http://www.freshfromflorida.com/content/download/11473/145134/TPPD.pdf. Accessed on 18 July 2014.

Fránová, J., Ludvíková, H., Paprštein, F., & Bertaccini, A. (2013). Genetic diversity of Czech '*Candidatus* Phytoplasma mali' strains based on multilocus gene analyses. *European Journal of Plant Pathology, 136*(4), 675–688.

Fukuta, S., Kato, S., Yoshida, K., Mizukami, Y., Ishida, A., Ueda, J., Kanbe, M., & Ishimoto, Y. (2003). Detection of tomato yellow leaf curl virus by loop-mediated isothermal amplification reaction. *Journal of Virological Methods, 112*(1–2), 35–40.

Halbert, S. E., Wilson, S. W., Bextine, B., & Youngblood, S. B. (2014). Potential planthopper vectors of palm phytoplasmas in Florida with a description of a new species of the genus *Omolicna* (Hemiptera: Fulgoroidea). *Florida Entomologist, 97*(1), 90–97.

Harrison, N., & Jones, P. (2003). Diseases of coconut. In R. C. Ploetz (Ed.), *Diseases of tropical fruit crops* (197–226 pp). Wallingford: CABI Publishing.

Harrison, N. A., Myrie, W., Jones, P., Carpio, M. L., Castillo, M., Doyle, M. M., & Oropeza, C. (2002a). 16S rRNA interoperon sequence heterogeneity distinguishes strain populations of

palm lethal yellowing phytoplasma in the Caribbean region. *Annals of Applied Biology, 141*(2), 183–193.
Harrison, N. A., Womack, M., & Carpio, M. L. (2002b). Detection and characterization of a lethal yellowing (16SrIV) group phytoplasma in Canary Island date palms affected by lethal decline in Texas. *Plant Disease, 86*(6), 676–681.
Harrison, N. A., Helmick, E. E., & Elliott, M. L. (2008). Lethal yellowing-type diseases of palms associated with phytoplasmas newly identified in Florida, USA. *Annals of Applied Biology, 153*(1), 85–94.
Harrison, N. A., Helmick, E. E., & Elliott, M. L. (2009). First report of a phytoplasma-associated lethal decline of *Sabal palmetto* in Florida, USA. *Plant Pathology, 58*(4), 792.
Harrison, N. A., Davis, R. E., Oropeza, C., Helmick, E. E., Narváez, M., Eden-Green, S., Dollet, M., & Dickinson, M. (2014). '*Candidatus* Phytoplasma palmicola', associated with a lethal yellowing-type disease of coconut (*Cocos nucifera* L.) in Mozambique. *International Journal of Systematic and Evolutionary Microbiology, 64*, 1890–1899.
Himeno, M., Neriya, Y., Minato, N., Miura, C., Sugawara, K., Ishii, Y., Yamaji, Y., Kakizawa, S., Oshima, K., & Namba, S. (2011). Unique morphological changes in plant pathogenic phytoplasma-infected petunia flowers are related to transcriptional regulation of floral homeotic genes in an organ-specific manner. *The Plant Journal, 67*(6), 971–979.
Hodgetts, J., Boonham, N., Mumford, R., Harrison, N., & Dickinson, M. (2008). Phytoplasma phylogenetics based on analysis of secA and 23S rRNA gene sequences for improved resolution of candidate species of '*Candidatus* Phytoplasma'. *International Journal of Systematic and Evolutionary Microbiology, 58*(8), 1826–1837.
Hodgetts, J., Boonham, N., Mumford, R., & Dickinson, M. (2009). Panel of 23S rRNA gene-based real-time PCR assays for improved universal and group-specific detection of phytoplasmas. *Applied and Environmental Microbiology, 75*(9), 2945–2950.
Hodgetts, J., Tomlinson, J., Boonham, N., González-Martín, I., Nikolić, P., Swarbrick, P., Yankey, E. N., & Dickinson, M. (2011). Development of rapid in-field loop-mediated isothermal amplification (LAMP) assays for phytoplasmas. *Bulletin of Insectology* (Supplementary), *64*, S41–S42.
Howard, F. W. (1980). Attractiveness of date and coconut palms to *Myndus crudus* and other homopterans. *Proceedings of the Florida State Horticultural Society, 93*, 199–201.
Howard, F. W. (1983). World distribution and possible geographic origin of palm lethal yellowing disease and its vectors. *FAO Plant Protection Bulletin, 31*, 101–113.
Howard, F. W. (1992). Lethal yellowing susceptibility of date palms in Florida. *Principes, 36*(4), 217–222.
Howard, F. W., & Barrant, C. I. (1989). Questions and answers about lethal yellowing diseases. *Principes, 33*, 163–171.
Howard, F. W., & Oropeza, C. (1998). Organic mulch as a factor in the nymphal habitat of Myndus crudus (Hemiptera: Auchenorrhyncha: Cixiidae). *Florida Entomologist, 81*, 92–97.
Howard, F. W., Thomas, D. L., Donselman, H. M., & Collins, M. E. (1979). Susceptibilities of palm species to mycoplasma organism-associated diseases in Florida. *FAO Plant Protection Bulletin, 27*, 109–117.
Howard, F. W., Norris, R., & Thomas, D. (1983). Evidence of transmission of palm lethal yellowing agent by a planthopper, *Myndus crudus* (Homoptera, Cixiidae). *Tropical Agriculture, 60*(3), 168–171.
Howard, F. W., Williams, D. S., & Norris, R. C. (1984). Insect transmission of lethal yellowing to young palms. *International Journal of Entomology, 26*(4), 331–338.
Howard, F. W., Atilano, R., & Williams, D. (1985). Experimental establishment of five date palm cultivars in Southern Florida. *The Date Palm Journal, 4*, 91–101.
Hren, M., Boben, J., Rotter, A., Kralj, P., Gruden, K., & Ravnikar, M. (2007). Real-time PCR detection systems for Flavescence dorée and Bois noir phytoplasmas in grapevine: Comparison with conventional PCR detection and application in diagnostics. *Plant Pathology, 56*(5), 785–796.
IRPCM. (2004). '*Candidatus* Phytoplasma', a taxon for the wall-less, non-helical prokaryotes that colonize plant phloem and insects. *International Journal of Systematic and Evolutionary Microbiology, 54*(4), 1243–1255.

Iwamoto, T., Sonobe, T., & Hayashi, K. (2003). Loop-mediated isothermal amplification for direct detection of *Mycobacterium tuberculosis* complex, *M. avium*, and *M. intracellulare* in sputum samples. *Journal of Clinical Microbiology, 41*(6), 2616–2622.

Jarausch, W., Lansac, M., & Dosba, F. (1999). Seasonal colonization pattern of European stone fruit yellows phytoplasmas in different prunus species detected by specific PCR. *Journal of Phytopathology, 147*(1), 47–54.

Jarausch, W., Peccerella, T., Schwind, N., Jarausch, B., & Krczal, G. (2004). Establishment of a quantitative real-time PCR assay for the quantification of apple proliferation phytoplasmas in plants and insects. *Acta Horticulturae, 657*, 415–420.

Kollar, A., Seemüller, E., Bonnet, F., Saillard, C., & Bove, J. (1990). Isolation of the DNA of various plant pathogenic mycoplasmalike organisms from infected plants. *Phytopathology, 80*(3), 233–237.

Krueger, R. R. (2015). Date palm status and perspective in United States of America. In J. M. Al-Khayri, S. M. Jain, & D. V. Johnson (Eds.), *Date palm genetic resources and utilization* (Africa and the Americas, Vol. 1, 544 p). Dordrecht: Springer, Science + Business Media.

Lee, I.-M., Hammond, R., Davis, R., & Gundersen, D. (1993). Universal amplification and analysis of pathogen 16S rDNA for classification and identification of mycoplasmalike organisms. *Phytopathology, 83*(8), 834–842.

Lee, I.-M., Gundersen, D. E., Hammond, R. W., & Davis, R. E. (1994). Use of mycoplasmalike organism (MLO) group-specific oligonucleotide primers for nested-PCR assays to detect mixed-MLO infections in a single host-plant. *Phytopathology, 84*(6), 559–566.

Lee, I.-M., Bertaccini, A., Vibio, M., & Gundersen, D. (1995). Detection of multiple phytoplasmas in perennial fruit trees with decline symptoms in Italy. *Phytopathology, 85*(6), 728–735.

Lee, I.-M., Gundersen-Rindal, D. E., Davis, R. E., & Bartoszyk, I. M. (1998). Revised classification scheme of phytoplasmas based on RFLP analysis of 16S rRNA and ribosomal protein gene sequences. *International Journal of Systematic Bacteriology, 48*(4), 1153–1169.

Lee, I.-M., Davis, R. E., & Gundersen-Rindal, D. E. (2000). Phytoplasma: Phytopathogenic mollicutes. *Annual Review of Microbiology, 54*(1), 221–255.

Lee, I.-M., Zhao, Y., Davis, R. E., Wei, W., & Martini, M. (2007). Prospects of DNA-based systems for differentiation and classification of phytoplasmas. *Bulletin of Insectology, 60*(2), 239–244.

Lee, I.-M., Bottner-Parker, K. D., Zhao, Y., Bertaccini, A., & Davis, R. E. (2012). Differentiation and classification of phytoplasmas in the pigeon pea witches'-broom group (16SrIX): an update based on multiple gene sequence analysis. *International Journal of Systematic and Evolutionary Microbiology, 62*, 2279–2285.

Makarova, O., Contaldo, N., Paltrinieri, S., Kawube, G., Bertaccini, A., & Nicolaisen, M. (2012). DNA barcoding for identification of '*Candidatus* Phytoplasmas' using a fragment of the elongation factor Tu gene. *PLoS One, 7*(12), e52092. doi:10.1371/journal.pone.0052092.

Martini, M., Botti, S., Marcone, C., Marzachì, C., Casati, P., Bianco, P. A., Benedetti, R., & Bertaccini, A. (2002). Genetic variability among flavescence dorée phytoplasmas from different origins in Italy and France. *Molecular and Cellular Probes, 16*(3), 197–208.

Martini, M., Lee, I.-M., Bottner, K. D., Zhao, Y., Botti, S., Bertaccini, A., Harrison, N. A., Carraro, L., Marcone, C., Khan, A. J., & Osler, R. (2007). Ribosomal protein gene-based phylogeny for finer differentiation and classification of phytoplasmas. *International Journal of Systematic and Evolutionary Microbiology, 57*(9), 2037–2051.

McCoy, R. E., Thomas, D. L., & Tsai, J. H. (1976). Lethal yellowing: A potential danger to date production. *Date Growers Institute Report, 53*, 4–8.

McCoy, R. E., Miller, M. E., & Williams, D. S. (1980a). Lethal yellowing in Texas *Phoenix* palms. *Principes, 24*, 179–180.

McCoy, R. E., Miller, M. E., Thomas, D. L., & Amador, J. (1980b). Lethal decline of Phoenix palms in Texas associated with mycoplasma-like organisms. *Plant Disease, 64*, 1038–1040.

McCoy, R. E., Thomas, D. L., & Tsai, J. H. (1982). Lethal yellowing: a potential danger to date production. *Date Palm Journal, 1*, 295–305.

McCoy, R. E., Howard, F. W., Tsai, J. H., Donselman, H. M., Thomas, D. L., Basham, R. A., Atilano, R. A., Eskafi, F. M., Britt, L., & Collins, M. E. (1983). *Lethal yellowing of palms* (Institute of Food and Agricultural Sciences Bulletin No. 834, 100 p). Florida: Institute of Food and Agricultural Sciences, University of Florida.

McCoy, R. E., Caudwell, A., Chang, C. J., Chen, T. A., Chiykowski, L. N., Cousin, M. T., Dale, J. L., De Leeuw, G. T. N., Golino, D. A., Hacket, K. J., Kirkpatrick, B. C., Marvitz, R., Petzold, H., Sinha, R. C., Sugiura, M., Whitcomb, R. F., Yang, I. L., Zhu, B. M., & Seemüller, E. (1989). Plant diseases associated with mycoplasma-like organisms. In R. F. Whitcomb & J. C. Tully (Eds.), *The mycoplasmas, (Volume 5), spiroplasmas, acholeplasmas, and mycoplasmas of plants and arthropods* (545–640 pp). San Diego: Academic.

Miller, M., Maxwell, N., & Amador, J. (1980). Lethal decline of *Phoenix canariensis* and *P. dactylifera* in the Rio Grande Valley. *Journal Rio Grande Horticultural Society, 34*, 89–95.

Mitrović, J., Kakizawa, S., Duduk, B., Oshima, K., Namba, S., & Bertaccini, A. (2011). The groEL gene as an additional marker for finer differentiation of '*Candidatus* Phytoplasma asteris'-related strains. *Annals of Applied Biology, 159*(1), 41–48.

Munyaneza, J. E., Crosslin, J. M., Upton, J. E., & Buchman, J. L. (2010). Incidence of the beet leafhopper-transmitted virescence agent phytoplasma in local populations of the beet leafhopper, *Circulifer tenellus*, in Washington state. *Journal of Insect Science, 10*, 18. doi:10.1673/031.010.1801.

Namba, S., Kato, S., Iwanami, S., Oyaizu, H., Shiozawa, H., & Tsuchizaki, T. (1993). Detection and differentiation of plant-pathogenic mycoplasmalike organisms using polymerase chain reaction. *Phytopathology, 83*(7), 786–791.

Nicolaisen, M., Contaldo, N., Makarova, O., Paltrinieri, S., & Bertaccini, A. (2011). Deep amplicon sequencing reveals mixed phytoplasma infection within single grapevine plants. *Bulletin of Insectology* (Supplementary), *64*, S35–S36.

Notomi, T., Okayama, H., Masubuchi, H., Yonekawa, T., Watanabe, K., Amino, N., & Hase, T. (2000). Loop-mediated isothermal amplification of DNA. *Nucleic Acids Research, 28*(12), e63. doi:10.1093/nar/28.12.e63.

Ntushelo, K., Harrison, N. A., & Elliott, M. L. (2012). Comparison of the ribosomal RNA operon from Texas *Phoenix* decline and lethal yellowing phytoplasmas. *European Journal of Plant Pathology, 133*(4), 779–782.

Ntushelo, K., Harrison, N. A., & Elliott, M. L. (2013a). Differences between the Texas phoenix palm phytoplasma and the coconut lethal yellowing phytoplasma revealed by restriction fragment length polymorphism (RFLP) analysis of the *NUSA* and *HFLB* genes. *African Journal of Biotechnology, 12*(25), 3934–3939.

Ntushelo, K., Harrison, N. A., & Elliott, M. L. (2013b). Palm phytoplasmas in the Caribbean basin. *Palms, 57*(2), 93–100.

Obura, E., Masiga, D., Wachira, F., Gurja, B., & Khan, Z. R. (2011). Detection of phytoplasma by loop-mediated isothermal amplification of DNA (LAMP). *Journal of Microbiological Methods, 84*(2), 312–316.

Ogle, L., & Harries, H. (2005). Introducing the vector: How coconut lethal yellowing disease may have reached the Caribbean. *Ethnobotany Research and Applications, 3*, 139–142.

Ong, K., & Mcbride, S. (2009). *Palm diseases caused by phytoplasmas in Texas*. Texas Plant Disease Diagnostic Laboratory, Department of Plant Pathology and Microbiology, Texas AgriLife Extension Service, Texas A&M System, Texas A&M University, USA. Available at: http://www.npdn.org/webfm_send/1065. Accessed on: 18 July 2014.

Oshima, K., Kakizawa, S., Nishigawa, H., Jung, H. Y., Wei, W., Suzuki, S., Arashida, R., Nakata, D., Miyata, S., Ugaki, M., & Namba, S. (2004). Reductive evolution suggested from the complete genome sequence of a plant-pathogenic phytoplasma. *Nature Genetics, 36*(1), 27–29.

Plavsic-Banjac, B., Hunt, P., & Maramorosch, K. (1972). Mycoplasmalike bodies associated with lethal yellowing disease of coconut palms. *Phytopathology, 62*, 298–299.

Ploetz, R. C., Ohr, H. D., Carpenter, J. B., & Pinkas, Y. (2003). Diseases of date. In R. C. Ploetz (Ed.), *Diseases of tropical fruit crops* (227–239 pp). Wallingford: CAB International.

Schaad, N. W., Frederick, R. D., Shaw, J., Schneider, W. L., Hickson, R., Petrillo, M. D., & Luster, D. G. (2003). Advances in molecular-based diagnostics in meeting crop biosecurity and phytosanitary issues. *Annual Review of Phytopathology, 41*(1), 305–324.

Schneider, B., & Seemüller, E. (1994). Presence of two sets of ribosomal genes in phytopathogenic mollicutes. *Applied and Environmental Microbiology, 60*(9), 3409–3412.

Schneider, B., Ahrens, U., Kirkpatrick, B. C., & Seemüller, E. (1993). Classification of plant-pathogenic mycoplasma-like organisms using restriction-site analysis of PCR-amplified 16S rDNA. *Journal of General Microbiology, 139*(3), 519–527.

Seemüller, E., & Schneider, B. (2004). '*Candidatus* Phytoplasma mali', '*Candidatus* Phytoplasma pyri' and '*Candidatus* Phytoplasma prunorum', the causal agents of apple proliferation, pear decline and European stone fruit yellows, respectively. *International Journal of Systematic and Evolutionary Microbiology, 54*(4), 1217–1226.

Seemüller, E., Kunze, L., & Schaper, U. (1984). Colonization behaviour of MLO and symptom expression of proliferation-diseased apple trees and decline-diseased pear trees over a period of several years. *Journal of Plant Disease and Protection, 91*(5), 525–532.

Sugio, A., Kingdom, H. N., Maclean, A. M., Grieve, V. M., & Hogenhout, S. A. (2011). Phytoplasma protein effector SAP11 enhances insect vector reproduction by manipulating plant development and defense hormone biosynthesis. *Proceedings of the National Academy of Sciences, 108*(48), E1254–E1263.

Tanne, E., Boudon-Padieu, E., Clair, D., Davidovich, M., Melamed, S., & Klein, M. (2001). Detection of Phytoplasma by polymerase chain reaction of insect feeding medium and its use in determining vectoring ability. *Phytopathology, 91*(8), 741–746.

Thomas, D. L. (1974). Possible link between declining palm species and lethal yellowing of coconut palms. *Florida State Horticultural Society, Florida Agricultural Experiment Stations Journal Series No. 5640*, USA, 502–504 pp.

Thomas, D. L. (1979). Mycoplasma like bodies associated with lethal declines of palms in Florida. *Phytopathology, 69*(9), 928–934.

Tomlinson, J. A., Boonham, N., & Dickinson, M. (2010). Development and evaluation of a one-hour DNA extraction and loop-mediated isothermal amplification assay for rapid detection of phytoplasmas. *Plant Pathology, 59*(3), 465–471.

Tsai, J. H., & Harrison, N. A. (2003). Lethal yellowing and lethal declines of palms. In G. Loebenstein & G. Thottappilly (Eds.), *Virus and virus-like diseases of major crops in developing countries* (597–606 pp). Dordrecht: Springer Science+Business Media.

Vázquez-Euán, R., Harrison, N., Narvaez, M., & Oropeza, C. (2011). Occurrence of a 16SrIV group phytoplasma not previously associated with palm species in Yucatan, Mexico. *Plant Disease, 95*(3), 256–262.

Weintraub, P. G. (2007). Insect vectors of phytoplasmas and their control – An update. *Bulletin of Insectology, 60*(2), 169–173.

Weintraub, P. G., & Beanland, L. (2006). Insect vectors of phytoplasmas. *Annual Review of Entomology, 51*(1), 91–111.

Wilson, M. R. (1987). The Auchenorrhyncha (Homoptera) associated with palms. In M. Wilson & L. Nault (Eds.), *Proceedings of the 2nd international workshop on leafhoppers and planthoppers of economic importance*, 28 July to 1 August 1986, Provo. 327–342 pp.

Win, N. K., Lee, S. Y., Bertaccini, A., Namba, S., & Jung, H. Y. (2013). *Candidatus* Phytoplasma balanitae' associated with witches' broom disease of *Balanites triflora. International Journal of Systematic and Evolutionary Microbiology (Pt 2), 63*, 636–640.

Zhang, J., Miller, S., Hoy, C., Zhou, X., & Nault, L. (1998). A rapid method for detection and differentiation of aster yellows phytoplasma-infected and-inoculative leafhoppers. *Phytopathology* (Supplementary), *88*, S84.

Zhao, Y., Wei, W., Lee, I.-M., Shao, J., Suo, X., & Davis, R. E. (2009). Construction of an interactive online phytoplasma classification tool, iPhyClassifier, and its application in analysis of the peach X-disease phytoplasma group (16SrIII). *International Journal of Systematic and Evolutionary Microbiology, 59*(10), 2582–2593.

Chapter 11
The Role of Semiochemicals in Date Pest Management

Victoria Soroker, Ally Harari, and Jose Romeno Faleiro

Abstract This chapter covers general concepts of semiochemical implementations in pest management and the work done on the identification and use of semiochemicals for the management of date palm pests. Within the semiochemicals, pheromones are usually species specific and non-toxic to the environment. This makes pheromones in particular valuable components of integrated pest management as indicators of pest presence and as behavior modifiers, interfering with the proliferation of target pest. Control of pest populations can be achieved by a number of methods based on attractants, such as mating disruption, mass trapping, attract-and-kill, push-pull, or attract-and-infect. For any of these control tactics knowledge of the chemical ecology of pests and the identification of cues involved are essential. Isolation and identification of semiochemicals is a complicated process requiring: solvent extraction or head space collection by solid-phase microextraction (SPME) or other means, identification of active components by coupled gas chromatography-electroantennography (GC-EAG) and finally identification of chemical composition by gas chromatography-mass spectrometry (GC-MS) and nuclear magnetic resonance (NMR). A summary of known attractants of date pests is also provided. Currently effective attractants, pheromones, and/or host volatiles are known only for some of the major date palm pests. Case studies are described on sap beetles, the red palm weevil, *Rhynchophorus ferrugineus*, and the lesser date moth, *Batrachedra amydraula*. For red palm weevil and sap beetles, implementation of pheromones in combination with plant volatiles as a tool for monitoring is well accepted but essential information for optimization of their use is still missing. Moreover, development of semiochemical application as a control strategy, for most cases, is in its infancy and will remain a major challenge in future years. For other pests of palm trees, pheromones and other semiochemicals remain to be identified. Future

V. Soroker (✉) • A. Harari
Department of Entomology, Agricultural Research Organization, The Volcani Center, Bet Dagan, Israel
e-mail: sorokerv@volcani.agri.gov.il; aharari@volcani.agri.gov.il

J.R. Faleiro
Food and Agriculture Organization of the United Nations (UTF/SAU/043/SAU), Date Palm Research Center, Al-Ahsa, Saudi Arabia
e-mail: jrfaleiro@yahoo.co.in

W. Wakil et al. (eds.), *Sustainable Pest Management in Date Palm: Current Status and Emerging Challenges*, Sustainability in Plant and Crop Protection,
DOI 10.1007/978-3-319-24397-9_11

challenges and perspectives for the implementation of semiochemicals towards sustainable pest control in date plantations are given.

11.1 Introduction

Date palm, *Phoenix dactylifera* L., has been cultivated for nearly 6000 years and is an important crop in many arid regions in the world. The selection of superior female palms and their propagation by offshoots were the early techniques developed to enhance crop productivity and fruit quality in date palm (Johnson 2011). This technology remains the most popular even after tissue culture technique for date palms was developed. Although much effort was invested over the years in breeding date palms for quality and productivity not much effort has been put into breeding date palms aimed at pest resistance (Howard 2001). Yet, date palms offer a good habitat with a reliable food source year around, for numerous insects and other arthropods, some of which are pests, feeding on palm leaves, fruits, stipe and roots. The majority of date palm insect pests belong to three main orders: Hemiptera, Coleoptera and Lepidoptera. Some members of these orders have reached the status of major pests, endangering yield, fruit quality or even palm survival itself, in the major date growing areas. Moreover, for some of the date palm pests, detection and management is challenging as most of the pests are difficult to detect and not easily accessible for control measures. These realities drive the growers towards prophylactic chemical treatments. The negative aspects of prophylactic chemical control, although relatively cheap, have long been realized especially if sustainability and environmental safety are considered.

The objectives of ideal pest control are to reduce losses from pests in ways that are effective, economically sound, ecologically compatible and therefore, sustainable. An integrated pest management (IPM) strategy combines biological, cultural, physical and chemical tools in a way that minimizes the cost, side effects and risks to the environment. Ideally, IPM is based on pest prevention with minor intervention. It also includes monitoring and setting action thresholds. A substantial volume of literature supports the significant role of chemical cues (semiochemicals) in IPM, while regulatory agencies generally consider them safe to use (Witzgall et al. 2010). Semiochemicals are well-known management tools especially for cryptic species. The identification of pheromones and other attractive semiochemical agents enable the timing of chemical applications of a large number of pests using monitor traps.

Still IPM programs for date plantations are uncommon and those based on semiochemicals are especially scarce. This chapter covers the general concepts of semiochemical implementation in pest management and the work that has been done on their identification and implementation for some of the main date palm pests. We also discuss prospects and constrains in the implementation of semiochemicals outlining the need for further development.

11.2 Semiochemicals and Their Potential Use in IPM

Chemical cues are the oldest form of communication in the biosphere. Most organisms, including insects rely on chemical cues for their behavior. These cues, generally termed as behavior modifying compounds (BMS) or semiochemicals, are essential for pest survival including location of habitat, food and mates, enabling them to escape competition and natural enemies and to avoid plant defenses. In the literature, semiochemicals are divided into several groups according to their source and their effects on the emitter and the receiver. Those that act among intraspecifics are termed pheromones, while those that affect separate species are called allelochemicals (Wyatt 2003). Chemical signals have several advantages over other modes of communication being highly species specific (in case of pheromones), not affected by light conditions, not blocked by mechanical obstacles and are able to transfer information over large distances (depending on the chemical properties of the message and the medium).

The use of semiochemicals in pest management was developed more than a hundred years ago, long before chemical identification and synthesis was possible. With rapid development of sensitive analytical techniques during the past 50 years or so, it has been possible to identify chemical signals/cues for thousands of insect species. Most of them are species specific cues and, particularly sex pheromones that attract members of the opposite sex. Others that attract both sexes of species are termed aggregation pheromones (Borden 1985). Often the attraction of aggregation pheromones is synergized by plant volatiles. So that combination of the pheromones and the plant volatiles attract significantly more conspecifics than each of them alone. For some insect pests, natural repellents produced by plants and/or other organisms are also known (El-Sayed 2014). Most identified insect pheromones are from the orders Lepidoptera and Coleoptera. Sex pheromones are used to successfully control Lepidoptera and aggregation pheromones have been developed to control Coleoptera (Witzgall et al. 2010).

The species specificity of pheromones makes them ideal for selective pest management. The use of semiochemicals and in particularly pheromones with or without host coattractants, is mainly to monitor and detect pest populations. Synthetic pheromone lures are powerful tools that indicate early stages of pest or invasion, thereby enabling proper timing of selected control measures. In addition, semiochemicals can be used as specific control tools in management programs, especially when the pest population is low or pesticide applications are not possible or not effective as when insecticides might have difficulty reaching the target pest. The tactics in these cases are mass attraction to lower pest numbers or mating disruption to stop reproduction. Mass elimination of the pest population may be achieved by mass trapping, attract-and-kill, attract-and-infect or push-pull methods. Success depends on efficient attraction of one or both the sexes, preferably females. Communication disruption causes disorientation and confusion preventing successful encounters between the sexes, thus reducing mating and reproduction.

11.2.1 Mass Trapping

Mass trapping is used to attract the pest to a source baited with food, synthetic components of the pheromone or their combination, to remove the captured individuals from the population. Most mass trapping efforts are aimed at sustainable pest management rather than the eradication of the pests. The objective of the methods adopted is to remove the pest from the crop before mating and economic damage occur. For effective removal of the pest the method requires (1) a trap designed to capture the target pest that is walking or flying (Cork et al. 2003), (2) a bait that attracts individuals from long distances (Byers 1993), and (3) a lure that overcomes natural attractants such as food, oviposition sites or the species pheromones (Jones 1998).

Mass trapping is more efficient when both sexes are attracted and when the strategy is implemented at low densities of the pest population, and in isolated areas such as in small and hilly plots. Significant successful cases were reported in control of some beetle pests using aggregation pheromone plus food bait. In tropical America *Rhynchophorus palmarum* (L.) vectors the red ring nematode disease in oil palm, where mass trapping of the weevils significantly reduced infestation by the weevil leading to a reduction in the disease incidence (Oehlschlager 2005).

In moths, however, traps baited with the species specific sex pheromones attract males only. For effective control, mass trapping of male moths requires the removal of large proportions of virgin males from the population, as one male may mate with more than one female. Practical applications aimed at moth pests are, therefore, largely unsuccessful in maintaining infestations below the economic injury level (Hathaway et al. 1985). Nevertheless, mass trapping alone has in some cases resulted in significant reduction in pest moth populations and corresponding damage (Codling moth: Emel'yanov and Bulyginskaya 1999). However, often this tactic requires the support of other control methods (Teich et al. 1979). Mass trapping of the gypsy moth, *Lymantria dispar* (L.), resulted in its successful eradication from Southeastern USA after an area-wide campaign was combined with limited sprays of *Bacillus thuringiensis* (*Bt*) (Dreistadt and Dahlsten 1989). Nonetheless, mating disruption was the preferred environmentally friendly tactic to control this pest (Reardon et al. 1998).

11.2.2 Attract-and-Kill

The attract-and-kill, also called lure-and-kill strategy is similar to mass trapping, where individuals are attracted and eliminated by adding a killing material to the bait. However, unlike mass trapping, individuals that get killed are not collected in a trap (device). This method requires a powerful attractant and also an efficient non-repelling killing agent (insecticide). Usually species specific pheromones, food baits or their combination serve as an attractant.

The success of this method depends on the potencies of the attractant, the killing agent, and the combination of both. The insects should be lured from a distance, without relying on a sporadic, random approach to the bait sites. Baits should also be distributed in a way that maximizes detection by all the targeted individuals (i.e. males, females or both). The dissemination of attractants should not lead to mating disruption, the bait should be dominant over the natural attractants (e.g. sex pheromone, host plant) and also the killing agent should be fast acting to avoid further mating or damage to the crop and must not repel the pest. If female produced pheromone is used as a lure to trap males, the concentration of bait points per hectare relative to the female density may greatly affect the success of the method, since bait droplets or particles compete with calling females.

Special technologies were developed to be used in lure-and-kill viz. a paste on which known concentrations of the attractant and the killing agent were combined in a bead. The bait was distributed by manual applicators at a specific density per hectare or were sprayed (i.e. hollow-fibers) from land or air. Apparently, lure-and-kill is much more effective with dipterans and coleopterans than with lepidopterans (El-Sayed et al. 2009).

The use of attract-and-kill has yielded mixed results. In several cases low or no reduction in damage or pest population was observed (Moraal et al. 1993; Downham et al. 1995; Trematerra et al. 1999; Angeli et al. 2000), while in others, local eradication of *Bactrocera dorsalis* (Hendel) (Steiner et al. 1965), *B. papayae* Drew and Hancock (Lloyd et al. 1998) and significant reduction of *B. cucurbitae* Coquillett (Cunningham and Steiner 1972) were recorded. One drawback of the attract-and-kill strategy is an adverse effect of the killing agent on non-target and beneficial insects (Michaud 2003;Uchida et al. 2007).

11.2.3 Push-Pull

The push-pull method is also called stimulo-deterrent diversionary strategy, which intends to keep pests away from plants by a combination of repellents situated out side the crop, that push them away from the host into attractive pheromone-lure/food baited traps, which pull the pest in (Cook et al. 2007; Hassanali et al. 2008). This strategy has proven useful for the management of several species of bark beetles (Borden et al. 2006).

11.2.4 Attract-and-Infect

Autoinoculation is the process in which an attractive semiochemical is used in conjunction with pathogenic microorganisms resulting in the transfer of pathogens from one member of a species to another stimulating epizootic disease. Various autoinoculating devices have been developed for the control of different insect groups e.g. aphids, coleopterans and lepidopterans (Francardi et al. 2013; Vega et al. 1995, 2007). The treatment should not only contain an attractive lure but also maintain the virulence of the infecting agent for an extensive period of time. It should restrain the insects long enough to be infected but also to allow them to leave the treatment location in a healthy enough condition to find a mate and transfer the pathogen. In the selection for pathogens, slow killing agents that are capable of horizontal transmission are chosen. Entomopathogenic fungi and viruses chosen from local strains are often preferred.

11.2.5 Mating Disruption

Mating disruption aims at interfering with the mating process of insect pests with the expected result of reduced oviposition (Cardé and Minks 1995). The mechanisms responsible for the success of the method are still not completely understood. In effect, mating disruption causes miss-orientation of the male to the pheromone sources (dispensers and females) and delays or reduces mating of females (Miller et al. 2010; Harari et al. 2015). Mating disruption can also be achieved by using specific pheromone antagonists or inhibitors (Reddy and Guerrero 2010).

Classical mating disruption is usually achieved by permeation of the air with the volatile sex pheromone. This goal may be achieved by spraying pheromone releasing microcapsules or hollow fibers, or by manually applying fewer points of pheromone sources that release larger quantities of the pheromone (Witzgall et al. 2010). The method is most effective in large plots or area wide control programs where immigration of gravid females is minimal.

Successful mating disruption applications to control moth pests are numerous. The reduction of European berry moth, *Lobesia botrana* (Denis and Schiffermüller), in vineyards, light-brown apple moth *Epiphyas postvittana* (Walker), and the codling moth, *Cydia pomonella* (L.) in apple (Witzgall et al. 2008; Reddy and Guerrero 2010) are some classical examples.

One important drawback of a classical mating disruption tactic is the high cost of the large amount of the synthetic sex pheromone needed to cover the infested areas. Nevertheless, in the last decade the mating disruption technique has increased almost exponentially both in crop varieties treated and land coverage (Witzgall et al. 2010).

11.3 Isolation and Identification of Semiochemicals

The process of identification of a chemical message is complicated as it usually involves a complex of chemicals, produced in minute amounts at specific environmental and physiological conditions of the organism. Classically, the process is divided into the three steps of collection, purification and identification. Each step is usually guided by appropriate bioassays, which are electrophysiological via GC coupled electro-antennogram and behavioral studies, as the behavior modifying function of the isolated compounds needs to be established. The common methods used were reviewed in detail by Millar and Sims (2000). Briefly, if the glandular source of the pheromone is known, then the organ can be extracted as is commonly done with female Lepidoptera, however, this method poses serious disadvantages. The main ones being, that the extraction usually yields non relevant compounds, not active pre-cursors or the actual compounds or the active compounds are present but in ratios that do not represent the natural composition emitted by active organisms. For these reasons entrainment or head space collection is usually preferable. The popular methods adopted are to pass a purified air over the organism/organ isolated in "aeration chamber", where the resulting exhaust air sample may either be directly analyzed by GC-MS or collected on appropriate absorbent such as Super Q, using columns, especially if large amounts are required or by solid phase micro-extraction (SPME) (Pawliszyn 2000). SPME is a solvent-free sample preparation technique in which the volatiles are trapped from the air surrounding the sample into a needle like coated silica fiber that is held within a syringe like device. The fiber is inserted in a hot inlet and the resulting compounds collected on the fiber are desorbed and analyzed by GC or GC-MS. A variety of fibers that differ in their absorption properties are available in the market. Moreover, commercial headspace and auto samplers are available, enabling continuous and real time identification of released pheromone components by SPME/GC/MS analysis as was recently shown (Levi-Zada et al. 2011). Identification of active components usually relies on MS and proton nuclear magnetic resonance (NMR). Another approach for identification of low concentrations of biologically active small signal molecules in a mixture can be through the recently developed NMR-spectroscopic approach viz. differential analysis by 2D NMR spectroscopy (DANS) (Robinette et al. 2011). This method was successfully used for the identification of the pheromone of a nematode, *Caenorhabditis elegans* (Maupas) (Pungaliya et al. 2009). In any case, the ultimate step for correct identification is when a synthetically produced compound induces the behavioral responses of the natural active compound(s).

11.4 Semiochemicals of Date Palm Pests

Despite of their significant role in the behavior of practically all arthropods, pheromones are currently known for only a few date palm pests (Table 11.1). The identified pheromones unsurprisingly belong to two well-studied insect orders: coleopterans, in which aggregation pheromones are produced by males, and lepidopterans, in which the species specific sex pheromone produced by female moths are mostly reported (El-Sayed 2014). Female produced multi-component sex pheromones are known for some moth date pests, but not for the raisin moth, *Cadra figulilella* (Gregson), where only one component was identified so far. Moreover, for the

Table 11.1 Semiochemicals of date palm insect pests

Order	Family	Species	Pheromones*	References	Allelochemicals**	References
Lepidoptera	Batrachedridae	Lesser date moth	(f)	Levi-Zada et al. (2011)	Not reported	
		Batrachedra amydraula	Z4, Z7-10Ac	Levi-Zada et al. (2013)		
			Z4-10Ac			
			Z4-10OH			
			Z5-10Ac			
	Pyralidae	Greater date moth	(m)	Levi-Zada et al. (2014)	Not reported	
		Aphomia (= *Arenipses*) *sabella*	Benzaldehyde			
			Sulcatol			
			Phenyl			
			Acetaldehyde			
			2-phenylpropenal			
			Geranyl acetone			
			Fuscamol			
	Pyralidae	Indian meal moth	(f)	Zhu et al. (1999)	(a)	Olsson et al. (2005)
		Plodia interpunctella[pp]	Z9, E12-14Ac		Cyclohexanone	
			Z9, E12-14Ald		2-phenylacetaldehyde	
			Z9, E12-14OH		Cyclohexanol	
			Z9-14Ac		2me5me-3-ethylpyrazine	
					Nonanal Vanillin	
					Ethyl vanillin	Tóth et al. (2002)
					3-methylbutan-1-ol Acetic acid	

		Raisin moth *Cadra (= Ephestia) figulilella* pp	(f)	Brady and Daley (1972) Kehat et al. (1992)	Not reported	
	f[pp]		Z9, E12-14Ald			
	Carob moth *Spectrobates* (= *Ectomyelois*) *ceratoniae*[pp]	(f)	Baker et al. (1991)	Ethyl hexanoate (a)	Cossé et al. (1994)	
		Z9, E11,13-14Ald		Ethyl alcohol (a+os)	Gothilf et al. (1975)	
		Z9, E11-14Ald		Acetaldehyde (a)		
		Z9-14Ald		2-phenylethanol (a)		
				Propan-1-ol (a+os)		
				Propan-2-ol (os)		
				2-methylpropan-1-ol (os)		
				Butan-1-ol (os)		
				3-Methylbutan-1-ol (os)		
Coleoptera	*Curculionidae*	Red palm weevil	(m)	Oehlschlager et al. (1995)	Ethyl acetate (a)	El-Sebay. (2003), Guarino et al. (2011), Gunawardena et al. (1998), Gunawardena and Herath (1995), Vacas et al. (2014), and Guarino et al. (2013, 2015)
		Rhynchophorus ferrugineus	4S, 5S-nonanol	Hallett et al. (1993)	Ethyl alcohol (a)	
			4S, 5S-nonanone	Gunawardena and Bandarage (1995)	Ethyl propionate (a)	
					Pentan-1-ol (a)	
					2-methoxy-4-vinylphenol (a)	
					Gamma-nonanoic lactone (a)	
					1	
					α-pinene (r)	
					1-octen-3-ol (r)	
					Geraniol (r)	

(continued)

Table 11.1 (continued)

Order	Family	Species	Pheromones*	References	Allelochemicals**	References
	Nitidulidae sap beetles	Dried fruit beetle	(m)	Bartelt et al. (1992b)	(a)	Phelan and Lin. (1991)
		Carpophilus hemipterus[pp] (U.S. and IS)	(2E, 4E, 6E, 8E)-3,5,7-trimethyl-2,4,6,8-decatetraene		Ethyl acetate	Dowd and Bartelt. (1991)
			(2E,4E,6E,8E)-3,5,7-trimethyl-2,4,6,8-undecatetraene (2E,4E,6E,8E)-7-ethyl-3,5 dimethyl-2,4,6,8-decatetraene		Ethyl alcohol	Smilanick et al. (1978)
			(2E,4E,6E,8E)-7-ethyl-3,5-dimethyl-2,4,6,8-undecatetraene		Acetaldehyde	Bartelt et al. (1992a)
					2me-1-propanol	
					3me-1-butanol	
					2me-1-butanol	
					Propanoic acid	
					Butyric acid	
					Methanol	
					Propan-2-ol	
					2-pentanol	
					2-pentanone	
					Heptan-1-ol	
					Methyl butyrate	
					Propanal	
					3-hydroxy-2-butanone	

		Corn sap beetle	(m)	Bartelt et al. (1995)	Not reported	
		C. dimidiatus[pp](IS)	(3E,5E,7E,9E)-6,8-diethyl-4-methyl-dodecatetraene			
			(3E,5E,7E,9E)-5,7-diethyl-9-methyl-tridodecatetraene			
		Confused sap beetle	(m)	Bartelt et al. (1993)	Acetaldehyde	Smilanick et al. (1978)
		C. mutilatus[pp](IS)	(3E,5E,7E)-5-ethyl-7-methyl-undecatriene		Ethyl acetate	
			(3E,5E,7E)-6-ethyl-4-methyl-decatriene		Ethyl alcohol	
		Pineapple beetle *Urophorus* (= *Carpophilus*) *humeralis*[pp] (IS)	Not reported		Ethyl alcohol	Bartelt and Wicklow (1999)
					4-ethylguaiacol	Zilkowski et al. (1999)
					2,5-diisopropylpyrazine	
					2-phenylethanol	
					Propyl acetate	
					Acetaldehyde	
					1-propanol	
					2-methylpropan-1-ol	
					3-methylbutan-1-ol	
		Epuraea (= *Haptoncus*) *luteola* (U.S. and IS)	Not reported		Probably the same as for some other *Carpophilus* sp.	Bartelt and Hossain (2010)

[pp] – polyphagous; *(m) male produced pheromone; (f) female produced pheromone; **(a) host associated attractant; (r) plant derived repellents; (os) oviposition stimulant; IS – Israel; U.S.-United States

greater date moth, *Aphomia* (= *Arenipses*) *sabella* Hampson only male specific volatiles were recently identified (Levi-Zada et al. 2014). This may suggest that the pheromone identification for these species is still partial. Some host coattractants (Table 11.1), were identified for both sap beetles and for the red palm weevil. Some similarity in the attraction to the host volatiles of both pests is not surprising; as both are attracted to fermenting plant material (see below).

To date, no pheromones or other semiochemicals of scale insects, mealybug pests (Homoptera) or date bugs (Heteroptera) damaging date palms have been identified. However, the identification of semiochemicals seems feasible, especially for scales and mealybugs, as sex pheromones of some related species have been identified e.g. *Dysmicoccus grassii* (Leonardi) (De Alfonso et al. 2012).

The section below will present case studies of semiochemicals of several date pests for which information on chemical cues is available (Table 11.1).

11.4.1 Lesser Date Moth (Lepidoptera: Batrachedridae)

The lesser date moth (LDM), (*Batrachedra amydraula* Meyrick) is an oligophagous moth developing on date palms. LDM is known to occur from Bangladesh to western Saudi Arabia, Yemen, Israel, Iraq, Iran and Pakistan, as well as most of North Africa (Blumberg 2008; Kakar et al. 2010). LDM is a key pest of date palms that causes over 50 % loss of crop due to larval feeding of green dates and in some varieties even up to 80 % loss (Michael 1970; Blumberg 2008). In other areas it is rarely a pest as it merely thins the fruit bunches (Talhouk 1991).

In the United Arab Emirates (UAE) and Israel, LDM produces three generations per year, which are active from March to August. During autumn and winter the larvae are believed to be dormant within cocoons, which are formed between the base of the terminal fronds and the remains of the dry bunches (Michael and Habib 1971). Presumably triggered by increasing day length, the larvae terminate diapause and pupate. The emerging adults lay eggs of the first generation in early spring.

Larvae that hatch during March-April feed on newly formed fruits. Fruits during this period are small and the larvae damage a few fruits before completing their development to adults (Blumberg 2008). Damage to fruit at the beginning of the season is particularly extensive. The first chemical spray is usually timed by observation of first generation larvae and damaged fruits, or it is applied prophylactically during the pollination process (Al-Samarraie et al. 1989; Blumberg 2008). This spray and the following chemical applications are usually too late as the damage to the fruit yield is already done. Recently, components of the female *B. amydraula* sex pheromone were identified (Levi-Zada et al. 2011, 2013) as (4Z,7Z)-4,7-decadien-1-yl acetate, (4Z)-decen-1-yl acetate, (5Z)-decen-1-yl acetate and decan-1-yl acetate. The moth can be efficiently trapped using either delta or Unitrap (Universal Moth Trap) (Fig. 11.1a). Delta traps can be easily self-made from plastic material with disposable plates painted with sticky glue-layer (Fig. 11.1b). Recently, these traps with two pheromone components provided by Allan Cam Oehlschlager,

Fig. 11.1 Pheromone loaded commercial Unitrap (Universal Moth Trap) (**a**) and self-made Delta trap (**b**) suitable for trapping and monitoring the lesser date moth in date plantations (Photos: Victoria Soroker)

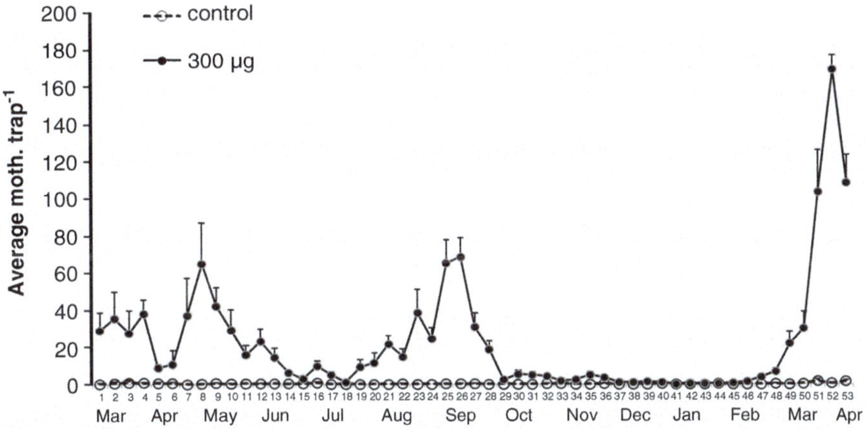

Fig. 11.2 Lesser date moth trapping from March 2012 to April 2013 in Medjool date plantation in Southern Arava, Israel using delta traps baited with lures with 300 µg of a two-component sex pheromone (4Z, 7Z)-4,7-decadien-1-yl acetate and 5Z-decen-1-yl acetate at a ratio of (1:2) (ChemTica Internacional, Costa Rica). The data are mean weekly trappings + SE of five traps (Soroker et al. unpublished)

ChemTica Internacional, Costa Rica were used to detect the emergence of the first generation of moths in a date plantation, enabling better timing of the most critical pest treatment (Victoria Soroker, Ally Harari andDafna Carmeli – unpublished). It also revealed that, unlike previously thought, moth activity in Israel is not limited to 3 months (April–June) but is extended from March onwards up to the harvest period in September (Fig. 11.2). In addition to monitoring and timing of insecticidal treatments this pheromone can be used as a control measure for mating disruption or mass trapping. The data indicated that mating disruption of LDM is feasible (Fig. 11.3), the approach, however, needs to be explored with a lure containing additional attractive compounds (Table 11.1).

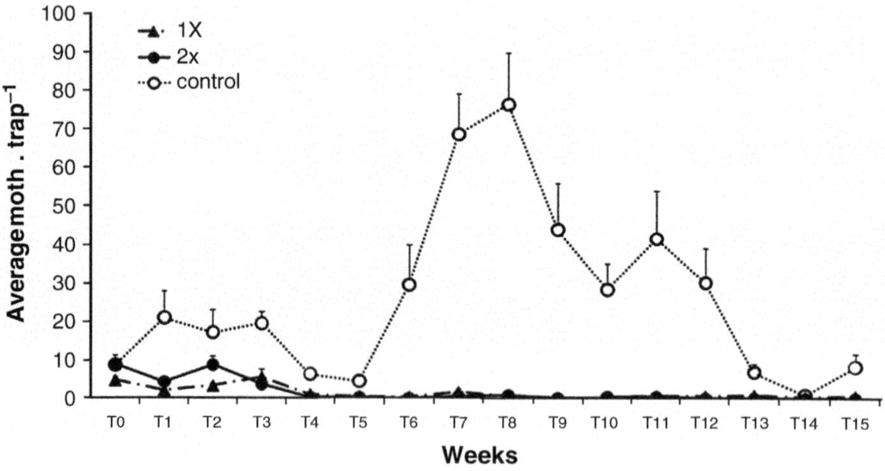

Fig. 11.3 Effect of two pheromone loads: 1X (500 mg × 15 palms/dun) and 2X of a two-component sex pheromone (4Z, 7Z)-4,7-decadien-1-yl acetate and 5Z-decen-1-yl acetate (at 1:2 ratio), on weekly catches of the lesser date moth in monitoring traps baited with lures with 300 μg of the same two -component pheromone (ChemTica Internacional, Costa Rica). Data are mean weekly catches + SE of five delta traps. Experiment conducted in Medjool date plantation in Southern Arava, Israel (Soroker et al. unpublished)

11.4.2 Sap Beetles (Coleoptera: Nitidulidae)

Sap beetles of the genus *Carpophilus* and *Epuraea* (= *Haptoncus*) *luteola* Erichson are pests of several agricultural crops throughout the world including dates. At least four species have been reported to occur abundantly in date orchards, both in the USA and in the Middle East. In date orchard, sap beetles are considered primary pests damaging fruits ripening on palms and in storage (Rafaeli et al. 2006). Damage is caused directly by feeding of larvae and adults but also indirectly by supporting the development of opportunistic pathogens that accelerate fruit rotting (Blumberg 2008). Adult sap beetles may also vector harmful microorganisms (Bartelt and Hossain 2010; Ambourn et al. 2005). However, the involvement of *Carpophilus* spp. in vectoring date pathogens is still unclear. Being largely polyphagous, beetles can migrate from adjacent crops. However, beetle populations often persist in the orchard year-round in the dates left on the ground after harvest and move to the fruits on trees when a new crop ripens. The problem is especially serious in plantations with mixed varieties, in which early varieties, such as Barhi, are grown together with late varieties, such as Deglet Noor.

Various methods have been used to control sap beetles. The main problem is that the beetles damage the fruits just before or during harvest, when toxic residues should be avoided, and use of conventional insecticides is not permitted. Although most of the studies on sap beetles have been reported on a wide range of crops, the gained knowledge can be used to design semiochemical based management in date palms.

It has been known for more than 50 years that both sexes of sap beetle are attracted to odors from overripe or decomposing fruit (Bartelt and Hossain 2010). Today it is clear that the attraction of sap beetles is mediated by a complex mixture of compounds from the host and conspecifics that operate in synergy. Imitation of the natural odor is a challenge. One popular substitute for natural food odor source is whole-wheat bread dough. In the species such as *Carpophilus mutilatus* Erichson and *C. hemipterus* (L.), pheromone or odor of wheat bread dough alone have very low attractiveness, while their combination is an extremely powerful attractant; for example, *C. hemipterus* pheromone combined with food odor attracted 75 times more beetles than either pheromone or dough alone (Bartelt and Hossain 2010). However, the use of fermenting mixture is a highly inconvenient and unreliable odor source.

Male-produced aggregation pheromones identified for several *Carpophilus* spp. are known to attract both males and females. Pheromone emission depends on male feeding and sexual experience, where feeding is known to stimulate pheromone production, while mating inhibits it (Bartelt and Hossain 2010). All identified pheromones are unsaturated hydrocarbons (e.g. tri- and tetraene hydrocarbons) (Bartelt et al. 1992b, 1994a; Bartelt 1997). These pheromone components are not entirely species-specific, and some pheromone components are shared by a few *Carpophilus* spp., although the combination of components and component ratios are usually species specific. There is cross-attraction between pheromones of some *Carpophilus* species including *C. humeralis* (F.), for which the pheromone has yet to be identified. However, aggregation pheromones of *Carpophilus* spp. did not have any effect on *E. luteola* (Blumberg et al. 1993; Bartelt et al. 1994a; Zilkowski et al. 1999). Bartelt et al. (1994a) suggested that mass trapping of sap beetles (particularly for *C. mutilatus*) in date plantations would be especially useful to trap and reduce much of the local beetle population during the months of April and May, and to protect the ripening fruits from the surviving local beetles and also from immigrants later in the season.

For establishing a trapping system in the field, information related to lure formulation, dispenser and trap design needs to be collected as well as the understanding of optimal trap distribution. The use of different traps and food odor sources was reviewed in Bartelt and Hossain (2010). However, easy use and the success of a variety of food odors to attract the beetles prevent standardization of the lure. The pheromone dispenser used is usually a rubber septa, baited with 500 μg of the pheromone per septa, but larger quantities can also be used (Bartelt and Hossain 2010). High pheromone doses were not observed to repel *Carpophilus* beetles. The pheromone dispenser, however, must be replaced every 2 weeks, making the operation rather costly and labor intensive.

In date plantations in Israel, Unitrap was successfully used to trap sap beetles (Fig. 11.4). Studies on trapping sap beetles in date orchards revealed that capture rates and species composition were significantly affected by the trap design and its suspension height on the date palms. In date plantations in Israel, in traps placed at near ground level (0.2 m), *E. luteola* was the most dominant species, whereas at fruit bunch level (16 m), *C. hemipterus* was abundant. In the USA, *C. hemipterus*

Fig. 11.4 Two unitraps set at different heights to compare the efficacy in monitoring sap beetles in date orchard. The position of the traps is marked by red arrows (Photo: Victoria Soroker)

and *C. mutilatus* were trapped more efficiently in California date farm experiments in traps set at 3 m rather than at 0.3 m above the ground. However, *C. humeralis* responding to the pheromone of *C. mutilatus* was caught more frequently in lower (0.3 m) traps (Bartelt et al. 1994b). Trapping each species separately is labor and timing consuming, but since cross attraction is documented between sap beetles, it could be an advantage to implement a trapping program against sap beetles in date palm using a combination of pheromones of several species. The lure combination of three sap beetle species was successfully implemented in stone fruit orchards in Australia in recent years. A commercial pheromone mixture (Great Lakes IPM), attractive to *C. hemipterus* and *C. mutilatus* supplemented with dates and yeast, was tested in organic date plantations in the Arava Region in Israel (Dafna Carmeli and

Ada Rafaeli – personal communication, 2014). This pheromone mixture trapped a large number of beetles from both species as well as others, including *E. luteolus* Erichson. However, best trap distribution as an efficient mass trapping tool in date plantation needs to be further developed.

Studies in stone fruit orchards revealed a problem in placing the traps on fruit trees, as beetles may cause damage before being caught. Based on studies in peach orchards, James et al. (2001), suggested a perimeter-based protection protocol, in which traps were set on regular intervals, several meters outside the orchard. Such an arrangement is expected to drive the pest out of the orchard and to intercept the immigrant beetles on their way into the orchard. Although this strategy sounds promising, it was found to be costly and not easily accepted by the growers as it created management problems.

Another interesting approach to reduce the cost of trap operations is attract-and-kill stations. This approach was developed recently for protection of stone fruit orchards using synthetic lure that included the pheromone and synthetic food attractant (Table 11.1). The latter was based on the odor of peach juice (Bartelt and Hossain 2010). Simultaneous area-wide application of attract-and-kill proved more successful in fruit protection than regular insecticide applications.

11.4.3 Red Palm Weevil (Coleoptera: Curculionidae)

Red Palm Weevil (RPW), *Rhynchophorus ferrugineus* (Olivier), is a key pest of coconut, *Cocos nucifera* L. in South and Southeast Asia. At present, the pest's geographical range includes Asia, Africa, Australasia, Southern Europe and the Americas (Giblin-Davis et al. 2013). Since the mid-1950s, the host range of RPW has significantly increased from only 4 palm species to 40 palm species worldwide (Nirula 1956; Anonymous 2011). The red palm weevil is known to cause significant damage to several economically important palm species, including date palm *Phoenix dactylifera* L. and *P. canariensis* Chabaud (Abraham et al. 1998; EPPO 2008; El-Sabea et al. 2009). In the date growing regions, RPW was first reported in 1985 at Rass-El-Khaima, UAE, and has spread quickly, mostly through infested planting material to all date producing countries in the middle east and major date palm growing countries in North Africa (Zaid et al. 2002; Faleiro 2006; Faleiro et al. 2012). Based on ecological niche modeling, Fiaboe et al. (2012) predicted that RPW will further expand its range.

The annual loss in the Gulf Cooperation Council countries, due to the eradication of severely infested palms is reported to range from US$ 1.74 to 8.69 million at 1 and 5 % infestation, respectively (El-Sabea et al. 2009). Early detection of RPW infested date palms is essential for limiting its spread and for the successful management of this lethal palm pest. However, most of the life cycle of the weevil is hidden within the palm stem. RPW is able to fly several kilometers to new host areas (Faleiro 2006). Recent computerized flight mill studies using field captured RPW in Saudi Arabia have indicated that most adult flights take place over short distances of less than 1Km (Hoddle et al. 2015).

RPW adults are long known to be attracted to palm volatiles emitted from freshly damaged parts of palms to gain access to stem tissue for oviposition. Traps baited with food attractants laced with insecticide were therefore traditionally used in the past to trap adult palm weevils (Kalshoven 1950; Hagley 1965; Kurian et al. 1979; Abraham et al. 1989). Later, the two-component male-produced aggregation pheromone was identified, with a major component 4-methyl-5-nonanol (ferruginol) and a minor component 4-methyl-5-nonanone (Hallett et al. 1993) (Table 11.1). Studies were carried out in Saudi Arabia to ascertain the response (attraction), using a custom made, 4-Choice Olfactometer® (Analytical Research Systems Inc., Gainesville, Florida, USA) of both male and female RPW adults to the aggregation pheromone (Ferrolure™). These studies revealed high attraction (>50 %) to the aggregation pheromone in newly emerged unmated male and female weevils, which decreased with age and mating (Faleiro and El-Shafie 2012).

Similar to other *Rhynchophorus* spp., in RPW there is a strong synergistic interaction between palm volatiles and the pheromone. A combination of food bait and the pheromone leads to trapping of significantly higher numbers of adult male and female weevils than either of the two alone (Hallett et al. 1999). It has been suggested for the related species *R. palmarum*, that the pheromone lure serves as a long range attractant that brings the adult weevils near the pheromone trap, while it is the food bait that acts at short range and encourages the weevil to enter the trap (Jaffé et al. 1993).

Trap captured female RPW adults from date plantations in Saudi Arabia were found to be young, gravid and fertile, which indicates that mass trapping of RPW can potentially reduce the buildup of this pest in the field (Abraham et al. 2001). In fact, RPW is often managed using an IPM based strategy (Al-Shawaf et al. 2012, 2013; Faleiro et al. 2010; Hoddle et al. 2013; Vidyasagar et al. 2000) involving traps baited with the synthetic aggregation pheromone lure and a food bait (dates) that are used both for monitoring and mass trapping the adult weevils; regular inspection of palms to detect infestations; preventive and curative chemical treatments; and eradication of severely infested palms besides identifying hidden breeding sites. In Israel, where RPW-IPM tactics were implemented between 1999 and 2001 in a 450 ha date plantation, a decrease in the number of weevils trapped by the end of 2001, with no infestation being found between 2002 and 2009 (Soroker et al. 2005, 2013) was observed. Further, a large study carried out during 1997 in the UAE on 1,466 farms that initially contained 349,342 palms, infestations decreased in 1998 by 64 % in the farms that received insecticide treatments and pheromone traps, as compared to a decrease of 36 % in the farms that received only chemical treatment (El-Ezaby et al. 1998; Oehlschlager 2006). In Oman, pheromone trapping of RPW in date plantations resulted in reduction of infestation from 24 % in 1998 to 3 % in 2003 (Al-Khatri 2004). In another report from the UAE, six date plantations where pheromone traps captured the highest numbers of weevils also exhibited the greatest reduction in infestation. In this study the average annual reduction in infestation over six farms was 71 % (Kaakeh et al. 2001; Oehlschlager 2005). Geographic Information System (GIS) techniques are often used by pest managers for decision making in area-wide pest management programs. Mapped traps and catches in a 234 ha date plantation provided the spatial and temporal information on RPW spread during 2009 and 2010 in Al-Sohemia-1 in the Al-Ahsa Region of Saudi Arabia (Massoud et al. 2012). GIS was also used in Israel by The Plant Protection Inspection Services (PPIS) authorities almost country wide between the years 2000 and 2013 (Soroker et al. 2013) for understanding the spatial and temporal spread of RPW.

Aspects pertaining to lure composition, trap design, trapping protocols, including trap placement and trapping density, in date plantations are crucial and are addressed below.

11.4.3.1 Lure Composition and Trap Design

The pheromone lure together with the food bait constitute important components of the RPW pheromone trap. Inclusion of food bait along with 1 L water in the RPW pheromone trap is known to significantly enhance captures (Hallett et al. 1993; Soroker et al. 2005; Abbas et al. 2006; Faleiro 2006; Al-Saoud 2013) that are female dominant. Trapping efficiency is sustained when the food bait is renewed every 7–15 days (Faleiro 2006). Dates, when used as food bait in RPW pheromone traps, resulted in the best captures as compared to other food baits (Faleiro and Satarkar 2005), and are the most widely used in RPW pheromone traps in date producing countries where RPW is a problem. Sugar or date molasses are also commonly used with or without dates (Soroker et al. 2005). Addition of non-repellent additives (propylene glycol) extends the effective life of the food bait from 2 to 7 weeks, without addition of water (Oehlschlager 2006). Ethyl acetate when added to RPW pheromone traps enhances weevil captures by a factor of between 2 to 5 (El-Sebay 2003; Oehlschlager 1998, 2005; Al-Shagag et al. 2008; Al-Saoud 2013). Field studies carried out in Saudi Arabia indicated that the amount of ethyl acetate released from food baited RPW pheromone traps was inversely proportional to weevil captures. The highest and significantly superior weevil captures (one weevil/trap) were recorded in traps that released the lowest amount of ethyl acetate (0.63 ml/day) as compared the highest release rate tested (19.63 ml/day) which recorded the least captures (0.5 weevils/trap) (Al-Dandan et al. 2014).

Saudi Arabian studies showed that 65 % increase in weevil captures could be achieved when a related ten carbon ketone was included in the lure (Abozuhairah et al. 1996; Oehlschlager 2006). Similarly, the activity of ferruginol was enhanced when combined with n-pentanol, reported by researchers in Sri Lanka (Gunawardena and Bandarage 1995). Some other potential co-attractants have also been identified (Table 11.1). Recently very promising results were acquired with mixtures of ethyl acetate and ethanol especially for female catches (Vacas et al. 2014). However, to date the most popular lure combination used contains pheromone, ethyl acetate and food bait in water. Monitoring and mass trapping of adult weevils has been widely practiced in area-wide RPW-IPM programs in date growing countries. However, the food bait, water and insecticide solution in these conventional food-baited RPW-pheromone traps has to be periodically replaced which is labor intensive and

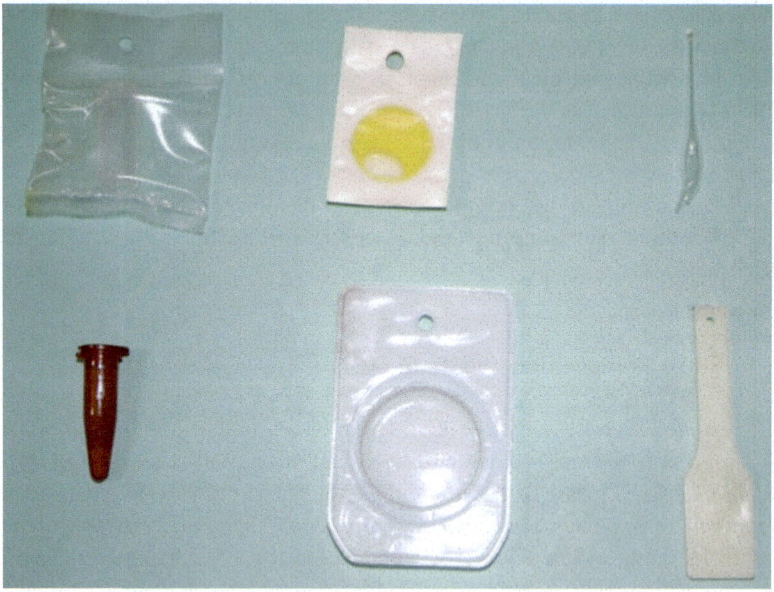

Fig. 11.5 Commonly used dispensers for RPWpheromone (Photo: Jose R. Faleiro)

impractical if a large number of traps are to be serviced. Within this context, a bait-free method (Hook RPW™) to attract-and-kill RPW adults was recently developed and tested at a trapping density of 250 attract-and-kill points/ha in RPW infested date plantations of Saudi Arabia and later in infested urban area in Israel (Victoria Soroker – unpublished). The results were inconsistent, while Hook RPW™ demonstrated the same level of attractiveness with or without the presence of food bait (El-Shafie et al. 2011) traps without food bait were significantly less attractive in urban RPW infested area of Israel (Victoria Soroker – unpublished).

For optimal attraction, pheromone release rate is also an important consideration. In oil palm plantations of Indonesia, the optimal release rate of the pheromone was three milligrams per day (Hallett et al. 1993). Ferruginol based lure (Ferrolure+) was found to sustain trapping efficiency at a lower release rate of 0.48 mg/day (Faleiro 2005). Today commercial ferruginol dispensers are available in several countries. These vary in shape and content including thermoplastic spatula, vials, sachets, glass ampoules etc. (Fig. 11.5). It is essential to select a lure that captures most weevils and lasts longest in the field under local climatic conditions.

Selection of the trap design is crucial. As RPW adults fly to the trap area, land on a surface and then walk into a trap, a large landing surface (ground) and also a rough trap surface is essential so that weevils easily crawl into the trap. Visual stimuli are apparently also important. Dark colored traps (red, black or brown) with a rough exterior, increased weevil captures (Hallett et al. 1999; Abuagla and Al-Deeb 2012; Al-Saoud 2013).

Fig. 11.6 Commonly used traps for RPW: bucket trap (**a**) and Picusan© trap (**b**) (Photos: Victoria Soroker)

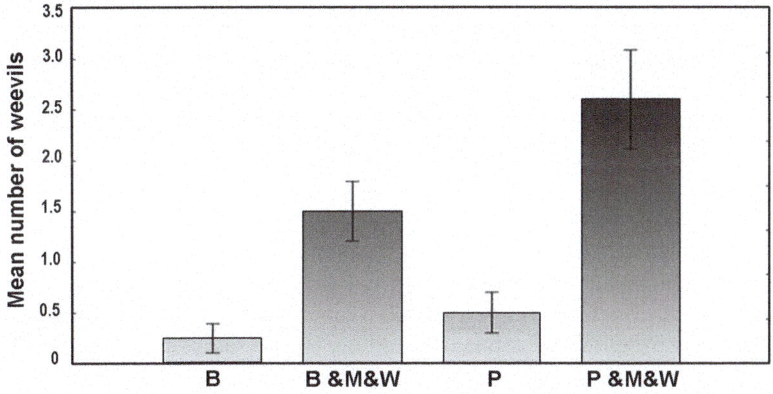

Fig. 11.7 Comparison of RPW catches in Picusan (P) and Bucket trap (B) baited with Ferrolure (ChemTica Internacional, Costa Rica), ethyl acetate or supplemented with molasses water solution, in bucket (B&W&M) or in Picusan (P&M&W) traps (Soroker et al. 2014)

In Saudi Arabia jute is attached to the exterior of the traps (Abraham et al. 1998), while in the UAE roughness is achieved by molding dimples on the outside of the trap surface (Oehlschlager 2006). Several commercial traps are currently available ranging from red traps with upper and side openings (Nardi et al. 2011), or black conical traps (Picusan©) (Fig. 11.6b) with a rough surface manufactured by a number of companies, varying slightly in their design. In Israel this design was found to be very efficient in trapping and retaining weevils, if the synthetic lure is supplemented with sugar molasses in water (Fig. 11.7) (Soroker et al. 2014). The same was also shown in experiments in Spain (Vacas et al. 2013).

Four-window bucket trap (5–20 L capacity) containing the pheromone lure along with fermenting food bait and with non-repelling insecticide to prevent escape of trapped weevils, is widely used in several countries to trap adults of RPW in both monitoring and mass trapping programs (Faleiro 2006). Abozuhairah et al. (1996) showed that repellency of insecticides to RPW decreases as follows; Deltamethrin™ < Carbaryl™ < Clhorpyriphos™ < Actellic™ < Endosulfan™ < Quinalphos™. Retaining captured weevils in RPW pheromone traps can also be achieved by providing adequate water in the trap. This method, although environment friendly is laborious.

11.4.3.2 Trap Placement and Trapping Density

To maintain field longevity of the lure it is recommended to place the trap in the shade. RPW trap placement in date plantations, for mass trapping and monitoring programs, are placed either on the ground (UAE, Israel, Spain), hung on palms 1 m above ground, or set on palm stumps 1.0 m high (Saudi Arabia: Faleiro 2006; Faleiro et al. 2011).

Adopting the appropriate trapping density in area-wide monitoring and mass trapping programs against RPW is vital not only for the success of the operation but also for the judicious use of resources without impairing the efficiency of the strategy. The agro-ecosystem, the intensity of the pest and resources available influences the decision on the number of traps to be set in the field and are particularly critical when RPW can migrate from adjacent ornamental palms such as *P. canariensis*. Initially a trap density of one trap per hectare in mass trapping programs was recommended (Oehlschlager 1994).

In RPW endemic date plantations in Saudi Arabia, effective mass trapping of the pest was achieved at a density of one trap per 1.5 ha, while a trap density of one trap per 100 ha effectively monitored RPW in pest-free areas (Abraham et al. 2000). In mass trapping programs, depending on the intensity of the pest in the field, trap density could range from 1 trap/ha, where weevil population is low, to 10 traps/ha in heavily infested plantations (Faleiro et al. 2011). In Israel, mass trapping the pest at a high trap density of 10 traps per hectare successfully managed RPW within 2 years in a 450 ha date plantation, while monitoring its activity in a 2,200 ha plantation was done with traps placed at a density of one trap per 3 ha (Soroker et al. 2005).

The potential negative effect of RPW pheromone traps on the infestation to the palm trees that are adjacent to the traps has been repeatedly raised, as weevils that are attracted but not captured by the trap may attack the nearby tree. In fact, in urban *P. canariensis*, the percentage of infested palms was high (72.7 %) when the traps were placed directly under the palms. This decreased considerably (18.8 %) when traps were set 5 m away from the palm with no significant additional decrease being observed when traps were set 50 m away from palms (Nardi et al. 2011).

11.5 Future Research

Currently, attractants for several important pests of date palm need to be identified, especially for the rhinoceros beetle, *Oryctes agamemnon* (Burmeister), the greater date moth, *A. sabella*, and the date stone beetle, *Coccotrypes dactyliperda* F., that can cause significant crop damages. Future attempts for semiochemical identification should be extended to homopteran pests as well, as recently there has been success in controlling this group of pests using pheromones.

Although, management of the red palm weevil and to a lesser extent, that of sap beetles has been implemented using semiochemicals, these strategies have to be further optimized. Mass trapping techniques could be optimized by evaluating the trap position and distribution pattern in the field, especially for the protection of organic orchards. Identification of effective plant volatiles that synergies the aggregation pheromones remain a challenge and an effort needs to be made towards development of synthetic lures that stand alone without the need to be supported with natural food baits.

The future advancement of synthetic semiochemical identifications could fuel the area-wide pest management in date plantations based on semiochemicals techniques such as: push-pull, attract-and-kill and attract-and-infect strategies. The management of RPW could involve a push-pull strategy, using RPW repellents that offer protection to the most vulnerable palm, along with pheromone trapping which could significantly supplement the present pheromone based control of RPW. In this context, a recent study from Italy and Israel indicated that 1-octen-3-ol and geraniol or α-pinene, by itself or together with methyl salicylate, could serve as a potential repellent against RPW (Guarino et al. 2013, 2015).

Another worthy approach can be "lure-and-infect" using autocontamination traps loaded with entomopathogenic fungi. In general, this approach is suitable for any pest of date palm for which attractants and effective pathogens are already known. In the case of RPW auto-contaminating traps have already been developed (Francardi et al. 2013) and several strains of effective entomopathogenic fungi have been isolated (Gindin et al. 2006; Dembilio et al. 2010).

Another important component of the IPM strategy in date palm is moth control in date orchards using pheromones. Development of monitoring programs, along with evaluating the efficacy of using mating disruption and mass trapping as control tactics need to be assessed.

In the long run it would be worthwhile to explore semiochemicals that are produced by palms in response to various pests and their impact on natural enemies, as currently there is no information on induced defenses of palms and date palms in particular. This would significantly contribute to the understanding of multitrophic interactions in the date palm ecosystem leading to new and sustainable semiochemical based IPM tactics for the management of insect pests of date palm.

11.6 Summary

Great concern over human and environmental health has resulted in a ban of some broad spectrum pesticides and strict regulations over residuals of the products. Therefore, new environmentally-friendly means and better targeting of the pests are eagerly sought. Semiochemicals have played an important role in IPM of several insect pests as monitoring, forecasting and management tools, besides being used for mating disruption of moth species and mass trapping of flies and beetles. There is also in general a consensus that use of semiochemicals as control agents, especially by means of mating disruption, mass trapping or attract-and-kill, are particularly effective when used in area-wide programs and low pest populations. Although direct comparisons between the three semiochemical based control tactics are scarce, several authors have stressed the importance of semiochemicals based IPM strategies to trap or avoid mating of the protandrous males (Hofer and Angst 1995; Charmillot et al. 2000) and also the need to maintain continued baits throughout the season (Ebbinghaus et al. 2001). Using a female attractant poses a risk of immigrating gravid females (Moraal et al. 1993), whereas, male immigration into the treated plot interferes with eliminating the males from the local population, in both mass trapping and attract-and-kill tactics. These risks are reduced in moths using mating disruption although here too high population density may increase the chance of random encounters between males and females.

Despite general progress in chemical ecology, and the potential, commercial exploitation of semiochemicals in date plantations is still limited and more date pests, are waiting for friendly control solutions. Much effort was invested in the last decade in the identification of host attractants and repellents as well as in the development of various implementation techniques. Taken together, these efforts are expected to bring to a broader implementation of semiochemicals in date farming system in the near future.

Acknowledgments We wish to thank Dr. Allan Cam Oehlschlager, ChemTica Internacional, Costa Rica for his advice and stimulating discussions on the use of semiochemicals in date plantations and especially for providing various formulations of LDM pheromones. We are also grateful to Yakov Nakache, Dafna Carmeli, and Nirit Ketner from Kibbutz Samar, Israel for their long term contribution in research and implementation of semiochemicals based IPM strategies in organic date palm orchards.

References

Abbas, M. S. T., Hanounik, S. B., Shahdad, A. S., & Al-Bagham, S. A. (2006). Aggregation pheromone traps, a major component of IPM strategy for the red palm weevil, *Rhynchophorus ferrugineus*, in date palms (Coleoptera: Curculionidae). *Journal of Pest Science, 79*, 69–73.

Abozuhairah, R. A., Vidyasagar, P. S. P. V., & Abraham, V. A. (1996). Integrated management of red palm weevil, *Rhynchophorus ferrugineus* F. in date palm plantations of the Kingdom of Saudi Arabia. In: *Proceedings of the XX International Congress of Entomology*, 25–31 August 1996. Florence (541 pp).

Abraham, V. A., Abdulla Koya, K. M., & Kurian, C. (1989). Integrated management of red palm weevil (*Rhynchophorus ferrugineus* Oliv.) in coconut gardens. *Journal of Plantation Crops* (Suppl.), *16*, 159–162.

Abraham, V. A., Al Shuaibi, M. A., Faleiro, J. R., Abozuhairah, R. A., & Vidyasagar, P. S. P. V. (1998). An integrated management approach for red palm weevil, *Rhynchophorus ferrugineus* Oliv., a key pest of date palm in the Middle East. *Journal of Agricultural and Marine Sciences, 3*, 77–84.

Abraham, V. A., Faleiro, J. R., Al-Shuaibi, M. A., & Prem Kumar, T. (2000). A strategy to manage red palm weevil *Rhynchophorus ferrugineus* Oliv. on date palm *Phoenix dactylifera* L. its successful implementation in Al-Hassa, Kingdom of Saudi Arabia. *Pestology, 24*, 23–30.

Abraham, V. A., Faleiro, J. R., Al Shuaibi, M. A., & Al-Abdan, S. (2001). Status of pheromone trap captured female red palm weevils from date gardens in Saudi Arabia. *Journal of Tropical Agriculture, 39*, 197–199.

Abuagla, A. M., & Al-Deeb, M. A. (2012). Effect of bait quantity and trap color on the trapping efficacy of the pheromone trap for the red palm weevil, *Rhynchophorus ferrugineus. Journal of Insect Science, 12*, 120. doi:10.1673/031.012.12002.

Al-Dandan, A. M., Al-Suleiman, Y., Al-Abdullah, E., Al-Shawaf, A. M., Al-Shagag, A., Al-Bagshi, M., Al-Saroj, S., Al-Bather, S., Ben Abdallah, A., & Faleiro, J. R. (2014). Red palm weevil captures in pheromone traps in relation to ethyl acetate release rates. In: *Proceedings of the fifth international date palm conference*, 16–18 March 2014. Abu Dhabi.

Al-Khatri, S. A. (2004). Date palm pests and their control. In: J. E. Peña (Ed.), *Proceedings of date palm regional workshop on ecosystem-based IPM for date palm in gulf countries* (pp. 84–88), 28–30 Mar 2004. Al-Ain.

Al-Samarraie, A. I., Al-Hafdh, E., Abdul-Majed, K., & Basumy, M. (1989). The chemical control of the lesser date moth, *Batrachedra amydraula. Pesticide Science, 25*, 227–230.

Al-Saoud, A. H. (2013). Effect of ethyl acetate and trap color on weevil captures in red palm weevil *Rhynchophorus ferrugineus* (Coleoptera: Curculionidae) pheromone traps. *International Journal of Tropical Insect Science, 33*, 202–206.

Al-Shagag, A., Al-Abbad, A. H., Al-Dandan, A. M., Ben Abdallah, A., & Faleiro, J. R. (2008). Enhancing trapping efficiency of red palm weevil pheromone traps with ethyl acetate. *Indian Journal of Plant Protection, 36*, 310–311.

Al-Shawaf, A. M., Al-Abdan, S., Al-Abbad, A. H., Ben Abdallah, A., & Faleiro, J. R. (2012). Validating area-wide management of *Rhynchophorus ferrugineus* (Coleoptera: Curculionidae) in date plantation of Al-Hassa. *Indian Journal of Plant Protection, 40*, 255–259.

Al-Shawaf, A. M., Al-Shagag, A., Al-Bagshi, M., Al-Saroj, S., Al-Bather, S., Al-Dandan, A. M., Ben Abdallah, A., & Faleiro, J. R. (2013). A quarantine protocol against red palm weevil *Rhynchophorus ferrugineus* (Olivier) (Coleoptera: Curculionidae) in date palm. *Journal of Plant Protection Research, 53*, 409–415.

Ambourn, A. K., Juzwik, J., & Moon, R. D. (2005). Seasonal dispersal of the oak wilt fungus by *Colopterus truncates* and *Carpophilus sayi* in Minnesota. *Plant Disease, 89*, 1067–1076.

Angeli, G., Ioriatti, C., & Finato, S. (2000). A new method for the control of codling moth. *Entomological Society of America, 56*, 63–66.

Anonymous. (2011). *The insect killing our palm trees EU efforts to stop the red palm weevil.* Available at: http://ec.europa.eu/food/plant/organisms/emergency/docs/111024_red_palm_weevil_en.pdf. Accessed on 8 June 2015.

Baker, T. C., Francke, W., Millar, J. G., Lofstedt, C., Hansson, B., Du, J. W., Phelan, P. L., Vetter, R. S., Youngman, R. S., & Todd, J. L. (1991). Identification and bioassay of sex pheromone components of carob moth, *Ectomyelois ceratoniae* (Zeller). *Journal of Chemical Ecology, 17*, 1973–1988.

Bartelt, R. J. (1997). Aggregation pheromones of *Carpophilus* spp. (Coleoptera: Nitidulidae): Review of chemistry and biology. *Recent Research Developments in Entomology, 1*, 115–129.

Bartelt, R. J., & Hossain, M. S. (2010). Chemical ecology of *Carpophilus* sap beetles (Coleoptera: Nitidulidae): And development of environmentally friendly method of crop protection. *Terrestrial Arthropod Reviews, 3*, 29–61.

Bartelt, R. J., & Wicklow, D. T. (1999). Volatiles from *Fusarium verticillioides* (Sacc.) Nirenb. and their attractiveness to nitidulid beetles. *Journal of Agricultural and Food Chemistry, 47*, 2447–2454.

Bartelt, R. J., Dowd, P. F., Vetter, R. S., Shorey, H. H., & Baker, T. C. (1992a). Responses of *Carpophilus hemipterus* (Coleoptera: Nitidulidae) and other sap beetles to the pheromone of *C. hemipterus* and host-related coattractants in California field tests. *Environmental Entomology, 21*, 1143–1153.

Bartelt, R. J., Weisleder, D., Dowd, P. F., & Plattner, R. D. (1992b). Male specific tetraene and triene hydrocarbons of *Carpophilus hemipterus*: Structure and pheromonal activity. *Journal of Chemical Ecology, 18*, 379–402.

Bartelt, R. J., Carlson, D. G., Vetter, R. S., & Baker, T. C. (1993). Male-produced aggregation pheromone of *Carpophilus mutilatus* (Coleptera: Nitidulidae). *Journal of Chemical Ecology, 19*, 107–118.

Bartelt, R. J., Vetter, R. S., Carlson, D. G., & Baker, T. C. (1994a). Responses to aggregation pheromones for five *Carpophilus* species (Coleoptera: Nitidulidae) in a California date garden. *Environmental Entomology, 23*, 1534–1543.

Bartelt, R. J., Vetter, R. S., Carlson, D. G., & Baker, T. C. (1994b). Influence of pheromone dose, trap height, and septum age on effectiveness of pheromones for *Carpophilus mutilatus* and *C. hemipterus* (Coleoptera: Nitidulidae) in a California date garden. *Journal of Economic Entomology, 87*, 667–675.

Bartelt, R. J., Weaver, D. K., & Arbogast, R. T. (1995). Aggregation pheromone of *Carpophilus dimidiatus* (F.) (Coleoptera: Nitidulidae) and responses to *Carpophilus* pheromones in South Carolina. *Journal of Chemical Ecology, 21*, 1763–1779.

Blumberg, D. (2008). Date palm arthropod pests and their management in Israel. *Phytoparasitica, 36*, 411–448.

Blumberg, D., Goldenberg, S., Bartelt, R. J., & Williams, R. N. (1993). Responses to synthetic aggregation pheromones, host-related volatiles, and their combinations by *Carpophilus* spp. (Coleoptera: Nitidulidae) in laboratory and field tests. *Environmental Entomology, 22*, 837–842.

Borden, J. H. (1985). Aggregation pheromones. In G. A. Kerkut & L. I. Gilbert (Eds.), *Comprehensive insect physiology, biochemistry and pharmacology* (Vol. 9, pp. 257–285). Oxford: Pergamon Press.

Borden, J. H., Birmingham, A. L., & Burleigh, J. S. (2006). Evaluation of the push-pull tactic against the mountain pine beetle using verbenone and non-host volatiles in combination with pheromone-baited trees. *Forestry Chronicle, 82*, 579–590.

Brady, U. E., & Daley, R. C. (1972). Identification of a sex pheromone from the female raisin moth, *Cadra figulilella. Annals of the Entomological Society of America, 65*, 1356–1358.

Byers, J. A. (1993). Simulation and equation models of insect population control by pheromone-baited traps. *Journal of Chemical Ecology, 19*, 1939–1956.

Cardé, R. T., & Minks, A. K. (1995). Control of moth pests by mating disruption: Successes and constraints. *Annual Review of Entomology, 40*, 559–585.

Charmillot, P. J., Hofer, D., & Pasquier, D. (2000). Attract and kill: A new method for control of the codling moth *Cydia pomonella. Entomologia Experimentalis et Applicata, 94*, 211–216.

Cook, S. M., Khan, Z. R., & Pickett, J. A. (2007). The use of push-pull strategies in integrated pest management. *Annual Review of Entomology, 52*, 375–400.

Cork, A., Alam, S. N., Rouf, F. M. A., & Talekar, N. S. (2003). Female sex pheromone of brinjal fruit and shoot borer, *Leucinodes orbonalis* (Lepidoptera: Pyralidae): Trap optimization and application in IPM trials. *Bulletin of Entomological Research, 93*, 107–113.

Cossé, A. A., Endriss, J. J., Millar, J. G., & Baker, T. C. (1994). Identification of volatile compounds from fungus-infected date fruit that stimulate upwind flight in female *Ectomyelois ceratoniae. Entomologia Experimentalis Applicata, 72*, 233–238.

Cunningham, R. T., & Steiner, L. F. (1972). Field trial of cue-lure + naled on saturated fiber board blocks for control of the melon fly by the male-annihilation technique. *Journal of Economic Entomology, 65*, 505–507.

De Alfonso, I., Hernandez, E., Velazquez, Y., Navarro, I., & Primo, J. (2012). Identification of the sex pheromone of the mealybug *Dysmicoccus grassii* Leonardi. *Journal of Agricultural and Food Chemistry, 60*, 11959–11964.

Dembilio, Ó., Quesada-Moraga, E., Santiago-Álvarez, C., & Jacas, J. A. (2010). Potential of an indigenous strain of the entomopathogenic fungus *Beauveria bassiana* as a biological control agent against the Red Palm Weevil, *Rhynchophorus ferrugineus*. *Journal of Invertebrate Pathology, 104*, 214–221.

Dowd, P. F., & Bartelt, R. J. (1991). Host-derived volatiles as attractants and pheromone synergists for driedfruit beetle, *Carpophilus hemipterus*. *Journal of Chemical Ecology, 17*, 285–308.

Downham, M. C. A., McVeigh, L. J., & Moawad, G. M. (1995). Field investigation of an attracticide control technique using the sex pheromone of the Egyptian cotton leaf-worm, *Spodoptera littoralis* (Lepidoptera: Noctuidae). *Bulletin of Entomological Research, 85*, 463–472.

Dreistadt, S. H., & Dahlsten, D. L. (1989). Gypsy moth eradication in Pacific Coast States: History and evaluation. *Bulletin of the Entomological Society of America, 35*, 13–19.

Ebbinghaus, D., Losel, P. M., Romeis, J., Cianciulli-Teller, M. G., Leusch, H., Olszak, R., Pluciennik, Z., & Scherkenbeck, J. (2001). Appeal: Efficacy and mode of action of attract and kill for codling moth control. *OILB/SROP Bulletin, 24*, 95–99.

El-Ezaby, F. A. A., Khalifa, O., & El-Assal, A. (1998). Integrated pest management for the control of red palm weevil, *Rhynchophorus ferrugineus* Oliv. in the United Arab Emirates, Eastern Region, Al Ain. In: *Proceedings of the 1st international conference on date palms* (pp. 269–281), 8–10 March. UAE. University, Al-Ain, UAE. Available at: http://www.pubhort.org/datepalm/datepalm1/datepalm1_23.pdf. Accessed on 29 Apr 2015.

El-Sabea, A. M. R., Faleiro, J. R., & Abo El Saad, M. M. (2009). The threat of red palm weevil *Rhynchophorus ferrugineus* to date plantations of the Gulf region of the Middle East: An economic perspective. *Outlook on Pest Management, 20*, 131–134.

El-Sayed, A. M. (2014). *The pherobase: Database of pheromones and semiochemicals*. (http://www.pherobase.com©) 2003–2014 The Pherobase. Accessed on 15 Oct 2014.

El-Sayed, A. M., Suckling, D. M., Byers, J. A., Jang, E. B., & Wearing, C. H. (2009). Potential of "lure and kill" in long-term pest management and eradication of invasive species. *Journal of Economic Entomology, 102*, 815–835.

El-Sebay, Y. (2003). Ecological studies on the red palm weevil, *Rhynchophorus ferrugineus* Oliv (Coleoptera: Curculionidae) in Egypt. *Egyptian Journal of Agricultural Research, 81*, 523–529.

El-Shafie, H. A. F., Faleiro, J. R., Al-Abbad, A. H., Stoltman, L., & Mafra-Neto, A. (2011). Bait-free attract and kill technology (Hook™ RPW) to suppress red palm weevil, *Rhynchophorus ferrugineus* (Coleoptera: Curculionidae) in date palm. *Florida Entomologist, 9*, 774–778.

Emel'yanov, V. A., & Bulyginskaya, M. A. (1999). Use of pheromones for control of codling moth *Laspeyresia pomonella* L. (Lepidoptera, Tortricidae) by elimination and disorientation of males. *Entomological Review, 79*, 539–546.

EPPO (European and Mediterranean Plant Protection Organization). (2008). Data sheets on quarantine pests. *Rhynchophorus ferrugineus. EPPO Bulletin, 38*, 55–59.

Faleiro, J. R. (2005). *Pheromone technology for the management of red palm weevil, Rhynchophorus ferrugineus (Olivier) (Coleoptera: Rhynchophoridae) – A key pest of coconut* (Technical Bulletin No. 4, 40 pp). Goa: ICAR Research Complex, India.

Faleiro, J. R. (2006). A review of the issues and management of the red palm weevil *Rhynchophorus ferrugineus* (Coleoptera: Rhynchophoridae) in coconut and date palm during the last one hundred years. *International Journal of Tropical Insect Science, 26*, 135–154.

Faleiro, J. R., & El-Shafie, H. A. F. (2012). *Olfactometer assays to evaluate the response of Rhynchophorus ferrugineus (Olivier) to the aggregation pheromone*. Presented at the Annual Meeting (Teleconference) of the Entomological Society of America; 13 November 2012 Knoxville, USA.

Faleiro, J. R., & Satarkar, V. R. (2005). Attraction of food baits for use in red palm weevil, *Rhynchophorus ferrugineus* Olivier pheromone traps. *Indian Journal of Plant Protection, 33*, 23–25.

Faleiro, J. R., Abdallah, B. A., & Ashok, K. J. (2010). Sequential sampling plan for area-wide management of *Rhynchophorus ferrugineus* (Olivier) in date palm plantations of Saudi Arabia. *International Journal of Tropical Insect Science, 30*, 145–153.

Faleiro, J. R., El-Saad, M. A., & Al-Abbad, A. H. (2011). Pheromone trap density to mass trap *Rhynchophorus ferrugineus* (Coleoptera: Curculionidae/rhynchophoridae/dryophoridae) in date plantations of Saudi Arabia. *International Journal of Tropical Insect Science, 31*, 75–77.

Faleiro, J. R., Ben Abdallah, A., El Bellaj, M., Al Ajlan, A. M., & Oihabi, A. (2012). Threat of red palm weevil, *Rhynchophorus ferrugineus* (Olivier) to date plantations of the Maghreb region in North Africa. *Arab Journal of Plant Protection, 30*, 274–280.

Fiaboe, K. K. M., Peterson, A. T., Kairo, M. T. K., & Roda, A. L. (2012). Predicting the potential worldwide distribution of the red palm weevil *Rhynchophorus ferrugineus* (Olivier) (Coleoptera: Curculionidae) using ecological niche modeling. *Florida Entomologist, 95*, 559–673.

Francardi, V., Benventi, C., Barzanti, G. P., & Rovers, P. F. (2013). Auto contamination trap with entomopathogenic fungi: A possible strategy in the control of *Rhynchophorus ferrugineus* (olivier) (Coleoptera Curculionidae). *REDIA, XCVI*, 57–67.

Giblin-Davis, R. M., Faleiro, J. R., Jacas, J. A., Peña, J. E., & Vidyasagar, P. S. P. V. (2013). Coleoptera: Biology and management of the red palm weevil, *Rhynchophorus ferrugineus*. In J. E. Peña (Ed.), *Potential invasive pests of agricultural crop species* (pp. 1–34). Wallingford: CABI Wallingford.

Gindin, G., Levski, S., Glazer, I., & Soroker, V. (2006). Evaluation of the entomopathogenic fungi *Metarhizium anisopliae* and *Beauveria bassiana* against the red palm weevil *Rhynchophorus ferrugineus. Phytoparasitica, 34*, 370–379.

Gothilf, S., Levy, E. C., Cooper, R., & Lavie, D. (1975). Oviposition stimulants of the moth *Ectomyelois ceratoniae*: The effect of short-chain alcohols. *Journal of Chemical Ecology, 1*, 457–464.

Guarino, S., Bue, P. L., Peri, E., & Stefano, C. (2011). Responses of *Rhynchophorus ferrugineus* adults to selected synthetic palm esters: Electroantennographic studies and trap catches in an urban environment. *Pest Management Science, 67*, 77–81.

Guarino, S., Peri, E., Bue, P. L., Germanà, M. P., Colazza, S., Anshelevich, L., Ravid, U., & Soroker, V. (2013). Assessment of synthetic chemicals for disruption of *Rhynchophorus ferrugineus* response to attractant-baited traps in an urban environment. *Phytoparasitica, 41*, 79–88.

Guarino, S., Colazza, S., Peri, E., Bue, P. L., Germanà, M. P., Kuznetsova, T., Gindin, G., & Soroker, V. (2015). Behaviour-modifying compounds for management of the red palm weevil (*Rhynchophorus ferrugineus* Olivier). *Pest Management Science*. doi:10.1002/ps.3966.

Gunawardena, N. E., & Bandarage, U. K. (1995). 4-methyl-5-nonanol (ferrugineol) as an aggregation pheromone of the coconut pest, *Rhynchophorus ferrugineus* F. (Coleoptera: Curculionidae): Synthesis and use in a preliminary field assay. *Journal of the Natural Sciences Council of Sri Lanka, 23*, 71–79.

Gunawardena, N. E., & Herath, H. (1995). Enhancement of the activity of ferrugineol by N-pentanol in an attractant baited trap for the coconut pest, *Rhynchophorus ferrugineus* (Coleoptera: Curculionidae). *Journal of National Science Council of Sri Lanka, 23*, 81–86.

Gunawardena, N. E., Kern, F., Janssen, E., Meegoda, C., Schafer, D., Vostrowsky, O., & Bestmann, H. J. (1998). Host attractants for red weevil, *Rhynchophorus ferrugineus*: Identification, electrophysiological activity, and laboratory bioassay. *Journal of Chemical Ecology, 24*, 425–437.

Hagley, E. (1965). Test of attractant for the palm weevil. *Journal of Economic Entomology, 58*, 1002–1003.

Hallett, R. H., Gries, G., Borden, J. H., Czyzewska, E., Oehlschlager, A. C., Pierce, H. D., Jr., Angerilli, N. P. D., & Rauf, A. (1993). Aggregation pheromones of two Asian palm weevils, *Rhynchophorus ferrugineus and R. vulneratus. Naturwissenschaften, 80*, 328–331.

Hallett, R. H., Oehlschlager, A. C., & Borden, J. H. (1999). Pheromone trapping protocols for the Asian palm weevil, *Rhynchophorus ferrugineus* (Coleoptera: Curculionidae). *International Journal of Pest Management, 45*, 231–237.

Harari, A. R., Zahavi, T., & Steinitz, H. (2015). Female detection of the synthetic sex pheromone contributes to the efficacy of mating disruption on the European grapevine moth, *Lobesia botrana*. *Pest Management Science, 71*, 16–322.

Hassanali, A., Herren, H., Khan, Z. R., Pickett, J. A., & Woodcock, C. M. (2008). Integrated pest management: The push-pull approach for controlling insect pests and weeds of cereals, and its potential for other agricultural systems including animal husbandry. *Philosophical Transactions of the Royal Society B, 363*, 611–621.

Hathaway, D. O., Tamaki, G., Moffitt, H. R., & Burditt, A. K. (1985). Impact of removal of males with sex-pheromone-baited traps on suppression of the peach-twig borer, *Anarsia lineatella* (Zeller). *Canadian Entomologist, 117*, 643–645.

Hoddle, M. S., Al-Abbad, A. H., El-Shafie, H. A. F., Faleiro, J. R., Sallam, A. A., & Hoddle, C. D. (2013). Assessing the impact of pheromone trapping, pesticide applications, and eradication of infested date palms for *Rhynchophorus ferrugineus* (Coleoptera: Curculionidae) management in Al-Ghowaybah, Saudi Arabia. *Crop Protection, 53*, 152–160.

Hoddle, M. S., Hoddle C. D., Faleiro, J. R., El-Shafie, H. A. F., Jeske, D. R., & Sallam A. A. (2015). How far can the red palm weevil (Coleoptera: Curculionidae) fly? Computerized flight mill studies with field-captured weevils. *Journal of Economic Entomology* 1–11. doi: 10.1093/jee/tov240

Hofer, D., & Angst, M. (1995). Control of pink bollworm in cotton with Sirene, a novel sprayable attract & kill formulation. *Proceedings of Beltwide Cotton Conference, 2*, 949–952.

Howard, F. W. (2001). Principles of insect pest control on palms. In R. M. Giblin & R. G. Abad (Eds.), *Insects on palms* (pp. 315–321). New York: CABI Publishing.

Jaffé, K., Sánchez, P., Cerda, H., Hernández, J. V., Jaffé, R., Urdaneta, N., Guerra, G., Martínez, R., & Miraz, B. (1993). Chemical ecology of the palm weevil *Rhynchophorus palmarum* L. (Coleoptera: Curculionidae): Attraction to host plants and to a male-produced aggregation pheromone. *Journal of Chemical Ecology, 19*, 1703–1720.

James, D. G., Vogele, B., Faulder, R. J., Bartelt, R. J., & Moore, C. J. (2001). Pheromone-mediated mass trapping and population diversion as strategies for suppressing *Carpophilus* spp. (Coleoptera: Nitidulidae) in Australian stone fruit orchard. *Agricultural and Forest Entomology, 3*, 41–47.

Johnson, D. V. (2011). Introduction: Date palm biotechnology from theory to practice. In M. Jain, J. Al-Khayri, & D. V. Johnson (Eds.), *Date palm biotechnology* (pp. 1–11). Dordrecht: Springer Science Business Media B.V.

Jones, O. T. (1998). Practical applications of pheromones and other semiochemicals (sections 11. Lure and kill). In P. Howse, I. Stevens, & O. T. Jones (Eds.), *Insect pheromones and their use in pest management* (pp. 280–300). London: Chapman and Hall.

Kaakeh, W., El-Ezaby, F., Abu Al-Nour, M. M., & Khamis, A. A. (2001). Management of the red palm weevil, *Rhynchophorus ferrugineus* Oliv., by a pheromone food-based trapping system. In: *Second international conference on date palms*, 25–27 March 2001. Al-Ain (325–343 pp).

Kakar, M. K., Nizamani, S. M., Rustamanu, M. A., & Khurhro, R. D. (2010). Periodical lesser date moth infestation on intact and dropped fruits. *Sarhad Journal Agriculture, 26*, 393–396.

Kalshoven, L. G. E. (1950). *The pests of crops in Indonesia* (701 p). Revised and translated by P. A. van der Laan, 1981, with G. H. L. Rothschild, Jakarta: P. T. Ichtiar Barn – van Hoeve.

Kehat, M., Eitam, A., Blumberg, D., Dunkelblum, E., Anschelevich, L., & Gordon, D. (1992). Sex pheromone traps for detecting and monitoring the raisin moth, *Cadra figulilella* in date palm plantations. *Phytoparasitica, 20*, 99–106.

Kurian, C., Sathiamma, B., Sukumaran, A. S., & Ponnamma, K. N. (1979). *Role of attractants and repellents in coconut pest control in India*. Paper presented at the 5th session of the FAO technical working party, Manila, The Phillipines.

Levi-Zada, A., Fefer, D., Anshelevitch, L., Litovsky, A., Gindin, G., & Soroker, V. (2011). Identification of the sex pheromone of the lesser date moth, *Batrachedra amydraula*, using SPME auto-sampling. *Tetrachedron Letters, 52*, 4550–4553.

Levi-Zada, A., Sadowsky, A., Dobrinin, S., Maayan, D., Ticuchinski, T., Fefer, D., Grinberg, A., & Blumberg, D. (2013). Reevaluation of sex pheromone of the lesser date moth, *Batrachedra amydraula*, using autosampling SPME-GC/MS and field bioassays. *Chemoecology, 23*, 13–20.

Levi-Zada, A., David, M., Fefer, D., Seplyarsky, V., Sadowsky, A., Dobrinin, S., Ticuchinski, T., Harari, D., Blumberg, D., & Dunkelblum, E. (2014). Circadian release of male-specific components of the greater date moth, *Aphomia (Arenipses) sabella*, using sequential SPME/GC/MS analysis. *Journal of Chemical Ecology, 40*, 236–243.

Lloyd, A., Leach, P., & Kopittke, R. (1998). Effects of exposure on chemical content and efficacy of male annihilation blocks used in the eradication of *Bactrocera papayae* in North Queensland. *General and Applied Entomology, 28*, 1–8.

Massoud, M. A., Sallam, A. A., Faleiro, J. R., & Al-Abdan, S. (2012). Geographic information system-based study to ascertain the spatial and temporal spread of red palm weevil *Rhynchophorus ferrugineus* (Coleoptera: Curculionidae) in date plantations. *International Journal of Tropical Insect Science, 32*, 108–115.

Michael, I. F. (1970). Economic importance and control of *Batrachedra amydraula* Meyr. (the lesser date moth) in the United Arab Republic. *Date Growers Institute Report, 47*, 9–10.

Michael, I. F., & Habib, A. A. (1971). Biology of *Batrachedra amydraula* Meyr, the lesser date moth. *Date Growers Institute Report, 48*, 6–8.

Michaud, J. P. (2003). Toxicity of fruit by baits to beneficial insects in citrus. *Journal of Insect Science, 3*, 1–9.

Millar, J. G., & Sims, J. J. (2000). Preparation, cleanup and preliminary fractionation of extracts. In J. G. Millar & K. F. Haynes (Eds.), *Methods in chemical ecology: Chemical methods* (pp. 2–37). Boston: Kluwer Academic Publishers.

Miller, J. R., Mcghee, P. S., Siegert, P. Y., Adams, C. G., Huang, J., Grieshop, M. J., & Gut, L. J. (2010). General principles of attraction and competitive attraction as revealed by large-cage studies of moths responding to sex pheromone. *Proceedings of the National Academy of Sciences of the U.S.A., 107*, 22–27.

Moraal, L. G., Van der Kraan, C., & Van der Voet, H. (1993). Studies on the efficacy of the sex attractant of *Paranthrene tabaniformis* Rott. (Lep., Sesiidae). *Journal of Applied Entomology, 116*, 364–370.

Nardi, S., Ricci, E., Lozzi, R., Marozzi, F., Ladurner, E., Chiabrando, F., Granchelli, L., Verdolini, E., Isidoro, N., & Riolo, P. (2011). Control of *Rhynchophorus ferrugineus* (Olivier, 1790) according to EU Decision 2007/ 365/EC in the Marche region (Central-Eastern Italy). *OEPP/EPPO Bulletin, 41*, 103–115.

Nirula, K. K. (1956). Investigations on the pests of coconut palm. Part 4. *Rhynchophorus ferrugineus. Indian Coconut Journal, 9*, 229–247.

Oehlschlager, A. C. (1994). *Use of pheromone baited traps in control of red palm weevil in the kingdom of Saudi Arabia* (17 pp). Consultancy report – Submitted to Ministry of Agriculture, Saudi Arabia.

Oehlschlager, A. C. (1998). Trapping of the date palm weevil. In: *FAO workshop on date palm weevil (Rhynchophorus ferrugineus) and its control*, 15–17 December 1998. Cairo.

Oehlschlager, A. C. (2005). Current status of trapping palm weevils and beetles. *Planter, 81*, 123–143.

Oehlschlager, A. C. (2006). *Mass trapping as a strategy for management of Rhynchophorus palm weevils* (143–180 pp). I Jornada Internacional sobre el Picudo Rojo de las Palmeras.

Oehlschlager, A. C., Prior, R. N. B., Perez, A. L., Gries, R., Gries, G., Pierce, H. D., Jr., & Laup, S. (1995). Structure, chirality, and field testing of a male-produced aggregation pheromone of Asian palm weevil *Rhynchophorus bilineatus* (Montr.) (Coleoptera: Curculionidae). *Journal of Chemical Ecology, 21*, 1619–1629.

Olsson, C. P. O., Anderbrant, O., Löfstedt, C., Borg-Karlson, A. K., & Liblikas, I. (2005). Electrophysiological and behavioral responses to chocolate volatiles in both sexes of the pyralid moths *Ephestia cautella* and *Plodia interpunctella. Journal of Chemical Ecology, 31*, 2947–2961.

Pawliszyn, J. (2000). Theory solid phase microextraction. *Journal of Chromatographic Science, 38*, 270–278.

Phelan, P. L., & Lin, H. (1991). Chemical characterization of fruit and fungal volatiles attractive to dried-fruit beetle, *Carpophilus hemipterus* (Coleoptera: Nitidulidae). *Journal of Chemical Ecology, 17*, 1253–1272.

Pungaliya, C., Srinivasan, J., Fox, B. W., Malik, R. U., Ludewig, A. H., Sternberg, P. W., & Schroeder, F. C. (2009). A shortcut to identifying small molecule signals that regulate behavior and development in *Caenorhabditis elegans. Proceeding of National Academy of Science of United States of America, 106*, 7708–7713.

Rafaeli, A., Kostukovshy, A., & Carmeli, D. (2006). Successful disinfestation of sap-beetle contaminations from organically grown dates using heat treatment. A case study. *Phytoparasitica, 34*, 204–212.

Reardon, R. C., Leonard, D. S., Mastro, V. C., Leonhardt, B. A., McLane, W., Talley, S., Thorpe, K. W., & Webb, R. E. (1998). *Using mating disruption to manage gypsy moth: A review* (85 pp, pp. 98–01). Morgantown Department of Agriculture Forest Service, FHTET.

Reddy, G. V. P., & Guerrero, A. (2010). New pheromones and insect control strategies. In G. Litwack (Ed.), *Vitamins and hormones* (pp. 493–520). Burlington: Elsevier Academic Press.

Robinette, S. L., Bruschweiler, R., Schoroeder, F. C., & Edison, A. S. (2011). NMR in metabolomic and natural products research: Two sides of the same coin. *Accounts of Chemical Research, 45*, 288–297.

Smilanick, J. M., Ehler, L. E., & Birch, M. C. (1978). Attraction of *Carpophilus* spp. (Coleoptera: Nitidulidae) to volatile compounds present in figs. *Journal of Chemical Ecology, 4*, 701–707.

Soroker, V., Blumberg, D., Haberman, A., Hamburger-Rishard, M., Reneh, S., Talebaev, S., Anshelevich, L., & Harari, A. R. (2005). Current status of RPW in date palm plantations in Israel. *Phytoparasitica, 33*, 97–106.

Soroker, V., Haberman, A., Nakache Y., Ketner, R., Ezra, N., Ben David T., Kuznetsova, T., Biton, S., & Cohen, Y. (2013). *History of RPW management in Israel 1999–2012*. Palm Pest Mediterranean Conference Nice, 16–18 January 2013.

Soroker, V., Barkan, S., Livne, Y., Nachache, Y., & Bar, Z. (2014). Comparison of red palm weevil trapping ability of bucket and Picusan traps. *Biobitaon, 8*, 10–14 (in Hebrew).

Steiner, L. F., Mitchell, W. C., Harris, E. J., Kozuma, T., & Fujimoto, M. S. (1965). Oriental fruit by eradication by male annihilation. *Journal of Economic Entomology, 58*, 96–964.

Talhouk, A. S. (1991). On the management of the date palm and its arthropod enemies in the Arabian Peninsula. *Journal of Applied Entomology, 111*, 514–520.

Teich, I., Neumark, S., & Jacobsen, M. (1979). Mass trapping of males of Egyptian cotton leafworm, *Spodoptera littoralis* and large scale synthesis procedure. In F. J. Ritter (Ed.), *Chemical ecology and odour communication in animals* (pp. 343–350). Amsterdam: Elsevier.

Tóth, M., Repasi, V., & Szöcs, G. (2002). Chemical attractants for females of pest pyralids and phycitids (Lepidoptera: Pyralidae, Phycitidae). *Acta Phytopathologica et Entomologica Hungarica, 37*, 375–384.

Trematerra, P., Sciarretta, A., & Tamasi, E. (1999). Control of codling moth, *Cydia pomonella*, with an attracticide (attract and kill) method. *Information of Fitopatology, 49*, 41–44.

Uchida, G. K., McInnis, D. O., Vargas, R. I., Kumashiro, B. R., Klungness, L. M., & Jang, E. (2003). Nontarget arthropods captured in cue-lure-baited bucket traps at area- wide pest management implementation sites in Kamuela and Kula, Hawaiian Islands. *Proceedings of the Hawaiian Entomological Society, 36*, 135–143.

Uchida, G. K., Mackey, B. E., McInnis, D. O., & Vargas, R. I. (2007). Attraction of *Bactrocera dorsalis* (Diptera: Tephritidae) and non-target insects to methyl eugenol bucket traps with different preservative fluids on Oahu, Hawaiian Islands. *Journal of Economic Entomology, 79*, 1580–1582.

Vacas, S., Primo, P., & Llopis, V. N. (2013). Advances in use of trapping system for *Rhynchophorus ferrugineus. Journal of Economic Entomology, 106*, 1739–1746.

Vacas, S., Payá, M. A., Primo, P., & Llopis, V. N. (2014). Identification of pheromone synergists for *Rhynchophorus ferrugineus* trapping systems from *Phoenix canariensis* palm volatiles. *Journal of Agricultural and Food Chemistry, 62*, 6053–6064.

Vega, F. E., Dowd, P. F., & Bartelt, R. J. (1995). Dissemination of microbial agents using an autoinoculating device and several insect species as vectors. *Biological Control, 5*, 545–552.
Vega, F. E., Dowd, P. F., Lacey, L. A., Pel, l. J. K., Jackson, D. M., & Klein, M. G. (2007). Dissemination of beneficial microbial agents by insects. In L. A. Lacey & H. K. Kaya (Eds.), *Field manual of techniques in invertebrate pathology* (pp. 127–146). Dordrecht: Springer.
Vidyasagar, P. S. P. V., Hagi, M., Abozuhairah, R. A., Al-Mohanna, O. E., & Al-Saihati, A. A. (2000). Impact of mass pheromone trapping on red palm weevil adult population and infestation level in date palm gardens of Saudi Arabia. *Planter, 76*, 347–355.
Witzgall, P., Stelinski, L., Gut, L., & Thomson, D. (2008). Codling moth management and chemical ecology. *Annual Review of Entomology, 53*, 503–522.
Witzgall, P., Kirsch, P., & Cork, A. (2010). Sex pheromones and their impact on pest management. *Journal of Chemical Ecology, 36*, 80–100.
Wyatt, T. D. (2003). *Pheromones and animal behaviour: Communication by smell and taste* (391 pp). Cambridge: Cambridge University Press.
Zaid, A., De Wet, P. F., Djerbi, M., & Oihab, A. (2002). Diseases and pests of date palm. In A. Zaid (Ed.), *Date palm cultivation* (FAO Plant Production and Protection Paper No. 156, Revision 1). Rome: Food and Agriculture Organization.
Zhu, J. W., Ryne, C., Unelius, C. R., Valeur, P. G., & Löfstedt, C. (1999). Re-identification of the female sex pheromone of the Indian meal moth, *Plodia interpunctella*: Evidence for a four-component pheromone blend. *Entomologia Experimentalis et Applicata, 92*, 137–146.
Zilkowski, B. W., Bartelt, R. J., Blumberg, D., James, D. G., & Weaver, D. K. J. (1999). Identification of host-related volatiles attractive to pineapple beetle *Carpophilus humeralis*. *Journal of Chemical Ecology, 25*, 229–252.

Chapter 12
Mite Pests of Date Palms

Mohamed W. Negm, Gilberto J. De Moraes, and Thomas M. Perring

Abstract The date palm, *Phoenix dactylifera*, is attacked by several mite species that frequently cause significant damage. The Banks grass mite, *Oligonychus pratensis* (Banks), and the Old World date mite, *Oligonychus afrasiaticus* (McGregor), (Acari: Tetranychidae) are considered major pests of date palms. The spider mite *Eutetranychus palmatus* Attiah (Tetranychidae), the red palm mite, *Raoiella indica* Hirst, and the red and black flat mite, *Brevipalpus phoenicis* (Geijskes), (Acari: Tenuipalpidae) are considered pests of minor importance to date palm. This chapter summarizes the present knowledge about the distribution, host range, damage, biology, seasonal incidence and management practices of these species. Strategies used for mites are centered on the use of chemical pesticides. Often, the extensive use of these products has resulted in secondary pest outbreaks, undermining the adoption of sustainable pest management programs. Alternative management tactics, including biological control, have not been widely adopted by date producers. Therefore, since more research is needed, future research priorities are also discussed.

12.1 Introduction

The date palm, *Phoenix dactylifera* L. (Arecaceae), is attacked by various arthropods. The first review of pests and diseases of date palm was presented by Carpenter and Elmer (1978), reporting about 54 insect and mite species. This number of species

---

M.W. Negm (✉)
Department of Plant Protection, Faculty of Agriculture, Assiut University,
Assiut 71526, Egypt
e-mail: waleednegm@yahoo.com

G.J. De Moraes
Departamento de Entomologia e Acarologia, Escola Superior de Agricultura
"Luiz de Queiroz" (ESALQ)-Universidade de São Paulo (USP), Piracicaba, SP, Brazil
e-mail: moraesg@usp.br

T.M. Perring
Department of Entomology, University of California, Riverside, CA 92521, USA
e-mail: thomas.perring@ucr.edu

W. Wakil et al. (eds.), *Sustainable Pest Management in Date Palm: Current Status and Emerging Challenges*, Sustainability in Plant and Crop Protection,
DOI 10.1007/978-3-319-24397-9_12

has increased worldwide, reaching 112 species in a recent review (El-Shafie 2012). General information about mite pests (Arachnida: Acari) of date palm was provided by Al-Deeb (2012) and Kinawy (2012). Some phytophagous mites reported from this host are cosmopolitan agricultural pests with extensive host ranges and records of plant damage (Jeppson et al. 1975). They belong mainly to the superfamily Tetranychoidea (Acari: Trombidiformes), which includes the Tetranychidae (spider mites) and Tenuipalpidae (flat mites), the most common and ecologically important families infesting date palm. The Banks grass mite, *Oligonychus pratensis* (Banks), the Old World date mite, *Oligonychus afrasiaticus* (McGregor) and the spider mite, *Eutetranychus palmatus* Attiah (Acari: Tetranychidae), are considered important pests of dates (Carpenter and Elmer 1978; Blumberg 2008). Among the flat mites, the red palm mite, *Raoiella indica* Hirst, and the red and black flat mite, *Brevipalpus phoenicis* (Geijskes), are the most damaging species (Jeppson et al. 1975; Childers et al. 2003; Flechtmann and Etienne 2004). The aim of this chapter is to summarize biological and ecological information about major mite pests reported on date palms, as well as about their management. This information includes distribution, host range, damage, biology, seasonal incidence and different management tactics. Future research priorities towards the management of these mite pests are also discussed.

12.2 Spider Mites (Tetranychidae)

Mites in the family Tetranychidae (Trombidiformes: Tetranychoidea) are known as spider mites because of their ability to spin silk webbing used to protect their colonies against the attack of natural enemies and to disperse. Spider mites are phytophagous and some of them are important pests of a wide range of host plants in most terrestrial ecological communities. All spider mites have five developmental stages, namely egg, larva, protonymph, deutonymph and adult (males and females). A non-feeding, quiescent period is always found at the end of each post-embryonic immature stage, and these are termed the larval chrysalis, protochrysalis and deutochrysalis. In addition to the three tetranychid species mentioned as most important pests of date palm, about nine other species of this family have been reported infesting this crop (Bolland et al. 1998; Knihinicki and Flechtmann 1999; Blumberg 2008; Majidi and Akrami 2013).

12.2.1 Banks Grass Mite

The Banks grass mite is the only mite pest of dates in North America. Known previously as the timothy mite and date mite, it is believed to be native to the USA (Stickney et al. 1950). According to Jeppson et al. (1975), this mite originally was found to damage sugarcane in Hawaii, but it was misidentified as *Oligonychus exsiccator* Zehntner. The first description was by Banks (1912), from mites collected on timothy grass, *Phleum pratense* L. (Poaceae), in the state of Washington, near Pullman, USA. He named the mite *Tetranychus pratensis* Banks. A year later, Ewing (1913) revised the description based on characteristics of the female tarsal claw and male aedeagus. In 1914, Banks collected what he thought was a new species from date palms near El Centro (Imperial County), California, USA describing it as *Tetranychus simplex* Banks (Banks 1914). A short time later, McGregor reviewed the tarsal claw and aedeagus descriptions of *T. pratensis* by Ewing (1913) and determined that this mite belonged to the genus *Paratetranychus*, thus giving rise to *Paratetranychus pratensis* Banks (McGregor 1919). Further collections by Ewing (1922) from date palms in the Coachella Valley (Riverside County), California were described as *Paratetranychus heteronychus* Ewing. McGregor reviewed Banks' earlier descriptions and concluded that all mites collected from date palms in Imperial and Riverside counties belonged to the same species (McGregor 1939). He placed *T. simplex* in the genus *Paratetranychus* and synonymized *P. heteronychus* creating the species *Paratetranychus simplex*. Pritchard and Baker (1955) combined *P. pratensis* and *P. simplex* into a single species and moved the mite to the genus *Oligonychus*, giving its current binomial name, *Oligonychus pratensis* (Banks).

Oligonychus pratensis has been found in North America, South America, Africa and Asia. In the USA it has been reported as a pest of grasses and dates in the states of Arizona (McGregor 1939; Bibby and Tuttle 1959; Tuttle and Baker 1964), Arkansas, South Dakota (USDA 1972), California, Washington (Banks 1914; McGregor and Stickney 1965), Colorado (Schweissing 1969), Florida, Louisiana (McGregor 1914; McGregor and Stickney 1965), Georgia (Flechtmann and Hunter 1971), Hawaii (Jeppson et al. 1975; Goff 1986), Idaho, Montana (Malcolm 1955), Kansas (Pritchard and Baker 1955; DePew 1960; McGregor and Stickney 1965), Missouri (Thewke and Enns 1970), Nebraska, Oklahoma (Owens et al. 1976), Nevada (Harvey 1954; McGregor and Stickney 1965; USDA 1972), New Mexico (Pritchard and Baker 1955; McGregor and Stickney 1965), Oregon (Krantz 1957; McGregor and Stickney 1965), Texas (Walter and Wene 1956; Dean 1957; McGregor and Stickney 1965; Ehler 1974), Utah (Pritchard and Baker 1955; McGregor and Stickney 1965) and Virginia (Walter and Wene 1956).

Oligonychus pratensis also has been reported from Brazil (Flechtmann and Baker 1975; Feres et al. 2009), Colombia (Figueroa 1919), Costa Rica (Ochoa et al. 1991; Aguilar and Murillo 2012), El Salvador (Baker and Pritchard 1962; Andrews and Poe 1980), Honduras (Ochoa et al. 1991), Mexico (Beer and Lang 1958; Baker and Pritchard 1962; Estebanes and Baker 1968) and Puerto Rico (Jeppson et al. 1975). In the Old World, *O. pratensis* has been reported from many parts of Africa (Jeppson et al. 1975) including Algeria (André 1932; Pasquier 1932; Perrot 1934; Athias-Henriot 1959), Egypt (Zaher et al. 1982; Omar et al. 2013), Madagascar (Helle et al. 1970), Mauritania (Mallamaire 1955), Mauritius (Moutia 1958; Jeppson et al. 1975), Morocco (Pereau-Leroy 1958), Sahara and North Africa (Calcat 1959; Strumpel 1969), Senegal (Gutierrez and Etienne 1981), South Africa (Ryke and Meyer 1958; Meyer and Ryke 1959) and Sudan (Siddig and Elbadry 1971; Khairi et al. 2010). Banks grass mite also has been identified in Asia from China

(Ma and Yuan 1980), India (Channabasavanna and Banu 1972; Akbar and Aheer 1994), Iraq (Hirst 1920; USDA 1972), Pakistan (Chaudhri et al. 1974), Russia (Roslavtseva and Butovskii 1991) and Thailand (Baker 1975).

Banks grass mite is reported from a wide range of host plants in ten plant families (Table 12.1). Many of the plants are from early descriptions of *O. pratensis*. While certain situations (for example climatic conditions, plant stress and proximity to heavily infested neighboring plants) could lead to damaging densities of *O. pratensis* on many hosts (Table 12.1), most of the plants are not infested on a regular basis. Of the ten families, economic losses are typically reported in just two; Arecaceae and Poaceae. Pest intervention is required on date palm, timothy grass, sorghum [*Sorghum bicolor* (L.) Moench], maize (*Zea mays* L.) and sometimes wheat (*Triticum aestivum* L.). Considerable research has been conducted on these plants to determine the best practices for managing the Banks grass mite.

12.2.1.1 *Oligonychus pratensis*, a Serious Pest of Dates in the New World

In the New World, commercial date production takes place in a very small region of the United States. Most dates are grown in the Coachella Valley and recent plantings of dates have taken place in the Bard/Winterhaven area of Imperial County, California and near Yuma, Arizona, USA. In 2012, dates were harvested from 8400 acres, producing a crop worth $41.6 million (USDA 2013). Two varieties make up most of this production, Deglet Noor and Medjool, and nearly all of the damage by *O. pratensis* is on Deglet Noor.

The Banks grass mite (Fig. 12.1) is one of two key pests on dates in the United States; the other pest is the carob moth, *Ectomyelois ceratoniae* Zeller. Malcolm (1955) described the biology of *O. pratensis*, which is similar to other species of tetranychid mites (Jeppson et al. 1975). They have a haplodiploid sex-determinant reproductive system, with arrhenotokous parthenogenesis producing haploid males from unfertilized eggs (Malcolm 1955). Males develop faster than females (Malcolm 1955; Tan and Ward 1977) and newly emerged males are attracted to and defend quiescent females. As soon as the females emerge, they are immediately mated by the tending male. Mating lasts 1.5–3 min and females typically mate multiple times (Malcolm 1955). The life stages of *O. pratensis* first were described by Malcolm (1955). Egg to adult development is closely linked with temperature (Malcolm 1955; Feese and Wilde 1977; Tan and Ward 1977; Congdon and Logan 1983; Perring et al. 1984a, b) and humidity (Perring et al. 1984a, b), particularly in the microenvironment of the plant boundary layer (Perring et al. 1986; Holtzer et al. 1988).

While the key damage is caused to the young developing date fruit, *O. pratensis* also feeds on the fronds (Gispert et al. 2001). Adults, immatures and eggs are present throughout the year on the fronds and grasses in the date garden, suggesting that feeding and reproduction occur year-round, i.e. there is no true diapause. The highest numbers of mites on the fronds occur from March through September, while numbers are low from October through February. Dates in this geographic area flower in March and April, and sampling found no mites present in the flowers (Gispert et al. 2001).

Table 12.1 Plants reported to be hosts of *Oligonychus pratensis* (Banks)

Family name	Scientific name	Common name	Reference
Arecaceae			
	Cocos nucifera L.	Coconut (agriculture)	Pritchard and Baker (1955)
	Phoenix canariensis Chabaud	(ornamental)	McGregor and Stickney (1965)
	Phoenix dactylifera L.	Date palm (agriculture)	Banks (1914), Ewing (1922), Pritchard and Baker (1955), McGregor and Stickney (1965), Elmer (1965), Elmer et al. (1968)
	Phoenix roebelenii O'Brien	Pygmy date palm (ornamental)	Feres et al. (2009)
	Washingtonia filifera (Lindley) Hermann Wendland	Desert fan palm, California palm (ornamental)	McGregor and Stickney (1965), Feres et al. (2009)
Cucurbitaceae			
	Citrullus lanatus (Thunberg) Matsumura and Nakai	(agriculture)	Tuttle et al. (1976)
	Sereno Watson	Coyote melon, coyote gourd	Tuttle and Baker (1964, 1968)
Cyperaceae (sedges)			
	Carex simulata Mackenzie	Analogue sedge (natural)	Tuttle and Baker (1968)
	Cyperus alternifolius L.	Umbrella papyrus (ornamental)	Tuttle and Baker (1964, 1968)
Euphorbiaceae			
	Euphorbia albomarginata Torrey and A. Gray	Rattlesnake weed (natural)	Tuttle and Baker (1968)
Malvaceae			
	Abelmoschus esculentus (L.) Moench	Okra (agriculture)	Chaudhri et al. (1974)

(continued)

Table 12.1 (continued)

Family name	Scientific name	Common name	Reference
Poaceae			
	Agrostis sp.	3 species grown in experimental plots	Malcolm (1955)
	Agrostis gigantea Roth = *Agrostis alba*	Redtop (grazing)	Tuttle and Baker (1968)
	Agropyron sp.	21 species grown in experimental plots	Malcolm (1955)
	Agropyron intermedium (=*Thinopyrum intermedium* (Host) Barkworth and Dewey)	Intermediate wheatgrass (grazing)	Malcolm (1955), McGregor and Stickney (1965)
	Alopecurus arundinaceus Poiret	Creeping meadow foxtail (grazing)	Malcolm (1955)
	Alopecurus pratensis L.	Meadow foxtail (grazing)	Malcolm (1955)
	Andropogon gerardii Vitman (=*furcatus*)	Big bluestem (grazing)	Malcolm (1955)
	Andropogon trachycaulum (Link) Malte (=*subsecundum*)	Wheatgrass (natural)	Malcolm (1955)
	Andropogon glomeratus (Walt.) Britton, Sterns and Poggenb.	, bushy bluestem (ornamental)	Prichard and Baker (1955), Tuttle and Baker (1964)
	Arundinaria simonii (Carr.)	Simons bamboo (ornamental)	Tuttle and Baker (1968)
	Arundo donax L.	Giant cane, arundo (weed)	McGregor and Stickney (1965)
	Aristida adscensionis L.	Six weeks threeawn (natural)	Tuttle and Baker (1968)
	Arrhenatherum elatius (L.) Beauvois ex J. Presl and C. Presl	False oat-grass (grazing)	Malcolm (1955)
	Species not provided	Bamboo (ornamental)	Goff (1986)
	Bothriochloa saccharoides (Swartz) Rydberg	Silver bluestem (grazing)	Tuttle et al. (1976)
	Bouteloua rothrockii Vasey	Rothrock's grama (natural)	Tuttle and Baker (1968)
	Bromus sp.	11 species grown in experimental plots	Malcolm (1955)
	Bromus anomalus Rupr. ex Fournier	Nodding brome	Tuttle and Baker (1968)

	Bromus arizonicus (Shear) Stebbins	Arizona brome (natural)	Tuttle and Baker (1968)
	Brachypodium pinnatum (L.) Beauvois	Heath false brome (natural)	Malcolm (1955)
	Chloris virgata L.	Southern sandspur (natural)	Tuttle and Baker (1968)
	Cenchrus pauciflorus L.	Coastal sandbur (weed)	McGregor and Stickney (1965)
	Chloris virgata Swartz	Feather fingergrass, feather windmill grass (natural)	McGregor and Stickney (1965), Tuttle and Baker (1968)
	Chondrosum (= Bouteloua) barbata (Lagasa) Clayton	Six-weeks grama (natural)	Tuttle and Baker (1964, 1968), McGregor and Stickney (1965)
	Cynodon dactylon (L.) Persoon	Bermuda grass (ornamental)	Pritchard and Baker (1955), Bibby and Tuttle (1959), Tuttle and Baker (1964, 1968), McGregor and Stickney (1965), Zaher et al. (1982)
	Dactylis glomerata L.	Cock's foot, orchard grass (natural)	Malcolm (1955)
	Digitaria sanguinalis (L.) Scopoli	Hairy crabgrass (weed)	McGregor and Stickney (1965)
	Echinochloa colona (L.) Link	Canada wild rye (natural)	Tuttle and Baker (1964, 1968), McGregor and Stickney (1965)
	Elymus canadensis L.	Slender wheatgrass (natural)	Malcolm (1955), Tuttle and Baker (1964, 1968)
	Elymus repens (L.) Gould	(natural)	Tuttle and Baker (1964, 1968)
	(Link) Gould ex Shinners = *Elymus pauciflorus*	Slender wheatgrass (natural)	Tuttle and Baker (1968)
	Elymus sp.	11 species grown in experimental plots	Malcolm (1955)
	Elymus sp.	Ryegrass (grazing)	McGregor and Stickney (1965)
	Elymus virginicus L.	Virginia wildrye (grazing)	Thewke and Enns (1970)
	Eragrostis cilianensis (Allioni) Vign. Ex Janchen	Stinkgrass (natural)	McGregor and Stickney (1965)
	Eriochloa lemmonii Vasey and Scribn	Canyon cupgrass (natural)	Tuttle and Baker (1964, 1968)

(continued)

Table 12.1 (continued)

Family name	Scientific name	Common name	Reference
	Festuca sp.	6 species grown in experimental plots	Malcolm (1955)
	Hilaria rigida (Thurb.) Benth. ex Scribn.	Big galleta (natural)	Tuttle and Baker (1968)
	Hordeum sp.	5 species grown in experimental plots	Malcolm (1955)
	Hordeum jubatum L.	Foxtail barley (ornamental)	Malcolm (1955), Tuttle and Baker (1968)
	Leptochloa filiformis (Persoon) Beauvois	Red sprangletop (natural)	Ochoa et al. (1991)
	Leptochloa uninervia (Presl.) Hitchc. and Chase	(weed)	Tuttle and Baker (1968)
	Lolium perenne L.	Perennial ryegrass (grazing)	Tuttle and Baker (1968)
	Lolium remotum Schrank	Remote ryegrass (weed)	Malcolm (1955)
	Oryza glaberrima Steudel	(agriculture)	Gutierrez and Etienne (1981)
	Oryza sativa L.	Asian rice (agriculture)	Ochoa et al. (1991)
	Panicum antidotale Retzius	Ghamur, blue panicgrass (grazing)	Tuttle and Baker (1968)
	Panicum capillare L.	Witchgrass (weed)	Tuttle and Baker (1968)
	Panicum maximum Jacquin	Fataque, (grazing)	Jeppson et al. (1975)
	Panicum sp.	Grasses	Prichard and Baker (1955), Tuttle and Baker (1964), McGregor and Stickney (1965)
	Paspalum dilatatum Poiret	Dallisgrass (weed)	Tuttle and Baker (1964, 1968)
	Phalaris arundinacea L.	(grazing)	Malcolm (1955)
	Phalaris minor Retzius	Little seed canary grass (grazing)	Tuttle and Baker (1964, 1968)
	Phleum sp.	4 species grown in experimental plots	Malcolm (1955)
	Phleum pratense L.	Timothy grass (grazing)	Banks (1912), Malcolm (1955)

	Phragmites australis (Cav.) Trin. ex Steud.	Carrizo, common reed (natural)	Tuttle and Baker (1964)
	Phragmites communis Trin.	Carrizo, common reed (natural)	Tuttle and Baker (1968)
	Poa sp.	7 species grown in experimental plots	Malcolm (1955)
	Poa nemoralis L.	Wood bluegrass (grazing)	Tuttle and Baker (1968)
	Poa pratensis L.	Kentucky bluegrass (grazing)	Malcolm (1955), Tuttle and Baker (1968)
	Saccharum officinarum L.	Sugarcane (agriculture)	McGregor (1950), Stickney et al. (1950), Pritchard and Baker (1955), Tuttle and Baker (1964), McGregor and Stickney (1965), Baker (1975), Flechtmann and Baker (1975), Goff (1986), Aguilar and Murillo (2012)
	Secale montanum Gussone	(natural)	Malcolm (1955)
	Setaria viridis (L.) Beauvois	Green foxtail (weed)	McGregor and Stickney (1965), Tuttle and Baker (1968), Ma and Yuan (1980)
	Sitanion hystrix (Elymus elymoides (Rafinesque) Swezey)	Squirreltail (grazing)	Tuttle and Baker (1968)
	Sorghum bicolor (L.) Moench	Sorghum	Walter and Wene (1956), Dean (1957), Bibby and Tuttle (1959), DePew (1960), Tuttle and Baker (1964, 1968), Ehler (1974), Baker (1975), Gutierrez and Etienne (1981)
	Sorghum halepense (L.) Persoon	Johnson grass (weed)	Tuttle and Baker (1964, 1968)
	Sporobolus cryptandrus (Torrey) Gray	Sand dropseed (natural)	McGregor and Stickney (1965)
	Sporobolus indicus (L.) R. Brown	Smutgrass (weed)	Prichard and Baker (1955), Tuttle and Baker (1964)
	Sporobolus junceus (Beauvois) Kunth	Pineywoods dropseed (natural)	Tuttle et al. (1976)
	Stipa nelsonii Scribner (=*williamsii*)	Williams needlegrass (natural)	Malcolm (1955)

(continued)

Table 12.1 (continued)

Family name	Scientific name	Common name	Reference
	Stipa X *Oryzopsis*	Cross in experimental plot	Malcolm (1955)
	Tripsacum dactyloides L.	Eastern gamagrass (grazing)	Malcolm (1955)
	Triticum aestivum L.	Wheat	Pritchard and Baker (1955), DePew (1960), Tuttle and Baker (1964), Jeppson et al. (1975)
	Urochloa mutica Peter Forsskål = *Brachiaria mutica*	Para grass, buffalo grass (weed)	Malcolm (1955), Pritchard and Baker (1955), Tuttle and Baker (1964), McGregor and Stickney (1965)
	Zea mays L.	Corn	Pritchard and Baker (1955), Walter and Wene (1956), Dean (1957), Bibby and Tuttle (1959), DePew (1960), Baker and Pritchard (1962), Tuttle and Baker (1964, 1968, 1976), McGregor and Stickney (1965), Ehler (1974), Baker (1975), Andrews and Poe (1980), Ochoa et al. (1991)
Rubiaceae			
	Bouvardia ternifolia (Cav.) Schltdl.	(ornamental)	Tuttle et al. (1976)
Salicaceae			
	Populus tremuloides Michaux	Quaking aspen, American aspen	Prichard and Baker (1955), McGregor and Stickney (1965)
Solanaceae			
	Calibrachoa parviflora (Jussieu) D'Arcy = *Petunia parviflora*	(ornamental)	Tuttle and Baker (1968)
	Solanum elaeagnifolium Cavanilles	Silverleaf nightshade (weed)	Tuttle et al. (1976)
Zygophyllaceae			
	Tribulus terrestris L.	Goathead, puncture vine (weed)	Tuttle and Baker (1968)

Many of the species in this table were cited in Migeon and Dorkeld (2006–2013). Some hosts in that site could not be confirmed by referring to the original literature. These include *Arenga* sp., *Bouteloua gracilis* (Willd. ex Kunth) Lag. ex Griffiths, *Cenchrus incertus, Dactyloctenium capitatum, Digitaria argillacea, Digitaria diversinervis, Echinochloa colona, Elymus hispidus, Elymus smithii, Eragrostis cilianensis, Leptochloa mucronata*, and *Pennisetum alopecuroides*

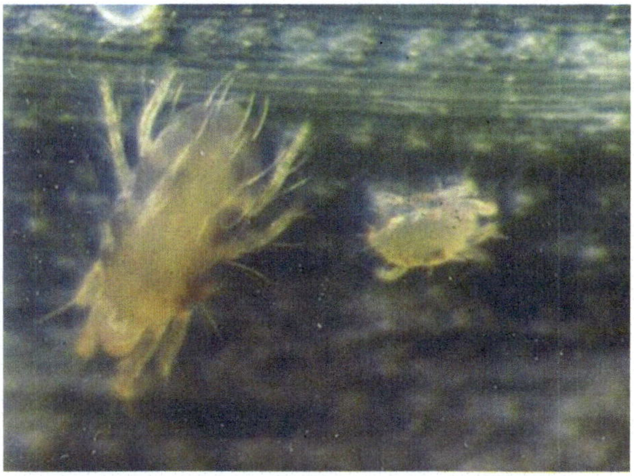

Fig. 12.1 Banks grass mite, *Oligonychus pratensis*: female (*left*) and larva (*right*) (Photo: Thomas M. Perring)

Fig. 12.2 Progression of *Oligonychus pratensis* infestation of date fruit Deglet Noor. Early infestation of young Kimri dates (**a**); Rapid spread in Kimri dates (**b**); Heavy webbing as dates enter Khalal stage (**c**); Completely infested date bunch (**d**) (Photos: Thomas M. Perring)

In April, during the early Kimri growth stage, *O. pratensis* females infest date fruit (Fig. 12.2a). They infest bunches by crawling from infested fronds or by aerial dispersal from fronds and grasses. The majority of new infestations in date bunches is thought to come from windborne mites (Thomas M. Perring – unpublished data). Margolies (1987) documented *O. pratensis* raising their posture which contributed to being picked up by the wind to be dispersed in the air. This behavior was displayed in response to declining food sources (Stiefel and Margolies 1992). At the same time, a higher proportion of offspring produced on these poor quality hosts were female, which facilitates the colonization of new hosts by mites (Margolies 1993; Palevsky et al. 2003).

Colony development is very rapid from April to June, when normal day-time temperatures in this area are typically above 39 °C. In addition, Pasquier (1932) hypothesized that population development was highest under low humidity and Perring et al. (1984a) determined that at 39 °C and low humidity, typical of date growing regions, the population doubling time is 48 h. Through this rapid population growth period, the Kimri dates also are increasing in size, providing optimal food resource and refuge for the developing spider mite colonies (Fig. 12.2b). The Banks grass mite produces copious amount of webbing and the date bunches become heavily webbed (Fig. 12.2c), providing protection to the interior part of the bunch where the population develops. This webbing (Fig. 12.2d) can interfere with acaricide penetration (Mauk et al. 2005).

The establishment by founding *O. pratensis* females leads to isolated colonies, which results in heavily infested bunches adjacent to bunches with few mites (Fig. 12.3). Mite colonies reach maximum abundance in mid-July and begin to decline with the onset of the Khalal or ripening stage. This reduction could be due to the thickening epidermis of the date as it begins to ripen or it may be caused by an increase in soluble carbohydrates in the ripening date. Perring et al. (1983) found that soluble carbohydrates in the leaves of maturing grain sorghum were negatively

Fig. 12.3 Date bunch heavily infested by *Oligonychus pratensis* (center), adjacent to other non- or lightly-infested bunches (Photo: Thomas M. Perring)

Fig. 12.4 *Oligonychus pratensis* feeding damage on Deglet Noor date fruit. Normal, non-damaged date is on the *left* (Photo: Thomas M. Perring)

correlated to *O. pratensis* densities and Palevsky et al. (2005) found that *O. afrasiaticus* declined on date fruits at the same time total sugars increased in the dates. While *O. pratensis* numbers decline on date fruit, the mite persists on date fronds and grasses in the date garden (Gispert et al. 2001), similar to *O. afrasiaticus* in date gardens in Israel (Palevsky et al. 2003). This insures a sufficient number of mites to re-infest date bunches the following April, as suggested by Lindgren and Vincent (1949).

12.2.1.2 Damage and Economic Losses

Oligonychus pratensis feeds on the epidermis of pre-Khalal dates. The mites disrupt the walls of the epidermal cells with their stylet-like chelicerae, removing the cell content, including the chlorophyll (Malcolm 1955; Krantz 1978). This damage to the fruit surface (Fig. 12.4) causes the infested area to take on a grayish-white, scarred quality (Carpenter and Elmer 1978) and severely damaged fruits may turn brown (Stickney et al. 1950). The fruits harden pre-maturely, shrivel and crack (Elmer 1965). Damaged fruits are down-graded in the packing houses, being unsuitable for fresh market consumption. Instead, they are used to make processed date products at a loss to the grower. Until 1998, when the acaricide hexythiazox was introduced into the date industry in California, *O. pratensis* caused annual losses and control costs between 1 and 2.5 million U.S. dollars.

12.2.1.3 Management

No sampling program or economic thresholds for the Banks grass mite have been adopted by the date industry in the USA. This is due to the historical reliance on acaricides for mite control applied mostly on a calendar basis. From the time *O. pratensis* was recognized as a pest in the early 1900s (Stickney et al. 1950),

Fig. 12.5 Applying sulfur dust to dates for *Oligonychus pratensis* control, Coachella Valley, circa 1996 (Photo: Thomas M. Perring)

a number of acaricides have been tested for its control on dates (Elmer 1965). Studies as early as 1922 demonstrated control of *O. pratensis* by nicotine and sulfur dust (Skinner 1927), and blowing sulfur dust up into the date bunches has been the industry standard since the mid-1950s (Vincent and Lindgren 1958; Carpenter and Elmer 1978; Nixon and Carpenter 1978). As recent as 1997, up to eight applications of sulfur were dusted into the date bunches (Fig. 12.5) during the dusting season (June through September) costing the industry 0.5 million U.S. dollars each year. This high amount of sulfur dust resulted in the date industry using an estimated 300 lb of pesticide per acre per year, contributing to the date industry being the second largest user of pesticides (by weight) in California. Not only was this practice costly, but the environmental concerns of dusting, the complaints of neighboring communities (see houses in background of Fig. 12.5) and field failures with sulfur prompted growers to seek alternative strategies.

Starting in 1997, the California Date Commission supported research to understand the biology and ecology of *O. pratensis* and develop a more integrated management program. Below is a summary of these studies.

Impact of Sulfur Dust on *O. pratensis* Predatory Mites, and Resistance

This study found similar numbers of *O. pratensis* on date fruit and fronds between sulfur-treated and non-treated palms on most weeks (Gispert et al. 2001). Interestingly, there were several weeks when significantly more mites were counted on the sulfur-treated palms. Subsequent research compared the toxicity of three sulfur formulations (stone, mined and petroleum distilled) and showed that the petroleum distilled sulfur (which was the formulation used most commonly in the date gardens) was the most toxic to *O. pratensis*. Further research found differential susceptibility at different date gardens indicating the development of resistance to sulfur. Additional studies found that mite susceptibility to sulfur declined as date gardens were treated with this product. Resistance ratios increased threefold in mites collected pre- and post-treatment with as few as two sulfur applications (Thomas M. Perring – unpublished data). Studies also demonstrated a negative impact of sulfur on the predatory mite, *Metaseiulus flumenis* Chant (Phytoseiidae), which is commonly cited in the literature on dates as *Galendromus mcgregori* (Chant) (Gispert et al. 2001), its junior synonym.

Hexythiazox and Additional Acaricides

At the same time research on the impact of sulfur dust was being conducted, studies evaluating efficacy of the insect growth regulator, hexythiazox were being conducted. This acaricide proved to be very effective for controlling mites with a one-time, early season application. Given the environmental concerns and non-target mortality with sulfur dusting, and armed with the knowledge that mites had developed resistance to sulfur, date growers were quick to adopt this new acaricide. A Section 18 Emergency Use registration from the United States Environmental Protection Agency was issued in 1998 (EPA 1998) and the product received full registration in 2003 (EPA 2003). While sulfur is still registered and used on a limited basis in organic date gardens, hexythiazox is the current industry standard for controlling *O. pratensis* in California (Mauk et al. 2005). Despite resistance to this material in several other mite agroecosystems (Van Leeuwen et al. 2009; Tirello et al. 2012), recent studies have shown no resistance to hexythiazox in the USA (Thomas M. Perring – unpublished data). Nevertheless, additional research has evaluated the acaricides, abamectin, buprofezin, etoxazole, hexythiazox, milbemectin, spirodiclofen, spiromesifen and spirotetramat in field trials and abamectin, bifenthrin, (fenpropathrin, hexythiazox, milbemectin, spirodiclofen, spiromesifen, spirotetramat and ultiflora in laboratory bioassays. These studies showed excellent mite control with etoxazole, milbemectin, spirodiclofen, spiromesifen and spirotetramat in field studies. In the laboratory bioassays, abamectin, bifenthrin, fenpropathrin, milbemectin and spirodiclofen were most effective against adult *O. pratensis*, while spirotetramat was the most effective material against immature mites (Thomas M. Perring – unpublished data).

Sampling Program for *O. pratensis* in Dates and Timing of Acaricides

The first step in the development of a sampling program is to identify a sampling unit. Sampling mites in dates presents a unique problem because mites are so small and date palms are so large; it is not practical to sample individual mites on the bunches. Therefore, studies were conducted to investigate the potential of using mite webbing, which is visible from the ground, as the sampling unit. This work is ongoing, but current results confirm that web ratings provide a reliable estimate of mite density. Web ratings are also correlated with subsequent mite damage to the dates. Some growers are relying on webbing assessment to identify specific bunches that require spot treatments with hexythiazox. This has saved them significant chemical and application costs.

Further research has related the development of mite webbing in bunches as a function of degree days (DD). For this research, *O. pratensis* temperature-developmental data from Perring et al. (1984b) were used to develop the model and temperature data were retrieved from the University of California CIMIS station in Thermal, California, USA. This study showed that bunches infested with a single female mite required 1520 DD to reach a web rating of 1 (=7 % of the bunch exterior covered with webbing). Date bunches infested with ten female mites required 1,285 DD to reach the same web rating. As it is difficult to determine when dates are infested (and thus determine when to start DD accumulation or the "biofix"), it was determined to use pollination of the date flowers as the biofix. In the heavily infested treatments (ten female mites), visible webbing was present 900 DD after pollination. This represents a conservative estimate for timing acaricide treatments and growers should not apply acaricides before this time. Further studies showed that applications made approximately 1100 DD after pollination resulted in less than 2 % damaged fruit. Thus 900–1100 DD is a reasonable time for growers to search dates for mite webbing and treat if webbing is found.

Studies with Natural Enemies of *Oligonychus pratensis*

Surveys in date gardens in the United States have identified only a few predator species. Based on abundance, predatory mites in the family Phytoseiidae are the most important. Gispert et al. (2001) found two species *M. flumenis* (Fig. 12.6) and

Fig. 12.6 *Metaseiulus flumenis* collected in date bunches infested with *Oligonychus pratensis*, Coachella Valley, California, 2012 (Photo: Thomas M. Perring)

Neoseiulus comitatus De Leon. *Metaseiulus flumenis* was found on date fronds and bunches, while *N. comitatus* was collected from Bermuda grass, *Cynodon dactylon* (L.) Persoon, in the garden. These mites are Type II phytoseiids (McMurtry and Croft 1997) and studies by Gispert et al. (2001) showed that they prey readily on *O. pratensis*.

Studies are underway to elucidate the biology of *M. flumenis* in an effort to determine why it is not more effective at regulating *O. pratensis* populations on dates in California. Other predatory insects found in date bunches, only after the spider mite colonies are well established, include *Stethorus* sp., *Orius insidiosus* (Say) and *Scolothrips sexmaculatus* (Pergande). The lack of predators for *O. pratensis*, compared to other agro-ecosystems (Dean 1957; Pickett and Gilstrap 1986), is likely due to the extreme high temperatures and low humidities found in the desert region, where date palm is grown.

As date production in the United States increases, there is a growing interest in sustainable practices for managing arthropod pests, including *O. pratensis*. Growers recognize that dependence on one acaricide, despite its effectiveness, is placing too much reliance on a single strategy. Thus they have supported the research described above and they are beginning to adopt practices like web-based sampling, spot acaricide applications and weed management in date gardens. Continued evaluation of alternative acaricides and studies on the biology of natural enemies will provide additional tools for managing *O. pratensis* in dates.

12.2.2 Old World Date Mite

The Old World date mite originally was described in the genus *Paratetranychus* Zacher, on date fruits collected from Biskra, Algeria (McGregor 1939). Pritchard and Baker (1955) transferred it to the genus *Oligonychus*. Baker and Pritchard (1960), Jeppson et al. (1975) and Zaher et al. (1982) reported this species in the subgenus *Reckiella* (Tuttle and Baker 1968).

The Old World date mite is considered the dominant acarine pest of date palm in North Africa and the Middle East. It has been reported from the following countries (Fig. 12.7): Afrotropical area: Chad, Mali, Mauritania, Niger (Munier 1952), Sudan (Baker and Pritchard 1960); Palearctic area: Algeria (André 1932; McGregor 1939), Bahrain (Abdul-Aziz M. Abdul-Karim, personal communication 2014), Egypt (Saleh and Hosny 1979; Zaher et al. 1982), Iran (Gentry 1965), Iraq (Buxton 1921; FAO 1966), Israel (Gerson et al. 1983), Libya (Martin 1958), Morocco (Pereau-Leroy 1958; Saba 1973), Oman (MOA 1989, 1992; Al-Zadjali et al. 2006), Saudi Arabia (El-Baker 1952), Tunisia (Pagliano 1951), United Arab Emirates (Gassouma 2005) and Yemen (El-Haidari 1981).

Oligonychus afrasiaticus is primarily a pest of date palm. However, it also has been reported from plants of other families: Celastraceae, *Catha edulis* (Vahl) Forssk. ex Endl. (Bolland et al. 1998) and Poaceae, *C. dactylon* (Zaher et al. 1982), *Imperata cylindrica* (L.) P. Beauv., *Lolium* sp. (Bolland et al. 1998), *Sorghum* sp. (Ben Chaaban et al. 2012) and *Zea mays* L. (Estebanes and Baker 1968).

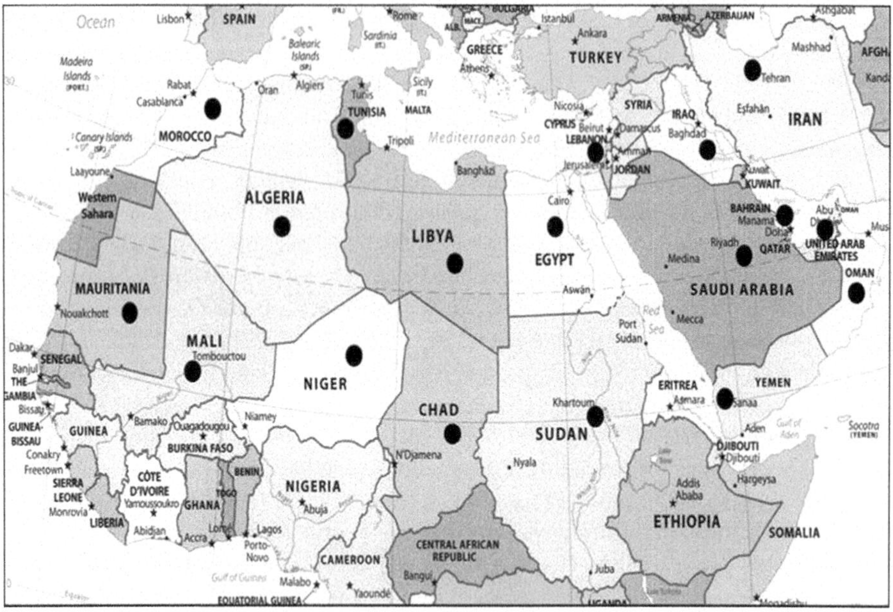

Fig. 12.7 World distribution of the Old World date mite, *Oligonychus afrasiaticus* (Mohamed W. Negm)

12.2.2.1 Biology and Seasonal Incidence

The biology of *O. afrasiaticus* first was studied by Hussain (1969). The egg is completely round and crystalline in color when first laid (Fig. 12.8), changing to waxy yellow as it matures. The larvae are light green when newly hatched, about 0.2 mm in length. Protonymphs and deutonymphs are similar in shape, except for the larger size of the latter. Adult females are oval, cream-colored and about 0.5 mm long, while males are smaller and with a posteriorly pointed body. *Oligonychus afrasiaticus* has six generations per year (Hussain 1969) and the duration of its life cycle depends mostly on temperature. The mean generation time has been determined as 6.0 and 14.0 days at 35 and 20 °C, respectively (Al-Sweedy et al. 2006). Age-specific fecundity schedules, life table parameters and seasonal abundance were studied in Iraq by Al-Swuidy (2003). A female-biased sex ratio, usually above 0.85, was reported on fruits of the varieties Deglet Noor, Barhee and Medjool (Palevsky et al. 2003).

Infestations occur in the hot dry summer season. Time of infestation varies depending on the variety. Initial infestation of the Medjool variety starts in early May, whereas in the Deglet Noor variety it starts in the first week of July and the

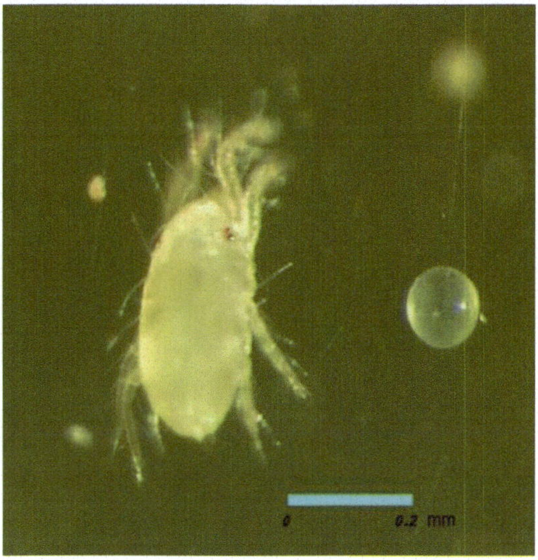

Fig. 12.8 Old World date mite, *Oligonychus afrasiaticus*: adult female (*left*) and deposited egg (*right*) (Photo: Palevsky et al. 2010; with permission)

Barhee variety is infested between late May and early July (Blumberg 2008). In Yemen, Saudi Arabia and Tunisia, populations reach the maximum level in mid-July to early August (Bass'haih 1999; Aldosari 2009; Ben Chaaban et al. 2012). The population decreases gradually in August and September (Ben Chaaban et al. 2011). Populations of *O. afrasiaticus* are rarely found during winter and spring; however, in the summer, numbers increase rapidly on fruit bunches and few can be found on fronds (Palevsky et al. 2003).

12.2.2.2 Damage and Losses

Similar to *O. pratensis*, *O. afrasiaticus* feeds mainly on date fruits. As the young green fruits appear, mites attack them causing the formation of many minute cracks and the production of gum-like exudates on the fruit surface. The mites remain active even at 45 °C and spin dense webbing onto which dust accumulates. In case of heavy infestations, webbing may cover the whole fruit bunch (Fig. 12.9).

Mites concentrate near the fruit calyx, where most eggs are laid. Heavy infestations can cause economic losses and the webbing makes the fruit unsuitable, even for processing (Fig. 12.10). In addition, this pest has a negative effect on the physicochemical characters of dates. Water and total soluble solid contents, particularly sugars, are significantly reduced (Ben Chaaban and Chermiti 2009). Infested dates of the Mijraf, Madini and Hamra varieties are smaller, malformed and do not ripe properly (Ba-Angood and Bass'haih 2000).

Fig. 12.9 Date palm fruit bunch covered by the webbing produced by the Old World date mite, *Oligonychus afrasiaticus* (Photo: Mohamed W. Negm)

Fig. 12.10 Date fruits damaged by the Old World date mite, *Oligonychus afrasiaticus* (Photo: Palevsky et al. 2010, with permission)

12.2.2.3 Management

Host Plant Resistance

Plant resistance to spider mites is reported in many crops (De Ponti 1985; Harman et al. 1996; Lopez et al. 2005). Date palm varieties vary in their degree of susceptibility to *O. afrasiaticus* infestation. The date varieties Cebiky and Sayer are resistant while the varieties Sokkary and Rothan are highly susceptible and Khodary has moderate resistance (Hussain 1974; Ali and Aldosari 2007). In Oman, the infestation with *O. afrasiaticus* to the varieties Hilali, Gibri and Khanazani starts in April, whereas other varieties are attacked later in the season (Elwan 2000). In Wadi Hadramout, Yemen, *O. afrasiaticus* attacks almost all date palm varieties; however, Hamra and Hajri seem less susceptible (Bass'haih 1999).

Biological Control

Efforts toward the biological control of *O. afrasiaticus* using mite and insect predators have been reported in some countries. In Iraq, Al-Jboory (2007) collected two phytoseiid mites (reported as *Amblyseius* sp. and *Euseius* sp.) in association with *O. afrasiaticus* on date palm fruits and pinnae. Othman et al. (2001) evaluated the possibility of controlling this pest by releases of a European strain of the phytoseiid mite, *Neoseiulus californicus* McGregor, in Tunisia. The results led the authors to recommend early releases in the season in order to obtain good control. This predator is commercially available in several countries, for the control of spider mites on different crops. However, it seems possible that phytoseiid mites locally-adapted to the hot semi-arid climatic conditions, under which date palm is cultivated, could provide better control of this pest than introduced species (Negm 2012). The efficacy of some local populations of phytoseiid species against *O. afrasiaticus* in Saudi Arabia was evaluated in laboratory studies [Al-Shammery (2010) for *Euseius scutalis* Athias-Henriot; Negm et al. (2014) for *Cydnoseius negevi* Swirski and Amitai (Fig. 12.11) and *Neoseiulus barkeri* Hughes]. Also, two predatory mites of the families Cheyletidae and Stigmaeidae have been evaluated by Abuyila et al. (2001) and Al-Shammery (2008), respectively.

Adults and larvae of *Stethorus* spp. and *Parastethorus* spp. (Coccinellidae) are reported to be specialist predators of spider mites (McMurtry et al. 1970; Biddinger et al. 2009). *Stethorus gilvifrons* Mulsant and *S. punctillum* Weise were reported as naturally occurring biocontrol agents of *O. afrasiaticus* in Iran (Afshari et al. 2007) and Algeria (Idder and Pintureau 2008), respectively.

Chemical Control

Pesticides have been used as the first choice for the control of *O. afrasiaticus* (Al-Doghairi 2004; Kamel et al. 2007; Aldosari 2009). Growers usually dust bunches with sulfur many times a year to control this pest (Palevsky et al. 2004).

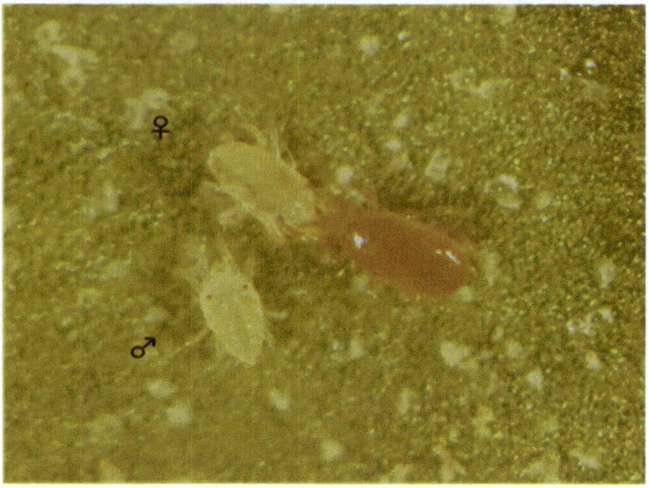

Fig. 12.11 The phytoseiid mite, *Cydnoseius negevi* (*right*) feeding on the Old World date mite, *Oligonychus afrasiaticus* (*left*) (Photo: Mohamed W. Negm)

This trend is likely to continue unless pest managers develop and implement alternative methods for the management of this pest. The acaricide hexythiazox, used for controlling *O. pratensis*, also provides good control of *O. afrasiaticus* in Israel (Palevsky et al. 2004). Aldosari (2009) evaluated the efficiency of the botanical compound (Baico) in comparison to other acaricides (Amitraz, Vertimec, Perpol and Salocide) and it appeared to be effective. Also, ethanolic extracts of demsisa, *Ambrosia maritima* L. (Asteraceae), duranta, *Duranta plumeria* (Verbenaceae) and cumin, *Cuminum cyminum* L. (Apiaceae) were recommended for the control of *O. afrasiaticus* (Fetoh and Al-Shammery 2011).

Other Tactics

Rainfall has been reported to limit the populations of this pest by destroying the webbing (Mohamed W. Negm – unpublished data). Spraying fruit bunches with a strong water stream can destroy the webbing and reduce mite populations, and this method could be useful for pest control especially in organic farming (Arbabi et al. 2010). Palevsky et al. (2004) evaluated different control measures to reduce mite damage, which did not yield satisfactory results.

Ground cover plants and weeds may harbor overwintering females of *O. afrasiaticus* during the absence of fruits (Palevsky et al. 2003). As a mechanical control practice, the removal of alternative host plants growing near date palms has been suggested in order to reduce the infestation of the next season.

Another control strategy, based on field observations, indicates that *O. afrasiaticus* infestations often are higher in palms growing near the margins of the orchard, along dusty roads. Thus paving roads or applying other practices that reduce dust levels also may reduce mite damage. Reducing dust has been shown to reduce mites in other agricultural systems (Godfrey 2011).

12.2.3 Spider Mite (*Eutetranychus palmatus*)

The spider mite, *E. palmatus* (Fig. 12.12) was described from date palms in Egypt (Attiah 1967). It also was reported from countries in the Palearctic region, including Iran (Kamali 1990), Israel (Gerson et al. 1983), Jordan (Palevsky et al. 2010) and Saudi Arabia (Al-Atawi 2011).

The host range of *E. palmatus* is limited to three plant families. In the Arecaceae, it has been reported from the date palm; the desert fan palm, *Washingtonia filifera* Lindley and Wendland; doum palm, *Hyphaene thebaica* L. Martius; Canary Island date palm, *Phoenix canariensis* Chabaud; and mountain date palm, *Phoenix loureiroi* Kunth (Ben-David et al. 2007). It also has been found on butternut squash, *Cucurbita moschata* Duchesne ex. Poiret in the Cucurbitaceae (Al-Atawi 2011) and on cotton, *Gossypium* sp., in the family Malvaceae (Bolland et al. 1998).

Fig. 12.12 Adult female of the spider mite, *Eutetranychus palmatus* (Photo: Palevsky et al. 2010; with permission)

12.2.3.1 Biology and Seasonal Incidence

Adult females of *E. palmatus* are reddish-brown and about 0.5 mm long. The body is round and somewhat flattened, and the legs are distinctly elongate. The eggs are round and pale brown. Unfortunately, little information is available about its biology (Blumberg 2008).

Its phenology and status as a pest of date fruit were studied by Palevsky et al. (2010). Mites can be found on the fronds throughout the year, reaching the highest levels from May to early June and then decline in July and August. Depending on variety and climate, the mites infest the fruits during June-July (Gerson et al. 1983), reaching the peak numbers on fruit strands in June (Palevsky et al. 2010).

12.2.3.2 Damage and Losses

Eutetranychus palmatus infests date palm fruits and fronds. Its damage is not as severe as that of *O. afrasiaticus* and *O. pratensis*, as it produces little webbing and it has not been observed to form large colonies. Damaged fruits are characterized by the presence of exudates, which then mix with mite exuviae, webbing and dust (Blumberg 2008). Moreover, irregular chlorotic scars of the feeding damaged areas appear along the middle to distal part of the fruit (Palevsky et al. 2010) (Fig. 12.13). These symptoms downgrade the marketable value of the dates, which then become inappropriate for exportation.

Fig. 12.13 Date fruits damaged by the spider mite, *Eutetranychus palmatus* (Photo: Palevsky et al. 2010; with permission)

12.2.3.3 Management

Date growers usually use synthetic acaricides for the control of *E. palmatus* when high densities are encountered. However, the application of fenbutatin oxide and hexythiazox did not lessen the chlorotic scarring caused by the mite at the early stage of infestation (Palevsky et al. 2010). Given that *E. palmatus* is found on date palms throughout the year, it is suggested that predatory mites, especially phytoseiids, reported in association with date palms (Palevsky et al. 2009; Ben Chaaban et al. 2011; Negm 2012; Negm et al. 2012a, b), as well as insect predators, may have an important role in regulating the populations of this pest.

12.3 Flat or False Spider Mites (Tenuipalpidae)

Mites of the family Tenuipalpidae (Trombidiformes: Tetranychoidea) commonly are known as flat mites or false spider mites, as they are generally compressed dorsoventrally. They do not produce silk webbing. Mesa et al. (2009) reported this family to consist of 891 species belonging to 34 genera. All tenuipalpids are phytophagous and they often are found on leaves near the midrib or other main veins. Certain species are found on the bark, in flower heads or in galls (Zhang 2003). Tenuipalpids are harmful to host plants directly, through their feeding, and indirectly, via transmission of plant viruses (Childers and Derrick 2003; Kitajima et al. 2003). They also are known to carry fungal spores and bacteria (Ochoa et al. 1994) that may infect the plant. The two largest flat mite genera, *Brevipalpus* Donnadieu and *Tenuipalpus* Donnadieu, include the most common species of economic importance. *Raoiella indica* Hirst and *Brevipalpus phoenicis* Geijskes are considered as pests of minor importance to date palm.

12.3.1 Red Palm Mite

The genus *Raoiella* was created by Hirst (1924) in the family Tetranychidae, with *R. indica*, described in the same publication as the type species. The genus subsequently was transferred to Pseudoleptidae, Trichadenidae and finally to Tenuipalpidae by Sayed (1950). Pseudoleptidae and Trichadenidae are considered today as synonyms of Tenuipalpidae.

12.3.1.1 Distribution and Host Range

Raoiella indica (Fig. 12.14) originally was described by Hirst (1924) from Coimbatore, Tamil Nadu, near the southern tip of India, from coconut, *Cocos nucifera* L. (Arecaceae). Since then, it has been reported from several Asian countries and islands in the Indian Ocean. In the first years of the twenty-first century, it dispersed west crossing the Atlantic Ocean and reached the Caribbean area, where

Fig. 12.14 Red palm mite, *Raoiella indica*: a colony showing eggs, nymphs, adults and exuviae (*top*), with detail of the droplets at the tip of dorsal setae (*bottom*) (Photo: Daniel C. Oliveira; with permission)

it was reported for the first time by Flechtmann and Etienne (2004). Since then, it has been found on other islands of the region as well as the American continent, where it has colonized from Southern Florida, USA to the tropical line in Brazil (Carrillo et al. 2011).

In Sudan, this mite first was found on date palms in 1938 (Pritchard and Baker 1958), being reported a few years later on the same crop in neighboring Egypt (Sayed 1942). It then was reported from date palm in Mauritius Island (Moutia 1958), Pakistan (Chaudhri et al. 1974), Iran (Arbabi et al. 2002; Majidi and Akrami 2013), Israel (Gerson et al. 1983), Sultanate of Oman and Tunisia (Zouba and Raeesi 2010).

Before its introduction to the New World, *R. indica* was known from only four hosts, namely coconut, date palm, areca nut *Areca catechu* L. and hurricane palm *Dictyosperma album* Bory H. Wendl. and Drude ex-Scheff. It recently was reported colonizing the pygmy date palm, *Phoenix roebelenii* O'Brien (Taylor et al. 2011). In the New World, it has been found on a much larger number of hosts in the family Arecaceae and other monocotyledonous families (Carrillo et al. 2012a; Gondim Jr. et al. 2012).

12.3.1.2 Biology and Seasonal Incidence

The life cycle of *R. indica* includes the same developmental stages reported for spider mites. Eggs of this species are ovoid, with one of the ends slightly wider than the other. They are reddish and smooth, about 0.1 mm long, bearing a narrow, white spiraled stipe distally. The post-embryonic stages are also reddish, but the adults are darker than the immatures. Adult females are rounded to ovoid, with a rounded posterior end, about 0.3 mm long × 0.2 mm wide; adult males are smaller and have a triangular posterior end. Under the microscope, the mites can be seen to have stout dorsal setae, many of which carry a droplet distally (Fig. 12.14). In the field, it has been noticed that particles of dust can stick to the droplets. It seems that when this happens predators have more difficulty attacking this mite.

Biological studies conducted under laboratory conditions, using coconut as the host, determined a relatively long immature phase, 3–4 weeks, which is usual for tenuipalpid mites. They also have been reported to have relatively low fecundity, 12–28 eggs (Moutia 1958; Jeppson et al. 1975; Lima et al. 2011). Despite these apparent limitations, it reaches very high population densities on coconut. Beard et al. (2012) indicated that these contradictory situations may be related to inadequate methodology used in the laboratory. These authors showed that *R. indica* feeds through the stomata and they suggested that when stomata close under stress, it makes it difficult for the mite to feed. On date palm leaves, Zaher et al. (1969) determined the immature phase to last about 21 days at 25–28 °C.

Some studies on the population dynamics of *R. indica* have been conducted (Lima et al. 2011; Taylor et al. 2011; Gondim Jr. et al. 2012). In two of these studies, Lima et al. (2011) and Gondim Jr. et al. (2012) observed the highest population densities to occur during periods of low rainfall. They suggested that more than 50–100 mm of monthly rainfall was detrimental to *R. indica* population increase. Also, as is valid for other mite species, rainfall could have an indirect effect (affecting natural enemies, cell turgidity and stomata opening and consequently food uptake) as well as a direct effect on the mite (washing mites off the leaves and increasing cell turgidity and consequently requiring more energy from the mite to get rid of excess ingested water). No studies have been conducted about a possible lower rainfall (or air humidity) threshold for the buildup of *R. indica* populations.

12.3.1.3 Damage and Losses

Beard et al. (2012) examined the leaf surface of host and non-host plants of *R. indica*, suggesting that the presence of wax and cutaneous materials immediately surrounding the stomatal opening could deter feeding by this mite. They reported that these were not present on coconut leaves, whereas they were present on the non-host *Sabal minor* Jacquin Persoon as well as on date palm leaves. In *S. minor* the particles were small and sunken below the leaf surface, clearly plugging the stomata. However, in date palm they were larger and only surrounded the stomata. These characteristics might be important in the high susceptibility of coconut in one extreme, no-susceptibility of *S. minor* in the other extreme, and date palm with intermediate susceptibility.

Most information about the symptoms of attack of *R. indica* refers to observations on coconut palms. Most serious damage is caused to plantlets, which often are killed by the mite. Older plants are less severely attacked, showing yellowish leaves under heavy mite densities. The worst symptoms appear in plants growing in soils with low fertility, with deficient drainage and with low content of organic matter (Sathiamma 1996).

On coconuts, this mite is found mostly on the lower leaf surface (Vasquez and Moraes 2013), while on dates they are reported to prefer the upper leaf surface, along the midrib (Zaher et al. 1969). This makes sense, given that stomata in coconut leaflets are practically restricted to the lower leaf surface, whereas in date palm they are about equally distributed on both leaf surfaces. In addition, coconuts are grown mostly in areas where rainfall is abundant, whereas dates are grown predominantly in areas where rainfall is low. El-Halawany et al. (2001) reported higher population levels of *R. indica* on fruits than on leaves of date palm in Egypt, but this information should be confirmed, given the need of this mite to feed through the stomata. Gerson et al. (1983) reported the occurrence of *R. indica* only on the lower leaf surface of the same host in Israel.

Raoiella indica has been reported as a minor seasonal pest of coconut in Southeastern India, as summarized by Taylor et al. (2011). However it was widely recognized as a pest of this crop after its introduction to the Caribbean area and the North American continent. Until now, there is no evaluation of the actual yield loss caused by the mite, but it is assumed to be high, because of the heavy symptoms. Few publications refer to *R. indica* as a pest of date palms (Zaher et al. 1969). It has been reported to not be a pest on this crop in Israel (Gerson et al. 1983; Blumberg 2008). However, high incidence of *R. indica* was reported on date fruits in all governorates of Egypt where date palms are grown (El-Halawany et al. 2001), but this report needs to be confirmed. Highest levels were reported between September and March, except when heavy rains occurred between November and January. It was mentioned as a minor date palm pest by Zaid et al. (2002) and also was briefly listed as a date palm pest in a review by El-Shafie (2012).

12.3.1.4 Management

Given the low pest status of this mite on dates, growers have sometimes used acaricides that have been evaluated on coconut. The USDA-APHIS investigated and concluded that the control of red palm mite using chemicals would be difficult and costly, especially when the palm trees are very tall (Hoy 2012). It may be cost effective in nursery situations. Studies also have been conducted to evaluate the tolerance of coconut varieties to the pest and the effect of different cultivation practices.

More recently, considerable effort has been dedicated to determine the role of natural enemies and their possible use as biological control agents on coconuts. An extensive review of this subject was published by Carrillo et al. (2012b). Most of the natural enemies found are mites of the family Phytoseiidae, although predaceous insects also have been reported in association with the pest. Most recent studies have indicated that the phytoseiid mite, *Amblyseius largoensis* Muma is the most common predator associated with *R. indica* (Peña et al. 2009; Gondim Jr. et al. 2012). However, as pointed out by Gondim Jr. et al. (2012), while *A. largoensis* often is the most dominant phytoseiid, its density is not dependent on *R. indica* as prey. But this does not mean that populations of this predator in long association with *R. indica* might not have evolved to prefer it over other possible food sources.

Populations of *A. largoensis* have been introduced from the Mauritius (Hoy 2012) and Réunion (Moraes et al. 2012) islands into the USA and Brazil, respectively, for possible field releases. Encouraging results have been obtained by Domingos et al. (2013), who reported the population introduced from Réunion Island to have significantly higher predation rate than the population of the same species collected in the Northern Brazilian State of Roraima. Recent laboratory investigations (D.C. Oliveira, E.G.F. de Morais and G.J. de Moraes – unpublished data), have confirmed the superior characteristics of the Réunion Island population. These encouraging results warrant further investigation under field conditions. However, doubts still remain as to what would be the result of an eventual release of the Réunion Island mites, given that a population of the same species is already present in the field. As expected, laboratory investigations (Navia et al. 2014) have shown reproductive compatibility between these two populations. No studies about the possible introduction of natural enemies for the control of the pest on date palm have been conducted.

12.3.2 Red and Black Flat Mite

This species has been known by various names depending on different host plants; on date palm, it is commonly referred to as the red and black flat mite. It has a worldwide distribution, affecting many crops in tropical and subtropical countries. Pritchard and Baker (1958) reported this mite on about 63 host plant genera. It is known now from more than 486 plant species worldwide including date palm (Childers et al. 2003; Hazarika et al. 2009). Important crops on which *B. phoenicis* occurs as a pest include citrus, coffee, date palm, papaya, passion fruit and tea (see Childers et al. 2003). On citrus, it is known as the citrus leprosis mite. It also has been reported as a serious tea pest in Asia, on which is known as the scarlet mite.

12.3.2.1 Biology and Seasonal Incidence

The immature stages of *B. phoenicis* are the same as reported for spider mites. Eggs are elliptical, bright red/orange in color when newly deposited, becoming darker as development proceeds. Eggs are deposited in groups by more than one female (Jeppson et al. 1975) and they are adhered to the plant surface. Larvae and nymphs are orange in color, with a pair of distinct lateral dark areas. Mite generations overlap in temperate climates. *Brevipalpus phoenicis* is thelytokous, usually producing only haploid females with two chromosomes; males are rarely produced, but seem non-functional reproductively (Helle et al. 2005). Weeks et al. (2001) indicated that haploid females of *B. phoenicis* were sustained as a result of an unidentified endosymbiotic bacterial infection. Thelytoky in *B. phoenicis* populations was studied genetically and discussed by Rodrigues et al. (2004) and Groot et al. (2005).

12.3.2.2 Damage and Losses

Symptoms of mite feeding vary according to the host. On date palm, it infests leaves and feeds on both sides, mainly along the midrib. At high infestation levels, damaged areas show brownish discoloration. Mites may not easily be detected because they are flattened, however, large populations can be seen as red spots and affected areas are dotted by white exuviae.

12.3.2.3 Management

Biological Control

Gerson (2008) reviewed the natural enemies affecting tenuipalpid mites in the field. These natural enemies mainly include predatory mites and entomopathogenic fungi. The predatory mites belong to families Cheyletidae, Cunaxidae, Eupalopsellidae, Stigmaeidae and mainly Phytoseiidae. The phytoseiid mite, *Euseius citrifolius* Denmark and Muma, collected from orange in Brazil, has been tested in the laboratory against *B. phoenicis* (Gravena et al. 1994). All mobile stages of *E. citrifolius* fed on *B. phoenicis* and preferred immature stages of the prey. This species is potentially useful as a biological control agent of *B. phoenicis*.

The entomopathogenic fungi, *Hirsutella thompsonii* Fisher, and *Metarhizium anisopliae* (Metchnikoff) Sorokin, appear to be effective pathogens of *B. phoenicis* in laboratory and field trials. *Hirsutella thompsonii* could be utilized as a microbial acaricide or as a natural biological control agent in humid areas (Childers et al. 2003). Rossi-Zalaf and Alves (2006) evaluated the efficiency of 52 isolates of the fungi, *Beauveria bassiana* (Balsamo) Vuillemin, *H. thompsonii*, *Lecanicillium lecanii* R. Zare and W. Gams, *L. muscarium* R. Zare and W. Gams, *M. anisopliae*, *Paecilomyces fumosoroseus* (Wize) A.H.S.Br. and G. Sm. and *P. lilacinus* (Thom) Samson, against *B. phoenicis* in the laboratory and found that *H. thompsonii* was the most effective.

Chemical Control

In Brazil, where this mite is the vector of the virus causing the disease known as citrus leprosis, synthetic chemical pesticides are used extensively for its control (Moraes and Flechtmann 2008). Unfortunately, *B. phoenicis* has developed resistance to some chemical compounds, so this may not be a sustainable tactic for controlling this tenuipalpid species.

12.4 Other Phytophagous Mites Found on Date Palm in the Field

The following species have been mentioned on date palms. However, they are not known to cause economic losses:

Family Eriophyidae
Acaphyllisa arabica Al-Atawi, Kamran and Flechtmann—Al-Atawi et al. (2014)
Tegonotus larii Kamali, Majidi and Akrami—Kamali et al. (2014)
Tetra iranica Kamali, Majidi and Akrami—Kamali et al. (2014)
Tumescoptes trachycarpi Keifer—Mohamed and El-Haidari (1968)
Family Phytoptidae
Mackiella phoenicis Keifer—Lindquist et al. (1996)
Retracrus johnstoni Keifer—El-Halawany et al. (2001)
Family Tarsonemidae
Polyphagotarsonemus latus (Banks)—El-Sanady and Mohamed (2013)
Steneotarsonemus spirifex (Marchal)—El-Sanady and Mohamed (2013)
Family Tenuipalpidae
Brevipalpus californicus (Banks)—Mashal and Albeidat (2006)
B. rica Chaudhri—Chaudhri et al. (1974)
Phyllotetranychus aegyptium Sayed—Zaher et al. (1969)
Phytoptipalpus phoenicis Alatawi and Kamran—Alatawi et al. (2015)
Tenuipalpus eriophyoides Baker—Majidi and Akrami (2013)
T. granati Sayed—Alatawi et al. (2015)
T. omani Moraes, Al-Shanfari and Silva—Moraes et al. (2011)
T. pareriophyoides Meyer and Gerson—Meyer and Gerson (1980)
T. trisegmentus Siddiqui and Chaudhri—Chaudhri et al. (1974)
T. yarensis Hasan, Bashir and Wakil—Hasan et al. (2003)
Family Tetranychidae
Bryobia rubrioculus (Scheuten)—Majidi and Akrami (2013)
Eutetranychus banksi (McGregor)—Bolland et al. (1998)
E. bredini Baker and Pritchard—Bolland et al. (1998)
E. orientalis Klein—Bolland et al. (1998)
Oligonychus calicicola Knihinicki and Flechtmann—Knihinicki and Flechtmann (1999)
O. senegalensis Gutierrez and Etienne—Blumberg (2008)
O. tylus Baker and Pritchard—Bolland et al. (1998)
Schizonobiella sp.—Majidi and Akrami (2013)
Tetranychus urticae Koch—Bolland et al. (1998)

12.5 Mites on Dates in Storage

Studies concerning the mite fauna of stored dried dates are very limited. Stickney et al. (1950) reported one species, *Tyrophagus lintneri* (Osborn), which is later proposed to be identical to *T. putrescentiae* (Schrank) (Klimov and OConnor 2010). However, its type specimens could not be located to allow a decision in this regard (Klimov and OConnor 2009), thus they are treated here as two separate species. Hoda et al. (1990) reported two predatory species from dates in stores, Acaropsellina sollers (Rohdendorf) (Cheyletidae) and Pyemotes tritici (LaGrèze-Fossat and Montagné). Another study was conducted in Egypt (Rezk 2000), reporting ten species; of these, four were phytophagous. Thus, five species have been reported from stored dates so far, namely *Suidasia* sp., *T. lintneri*, *T. putrescentiae* (Acaridae); *Carpoglyphus lactis* L. (Carpoglyphidae); and *Blomia freemani* Hughes (Glycyphagidae).

12.6 Future Research

The control of mite pests on date palm depends mostly on the use of synthetic acaricides. However, mite resistance to these products and concerns about the impact of pesticides on the environment and human health clearly identifies the need for a more integrated management approach. Biological control is an important component, yet the broad-scale use of natural enemies is in its infancy. Currently, there are just a handful of biological control agents that have been studied and more research is needed to determine the conditions under which these will be economically viable. For example, further surveys in the climatic regions infested by pest mites should be conducted, with the goal of identifying natural enemies that are already adapted to the extreme environmental conditions under which dates grow. As organisms are located, they must be properly identified, and while morphological characteristics are useful, there are more options for using molecular techniques.

In addition to biological control, we must continue searching for other methods to reduce pest mite densities. Removing overwintering sites for mites, washing them from the date bunches with water, reducing dust in the date gardens and developing sampling programs which can be used for more judicious use of acaricides are some of the ideas presented in this chapter. Hopefully these strategies will inspire additional research along these lines.

12.7 Summary

The number of mite species considered to be major pests of date palms is small, consisting of two *Oligonychus* species, *O. afrasiaticus* and *O. pratensis*. Both mites seriously affect yield and quality of dates, the first in North Africa and the

Middle East, and the second in North America. Producers rely primarily on acaricides to control these pests. Other mite species have been reported from date palms. Among these is *R. indica*, which is known to cause considerable damage to coconut; however it has not caused consistent damage to date palm. While there are relatively few mite pests on dates, the present scenario could change in different ways. First, mites currently identified in one region could be introduced into new areas. For example, the introduction of *O. afrasiaticus* to North America would compound the *O. pratensis* situation that currently exists. This would require an expansion of the biological control research to include the new introduction. Along the same line, when introduced into a new area, non-pest mite species may be able to cause major problems if introduced without their natural enemies. In addition use of acaricides to control present-day pests should be done in economically and environmentally sound ways. The judicious use of acaricides will delay the onset of resistance and give natural enemies the best possible chance of reducing densities of actual and potential pests.

Acknowledgments We would like to thank the Acarology Laboratory, King Saud University, Saudi Arabia for providing helpful information and Dr. Eric Palevsky for his permission to use images. Studies presented on *O. pratensis* were supported, in part, by the California Date Commission, California, USA.

References

Abuyila, O. A., Kerra, H. M., & Albahi, A. (2001). Biocontrol of date palm dust mite, *Oligonychus afrasiaticus* (Acarina: Tetranychidae) using predatory mite, *Eutogenes punctata* (Acarina: Cheyletidae: Prostigmata). *El Mokhtar Journal of Science, 8*, 1–8.

Afshari, M., Mossadegh, S., Soleyman-Nejadian, E., & Kamali, K. (2007). Geographical distribution and host plants of *Stethorus gilvifrons* (Mulsant) (Col.: Coccinellidae) and its biology under laboratory conditions in Khuzestan province. *Journal of Agricultural Sciences and Natural Resources, 14*, 201–210.

Aguilar, H., & Murillo, P. (2012). New hosts and records of plant feeding mites for Costa Rica: Interval 2008–2012. *Agronomia Costarricense, 36*, 11–28.

Akbar, S., & Aheer, G. M. (1994). Mite fauna of summer vegetables in Punjab. *Pakistan Journal of Zoology, 26*, 339–345.

Alatawi, F. J., Kamran, M., & Negm, M. W. (2015). False spider mites (Acari: Tenuipalpidae) associated with date palm orchards in Saudi Arabia: Description of two new species and new records. *Systematic and Applied Acarology, 20*, 809–830.

Al-Atawi, F. J. (2011). Phytophagous and predaceous mites associated with vegetable crops from Riyadh, Saudi Arabia. *Saudi Journal of Biological Sciences, 18*, 239–246.

Al-Atawi, F. J., Kamran, M., & Flechtmann, C. H. W. (2014). Eriophyoid mites (Prostigmata: Eriophyoidea) associated with date palms: New record and a new species of the genus *Acaphyllisa* from Saudi Arabia. *International Journal of Acarology, 40*, 353–357.

Al-Deeb, M. A. (2012). Date palm insect and mite pests and their management. In A. Manickavasagan, M. Mohamed Essa, & E. Sukumar (Eds.), *Date: Production, processing, food, and medicinal values* (pp. 113–128). Boca Raton: CRC Press, Taylor and Francis Group.

Al-Doghairi, M. A. (2004). Effect of eight acaricides against the date dust mite, *Oligonychus afrasiaticus* (McGregor) (Acari: Tetranychidae). *Pakistan Journal of Biological Sciences, 7*, 1168–1171.

Aldosari, S. A. (2009). Occurrence of dust mite, *Oligonychus afrasiaticus* McG. on fruits, leaflets of some date palm trees and evaluation the efficiency of botanical compound, (biaco) as compared with some acaricides. *Assiut University Bulletin Environmental Research, 12*, 69–77.

Ali, A. G., & Aldosari, S. A. (2007). Susceptibility of date palm fruit cultivars to the natural infestation by *Oligonychus afrasiaticus* McG. (Acari: Tetranychidae) in relation to their chemical composition. *Assiut University Bulletin Environmental Research, 10*, 1–7.

Al-Jboory, I. J. (2007). Survey and identification of the biotic factors in the date palm environment and its application for designing IPM-program of date palm pests in Iraq. *Aden Journal of Natural and Applied Sciences, 11*, 1–28.

Al-Shammery, K. A. (2008). *Biological studies on some plant parasitic mites and their control* (152 pp). Ph.D. dissertation, Riyadh University for Girls, Saudi Arabia.

Al-Shammery, K. A. (2010). Different biological aspects of the predaceous mite *Euseius scutalis* (Acari: Gamasida: Phytoseiidae) and the effects due to feeding on three tetranychid mite species in Hail, Saudi Arabia. *Asian Journal of Biological Sciences, 3*, 77–84.

Al-Sweedy, T. M., Al-Jboory, I. J., & Ahmad, T. R. (2006). Age specific fecundity of old world date mite *Oligonychus afrasiaticus* (McGregor) (Acari: Tetranychidae). *Arab Journal of Plant Protection, 24*, 14–19.

Al-Swuidy, T. M. (2003). *Heat accumulation, fecundity schedules, and life tables structure of old world date mite (Ghobar mite) Oligonychus afrasiaticus (McGregor) (Acari: Tetranychidae)* (94 pp). M.Sc. thesis, University of Baghdad, Iraq.

Al-Zadjali, T. S., Abd-Allah, F. F., & El-Haidari, H. S. (2006). Insect pests attacking date palms and dates in Sultanate of Oman. *Egyptian Journal of Agricultural Research, 84*, 51–59.

André, M. (1932). Contribution à l'étude du "Bou-Faroua" tétranyque nuisible au dattier en Algérie. *Bulletin de la Société d'histoire naturelle d'Afrique du Nord, 23*, 301–338.

Andrews, K. L., & Poe, S. L. (1980). Spider mites of El Salvador, Central America (Acari: Tetranychidae). *Florida Entomologist, 63*, 502–505.

Arbabi, M., Golmohammdad, Z. K. N., & Askari, M. (2002). Plant mite fauna of Sistan-Baluchestan and Hormozgan Provinces. *Journal of the Entomological Society of Iran, 22*, 87–88.

Arbabi, M., Asgari, M., Fasihi, M. T., Golmohammadzadeh-Khiaban, N., Damghani, M. R., Latifiean, M., & Babai, M. (2010). Evaluation of water spray application for organic control of date palm spider mite *Oligonychus afrasiaticus* (McGregor) (Acari: Tetranychidae) of date palm orchards in southern parts of Iran. *Journal of Entomological Research, 1*, 269–277.

Athias-Henriot, C. (1959). Acarologie appliquée et agronomie algérienne. *Acarologia, 1*, 181–200.

Attiah, H. H. (1967). The genus *Eutetranychus* in U.A.R. with descriptions of three new species (Acarina: Tetranychidae). *Bulletin de la Société Entomologique d'Egypte, 51*, 11–16.

Ba-Angood, A. S., & Bass'haih, G. S. (2000). A study on the effect of date palm dust mite *Oligonychus afrasiaticus* (McGregor) (Acarina: Tetranychidae) on the physiochemical characters of three different date varieties in Wadi Hadhramout, Yemen. *Arab Journal of Plant Protection, 18*, 82–85.

Baker, E. W. (1975). Plant-feeding mites of Thailand (Tetranychidae, Tenuipalpidae, and Tuckerellidae). *Plant Protection Service Technical Bulletin, 35*, 1–43.

Baker, E. W., & Pritchard, A. E. (1960). The tetranychoid mites of Africa. *Hilgardia, 29*, 455–574.

Baker, E. W., & Pritchard, A. E. (1962). Arañas rojas de América Central (Acarina: Tetranychidae). *Revista de la Sociedad Mexicana de Historia Natural, 23*, 309–340.

Banks, N. (1912). New American mites. *Proceedings of the Entomological Society of Washington, 14*, 96–99.

Banks, N. (1914). New acarina. *Pomona Journal of Entomology and Zoology, 6*, 55–66.

Bass'haih, G. (1999). *Studies on the occurrence of the date palm dust mite Oligonychus afrasiaticus (McGregor) (Acarina: Tetranychidae) and its natural enemies on different date palm varieties in Wadi Hadramout* (55 pp). M.Sc. thesis, University of Aden, Yemen.

Beard, J. J., Ochoa, R., Bauchan, G. R., Welbourn, W. C., Pooley, C., & Dowling, A. P. G. (2012). External mouthpart morphology in the Tenuipalpidae (Tetranychoidea): *Raoiella* a case study. *Experimental and Applied Acarology, 57*, 227–255.

Beer, R. E., & Lang, D. L. (1958). The Tetranychidae of Mexico (Acarina). *University of Kansas Science Bulletin, 38*, 1231–1259.

Ben Chaaban, S., & Chermiti, B. (2009). Characteristics of date fruit and its influence on population dynamics of *Oligonychus afrasiaticus* in the southern of Tunisia. *Acarologia, 49*, 29–37.

Ben Chaaban, S., Chermiti, B., & Kreiter, S. (2011). *Oligonychus afrasiaticus* and phytoseiid predators'seasonal occurrence on date palm *Phoenix dactylifera* (Deglet Noor cultivar) in Tunisian oases. *Bulletin of Insectology, 64*, 15–21.

Ben Chaaban, S., Chermiti, B., & Kreiter, S. (2012). Effects of host plants on distribution, abundance, developmental time and life table parameters of *Oligonychus afrasiaticus* (McGregor) (Acari: Tetranychidae). *Papéis Avulsos de Zoologia (São Paulo), 52*, 121–132.

Ben-David, T., Melamed, S., Gerson, U., & Morin, S. (2007). ITS2 sequences as barcodes for identifying and analyzing spider mites (Acari: Tetranychidae). *Experimental and Applied Acarology, 41*, 169–181.

Bibby, F. F., & Tuttle, D. M. (1959). Notes on phytophagous and predatory mites of Arizona. *Journal of Economic Entomology, 52*, 186–190.

Biddinger, D. J., Weber, D. C., & Hull, L. A. (2009). Coccinellidae as predators of mites: Stethorini in biological control. *Biological Control, 51*, 268–283.

Blumberg, D. (2008). Date palm arthropod pests and their management in Israel. *Phytoparasitica, 36*, 411–448.

Bolland, H. R., Gutierrez, J., & Flechtmann, C. H. W. (1998). *World catalogue of the spider mite family (Acari: Tetranychidae)* (392 pp). Leiden: Koninklijke Brill NV.

Buxton, M. A. (1921). Insect pests of dates and the date palm in Mesopotamia and elsewhere. *Bulletin of Entomological Research, 21*, 287–304.

Calcat, A. (1959). Diseases and pests of date palm in the Sahara and North Africa. *FAO Plant Protection Bulletin, 8*, 5–10.

Carpenter, J. B., & Elmer, H. S. (1978). *Pests and diseases of the date palm* (U.S. Department of Agriculture Handbook 527, 42 pp). Washington, DC: Department of Agriculture, Science and Education Administration,.

Carrillo, D., Navia, D., Ferragut, F., & Peña, J. E. (2011). First report of *Raoiella indica* (Acari: Tenuipalpidae) in Colombia. *Florida Entomologist, 94*, 370–371.

Carrillo, D., Amalin, D., Hosein, F., Roda, A., Duncan, R. E., & Peña, J. E. (2012a). Host plant range of *Raoiella indica* (Acari: Tenuipalpidae) in areas of invasion of the New World. *Experimental and Applied Acarology, 57*, 271–289.

Carrillo, D., Frank, J. H., Rodrigues, J. C., & Peña, J. E. (2012b). A review of the natural enemies of the red palm mite, *Raoiella indica* (Acari: Tenuipalpidae). *Experimental and Applied Acarology, 57*, 347–360.

Channabasavanna, G. P., & Banu, K. (1972). Plant feeding mites of India, Part 3. Spider mites of the species *Oligonychus* (Acarina: Tetranychidae). *Mysore Journal of Agricultural Sciences, 6*, 163–168.

Chaudhri, W. M., Akbar, S., & Rasool, A. (1974). *Taxonomic studies of the mites belonging to the families Tenuipalpidae, Tetranychidae, Tuckerellidae, Caligonellidae, Stigmaeidae and Phytoseiidae* (PL-480 Project on Mites, 250 p). Faisalabad: University of Agriculture.

Childers, C. C., & Derrick, K. S. (2003). *Brevipalpus* mites as vectors of unassigned rhabdoviruses in various crops. *Experimental and Applied Acarology, 30*, 1–3.

Childers, C. C., Rodrigues, J. C. V., & Welbourn, W. C. (2003). Host plants of *Brevipalpus californicus, B. obovatus*, and *B. phoenicis* (Acari: Tenuipalpidae). *Experimental and Applied Acarology, 30*, 29–105.

Congdon, B. D., & Logan, J. A. (1983). Temperature effects on development and fecundity of *Oligonychus pratensis*. *Environmental Entomology, 12*, 359–362.

De Ponti, O. M. B. (1985). Host-plant resistance and its manipulation through plant breeding. In W. Helle & M. W. Sabelis (Eds.), *Spider mites: Their biology, natural enemies and control* (pp. 395–404). Amsterdam: Elsevier.

Dean, H. A. (1957). Predators of *Oligonychus pratensis* (Banks), Tetranychidae. *Annals of the Entomological Society of America, 50*, 164–165.

de Moraes, G. J., & Flechtmann, C. H. W. (2008). *Manual de acarologia. Acarologia básica e ácaros de plantas cultivadas no Brasil* (288 pp). Ribeirão Preto: Holos Editora.

de Moraes, G. J., Al-Shanfari, A., & Da Silva, R. V. (2011). A new flat mite (Acari: Prostigmata: Tenuipalpidae) from date palm in the Sultanate of Oman. *Zootaxa, 2962*, 63–68.

de Moraes, G. J., de Castro, T. M. M. G., Kreiter, S., Quilici, S., Gondim, M. M. C., Jr., & de Sá, L. A. N. (2012). Search for natural enemies of Raoiella indica Hirst in Réunion Island (Indian Ocean). *Acarologia, 52*, 129–134.

DePew, L. J. (1960). Control of *Oligonychus pratensis* attacking winter wheat in Western Kansas. *Journal of Economic Entomology, 53*, 1061–1063.

Domingos, C. A., Oliveira, L. O., de Morais, E. G. F., Navia, D., de Moraes, G. J., & Gondim, M. G. C., Jr. (2013). Comparison of two populations of the pantropical predator *Amblyseius largoensis* (Acari: Phytoseiidae) for biological control of *Raoiella indica* (Acari: Tenuipalpidae). *Experimental and Applied Acarology, 60*, 83–93.

Ehler, L. E. (1974). *A review of the spider mite problem on corn and grain sorghum in West Texas* (Texas Agricultural Experiment Station Bulletin No. 1149, 15 pp).

El-Baker, A.-J. (1952). *Report to the government of Saudi Arabia on date cultivation* (FAO Report No. 31, TA 90/Rev. 1 Group 5. Rome, Italy, 25 pp).

El-Haidari, H. S. (1981). New records of mites and insects infesting date palms in the near east and north Africa. *Date Palm Journal, 1*, 133–136.

El-Halawany, M. E., Abdel-Samad, M. A., & El-Naggar, M. E. (2001). Mites inhabiting date palms. In: *Second international conference on date palms*, March 25–27, 2001, Al-Ain, UAE (pp. 366–373).

Elmer, H. S. (1965). Banks grass mite, *Oligonychus pratensis*, on dates in California. *Journal of Economic Entomology, 58*, 531–534.

Elmer, H. S., Carpenter, J. B., & Klotz, L. J. (1968). Pests and diseases of the date palm. *FAO Plant Protection Bulletin, 16*, 77–91.

El-Sanady, M. A., & Mohamed, A. A. (2013). Biodiversity and seasonal abundance of mites associated with two varieties of date palm in Giza and Sohag governorates, Egypt. *Acarines, 7*, 57–62.

El-Shafie, H. A. F. (2012). Review: List of arthropod pests and their natural enemies identified worldwide on date palm, *Phoenix dactylifera* L. *Agriculture and Biology Journal of North America, 3*, 516–524.

Elwan, A. A. (2000). Survey of the insect and mite pests associated with date palm trees in Al-Dakhliya region, Sultanate of Oman. *Egyptian Journal of Agricultural Research, 78*, 653–664.

EPA (Environmental Protection Agency). (1998). Hexythiazox, pesticide tolerances for emergency exemption. *Federal Register, 63*(197), 54595–54599. Available at: http://www.gpo.gov/fdsys/granule/FR-1998-10-13/98-27397. Accessed on 17 Apr 2015.

EPA (Environmental Protection Agency). (2003). Hexythiazox, pesticide tolerance. *Federal Register, 68*(43), 10370–10377. Available at: http://www.federalregister.com/Browse/Document/usa/na/fr/2003/3/5/03-5194. Accessed on 15 Apr 2015.

Estebanes, M. L., & Baker, E. W. (1968). Arañas rojas de Mexico (Acarina: Tetranychidae). *Anales de la Escuela Nacional de Ciencias Biologicas, 15*, 61–133.

Ewing, H. E. (1913). The taxonomic value of the characters of the male genital armature in the genus *Tetranychus* Dufor. *Annals of the Entomological Society of America, 6*, 453–460.

Ewing, H. E. (1922). The new species of peculiar and injurious spider mites. *Proceedings of the Entomological Society of Washington, 24*, 104–108.

FAO. (1966). *Report of the second Food and Agriculture Organization technical conference on the improvement of date production and processing*, October 16–25, 1966, Baghdad, Iraq, Meeting Report PL/1965/16. Rome, Italy, 25 p.

Feese, H., & Wilde, G. (1977). Factors affecting survival and reproduction of the Banks grass mite, *Oligonychus pratensis. Environmental Entomology, 6*, 53–56.

Feres, R. J. F., Vieira, R. M., Daud, R. D., Pereira, E. G. J., Oliveira, G. F., & Dourado, C. L. (2009). Ácaros (Arachnida, Acari) de plantas ornamentais na região noroeste do estado de São Paulo, Brasil: inventário e descrição dos sintomas causados pelos fitófagos. *Revista Brasileira de Entomologia, 53*, 466–474.

Fetoh, B. E. A., & Al-Shammery, K. A. (2011). Acaricidal ovicial and repellent activities of some plant extracts on the date palm dust mite, *Oligonychus afrasiaticus* McG. (Acari: Tetranychidae). *International Journal of Environmental Science and Engineering, 2*, 45–52.

Figueroa, J. (1919). Cultivation of clover. *Revista National de Agricultura, Bogota, 9–12*, 560–604.

Flechtmann, C. H. W., & Baker, E. W. (1975). A report on the Tetranychidae (Acari) of Brazil. *Revista Brasileira de Entomologia, 19*, 111–122.

Flechtmann, C. H. W., & Etienne, J. (2004). The red palm mite, *Raoiella indica* Hirst, a threat to palms in the Americas (Acari: Prostigmata: Tenuipalpidae). *Systematic and Applied Acarology, 9*, 109–110.

Flechtmann, C. H. W., & Hunter, P. E. (1971). The spider-mites (Prostigmata: Tetranychidae) of Georgia. *Journal of the Georgia Entomological Society, 6*, 16–30.

Gassouma, M. S. (2005). *Pests of the date palm*. Available at: http://ecoport.org/ep? SearchType= slideshowViewandslideshowId=133andcheckRequired=Y. Accessed on 15 Apr 2015.

Gentry, J. W. (1965). *Crop insects of Northeast Africa-Southwest Asia* (U.S. Department of Agriculture, 273 pp).

Gerson, U. (2008). The Tenuipalpidae: An underexplored family of plant-feeding mites. *Systematic and Applied Acarology, 13*, 82–101.

Gerson, U., Venezian, A., & Blumberg, D. (1983). Phytophagous mites on date palms in Israel. *Fruits, 38*, 133–135.

Gispert, C., Farrar, C., & Perring, T. M. (2001). Seasonal abundance and the banks grass mite *Oligonychus pratensis* (Banks) (Prostigmata: Tetranychidae) and a predatory mite, and their response to sulfur treatment on commercial date palms *Phoenix dactylifera* L. in Southern California. In R. B. Halliday, D. E. Walter, H. C. Proctor, R. A. Norton, & M. J. Colloff (Eds.), *Proceedings of the 10th International Congress of Acarology* (pp. 403–408). Melbourne: CSIRO Publishing.

Godfrey, L. D. (2011). *Pest notes: Spider mites*. UC ANR Publication 7405, UC Statewide Integrated Pest Management Program, University of California, Davis, California, USA. Available at: http://www.ipm.ucdavis.edu/PMG/PESTNOTES/pn7405.html. Accessed on 15 Dec 2014.

Goff, L. M. (1986). Spider mites (Acari: Tetranychidae) in the Hawaiian Islands. *International Journal of Acarology, 12*, 43–49.

Gondim, M. G. C., Jr., Castro, T. M. M. G., Marsaro, A. L., Jr., Navia, D., Melo, J. W. S., Demite, P. R., & de Moraes, G. J. (2012). Can the red palm mite threaten the Amazon vegetation? *Systematics and Biodiversity, 10*, 527–535.

Gravena, S., Benetoli, I., Moreira, P. H. R., & Yamamoto, P. T. (1994). *Euseius citrifolius* Denmark and Muma predation on citrus leprosis mite *Brevipalpus phoenicis* (Geijskes) (Acari: Phytoseiidae: Tenuipalpidae). *Anais da Sociedade Entomologica do Brasil, 23*, 209–218.

Groot, T. V. M., Janssen, A., Pallini, A., & Breeuwer, J. A. J. (2005). Adaptation in the asexual false spider mite *Brevipalpus phoenicis*: Evidence for frozen niche variation. *Experimental and Applied Acarology, 36*, 165–176.

Gutierrez, J., & Etienne, J. (1981). Quelques données sur les acariens Tetranychidae attaquant les plantes cultivées au Sénégal. *Agronomie Tropicale, 36*, 391–394.

Harman, J., Paul, P., Craig, R., Cox-Foster, D., Medford, J., & Mumma, R. O. (1996). Development of a mite bioassay to evaluate plant resistance and its use in determining regeneration of spider mite resistance. *Entomologia Experimentalis et Applicata, 81*, 301–305.

Harvey, T. L. (1954). Chemical control of a spider mite on wheat in Eastern New Mexico. *Journal of Economic Entomology, 47*, 593–597.

Hasan, M., Bashir, F., & Wakil, W. (2003). A new species of subgroup *eriophyoides* (*Tenuipalpus*: Tenuipalpidae: Acarina) from Punjab, Pakistan. *Acarologia, 43*, 359–361.

Hazarika, L. K., Bhuyan, M., & Hazarika, B. N. (2009). Insect pests of tea and their management. *Annual Review of Entomology, 54*, 267–284.

Helle, W., Gutierrez, J., & Bolland, H. R. (1970). A study on sex-determination and karyotype evolution in Tetranychidae. *Genetica, 41*, 21–32.

Helle, W., Bolland, H. R., & Gutierrez, J. (2005). Minimal chromosome number in false spider mites (Tenuipalpidae). *Cellular and Molecular Life Sciences, 28*, 707.

Hirst, S. (1920). Revision of the English species of red spider (Genera *Tetranychus* and *Oligonychus*). *Proceedings of the Zoological Society of London, 4*, 49–60.

Hirst, S. (1924). On some new species of red spiders. *Annals and Magazine of Natural History, London, Series, 9*(14), 522–523.

Hoda, F. M., El-Naggar, M. E., Taha, H. A., & El-Beheiry, M. M. (1990). Prostigmatid mites associated with stored products. *Agricultural Research Review, 68*, 77–85.

Holtzer, T. O., Norman, J. N., Perring, T. M., Berry, J. S., & Heintz, J. C. (1988). Effects of microenvironment on the dynamics of spider-mite populations. *Experimental and Applied Acarology, 4*, 247–264.

Hoy, M. A. (2012). Overview of a classical biological control project directed against the red palm mite in Florida. *Experimental and Applied Acarology, 57*, 381–393.

Hussain, A. A. (1969). Biology of *Paratetranychus afrasiaticus* McG. infesting date palm in Iraq. *Bulletin of the Entomological Society of Egypt, 33*, 221–225.

Hussain, A. A. (1974). *Date palms and dates with their pests in Iraq* (166 p). University of Baghdad, Ministry of Higher Education and Scientific Research, Iraq.

Idder, M. A., & Pintureau, B. (2008). Efficacité de la coccinelle *Stethorus punctillum* (Weise) comme predateur de l'acarien *Oligonychus afrasiaticus* (McGregor) dans les palmeraies de la region d'Ouargla en Algerie. *Fruits, 63*, 85–92.

Jeppson, L. R., Keifer, H. H., & Baker, E. W. (1975). *Mites injurious to economic plants* (614 pp). Berkeley: University of California Press.

Kamali, K. (1990). A checklist of plant mites (Acari) of Khuzestan, Southwestern Iran. *Scientific Journal of Agriculture, 13*, 73–83.

Kamali, H., Majidi, M., & Akrami, M. A. (2014). Two new eriophyid mite species (Acari, Prostigmata, Eriophyidae) on date trees from Iran. *Persian Journal of Acarology, 3*, 321–329.

Kamel, A., Al-Dosary, S., Ibrahim, S., & Ahmed, M. A. (2007). Degradation of the acaricide abamectin, flufenoxuron and amitraz on Saudi Arabian dates. *Food Chemistry, 100*, 1590–1593.

Khairi, M. M. A., Elhassan, M. I., & Bashab, F. A. (2010). The status of date palm cultivation and date production in Sudan. *IV International Date Palm Conference Book Series: Acta Horticulturae, 882*, 37–42.

Kinawy, M. (2012). *Date palm pests in Oman*. Muscat: Royal Gardens and Farms, Royal Court Affairs.

Kitajima, E. W., Chagas, C. M., & Rodrigues, J. C. (2003). *Brevipalpus*-transmitted plant virus and virus-like diseases: Cytopathology and some recent cases. *Experimental and Applied Acarology, 30*, 135–160.

Klimov, P. B., & OConnor, B. M. (2009). Conservation of the name *Tyrophagus putrescentiae*, a medically and economically important mite species (Acari: Acaridae). *International Journal of Acarology, 35*, 95–114.

Klimov, P. B., & OConnor, B. M. (2010). *Acarus putrescentiae* Schrank, 1781 (currently *Tyrophagus putrescentiae*; Acariformes, Acaridae): Proposed conservation of usage by designation of a replacement neotype. *Bulletin of Zoological Nomenclature, 67*, 24–27.

Knihinicki, D. K., & Flechtmann, H. W. (1999). A new species of spider mite, *Oligonychus calicicola* (Acari: Tetranychidae), damaging date fruit, *Phoenix dactylifera* L. (Arecaceae), in Australia. *Australian Journal of Entomology, 38*, 176–178.

Krantz, G. W. (1957). The timothy mite *Oligonychus pratensis* (Banks). *Oregon Seed Growers League Proceedings, 17*, 75–76.

Krantz, G. W. (1978). *A manual of acarology* (2nd ed., 509 pp). Corvallis Oregon: Oregon State University Bookstores.

Lima, M. R., Rodríguez, H., González, A. I., & González, M. (2011). Management strategy of *Raoiella indica* Hirst (Acari: Tenuipalpidae) in Cuba. In: G. J. de Moraes, & H. Proctor (Eds), Acarology XIII, proceedings of the International Congress, August 23–27, 2010, Recife, Pernambuco, Brazil. *Zoosymposia, 6*, 152–159.

Lindgren, D. L., & Vincent, L. E. (1949). Investigations on the life history and control of date insects and the date mite. *Date Growers' Institute Annual Report, 26*, 21–24.

Lindquist, E. E., Sabelis, M. W., & Bruin, J. (Eds.). (1996). *Eriophyoid mites: Their biology, natural enemies and control. World crop pests* (Vol. 6, 790 pp). Amsterdam: Elsevier Science Publishers.

Lopez, R., Levi, A., Shepard, B. M., Simmons, A. M., & Jackson, D. M. (2005). Sources of resistance to two spotted spider mite (Acari: Tetranychidae) in *Citrullus* spp. *HortScience, 40*, 1661–1663.

Ma, E. P., & Yuan, Y. L. (1980). New species and new records of tetranychid mites from China I. (Acarina: Tetranychidae). *Acta Zootaxonomica Sinica, 5*, 42–45.

Majidi, M., & Akrami, M. A. (2013). Mites associated with the date palm (*Phoenix dactylifera* L.) in Larestan (Fars province), Southern Iran. *Persian Journal of Acarology, 2*, 335–339.

Malcolm, D. R. (1955). *Biology and control of the timothy mite, Paratetranychus pratensis (Banks)* (Washington Agricultural Experiment Station Bulletin No. 17, 35 pp).

Mallamaire, A. (1955). Les tetranyques en Afrique occidentale. *Bulletin de la Protection des Végétaux, 1*, 101–109.

Margolies, D. C. (1987). Conditions eliciting aerial dispersal behavior in Banks grass mite *Oligonychus pratensis* (Acari: Tetranychidae). *Environmental Entomology, 16*, 928–932.

Margolies, D. C. (1993). Genetic variation of aerial dispersal behavior in the Banks grass mite. *Experimental and Applied Acarology, 17*, 461–471.

Martin, H. (1958). Ravageurs et maladies du palmier-dattier en Libye. *Bulletin Phytosanitaire de la FAO, 6*, 120–123.

Mashal, M., & Albeidat, B. (2006). A survey on insect pests of date palm trees in Jordan (research note). *Jordan Journal of Agricultural Sciences, 2*, 94–104.

Mauk, P. A., Perring, T. M., & Shea, T. (2005). Management of Banks grass mite in dates. *University of California Cooperative Extension, Topics in Subtropics Newsletter, 3*, 2.

McGregor, E. A. (1914). Four new tetranychids. *Annals of the Entomological Society of America, 7*, 354–364.

McGregor, E. A. (1919). The red spiders of America and a few European species likely to be introduced. *Proceeding of the United States Natural History Museum, 56*, 641–679.

McGregor, E. A. (1939). The specific identity of the American date mite: Description of two new species of *Paratetranychus. Proceedings of the Entomological Society of Washington, 41*, 247–256.

McGregor, E. A. (1950). Mites of the family Tetranychidae. *American Midland Naturalist, 44*, 257–420.

McGregor, E. A., & Stickney, F. (1965). Distribution and host plants of *Oligonychus pratensis* (Banks). *Proceedings of the Entomological Society of Washington, 67*, 28.

McMurtry, J. S., & Croft, B. A. (1997). Life styles of phytoseiid mites and their roles in biological control. *Annual Review of Entomology, 42*, 291–321.

McMurtry, J. A., Huffaker, C. B., & van de Vrie, M. (1970). Ecology of tetranychid mites and their natural enemies: A review. I. Tetranychid enemies: Their biological characters and the impact of spray practices. *Hilgardia, 40*, 331–390.

Mesa, N. C., Ochoa, R., Welbourn, W. C., Evans, G. A., & de Moraes, G. J. (2009). A catalog of the Tenuipalpidae (Acari) of the world with a key to genera. *Zootaxa, 2098*, 1–185.

Meyer, M. K. P., & Gerson, U. (1980). Some false spider mites (Prostigmata: Tenuipalpidae) from Israel. *Israel Journal of Entomology, 15*, 67–81.

Meyer, M. K. P., & Ryke, P. A. J. (1959). A revision of the spider-mites (Acarina: Tetranychidae) of South Africa with descriptions of a new genus and new species. *Journal of the Entomological Society of South Africa, 22*, 330–366.

MOA (Oman Ministry of Agriculture). (1989). *Plant crops control guide* (A. M. A. Al-Majing, A. M. A. Moktar, & M. M. Al-Shishtawy, Eds.) Sultanate of Oman (120 pp).

MOA (Oman Ministry of Agriculture). (1992). *Agricultural research annual report*. Sultanate of Oman (179 pp).

Mohamed, I. I., & El-Haidari, H. (1968). Contribution to the knowledge of the mite fauna of Iraq. *Bulletin de la Societé Entomologique d'Egypte, 1*, 127–128.

Moutia, L. A. (1958). Contribution to the study of some phytophagous acarina and their predators in Mauritius. *Bulletin of Entomological Research, 49*, 59–75.

Munier, P. (1952). *L'Assaba, essai monographique* (pp. 1–72). Saint-Louis: Centre IFAN, Maurit Etud. Maurit.

Navia, D., Domingos, C. A., Mendonça, R. S., Ferragut, F., Rodrigues, M. A. N., Morais, E. G. F., Tixier, M. S., & Gondim, M. G. C., Jr. (2014). Reproductive compatibility and genetic and morphometric variability among populations of the predatory mite, *Amblyseius largoensis* (Acari: Phytoseiidae), from Indian Ocean Islands and the Americas. *Biological Control, 72*, 17–29.

Negm, M. W. (2012). *Phytoseiid mites in Saudi Arabia with evaluation of two species as predators of the old world date mite, Oligonychus afrasiaticus* (151 pp). PhD dissertation, King Saud University, Riyadh, Saudi Arabia.

Negm, M. W., Alatawi, F. J., & Aldryhim, Y. N. (2012a). A new species of *Neoseiulus* Hughes, with records of seven species of predatory mites associated with date palm in Saudi Arabia (Acari: Phytoseiidae). *Zootaxa, 3356*, 57–64.

Negm, M. W., Alatawi, F. J., & Aldryhim, Y. N. (2012b). Incidence of predatory phytoseiid mites in Saudi Arabia: New records and a key to the Saudi Arabian species (Acari: Mesostigmata: Gamasina). *Systematic and Applied Acarology, 17*, 261–268.

Negm, M. W., Alatawi, F. J., & Aldryhim, Y. N. (2014). Biology, predation and life table of *Cydnoseius negevi* and *Neoseiulus barkeri* (Acari: Phytoseiidae) on the old world date mite, *Oligonychus afrasiaticus* (Acari: Tetranychidae). *Journal of Insect Science, 14*, 1–6.

Nixon, R. W., & Carpenter, J. B. (1978). *Growing dates in the United States* (United States Department of Agriculture Information Bulletin No. 207, 63 pp).

Ochoa, R., Aguilar, H., & Vargas, C. (1991). *Acaros fitofagos de America Central: guia ilustrada* (Serie Técnica. Manual Técnico, 251 pp). Turrialba:, CATIE.

Ochoa, R., Aguilar, H., & Vargas, C. (1994). *Phytophagous Mites of Central America: An Illustrated Guide* (CATIE. Serie Tecnica, Manual Tecnico No. 6, English Edition, 234 p).

Omar, N. A., Mohamed, O. M. O., & El-Sanady, M. A. (2013). Population fluctuations of *Oligonychus pratensis* (Banks) and its predator *Amblyseius zaheri* Yousef & El-Brollosy in Egypt (Tetranychidae, Phytoseiidae). *Research Journal of Agriculture and Biological Sciences, 9*, 89–95.

Othman, K., Rhouma, A., Belhadj, R., Alimi, E., Fallah, H., Kreiter, P., Lenfant, C., & Brun, J. (2001). Lutte biologique contre un acarien ravageur des dattes: essai d'utilisation de *Neoseiulus californicus* contre *Oligonychus afrasiaticus* dans les palmeraies du Djerid (Sud tunisien). *Phytoma, 540*, 30–31.

Owens, J. C., Ward, C. R., & Teetes, G. L. (1976). Current status of spider mites in corn and sorghum. In *Proceedings of the 31st annual corn and sorghum research conference*, Washington DC, USA (pp. 38–64).

Pagliano, T. (1951). *Les ennemis des vergers, des olivettes et des palmeraies* (2nd ed., 366 pp). Office Exp. et Vulgarisation Agricculture Société, Editions Françaises, Afrique du Nord.

Palevsky, E., Ucko, O., Peles, S., Yablonski, S., & Gerson, U. (2003). Species of *Oligonychus infesting* date palm cultivars in the Southern Arava Valley of Israel. *Phytoparasitica, 31*, 144–153.

Palevsky, E., Ucko, O., Peles, S., Yablonski, S., & Gerson, U. (2004). Evaluation of control measures for *Oligonychus afrasiaticus* infesting date palm cultivars in the Southern Arava Valley of Israel. *Crop Protection, 23*, 387–392.

Palevsky, E., Borochov-Neori, H., & Gerson, U. (2005). Population dynamics of *Oligonychus afrasiaticus* in the Southern Arava Valley of Israel in relation to date fruit characteristics and climatic conditions. *Agricultural and Forest Entomology, 7*, 283–290.

Palevsky, E., Gal, S., & Ueckermann, E. (2009). Phytoseiidae from date palms in Israel with descriptions of two new taxa and a key to the species found on date palms worldwide (Acari: Mesostigmata). *Journal of Natural History, 43*, 1715–1747.

Palevsky, E., Lotan, A., & Gerson, U. (2010). Evaluation of *Eutetranychus palmatus* (Acari: Tetranychidae) as a pest of date palms in Israel. *Israel Journal of Plant Sciences, 58*, 43–51.

Pasquier, R. (1932). Un dangereux ennemi du dattier: le oufaroua. *Revue d'horticulture et d'agriculture de l'Afrique du Nord, 36*, 36–40.

Peña, J. E., Rodigues, J. C., Roda, A., Carrillo, D., & Osborne, L. S. (2009). Predator-prey dynamics and strategies for control of the red palm mite (*Raoiella indica*) (Acari: Tenuipalpidae) in areas of invasion in the Neotropics. *Integrated Control of Plant Feeding Mites, IOBC/wprs Bulletin, 50*, 69–79.

Pereau-Leroy, P. (1958). *Date palms in Morocco* (142 pp). Institut Français de Recherches Fruitières Outre-Mer.

Perring, T. M., Archer, T. L., Krieg, D. L., & Johnson, J. W. (1983). Relationships between the Banks grass mite (Acariformes: Tetranychidae) and physiological changes of maturing grain sorghum. *Environmental Entomology, 12*, 1094–1098.

Perring, T. M., Holtzer, T. O., Kalisch, J. A., & Norman, J. M. (1984a). Temperature and humidity effects on ovipositional rates, fecundity, and longevity of adult female Banks grass mites (Acari: Tetranychidae). *Annals of the Entomological Society of America, 77*, 581–586.

Perring, T. M., Holtzer, T. O., Toole, J. L., Norman, J. N., & Myers, G. L. (1984b). Influences of temperature and humidity on pre-adult development of the Banks grass mite (Acari: Tetranychidae). *Environmental Entomology, 13*, 338–343.

Perring, T. M., Holtzer, T. O., Toole, J. L., & Norman, J. N. (1986). Relationships between corn-canopy microenvironments and Banks grass mite (Acari: Tetranychidae). *Environmental Entomology, 15*, 79–83.

Perrot, E. (1934). *Le dattier dans les oasis des Zibans et de l'Oued Rhir (Biskra-Touggourt)*. Notice. Centre de Documentation Technique et Economique sur les Plantes Medicinales 41, 70.

Pickett, C. H., & Gilstrap, F. E. (1986). Natural enemies associated with spider mites (Acari: Tetranychidae) infesting corn in the high plains region of Texas. *Journal of the Kansas Entomological Society, 59*, 524–536.

Pritchard, A. E., & Baker, E. W. (1955). *A revision of the spider mite family Tetranychidae* (Pacific Coast Entomological Society Memoirs, Series 2, 472 pp).

Pritchard, A. E., & Baker, E. W. (1958). The false spider mites (Acarina: Tenuipalpidae). *University of California Publications in Entomology, 14*, 175–274.

Rezk, H. A. (2000). Mites associated with stored dried-dates in Egypt and the role of *Blattisocius keegani* Fox as a biological control agent. *Alexandria Journal of Agricultural Research, 45*, 179–191.

Rodrigues, J. C. V., Gallo-Meagher, M., Ochoa, R., Childers, C. C., & Adams, B. J. (2004). Mitochondrial DNA and RAPD polymorphisms in the haploid mite *Brevipalpus phoenicis* (Acari: Tenuipalpidae). *Experimental and Applied Acarology, 34*, 275–290.

Roslavtseva, S. A., & Butovskii, R. O. (1991). Effects of pyrethroids on harmful and beneficial mites. *Agrokhimiya, 9*, 135–144.

Rossi-Zalaf, L. S., & Alves, S. B. (2006). Susceptibility of *Brevipalpus phoenicis* to entomopathogenic fungi. *Experimental and Applied Acarology, 40*, 37–47.
Ryke, P. A. J., & Meyer, M. K. P. (1958). Spider mites and false spider mites with notes on South African species and the basic concepts regarding their biological control. *South African Journal of Agricultural Science, 1*, 357–373.
Saba, F. (1973). Les acariens nuisibles aux plantes cultives au Maroc. *Al Awania, 49*, 69–97.
Saleh, M. R. A., & Hosny, M. M. (1979). Observation on *Oligonychus* spp. occurring on date bunches (Acari: Tetranychidae). *Ain Shams University Research Bulletin, 1114*, 1–8.
Sathiamma, B. (1996). Observations on the mite fauna associated with the coconut palm in Kerala, India. *Journal of Plantation Crops, 24*, 92–96.
Sayed, M. T. (1942). Contribution to the knowledge of the Acarina of Egypt: I The genus *Raoiella* Hirst Pseudotetranychidae -Tetranychidae). *Bulletin of the Society Fouad Ier Entomology, 26*, 81–91.
Sayed, M. T. (1950). On the taxonomy of tetranychid and allied genera, a new family and two new sub-families in Acarina. In *Proceedings of the eighth International Congress of Entomology, Stockholm, Sweden* (pp. 1012–1017).
Schweissing, F. C. (1969). *Cultural practice for possible use in reducing spider mite damage in corn* (Colorado State University, Agriculture Experiment Station Progress Report, Colorado, USA, pp. 69–72).
Siddig, M. A., & Elbadry, E. A. (1971). Biology of the spider mite, *Eutetranychus sudanicus*. *Annals of the Entomological Society of America, 64*, 806–809.
Skinner, H. B. (1927). *Report of the date palm inspector for 1923 and 1924. 15th and 16th report on Arizona Agriculture and Horticulture* (pp. 50–58). Arizona: Arizona Commission of Agriculture and Horticulture.
Stickney, F. S, Barnes, D. F., & Simmons, P. (1950). *Date palm insects in the United States* (USDA Circular No. 846, 57 pp).
Stiefel, V. L., & Margolies, D. C. (1992). Do components of colonization-dispersal cycles affect the offspring ratio of Banks grass mite (*Oligonychus pratensis*)? *Entomologia Experimentalis et Applicata, 64*, 161–166.
Strumpel, H. (1969). The animal pests of the date palm *Phoenix dactylifera* L. and its fruit in North Africa. *Zeitschrift für Angewandte Entomologie, 64*, 232–240.
Tan, F. M., & Ward, C. R. (1977). Laboratory studies on the biology of the Banks grass mite. *Annals of the Entomological Society of America, 70*, 534–536.
Taylor, B., Rahman, M., Murphy, S. T., & Sudheendrakumar, V. V. (2011). Exploring the host range of the red palm mite (*Raoiella indica*) in Kerala, India. *Zoosymposia, 6*, 86–92.
Thewke, S. E., & Enns, W. R. (1970). *The spider-mite complex (Acarina: Tetranychoidea) in Missouri* (University of Missouri Museum Contribution, Monograph No. 1, Missouri, USA, 106 pp).
Tirello, P., Pozzebon, A., Cassanelli, S., Van Leeuwen, T., & Duso, C. (2012). Resistance to acaricides in Italian strains of *Tetranychus urticae*: Toxicological and enzymatic assays. *Experimental and Applied Acarology, 57*, 53–64.
Tuttle, D. M., & Baker, E. W. (1964). The spider mites of Arizona (Acarina: Tetranychidae). *University of Arizona Agricultural Experiment Station Technical Bulletin, 158*, 1–41.
Tuttle, D. M., & Baker, E. W. (1968). *Spider mites of Southwestern United States and a revision of the family Tetranychidae* (143 pp). Tucson: The University of Arizona Press.
Tuttle, D. M., & Baker, E. W. (1976). New records and species of Tetranychidae and Tenuipalpidae (Acarina) from Utah and Idaho. *Great Basin Naturalist, 36*, 57–64.
Tuttle, D. M., Baker, E. W., & Abbatiello, M. (1976). Spider mites of Mexico (Acarina: Tetranychidae). *International Journal of Acarology, 2*, 1–102.
USDA. (1972). Banks grass mite situation in the U.S. – 1971. *United States Department of Agriculture Cooperative Economic Insect Report, 22*, 218–223.

USDA. (2013). United States Department of Agriculture Noncitrus fruits and nuts, 2012 preliminary summary. Available at: http://usda.mannlib.cornell.edu/usda/current/NoncFruiNu/NoncFruiNu-01-25-2013.txt. Accessed on 18 Apr 2015.

Van Leeuwen, T., Vontas, J., Tsagkarakou, A., & Tirry, L. (2009). Mechanisms of acaricide resistance in the two-spotted spider mite *Tetranychus urticae*. In I. Ishaaya & A. R. Horowitz (Eds.), *Biorational control of arthropod pests* (pp. 347–393). Dordrecht: Springer.

Vasquez, C., & de Moraes, G. J. (2013). Geographic distribution and host plants of *Raoiella indica* and associated mite species in Northern Venezuela. *Experimental and Applied Acarology, 60*, 73–82.

Vincent, L. E., & Lindgren, D. L. (1958). Control of the date mite, *Oligonychus pratensis* (Banks) in California. *Date Growers' Institute Annual Report, 35*, 15–17.

Walter, E. V., & Wene, G. P. (1956). *Oligonychus pratensis*, a new mite on corn. *Journal of Economic Entomology, 49*, 265–266.

Weeks, A. R., Marec, F., & Breeuwer, J. A. J. (2001). A mite species that consists entirely of haploid females. *Science, 292*, 2479–2482.

Zaher, M. A., Wafa, A. K., & Yousef, A. A. (1969). Biological studies on *Raoiella indica* Hirst and *Phyllotetranychus aegyptiacus* Sayed infesting date palm trees in the U.A.R. (Acarina-Tenuipalpidae). *Zeitschrift Für Angewandte Entomologie, 63*, 406–411.

Zaher, M. A., Gomaa, E. A., & El-Enany, M. A. (1982). Spider mites of Egypt (Acari: Tetranychidae). *International Journal of Acarology, 8*, 91–114.

Zaid, A., Wet, P. F. de, Djerbi, M., Oihabi, A. (2002). Diseases and pests of date palm. In A. Zaid (Ed), *Date palm Cultivation* (FAO Plant Production and Protection Paper, 156). http://www.fao.org/docrep/006/y4360e/y4360e00.htm#Contents. Accessed on 11 Aug 2014.

Zhang, Z.-Q. (2003). *Mites of greenhouses: Identification, biology and control* (244 pp). Willingford: CABI Publishing.

Zouba, A., & Raeesi, A. (2010). First report of *Raoiella indica* Hirst (Acari: Tenuipalpidae) in Tunisia. *The African Journal of Plant Science and Biotechnology, 4*, 100–101.

Chapter 13
Post-harvest Processing of Dates: Drying, Disinfestation and Storage

Hagit Navarro and Shlomo Navarro

Abstract Progressive demand for better date quality led to the development of solar drying technology. Equilibrium relative humidity (ERH) and water activity (a_w) are the most reliable criteria indicating the ability to safely store dates. ERH <65 % at 26 °C serves as a critical limit for date storage at the commercial level and ensures resistance to deterioration by microflora. The relationship between MC and ERH for Medjool, Khadrawy, Zahidi and Deglet Noor is established based on different studies. A solar dryer employing mechanical distribution fans was successfully implemented to integrate thermal disinfestation technology. It is based on the principle of transfer of hot air through a tunnel containing 3 tonnes of dates exposed for 3 h at 50 °C. By employing thermal disinfestation technology to replace methyl bromide (MeBr) fumigation, no effect on the organoleptic properties of dates was observed, while complete mortality of insects and high disinfestation levels for all dry date cultivars were achieved. The technology is currently implemented in all Israeli date packing houses to replace the MeBr fumigation. Vapormate™ is a fumigant that contains 16.7 wt.% ethyl formate in liquid CO_2. It is registered in Israel to control date pests, such as Nitidulid beetles, at dosage of 420 g/m^3 for 12 h at temperatures above 24 °C, as an alternative to MeBr for dates stored in bulk. The storage atmosphere containing a concentration of 60–80 % CO_2 in a gas tight chamber was found to be effective in controlling pests and preserving the quality of dates and moisture content (MC) for 4.5 months storage period.

13.1 Introduction

Date palms are mainly grown in North Africa and the Middle East. The largest date production (about 90 %) is concentrated in the countries Algeria, Egypt, Oman, Saudi Arabia, Iran, Iraq, UAE, Sudan, Pakistan and Morocco. The world date production was 1.8 million tons during 1961 which gradually increased to 2.8 million

H. Navarro (✉) • S. Navarro
Green Storage Limited, Argaman 5, Rishon Lezion, Israel
e-mail: hnavarro@green-storage.co.il; snavarro@013.net

W. Wakil et al. (eds.), *Sustainable Pest Management in Date Palm: Current Status and Emerging Challenges*, Sustainability in Plant and Crop Protection,
DOI 10.1007/978-3-319-24397-9_13

tonnes during 1985 and nearly 8.0 million tonnes in 2012 (FAOSTAT 2012). Most of the harvested dates are locally consumed. Only during recent years the date palm has been introduced as modern plantations in Australia, South America and South Africa, USA and Israel (Zaid 2002).

Date fruit is oblong in shape with a single seed, and becomes sweet when ripe. Since the fruits are rich in carbohydrates and other nutrients, they are considered high energy food. The Food and Agriculture Organization (FAO) publication "Date Palm Cultivation" provides an update of the technical information included in the two earlier FAO books: "Dates Handling, Processing and Packaging" by Dowson and Aten (1962) and "Date Production and Protection" by Dowson (1982).

13.2 Date Varieties

Based on appearance and texture, dates can be primarily categorized into four types: Fresh (fruits are consumed immediately after harvesting, typical of the variety, Barhee); Wet (fruits mature and stored at low temperature under refrigeration, typical variety Hayany); semi-dry (typical varieties Medjool and Deglet Noor) and Dry (typical varieties Ameri, Thoory, Halawy, Khadrawy and Zahidi) (Dowson 1982; Glasner et al. 2002).

13.3 Treatments Before Storage

During maturation, dates pass through the following stages: Khalal stage (yellow or red colored physiologically mature dates with sweet taste bearing moisture content (MC) of 50–85 %, Rutab stage (moderately brown in color, bearing 30–45 % MC) and Tamar stage (dark brown to amber color, with 10–25 % MC). The later stage is also considered as nonperishable (Glasner et al. 2002).

Usually, over a period of several weeks, a number of pickings have to be made. In Saudi Arabia and California (USA), the growers mostly use several mechanized means (extension ladders, harvesting saddles or cat walk equipped mobile towers) to harvest the dates. High quality dates are harvested by hand, while others are harvested as whole bunches (Morton 1987).

Since tree nuts and dried fruits are relatively high value products and used as confectionary ingredients or as snack foods, their successful marketing requires strict quality control measures. All modern date handling practices are designed to meet the buyer's demands with definite quality standards. In order to maintain quality, the processing facilities must produce fruits with the same properties consistently (Glasner et al. 2002). Since dates are considered a dry to semi-dry commodity, drying to control their moisture content must be done before their storage in order to maintain high quality.

Generally, field pest population development does not take place during storage (Johnson 2004), but these are pre-harvest insect pests that continue to cause quality deterioration and damage by direct feeding on the fruit (Simmons and Nelson 1975). The existence of these pests causes phytosanitary constraints for processors and cause continuous quality deterioration, therefore, they are considered post-harvest pests.

Moreover, aflatoxin producing molds (*Aspergillus* spp.) can gain access at sites of feeding damage caused by the insect pests (Campbell et al. 2003). Therefore, preliminary disinfestation of fresh product is necessary before storage to reduce such damage. Among a number of devastating insect pests of stored dates, the Indian meal moth, *Plodia interpunctella* (Hübner) may be the most serious (Simmons and Nelson 1975). A number of date cultivars are also vulnerable to attack by nitidulid beetles. The prolonged storage of ripen dates may allow several generations of pre-existing nitidulid beetles to develop, which results in heavy infestation of stored dates.

These beetles are strong flyers, capable of flying up to 4 km. The adult life span of Nitidulid beetles ranges from 6 to 12 months. They damage the ripening fruits by entering through the calyx end, and feeding on the date pulp starting on the palm, continue on mature fallen date fruits and even during storage. After the larvae hatch in the fruit, they feed and develop rapidly, contaminating the fruit with their molted skins and excretions. Usually, infestation by nitidulid beetles also provides entry points to the pathogens, which ultimately cause fermentation followed by rotting, and make the dates unmarketable (Blumberg 2008).

Since stored product pests reproduce during storage, repeated treatments are necessary to control the pests of stored dates particularly during long-term storage. Currently, fumigation is the main method of pest control in stored dates (Johnson 2004) that needs to be carried out immediately after the dates reach the packing stations.

In 2015, the fumigant methyl bromide (MeBr) will be withdrawn under the Montreal Protocol (MP) eliminating use of ozone depleting substances (UNEP 2006). MeBr has already been banned in Non-Article 5 countries. Therefore, a thermal disinfestation technology was developed as an alternative control measure to MeBr.

13.3.1 Solar Drying in Greenhouse

Amongst the preliminary actions must be taken into account for the safe storage of dates, the 'complete maturation' of all fruits is the significant one. Dates are known to ripen in stages at different times on the palm. The high quality dates are harvested by hand, and cutting off whole bunch for rest of the dates is a common practice. After harvesting, the drying process is carried out for semi-dry and dry cultivars (Navarro 2006).

In dates, the enzyme invertase converts disaccharide into glucose and fructose (monosaccharides), this enzymatic activity involved in converting the sugars, results in up to 90 % of the cellulose to break down by increasing the soluble pectin concentration that makes the fruit texture softer and juicier (Kanner et al. 1999). The activity of cellulase and polygalacturonase, level of glucose and fructose, relationship between water activity (a_w) [a_w; the proportion of water vapor pressure in a product to that of pure water the same temperature (Dowson and Aten 1962)] and moisture content (MC) are the main factors affecting the drying of date fruit.

Kanner et al. (1999) tested the enzymatic activities in Medjool cultivar. They observed that 0.65 a_w is safe for storage and effectively controls bacterial and mould growth with MC depending on the amount of sugars in the fruit. During an investigation dealing with the effect of sugar level on a_w and MC of Medjool date, it was found that dates absorbed more water at higher sugar levels for the same a_w value. At 0.75 a_w MC was 18 % and 20 % for dates containing 65 % reduced sugars and 24 % MC for dates containing 80 % reduced sugars (Kanner et al. 1999).

Due to difficulty of separating the dates one by one on the branches, incomplete ripen dates are harvested together with ripen dates. Those immature dates need to be separated from matured dates. In addition, sometimes, under unfavorable temperatures prevailing towards the end of the harvest season, ripening of dates on palms is not entirely completed; the maturation can be accomplished artificially by heat treatments. During those heat treatments, the temperature and time of exposure (at a particular temperature) are correlated and play an important role in the ripening of dates (Navarro 2006).

The temperature levels of 30 and 50 °C were required for maturation of Medjool dates in 156 and 21 h, respectively. The high temperature (55 °C) affected the date quality by causing a significant blistering (separation of skin from the fruit) and change in color of the fruit. The relationship between time and temperature for the maturation of dates can be expressed by an equation that describes the constant value (Navarro 2006):

$$K = T.\ \ln(t)$$

Where:

T = temperature (°C); t = time (hour)

With the calculated value (K) of 152.2 for ripening dates of Medjool cultivar (Navarro 2006) and based on the determination of maturation temperature the time required for the maturation of dates can be determined as follows:

$$\ln(t) = 2K / ((\Sigma(tmax) / n) + 30)$$

Where:

$\Sigma(tmax)/n$ = mean highest temperature within the maturation pallet on daily basis

For instance, when 40 °C is the mean highest daily temperature, then according to the equation, the maturation time for Medjool dates will be 77.4 h. This time (77.4 h) is the cumulative time period when the temperature of maturation is >30 °C.

Using solar energy for maturation, the dates placed in pallets are exposed to the solar heat and the energy requirements for field maturation can be determined by using the above equation. In the same regard, the use of shrink films to cover the trays containing immature Medjool dates in the field, gave excellent results (Navarro 2006). Moreover, leaving a ventilation window in the shrink film covering the date containing pallets prevents the condensation that further facilitates the ripening process. Since dates possess hygroscopic characteristics, they equilibrate the MC with the relative humidity of the ambient air by losing or gaining moisture from the outside environment through this ventilation window. This association can be expressed as equilibrium moisture content (EMC) or equilibrium relative humidity (ERH). EMC can be defined as the levels of MC in terms of ERH with which the product is in balanced position. While, ERH is the level of air RH surrounding the product with consistent MC on wet basis. The a_w helps in determining the intensity of infestation level caused by microorganisms. *Aspergillus restrictus* and a number of species of *Aspergillus glaucus* group can grow slowly at a_w of 0.65 or above. The water content may vary in different products having same a_w (Lacey et al. 1980) and generally commodities bearing high sugar content have higher MC than other stored products at the same a_w.

Immature dates possess less sugar content, hence results in an increased a_w. Therefore, dates with high MC before storage results in deterioration of fruits due to microflora activity. The a_w expresses the available water for metabolic activities under varying sugar concentrations (Kanner et al. 1999), therefore, a_w (or ERH) is a more practical parameter compared to the MC. Thus, a_w plays a significant role in decision making process for the storability of the produce. At 26 °C with 65 % RH drying dates to a MC below 21 % is not possible; e.g. Medjool dates exposed to air at 26 °C with 65 % RH can only be dried to a MC of 21 % (desorption) (Navarro 2006).

The activity of cellulase enzyme is determined by a_w of the fruit; at 0.87 a_w, which is typical for the Khalal stage, when progressed to Rutab stage with decreasing a_w, the enzymatic activity also decreased and stopped at 0.63 a_w (Kanner et al. 1999).

Elastic and soft texture dates are more in demand, therefore, fresh dates having MC 24–26 %, can be preserved after packing in cold storage. The growth of microbial agents like mold, bacteria and yeast is also inhibited in dates having ERH less than 65 %. In order to prevent ripening fruit from over drying on trees, they must be harvested when they have sufficient water content (Navarro 2006). Each cultivar has its own capacity of binding water, Figs. 13.1, 13.2, and 13.3 show this relationship between MC and ERH for date cultivars Khadrawy, Zahidi and Deglet Noor, respectively (Rindner et al. 2001; Navarro 2006).

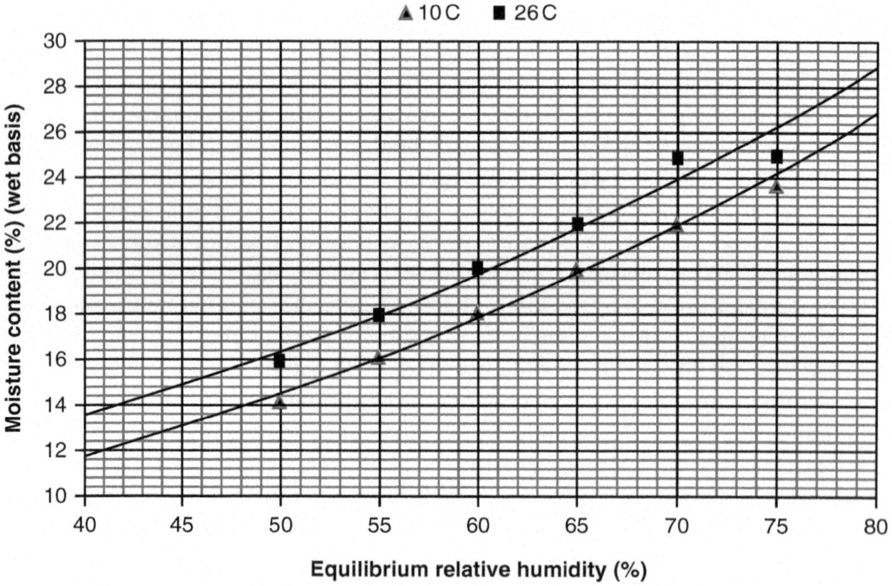

Fig. 13.1 Equilibrium relative humidity of Khadrawy date variety at different levels of relative humidity (50, 55, 60, 65 and 75 %) and at 26 and 10 °C (Rindner et al. 2001; with permission)

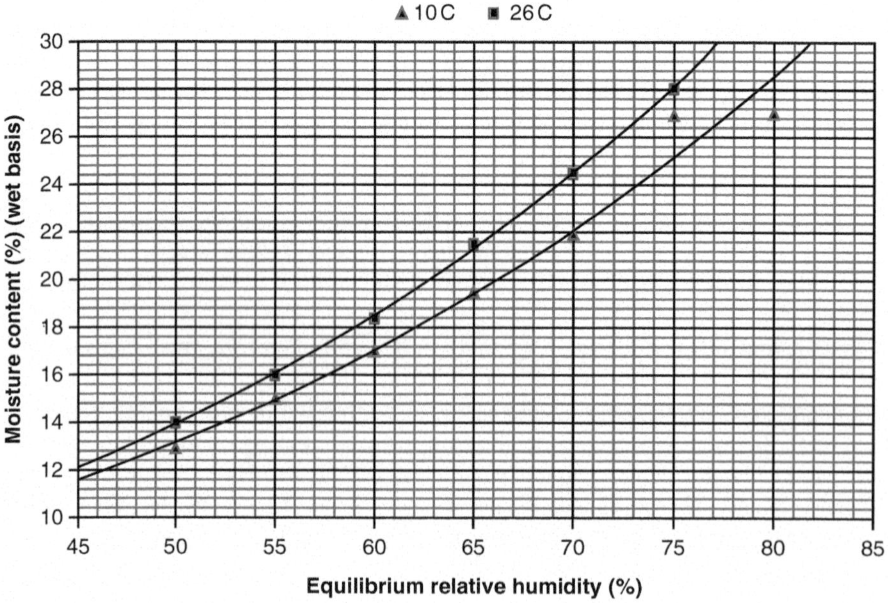

Fig. 13.2 Equilibrium relative humidity of Zahidi date variety at different levels of relative humidity (50, 55, 60, 65 and 75 %) and at 26 and 10 °C (Rindner et al. 2001; with permission)

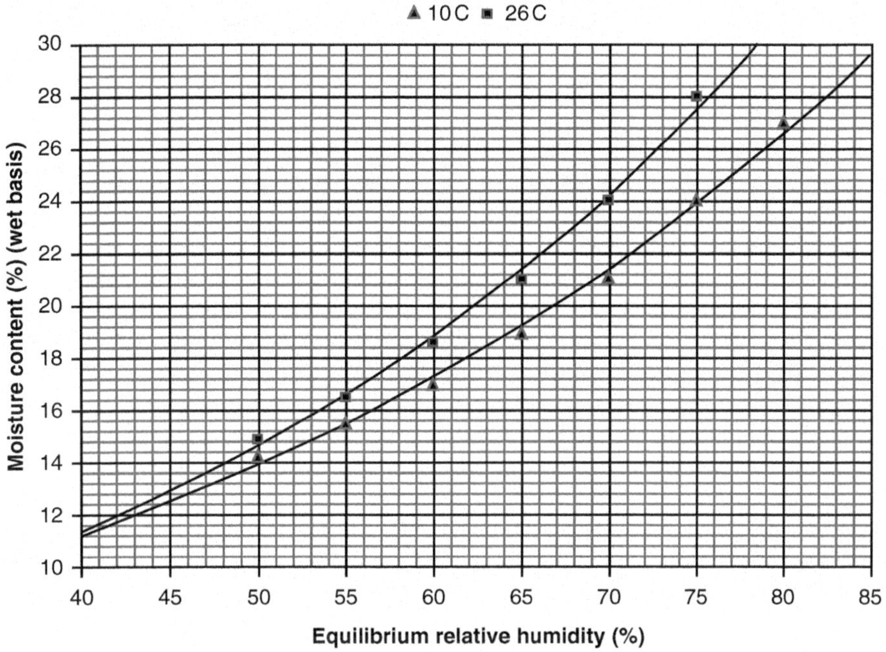

Fig. 13.3 Equilibrium relative humidity of Deglet Noor date variety at different levels of relative humidity (50, 55, 60, 65 and 75 %) and at 26 and 10 °C (Rindner et al. 2001; with permission)

Under climatic conditions where sunshine is adequate and ambient humidity is low, harvested dates may be dried outdoors as a whole date or cut into two halves.

Solar and mechanical drying methods are also used to dry the dates. In conventional sun drying method, date trays are directly exposed to sun radiations in the open. Although, this is a low cost method but the dust deposition by wind on the dates diminish their shine, quality and visual appearance. Additionally, washing done to remove the adhered dust on dates not only causes blistering (separation of skin from the pulp) but also exposes the fruits to insects, especially Nitidulid spp., and finally, if the grower cannot measure a_w, it is difficult to avoid over drying that would substantially lower the fruit quality. As an alternate to this conventional drying method, date pallets covered with shrink film exposed to sun radiation at a height up to 2 m is a very useful method that also allows aeration and avoid condensation. Such solar dryers are very simple and useful since drying is possible close to the date orchard where electrical power is not accessible. Navarro (2006) indicates average MC of sun dried pallets at 65 % ERH, which reveals that when dates are positioned at the top of dryer it takes about 7 days for dates with values of about 76 % ERH to reach 65 % while at the bottom layer it takes 8 days (Navarro et al. 2000).

The mechanical drying method is hot air flow through a tunnel to avoid the change in color and texture deterioration of dried fruits and nuts, where moderate temperature (35–55 °C) is required for drying process. In Israel, for drying Medjool dates, the recommended temperature is to maintain at 45 °C to avoid the blistering effect. Mechanical drying is the fastest and most accurate of all the above mentioned methods. It takes approximately 2 days for the dates to reach their desired ERH but it is also the most energy cost effective method. In this method, dates are placed on trays in one layer. After drying is completed by mechanical means the dates need to be packed into boxes manually which is time consuming and costly.

In the Jordan Valley of Israel, the hybrid solar-mechanical drying method was developed and successfully implemented taking advantage of the hot climatic conditions including adequate sunshine and low atmospheric humidity suitable for drying of harvested dates. In this method, the dates were placed in trays (in one layer containing 3 kg) or in PVC holed boxes (containing 12–14 kg) on pallets 2 m high. The principle in this method is making the hot air flow through the dates which are covered with polyethylene liner that makes the hot air scatter inside the tunnel. At the end of each tunnel, a fan sucks and circulates the hot air entering the tunnel. The average airflow must be 1.4–2.1 m/s on the surface of the dates or fan airflow rate at 12,600 m^3/h (Navarro et al. 2000). In 2003 the drying facility was a greenhouse converted to a solar dryer that measured 10 m wide × 40 m long × 3 m high covered with polyethylene, prepared specially for commercial drying of dates on a large scale (Navarro et al. 2004a). The greenhouse was designed in such a way that up to 12 stacked rows of dates were placed parallel across the greenhouse and drying ducts were formed by covering side and top with polyethylene liners. Ten pallets were positioned in each row, arranged in five pallets lengthwise and two pallets across. Every pallet (2.1 m high) bears trays stacked 20 layers high with 5 trays (40 × 60 × 10 cm high) per layer, in this way every pallet holds 100 trays. Each tray contained 3 kg Medjool dates, in a single layer. Therefore, a standard drying duct consisted of ten pallets (every pallet holds 300 kg fruit) containing 3 tonnes dates. The drying process was controlled using less energy than conventional drying based on LPG or using electricity as an energy source.

13.3.2 Thermal Disinfestation Method: Principles and Practices

The conventional method for disinfestation of raw dates is based on fumigation and is most suited for small fumigation chambers. Dates received at the packing stations are first fumigated, then cold stored, dried and re-stored. Methyl bromide (MeBr) treatments are useful to limit insect infestation in dates particularly when the drying facility is small. It is important to mention that storing the dates at low temperatures kills the insects inside the fruit. Fumigation using MeBr before storage improves quality.

In all date cultivars, field infestations of Nitidulid beetles serve as source to carry the infestation into storage. Recently, the fumigant MeBr controlled this problem successfully. MeBr; however, has been phased out since January 2005 under the Montreal Protocol for developed countries (Non-Article 5), and will be phased out for the developing countries (Article 5) by 2015, excluding treatments before shipment and quarantine (TEAP and MB 2003). Instead of using fumigants, heat treatment is now considered the most efficient way of controlling nitidulid beetles in stored dates.

According to Fields (1992), exposure to temperatures of 50 °C for 1 h or more is effective in controlling stored product insects. Target temperatures in the range of 50–60 °C are standard temperatures for heat treatments (Imholte and Imholte-Tauscher 1999). During actual field treatment, at a commercial drying facility distribution of heat is uneven and because of the differences in the specific heat of the materials, the composition and thickness of their mass where insects harbor, some areas of the fruit cannot receive the target temperatures or do not get treated for the required time to attain complete mortality (Roesli et al. 2003).

Exposure to 49 °C for 4–20 min was needed to obtain 90 % mortality of adult Nitidulids, depending upon the relative humidity, as reported through laboratory observations made by Lindgren and Vincent (1953) on dried fruits. Under controlled conditions the adults of *Carpophilus hemipterus* (L.), the dried fruit beetle, exhibited tolerance towards heat and took times ranging from 25 to 60 min to achieve complete mortality of the target insect pest (Al-Azawi et al. 1984). In stored dates, heat treatment at 60 °C for 33 min was required for the control of all the developmental stages of *Cadra cautella* Walker (Al-Azawi et al. 1983). For dates, the feasible temperature lies in the range of 45 and 55 °C, and temperatures beyond 60 °C may exert undesirable drying effect, blistering of fruit and even biochemical changes in the fruit. Treatment at lower temperatures would take extended periods to attain complete mortality of the pest.

Finkelman et al. (2006) observed the temperature effect (four temperatures 45, 50, 55 and 55 °C) on the mortality and disinfestation levels of *C. hemipterus* mobile stages under laboratory conditions. The target temperatures were employed for 2 h at feeding sites and 100 % mortality of *C. hemipterus* larvae was observed at 50 °C, with maximum number of larvae left (92.3 %) the commodity (disinfestation).

However, some previous findings revealed <90 % disinfestation when dates were fumigated with MeBr (Donahaye et al. 1991, 1992). Promising results were also recorded in controlling adults and larvae of *C. hemipterus* when exposed at 50 °C in a commercial drying facility (Finkelman et al. 2006).

On a commercial scale, natural infestations were examined by inserting the infested dates into drying ducts where infestations were examined after 3 h of exposure to heated air (50 °C). To achieve the desired temperature for controlling stored pests of dates, the exposure period of 1–2 h was required. Since these semi-commercial trials provided similar level of disinfestation as with the MeBr fumigation, the drying facility was considered a promising alternative for controlling stored dates infestation. The thermal disinfestation strategy also satisfied the needs of processors for disinfestation of Medjool date cultivar (Finkelman et al. 2006). In this case, after the initial disinfestation, at 50 °C the drying time at 45 °C can be extended up to few days depending on the initial MC of dates. After heat treatment for disinfestations of dates, they are dried and then stored either as cold or at ambient temperatures depending on the MC and cultivar of the dates. The thermal disinfestation in combination with drying technology was examined at a commercial drying facility with the highest disinfestations observed at 50 °C while most of the date verities get dried between 45 and 50 °C (Navarro et al. 2003). The development of commercial heat disinfestation application using solar dryers in Israel is the outcome of above findings (Navarro et al. 2004a, b; Finkelman et al. 2006). Additional investigation was needed to explore the employment of heat treatment for the other date cultivars. For this purpose, trials were conducted on Deglet Noor in branches (Navarro et al. 2009). In this study involving thermal disinfestation process of Deglet Noor dates, the results from both laboratory and field tests were encouraging. In commercial scale trials, the use of Dolev type crates (containing 100–150 kg dates in branches) indicated the successful application of increased temperature (50 °C) to Deglet Noor dates in branches. Furthermore, no change in the color of branches and dates was observed both in laboratory and field trials. The heat treatment of Zahidi and Halawy cultivars in bulk using Dolev type crates (containing 200–250 kg) has also been tested. During these trials, in contrast to Medjool cultivar with amount of 3 tons per tunnel, the total amount of dates used for Zahidi and Halawy cultivar was approximately 1,600 kg per tunnel. However, it was found that at the bottom side of the tunnel the bulk of dates took an additional 2 h to attain the target temperature (50 °C) compared to the rest of tunnel sections where hot air with an average speed of 1–1.7 m/s raised the date's temperature up to 55 °C in half an hour. The RH measured in this trial reached equilibrium at 20–25 % gradually at both ends of the tunnel, affecting the MC of dates (Navarro et al. 2010).

Based on these findings, the successful development of protocols using thermal treatment with regard to dates' bulk resistance to hot air and its uneven distribution in bulk dates, lead to further expansion of this technology to other cultivars like Khadrawy, Halawy, Zahidi, Derhi and Ameri. The technology involves transferring the hot air through a tunnel to the dates placed in factory boxes (12 kg). Field trials were carried out by disinfesting dates in a commercial chamber made for these cultivars to test even distribution of hot air and its effect on the quality of dates. For this purpose, the bulk of dates wrapped into 13 layers pallets were placed in boxes (each weighing 12 kg) inside the tunnel, and each tunnel contained 3,120 kg of dates in total. In boxes where the airflow through the bulk of dates was at the rate of 1.4 m/s, target temperature (50 °C) was achieved within 3 h, while at higher airflow the same temperature level was achieved within 1–2 h of heating. The temperature increase at this rate successfully disinfested the dates with no changes in their color and taste, however a slight reduction in MC (1–2 %) of dates was observed. Therefore, it is recommended that for date cultivars like Deglet Noor, the exposure period should not be longer than 3 h after attaining the target temperature for disinfestation, and some preliminary trials have also revealed that this exposure time is safe to retain the optimum MC of the dates (Navarro et al. 2009).

Finkelman et al. (2006) reported that under laboratory conditions the optimal temperature for date disinfestation was 50 °C achieved in 2 h. However, for successful disinfestations in packing houses and in order to let the temperature distribute uniformly in the disinfestation dryer, the exposure time was extended to 3 h (Navarro et al. 2003, 2004a, b). During these investigations, the actual environmental factor (temperature or low humidity) conferring mortality to *C. hemipterus* could not be identified, but it was seen that adults of *C. hemipterus* were more sensitive to high temperatures compared to its larvae (Navarro et al. 1989).

During some other investigations, the effect of thermal treatment on 1 day old eggs of *C. hemipterus* tested at 50 °C for 3 h exposure revealed that not a single viable egg was there after 48 and 72 h of treatment (Navarro et al. 2013). According to some studies, insects have metabolically adapted thermally challenging environments. Among a number of findings, it was seen that insect elicit heat shock proteins which provide resistance against heat (Fields 1992). However, further studies are required to investigate the thermotolerance impact on these proteins.

Dates that are brought to the packing houses exhibit varying levels of humidity e.g. the cultivar Deglet Noor when brought to packing house might be very dry (approximately 40 % RH) compared to the Medjool cultivar which in most cases contains approximately 85 % RH when harvested (Kanner et al. 1999). The possible impact of relative humidity levels in interaction with temperatures on the storing dates tested against *C. hemipterus* larvae revealed that highest larval mortalities were observed at 50 °C than at 29 °C with all tested humidity regimes (15, 65 and 95 % RH). In order to predict larval mortality in relation to the exposure duration (min) and relative humidity (%), a stepwise regression model analysis exhibited that the exposure parameter significantly affected the larval mortality [$F_{(1,70)}=289.168$, $P<0.01$]. In a multiple step regression performed for short exposure times (up to 30 min) humidity was identified as a significant variable [$F_{(2,33)}=50.793$, $P<0.01$] (Navarro et al. 2013). In a related study, the exposure time of 30 min completely killed all the developmental stages of *C. hemipterus* at 50 °C and 70 % RH (Al-Azawi et al. 1984).

13.3.2.1 Environmental Aspects of Thermal Disinfestation and Drying

Thermal disinfestation technology may be applied based either on liquid petroleum gas (LPG), diesel oil, solar energy or combined energy. All of the above mentioned disinfestation methods have different impacts on the environment. Nevertheless, the choice of method is influenced by its economic feasibility and its impact on the environment (e.g. external costs). The environmental impact of the energy used in different disinfestation methods is evaluated on the basis of their carbon footprint calculation. *Carbon footprint* is "a measure of the exclusive total amount of carbon dioxide emissions that is directly and indirectly caused by an activity or is accumulated over the life stages of a product" (Wiedmann and Minx 2008).

Carbon footprint serves as a common denominator that allows comparison between similar products of different processes that are generated and involved in global warming. This concept has become popular in recent years and is used for comparison. In contrast to life cycle analysis (LCA) which compares the entire environmental impacts, is actually a more complicated analysis for different products that might skip the concept of the common denominator of a product influencing the environment, and since the carbon footprint is popular, and has the potential to increase the green environmental awareness of consumers (Weidema et al. 2008).

13.3.3 Ethyl Formate Mixed with CO_2

Ethyl formate (EF) is an effective alternative to conventional fumigants and is commercially available under different brand names. EF is a natural component found in apples, honey, orange juice, wine and pears, and commonly used as an artificial fragrance, synthetic flavor in food industry and more importantly as a food additive it is registered as GRAS (Generally Regarded As Safe). A commercial product of EF contains a mixture of 16.7 % ethyl formate and 83.3 % carbon dioxide (w/w) (Ryan and Bishop 2003). The incorporation of CO_2 reduces its inflammability of the EF and their combination enhances the speed of kill the target insects (Haritos et al. 2006). Laboratory and field studies carried out to investigate its fumigant properties against insect pests of stored commodities and dried fruits are reported (Muthu et al. 1984). The laboratory bioassays performed against *C. maculatus* Murray and *C. hemipterus* depicted 100 % mortality of exposed insects when the fumigant was applied at the rate of 420 mg/L for 12 h at 30 °C. The treatment also removed 69.6 % infestation of *Carpophilus* spp. from stored dates (Donahaye et al. 1992). The product effectiveness (at 420 g/m^3) as larvicide and disinfestant was investigated for the exposure period of 12 h at five different temperature levels (16, 18, 24, 26 and 28 °C). The highest larval disinfestation >88 % was observed at 26 and 28 °C, while the complete larval mortality was achieved at temperatures ranging from 18 to 28 °C. In the same way, the fumigant applied at the same rate killed all the exposed eggs of *C. hemipterus* at 12 h exposure at both exposure temperatures i.e. 24 and 30 °C (Donahaye et al. 1992).

Semi-commercial trials using flexible fumigation chambers (9.5 m^3) and commercial trials employing rigid fumigation chambers (96.5 m^3) were also carried out. After treatment of naturally infested Halawy dates in a flexible chamber all the dates were completely disinfected from pest population, while 95 % beetles were found dead outside the fruits. In commercial rigid fumigation chamber, a complete disinfestation was observed at all the four locations and date samples from the same location were also free from insect contamination (Navarro et al. 1998). The product is now commercially registered as Vapormate™ and can be successfully used against the insect pests of stored grains, horticultural products and quarantine treatment of bananas in Australia and New Zealand (Krishna et al. 2002).

13.3.4 Controlled Atmosphere (CA) Storage of Dates

In Israel, as a general practice dates are stored at −18 °C, this storage temperature is particularly suitable for the soft fruit cultivars but at the same time it is also an energy consuming approach. Very limited research has been conducted regarding the influence of controlled atmospheres (CA) on quality of dates. Under laboratory conditions, Baloch et al. (2007) examined the stability of Dhakki dates stored in tinplate cans at 0.52–0.75 a_w under oxygen, air or nitrogen and 40 °C for the period of 4 months. They concluded that in order to maintain freshness of the produce with extended shelf life, Dhakki dates should be stored under inert atmosphere with a water activity close to 0.61 ± 0.01 a_w (corresponds to 24–25 % equilibrium MC). The water activity and storage atmosphere greatly affect the quality of Dhakki dates (darkening of fruit and other linked changes that deteriorate the fruit quality). Investigations showed that some quality parameters like darkening, pH and titratable acidity (is a simple measure of the amount of acid 'anions' in a juice) of dates were significantly reduced under the storage environment containing nitrogen and water activity of 0.52 a_w. The results revealed that oxygen free atmosphere and water activity ranging between 0.60 and 0.61 is best suited for the storage rendering enhanced shelf life of Dhakki dates (Baloch et al. 2007).

Stored dates under controlled atmosphere must be immediately marketed after exposure to air. The elevated CO_2 concentration significantly reduces the fungal growth (fungistatic), but when dates are exposed to ambient air, the fungal spores resume growth, especially under elevated humidity and temperature levels (Kader and Hussein 2009). Tamar dates under nitrogen packed conditions eliminates the possibility of insect infestation and darkening at Khalal stage (Kader and Hussein 2009). In case of Barhee dates, the storage temperature of 0 °C at 90–95 % RH in 20 % CO_2 enriched air successfully maintained the date quality up to 26 weeks compared to 7 weeks storage at ambient air (Kader and Hussein 2009). Some other studies have also shown that the factors like sugar formation and browning of dates responsible for shelf life are significantly delayed in the presence of CO_2 enriched environment in comparison with the storage at −18 °C (Navarro et al. 1988, 1992, 1995). Based on these findings, specially designed gastight plastic liners for CA storage of dates are used in Israel (Navarro et al. 1998), a flexible structure of 151 m^3 was used and tested for gas tightness using the pressure decay method (Annis et al. 1991; Navarro 1997).

Investigations conducted in Israel showed that in same size storage facilities, half-life pressure decay time (time elapse to half the pressure of a pressurized structure using pressurized air compared to ambient pressure) for less than 10 min was considered as critical, while for maintaining constant concentration of CO_2 a less gastight structures must be used (Fig. 13.4).

Usually, high glucose and fructose contents with very little or no sucrose are present in the soft fruit date cultivars, while sucrose content is high in firm and dry date cultivars. The sugar content in unripe fruit is sucrose which gets converted in the latter stages. During the latter stages of development, the activity of invertase

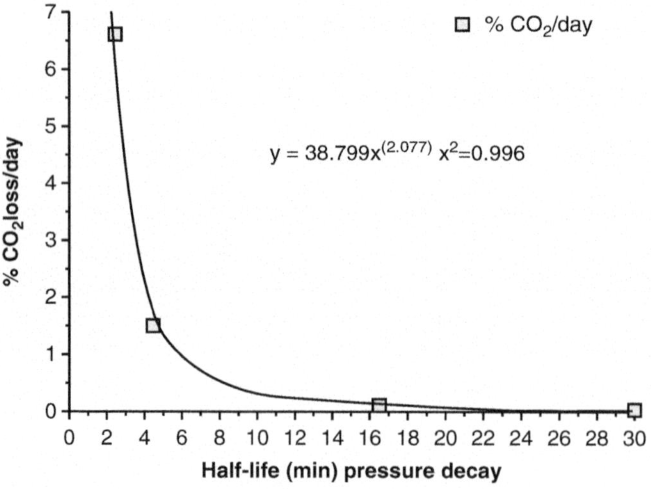

Fig. 13.4 Degree of gas tightness expressed in half-life (min) pressure decay and daily percentage carbon dioxide loss from the 151 m^3 capacity storage enclosure (Navarro et al. 1997)

was found to be the maximum in soft fruits compared to the dry date cultivars (Kanner et al. 1978). In ripe fruits of soft date cultivars which carry plenty of reducing sugars, a "caramel" like taste develops due to the non-enzymatic browning (Kanner 1967). The accumulation of sugar and changes in color determines the quality of dates during storage (Rygg 1956). During the storage period of 4.5 months no significant difference was observed in sugar formation and sloughing of peel on dates (Navarro et al. 1997). These results are in accordance with the previous studies performed for Halawy, Zahidi and Ameri cultivars stored under CO_2 enriched CA which revealed no adverse effect on increase in sugars, even after a long storage period of 8 months at ambient temperatures (Navarro et al. 1992, 1995).

13.4 Post-harvest Handling and Storage

Investigations showed that exhale of gases from dates take place at very low rate, for instance, less than 5 mg and 2 mg CO_2 per kg/h of dates at 20 °C at Khalal, and Rutab and Tamar stage, respectively. Similarly, very low 0.1 μl/kg/h ethylene is produced at Khalal stage with no ethylene production during the Rutab and Tamar stages (Yahia 2004). The unripe dates picked at early stages require post-harvest ripening (Yahia 2004). During post-harvest handling, the soft and semi-dry cultivars are dehydrated to remove extra moisture, while hard or dry cultivars are hydrated to soften their texture. In addition to the storage environment, the date moisture plays a significant role in maintaining the quality of fruit which may be deteriorated by certain physiological and pathological disorders in case of increased MC of the stored dates (Yahia 2004; Baloch et al. 2007). The storage of dates at Khalal stage is carried out at 0 °C and 85–95 % RH to avoid the excessive water loss. These storage conditions are also helpful in retaining the flavor and quality parameters of Rutab cultivar (Kader and Hussein 2009). The packaging of dates in boxes lined with plastic or in moisture barrier plastic bags is also effective in preventing moisture loss and storing dates for extended periods.

Stored dates quality can also be maintained by storing at low temperatures as it postpones insect infestation, reduces the growth and incidence of yeasts and molds, prevents development of syrupiness (formed due to formation of reducing sugar from sucrose) and souring of highly moist dates (Kader and Hussein 2009). Tamar dates can be stored at optimum temperature of 0 °C for 6–12 months, which may vary among cultivars (soft dates, like Barhee and Hiani have reduced shelf life compared to the semi-soft dates, like Halawy and Deglet Noor). For long term storage, a temperature level of −15.7 °C and below is recommended. Dry date cultivars with 20 % or less MC can be kept for longer periods at higher temperatures. Kader and Hussein (2009) found that storage temperatures of −18 °C for more than 12 months, 0 °C for 12 months, 4 °C for 8 months, or 20 °C for 30 days all with 65–75 % RH did not affect the quality of dry dates during entire storage period. In addition to the optimum storage temperature and RH, post-harvest treatments like modified atmosphere packaging and edible coating can significantly contribute to increase the shelf life of dates and other fruits (Kader and Hussein 2009).

13.5 Future Research

Thermal disinfestation is considered as an effective alternative to MeBr fumigation, however, this method is limited to Medjool cultivar in trays of 3 kg and in boxes of 12 kg capacity for Halawy, Khadrawy, Zahidi, Ameri and Derei cultivars. This technology is applied in handling and drying of date cultivars in solar dryers, however, an additional research is required to elaborate the likelihood of other handling and drying methods for heat disinfestation of dates. Controlled atmosphere storage method may replace expensive cold storage of dates and further research would be needed to make this technology more cost-effective and broadly applicable.

13.6 Summary

The use of solar energy in shrink covered pallets for ripening of immature Medjool dates was proved to be a feasible option. Ripening times can be determined by applying the equations developed for Medjool dates. For marketing, the acceptable moisture limit for the dates is 65 % ERH at 26 °C, whereas, the length of storage period depends upon the actual ERH or a_w of the stored dates. Temperature plays a vital role in determining the ERH for each date cultivar. Hence, for each cultivar the ERH should specify for which particular temperature the isotherm is valid.

The most common approach for drying of dates is solar drying (spreading date trays in the open field) and mechanical drying. For drying of dates, a solar dryer was constructed by wrapping the shrink film around the pallets. Dates dry within about 7–8 days in pallet size solar dryers which have successfully replaced the traditional practices of spreading date trays in the field. A hybrid mechanical and solar dryer is also effectively used in the Jordan Valley of Israel. These dryers are cost effective as have an ability to integrate thermal disinfestation and drying of dates together thus save energy.

The drying process (at 45 °C) can be followed after thermal disinfestation at 50 °C, that was found as an alternative to MeBr treatment. This approach is used in already installed drying facility without any modification of the date handling trays for Medjool dates (3 kg) and factory boxes (12 kg) used for other date cultivars. As far as the disinfestation of dry dates (dates with MC suitable for storage) is concerned, the dates are exposed to 50 °C for 3 h after attaining the desired temperature during which time, moisture loss was found to be insignificant.

For bulk stored dates, another date treatment method registered in Israel is the use of fumigant Vapormate™. This fumigant has shown to be very successful in both controlling the nitidulid beetles and disinfesting the dates at dosage of 420 g/m^3 for 12 h at temperatures above 24 °C.

Storing dates under controlled atmospheres instead of cold storage is feasible, and requires a daily supply of 0.8 kg of CO_2 to maintain 60–80 % CO_2 concentration in a hermetic cube or a rigid sealed structure. The half-life pressure decay time not less than 10 min is also needed to avoid further loss of CO_2. The date storage in CA containing CO_2 has been found very effective in insect pest control and maintaining quality of dates during long term storage.

References

Al-Azawi, A. F., El-Haidari, H. S., Aziz, F. M., & Murad, A. K. (1983). Effect of high temperatures on fig moth *Ephestia cautella* Walker (Lepidoptera: Pyralidae) in Iraq. *Date Palm Journal, 2*, 79–85.

Al-Azawi, A. F., El-Haidari, H. S., Aziz, F. M., Murad, A. K., & Al-Saud, H. M. (1984). The effect of high temperatures on the dried fruit beetle *Carpophilus hemipterus* (L.) a pest of stored dates in Iraq. *Date Palm Journal, 3*, 327–336.

Annis, P. C., Van, S., & Graver, J. E. (1991). *Suggested recommendations for the fumigation of grain in the ASEAN region. Part 2. Carbon dioxide fumigation of bag-stacks sealed in plastic enclosures: An operations manual* (58 pp). Malaysia/Canberra: ASEAN Food Handling Bureau Kuala Lumpur/ACIAR.

Baloch, A. K., Baloch, W. A., Baloch, M. K., & Saleem, S. A. (2007). Shelf stability of Dhakki dates as influenced by water activity and headspace atmosphere. In: A. Zaid, V. Hegarty, H. H. S. Al Kaabi (Eds.), *III international date palm conference, Abu Dhabi, UAE* (ISHS Acta Horticulturae 736, 575–586).

Blumberg, D. (2008). Review: Date palm arthropod pests and their management in Israel. *Phytoparasitica, 36*, 411–448.

Campbell, B. E., Molyneux, R. J., & Schatzki, T. F. (2003). Current research on reducing pre- and postharvest aflatoxin contamination of U.S. almond, pistachio, and walnut. *Journal of Toxicology – Toxin Reviews, 22*, 225–266.

Donahaye, J. E., Navarro, S., Rindner, M., & Dias, R. (1991). The influence of different treatments causing emigration of nitidulid beetles. *Phytoparasitica, 19*, 273–282.

Donahaye, E., Navarro, S., Rindner, M., & Dias, R. (1992). An artificial feeding site to investigate emigration of Nitidulid beetles from dried fruits. *Journal of Economic Entomology, 85*, 1990–1993.

Dowson, V. H. W. (1982). *Date production and protection with special reference to North Africa and the near east* (FAO Technical Bulletin No. 35, 294 pp). Rome: FAO.

Dowson, V. H. W., & Aten, A. (1962). *Dates handling, processing and packing* (FAO Agricultural Development Paper No. 72, 392 pp). Rome: FAO.

FAOSTAT. (2012). *Food and agricultural commodities production*. Available at: http://faostat.fao.org/faostat/collections?version=ext&hasbulk=0&subset=agriculture. Accessed on 21 May 2015.

Fields, P. G. (1992). The control of stored-product insects and mites with extreme temperatures. *Journal of Stored Products Research, 28*, 89–118.

Finkelman, S., Navarro, S., Rindner, M., & Dias, R. (2006). Use of heat for disinfestation and control of insects in dates: Laboratory and field trials. *Phytoparasitica, 34*, 37–48.

Glasner, B., Botes, A., Zaid, A., & Emmens, J. (2002). Chapter IX: Date harvesting, packinghouse management and marketing aspects. In A. Zaid (Ed.), *Date palm cultivation* (FAO Plant Production and Protection Paper No. 156, Revision 1). Rome: Food and Agricultural Organization of The United Nations.

Haritos, V., Damcevski, K., & Dojchinov, G. (2006). Improved efficacy of ethyl formate against stored grain insects by combination with carbon dioxide in a dynamic application. *Pest Management Science, 62*, 325–333.

Imholte, T. J., & Imholte-Tauscher, T. (1999). *Engineering for food safety and sanitation*. Woodinville: Technical Institute of Food Safety.

Johnson, J. A. (2004). Dried fruit and nuts: United States of America. In K. Hodges & G. Farrell (Eds.), *Crop post-harvest science and technology, Vol. 2, durables* (pp. 226–235). Oxford: Blackwell Science.

Kader, A. A., & Hussein, A. M. (2009). *Harvesting and postharvest handling of dates* (iv, 15 p). Aleppo: International Center for Agricultural Research in the Dry Areas (ICARDA).

Kanner, J. (1967). *Rehydration of over dried dates and biochemical changes involved in the rehydrated fruits*. MSc thesis, Faculty of Agriculture, Hebrew University, Jerusalem. (In Hebrew with English summary).

Kanner, J., Elmaleh, H., Reuveni, O., & Ben-Gera, I. (1978). Invertase (β-fructofuranosidase) activity in three date cultivars. *Journal of Agricultural Food and Chemistry, 26*, 1238–1240.

Kanner, J., Navarro, S., Donahaye, E., Ben-Shalom, N., Shoval, N., Granit, R., Rindner, M., Azrieli, A., & Pinto, R. (1999). *Development of technological systems to improve the quality of Medjool dates for export* (Annual Report 1999 of Research Project No. 416-0435-98, 23 pp). Israel: Ministry of Agriculture (In Hebrew).

Krishna, H., Ryan, R., Munez, A., Hirst, G., Yoshihara, H., & Barton, S. (2002). *Development of Vapormate™ for preshipment disinfestation of export bananas in the Philippines* (Crop & Food Research Confidential Report No. 007). Palmerston North: Crop and Food Research.

Lacey, J., Hill, S. T., & Edwards, M. A. (1980). Micro-organisms in stored grains: Their enumeration and significance. *Tropical Stored Products Information, 38*, 19–32.

Lindgren, D. L., & Vincent, L. E. (1953). Nitidulid beetles infesting California dates. *Hilgardia, 22*, 97–117.

Morton, J. F. (1987). Date. In: *Fruits of warm climates* (pp. 5–11). Miami: Julia F. Morton. Available at: http://www.hort.purdue.edu/newcrop/morton/Date.html. Accessed on 5 June 2015.

Muthu, M., Rajenderan, S., Krishnamurthy, T. S., Narasimhan, K. S., Rangaswamy, J. R., Jayaram, M., & Majumder, S. K. (1984). Ethyl formate as a safe general fumigant. In B. E. Ripp, H. J. Banks, E. J. Bond, D. Calverley, E. G. Jay, & S. Navarro (Eds.), *Controlled atmosphere and fumigation in grain storages* (pp. 367–381). Amsterdam: Elsevier.

Navarro, S. (1998). Pressure tests for gaseous applications in sealed storages: Theory and practice. In J. Zuxun, L. Quan, L. Yongsheng, T. Xianchang, & G. Lianghua (Eds.), *Proceedings of 7th international working conference on stored-product protection* (Vol. 1, 385–390 pp), October 14–19, 1998, Beijing, China. Chengdu: Sichuan Publishing House of Science and Technology.

Navarro, S. (2006). Postharvest treatment of dates. In Y. Sarig (Ed.), Quality. *Stewart Postharvest Review, 2*, 1–9. Published online 01 April 2006. 2006 Stewart Postharvest Solutions (UK) Ltd. doi:10.2212/spr.2006.2.1

Navarro, S., Paran, G., Azrieli, A., & Donahaye, E. (1988). *Influence of carbon dioxide on color preservation of Deglet noor dates* (Annual Report of Project No. 564-001-87, 10 pp). Bet-Dagan: Agricultural Research Organization (In Hebrew).

Navarro, S., Donahaye, E., Dias, R., & Jay, E. (1989). *Integration of modified atmospheres for disinfestation of dried fruits* (Final report of BARD Project No. I-1095-86, 86 pp). Bet Dagan: BARD.

Navarro, S., Donahaye, E., Rindner, M., Dias, R., & Azrieli, A. (1992). Integration of controlled atmosphere and low temperature for disinfestation and control of dried fruit beetles. In S. Navarro & E. Donahaye (Eds.), *International conference on controlled atmosphere fumigation in grain storages* (pp. 389–398). Jerusalem: Caspit Press Limited.

Navarro, S., Donahaye, E. J., Caliboso, F. M., & Sabio, G. C. (1995). *Application of modified atmospheres under plastic covers for prevention of losses in stored grain* (Final Report of CDR Project No. C7-053, August 1990 – November 1995, U.S., 32 pp). Israel: Agency for International Development.

Navarro, S., Donahaye, E., Rindner, M., & Azrieli, A. (1997). Storage of dried fruits under controlled atmospheres for quality preservation and control of nitidulid beetles. In U. Aksoy, L. Ferguson, & S. Hepaksoy (Eds.), *Proceedings of first international symposium on Fig, Ege University, Bornova, Izmir, Turkey* (ISHS Acta Horticulturae Number 480, pp. 221–226). Available at: http://www.ishs.org/ishs-article/480_38

Navarro, S., Donahaye, E., Rindner, M., & Azrieli, A. (1998). Control of nitidulid beetles in dried fruits by modified atmospheres. In C. Adler (Ed.), *Proceedings of the Working Group Integrated protection of stored products*. August 31–September 9, 1997 Zürich, Switzerland, *IOBC/WPRS Bulletin, 21*(3) 159–163.

Navarro, S., Donahaye, E., Kanner, Y., Rindner, M., Azrieli, A., Dias, R., Hovevey-zion, D., Nir, G., Zilberstain, B., Klayman, Y., Yakutieli, A., Baras, H., & Barak, M. (2000). *Solar drying of dates Medjool var. during harvest* (Annual Report 2000 of Research Project No. 417-0380-99, 18 pp). Israel: Ministry of Agriculture.

Navarro, S., Finkelman, S., Rindner, M., & Dias, R. (2003). Effectiveness of heat for disinfestation of nitidulid beetles from dates. In: G. L. Obenauf, & R. Obenauf (Eds.), *Annual international research conference on methyl bromide alternatives and emissions reductions*, November 2–6, 2003, San Diego 69 (pp. 1–2).

Navarro, S., Finkelman, S., Rindner, M., & Dias, R. (2004a). Heat treatment for disinfestation of nitidulid beetles from dates. In T. Batchelor & F. Alfarroba (Eds.), *Proceedings of fifth international conference on alternatives to methyl bromide*, September 27–30, 2004, Lisbon (pp. 297–300).

Navarro, S., Finkelman, S., Rindner, M., & Dias, R. (2004b). Emigration and control of nitidulid beetles from dates using heat. In S. Navarro, C. Adler, M. Schöller, M. Emekçi, A. G. Ferizli, & L. S. Hansen (Eds.), *Proceedings of the working group on integrated protection of stored products*, September 16–19, 2003, Kusadasi, Turkey (IOBC/wprs Bulletin 27, pp. 219–225).

Navarro, S., Finkelman, S., Ashbell, G., & Rimon, D. (2009). *Development of heat disinfestations technology for dates varieties in packing houses as alternatives to the use of methyl bromide* (58 pp). Report to Israel Ministry of Environmental Protection, Israel. Available at: http://www.sviva.gov.il/English/ResourcesandServices/Publications/ResearchReportsPublications/Pages/default.aspx. Accessed on 6 June 2015. (In Hebrew with English Abstract)

Navarro, S., Finkelman, S., Ashbell, G., & Rimon, D. (2010). *Development of heat disinfestations technology for dates varieties in packing houses as alternatives to the use of methyl bromide* (2nd Year Technical Report of Project, 64 pp). Israel: Ministry of Environment.

Navarro, S., Finkelman, S., & Navarro H. (2013). Mortality of *Carpophilus hemipterus* eggs and larvae under the influence of thermal disinfestations. In *IOBC/WPRS conference on integrated protection of stored products*, July 1–4, 2013 (IOBC/wprs Bulletin, 98, pp. 159–165).
Rindner, M., Navarro, S., & Azrieli, A. (2001). Critical moisture percents for safe storage of four dates cultivars. *Alon Hanotea, 9*(10), 433–436 (In Hebrew).
Roesli, R., Subramanyam, B., Fairchild, F. J., & Behnke, K. C. (2003). Trap catches of stored-product insects before and after heat treatment in a pilot feed mill. *Journal of Stored Products Research, 39*, 521–540.
Ryan, R., & Bishop, S. (2003). Vapormate™: Non-flammable ethyl formate/liquid carbon dioxide fumigant mixture. In E. J. Wright, M. C. Webb, & E. Highley (Eds.), *Proceedings of the Australian postharvest technical conference on stored grain in Australia 2003, June 25–27, 2003* (pp. 190–192). Canberra: CSIRO Stored Grain Research Laboratory.
Rygg, G. L. (1956). Effect of temperature and moisture content on the rate of the deterioration in Deglet Noor dates. *Annual Report of the Date Grower's Institute, 33*, 8–11.
Simmons, P., & Nelson, H. D. (1975). *Insects on dried fruits* (USDA Agricultural Handbook No. 464, 26 pp). Washington, DC: United States Department of Agriculture, Agricultural Research Service.
TEAP & MB (Technology and Economic Assessment Panel and Methyl Bromide Technical Options Committee). (2003). *Handbook on critical use nominations for methyl bromide*. Washington, DC: Environmental Protection Agency. Available at: http://www.epa.gov/spd-publc/mbr/mbtochandbook.pdf. Accessed on 20 May 2015.
UNEP (United Nations Environmental Programme). (2006). *Handbook for the Montreal protocol on substances that deplete the ozone layer* (7th ed.). Nairobi: UNEP Ozone Secretariate. Available at: http://ozone.unep.org/Publications/Handbooks/MP_Handbook_2006.pdf. Accessed on 20 May 2015.
Weidema, B. P., Thrane, M., Christensen, P., Schmidt, J., & Løkke, S. (2008). Carbon footprint – A catalyst for life cycle assessment? *Journal of Industrial Ecology, 12*, 3–6.
Wiedmann, T., & Minx, J. (2008). A definition of 'carbon footprint'. In C. C. Pertsova (Ed.), *Ecological economics research trends* (pp. 1–11). Hauppauge: Nova Science Publishers.
Yahia, E. M. (2004). Date. In K. C. Gross, C. Y. Wang, & M. Saltveit (Eds.), *The commercial storage of fruits, vegetables, and florist and nursery stocks* (Agriculture Handbook No. 66 (HB-66), 5 pp). Washington, DC: USDA-ARS.
Zaid, A. (2002). *Date palm cultivation* (FAO Plant Production and Protection Paper No. 156 Revision 1). Rome: FAO.

Index

W. Wakil et al. (eds.), *Sustainable Pest Management in Date Palm: Current Status and Emerging Challenges*, Sustainability in Plant and Crop Protection,
DOI 10.1007/978-3-319-24397-9

B

E

N

The manufacturer's authorised representative in the EU is Springer
Nature Customer Service Centre GmbH, Europaplatz 3, 69115 Heidelberg,
Germany. If you have any concerns regarding our products, please
contact ProductSafety@springernature.com

Printed and bound by CPI Group (UK) Ltd, Croydon, CR0 4YY
05/10/2026
02228133-0010